3

BRITTLE MATRIX COMPOSITES
2

Proceedings of the Second International Symposium on Brittle Matrix Composites (BMC2) held at Cedzyna, Poland, 20–22 September 1988.

BRITTLE MATRIX COMPOSITES 2

Edited by

A. M. BRANDT

*Institute of Fundamental Technological Research,
Polish Academy of Sciences, Warsaw, Poland*

and

I. H. MARSHALL

*Department of Mechanical and Production Engineering,
Paisley College of Technology, Scotland*

ELSEVIER APPLIED SCIENCE
LONDON and NEW YORK

ELSEVIER SCIENCE PUBLISHERS LTD
Crown House, Linton Road, Barking, Essex IG11 8JU, England

Sole Distributor in the USA and Canada
ELSEVIER SCIENCE PUBLISHING CO., INC.
655 Avenue of the Americas, New York, NY 10010, USA

WITH 109 TABLES AND 432 ILLUSTRATIONS

© 1989 ELSEVIER SCIENCE PUBLISHERS LTD

British Library Cataloguing in Publication Data

International symposium on brittle matrix composites.
(2nd: 1988: Cedzyna, Poland)
1. Brittle metals. Fracture
I. Brandt, A. M. (Andrej Marek)   II. Marshall, I. H. (Ian H.)
620.1'126

ISBN 1-85166-360-6

Library of Congress CIP data applied for

Printed in Great Britain at the University Press, Cambridge

Preface

The papers contained herein were presented at the Second International Symposium on Brittle Matrix Composites (BMC2) held at Cedzyna, Poland, 20–22 September 1988. The colloquium was organised and sponsored by the Institute of Fundamental Technological Research, Polish Academy of Sciences, in association with Kielce Technical University and the Civil Engineering Committee of the Polish Academy of Sciences. The Symposium forms a logical progression from BMC1 which was held in Jablonna, Poland, in November 1985.

As the title implies, the Symposium addressed a variety of problems ranging from concrete to ceramics and so on; others which become embrittled under extraneous conditions were also addressed. Participants from twenty countries (Belgium, Bulgaria, People's Republic of China, Czechoslovakia, Denmark, Finland, France, Federal Republic of Germany, German Democratic Republic, Hungary, Japan, North Korea, The Netherlands, Poland, Portugal, the Soviet Union, Sweden, the United Kingdom, the United States and Vietnam) contributed to the success of the event.

For logistical reasons, apart from five invited Plenary papers, the remaining sixty-nine contributions were, broadly speaking, subdivided into two Parallel Sessions. One branch dealt with Cement Based Composites, the other with Polymers and Ceramics. Except for the inherent disadvantages of Parallel Sessions, this was probably the most efficient subdivision possible.

With such a diverse geographical representation, it was not surprising that a healthy discourse of ideas permeated throughout the three days. This is not to be confused with disagreement; rather it should be considered as a lively exchange of scientific knowledge. Such exchanges are generally only possible during smaller conferences rather than the 'mega' variety and undoubtedly contribute significantly to their value.

Clearly much is happening on the brittle matrix front, ranging from new ceramic materials with tremendous potential to an increased awareness of the tensile properties of concrete, probably fuelled by an increasing application of fracture mechanics in a field traditionally dominated by static test results.

From the contents of this book, the reader will be left in no doubt as to the potential of a wide range of new or improved materials of construction, all characterised by their relatively brittle matrix properties. There is no doubt that these materials will play an increasingly important role in many aspects of engineering.

As always, an international symposium can only succeed in making a contribution to knowledge through the considerable efforts of a number of willing and, indeed, enthusiastic individuals. In particular, thanks are due to Andrzej Burakiewicz, Michal Glinicki, Janusz Kasperkiewicz and Janusz Potrzebowski.

The considerable efforts of the International Advisory Panel are greatly appreciated:

R. BARES (*Prague, Czechoslovakia*)
N. CLAUSSEN (*Hamburg, FRG*)
J. W. DOUGILL (*London, UK*)
C. GAULT (*Limoges, France*)
HUANG YIUAN-YUAN (*Shanghai, PR China*)
H. KRENCHEL (*Lyngby, Denmark*)
K. KROMP (*Stuttgart, FRG*)
Y. OHAMA (*Koriyama, Japan*)
R. PAMPUCH (*Cracow, Poland*)
S. P. SHAH (*Evanston, USA*)
S. SŌMIYA (*Tokyo, Japan*)
B. W. STAYNES (*Brighton, UK*)
P. STROEVEN (*Delft, The Netherlands*)
R. N. SWAMY (*Sheffield, UK*)
Y. V. ZAITSEV (*Moscow, USSR*)
ZHAO GUOFAN (*Dalian, PR China*)

A special word of thanks goes to Mariola Rejmund, our most able Symposium Secretary.

Finally, a grateful word of thanks is due to all Session Chairmen, authors and participants at the Symposium; without their efforts the Symposium would not have been possible.

As always, a final thanks should go to our respective families for their support during the Symposium and forbearance of the time demands necessary to make it possible.

A. M. BRANDT
I. H. MARSHALL

Contents

Session III: Ceramics—Properties and Applications

(*Chairman:* K. KROMP, *Max-Planck-Institut für Metallforschung, Stuttgart, FRG; Co-Chairman:* J. PIEKARCZYK, *Institute of Materials Science, Cracow, Poland*)

Session IV: Analytical and Numerical Methods in Brittle Matrix Composites

(*Chairman:* P. HAMELIN, *Institut National des Sciences Appliquées de Lyon, Villeurbanne, France; Co-Chairman:* L. HEBDA, *Keilce Technical University, Poland*)

Session V: Ceramics—Strength and Failure

(*Chairman:* G. ONDRACEK, *Rheinisch-Westfälische Technische Hochschule, Aachen, FRG; Co-Chairman:* L. RADZISZEWSKI, *Kielce Technical University, Poland*)

Session VI: Cement Based Composites—Probability and Damage

(*Chairman:* K. TOMATSURI, *Taisei Corporation, Yokohama, Japan; Co-Chairman:* M. GLINICKI, *Institute of Fundamental Technological Research, Warsaw, Poland*)

Session VII: Ceramics—Structure and Technology

(*Chairman:* N. CLAUSSEN, *Technische Universität Hamburg-Harburg, FRG; Co-Chairman:* A. BUCZYNSKI, *Warsaw Technical University, Poland*)

Session VIII: Plenary Paper

(*Chairman:* S. SCHMAUDER, *Max-Planck-Institut für Metallforschung, Stuttgart,
FRG; Co-Chairman:* L. RADZISZEWSKI, *Kielce Technical University, Poland*)

Session IX: Cement Based Composites—Micro and Macro Observations

(*Chairman:* P. STROEVEN, *Delft University of Technology, The Netherlands;
Co-Chairman:* R. BABUT, *Institute of Fundamental Technological Research,
Warsaw, Poland*)

Session XVII: Polymer Composites—Test Methods and Theoretical Models

(*Chairman:* C. GAULT, *Ecole Nationale Supérieure de Céramique Industrielle, Limoges, France; Co-Chairman:* P. LUKOWSKI, *Warsaw Technical University, Poland*)

Session XVIII: Test Methods of Brittle Composites

(*Chairman:* CHEN ZHI YUAN, *Tongji University, Shanghai, PR China; Co-Chairman:* A. BURAKIEWICZ, *Institute of Fundamental Technological Research, Warsaw, Poland*)

Session XIX: Polymer Composites—Mechanical Properties

(*Chairman:* V. WEISS, *Czech Technical University, Prague, Czechoslovakia; Co-Chairman:* A. SWIETLOW, *Warsaw Technical University, Poland*)

ELEMENTARY FAILURE MECHANISMS IN THERMALLY STRESSED MODELS
OF FIBER REINFORCED COMPOSITES

K.P. HERRMANN and F. FERBER
Laboratorium für Technische Mechanik
Paderborn University
D-4790 Paderborn, Pohlweg 47-49
Federal Republic of Germany

ABSTRACT

Elementary failure mechanisms like matrix and interface cracks, respectively, arising in self-stressed models of fibrous composites due to a steady cooling process are investigated by numerical as well as experimental methods of fracture mechanics. Further, by introduction of an appropriate nomenclature for special variations of the model geometries used in the cooling experiments thermal stress states originated in the uncracked as well as cracked composite structures, respectively, have been determined by using the finite element method. Moreover, by applying the substructure technique as well as by implementation of a maximum strain energy release rate criterion a simulation of a thermal crack propagation known from associated cooling experiments could be performed. Furthermore, fracture mechanical data like crack opening displacements, strain energy density variations and strain energy release rates, respectively, were calculated where these quantities have been used for the prediction of crack initiation as well as of further crack growth of branched thermal crack systems. Finally, by applying the method of caustics in transmission and reflection, respectively, opening-mode stress intensity factors K_I at the tips of propagating thermal matrix cracks were determined. A comparison of these experimental fracture mechanical quantities with associated finite element calculations obtainable by using the local as well as the global energy method, respectively, showed a very good agreement in the region of stable crack propagation.

INTRODUCTION

The existence of different failure mechanisms in the low- and high-fiber concentration ranges of unidirectionally fiber reinforced composites is well known from the experimentally observed failure behaviour of mechanically and/or thermally loaded fibrous composites. Further, elementary failure mechanisms like fiber breaking, debonding as well as matrix cracking due to special loading and overloading, respectively, were simulated by appropriate computer experiments [1]. The corresponding activation criteria concerning the initiation of such damage mechanisms were obtained by an analysis of the stress redistribution due to arising microcracks in the fibrous composites. A comparison of the numerical results with experimental data for a Carbon fiber/Aluminum matrix composite showed a sufficient coincidence. An analytical solution for the determination of the stress- and displacement fields in a unidirectionally reinforced composite containing an arbitrary number of broken fibers as well as a plastically deformed matrix material has been given in reference [2]. The corresponding compound material model basing on a shear-lag-assumption as well as on a shear stress fracture criterion allowed the prediction of the characteristic strength and failure properties of a Boron fiber/Aluminum matrix composite structure in agreement with experimental results. Micromechanical processes connected with crack growth in fibrous composites were also investigated in reference [3] by introduction of a local heterogeneous process zone around the crack tip whereas outside of this zone the heterogeneous material is considered as an anisotropic elastic continuum. By using this compound material model essential features of the crack growth in fibrous composites under monotonic increasing load could be studied. Furthermore, damage mechanics as well as the non-destructive testing and evaluation of composite laminates were considered in reference [4]. Thereby by using special damage mechanisms connected with the microstructure of the material a specific procedure can be applied for the description of a developing material damage. Further, the fatigue effect can be described in terms of a subcritical element damage, e.g. as matrix cracking, debonding and interface crack initiation which can be included in the associated damage mechanics as local stress redistributions. Additional literature concerning the application of the micromechanical and macromechanical methods, respectively, onto the strength and fracture behaviour of composites can be found in the references [5-7]. Moreover, thermal cracking of unidirectionally reinforced fibrous composites was investigated in a series of papers [8-12] where by applying the micromechanical method of composite mechanics the interaction of existing matrix and interface cracks, respectively, with the microstructure of fibrous composites submitted to thermal loading was considered.

NUMERICAL STRESS ANALYSIS

Statement of the Problem

In this paper, in continuation of investigations reported in reference [12] branched thermal crack systems consisting of a combination of curved matrix cracks and interface cracks, respectively, arising in self-stressed plane models of fibrous composites due to a steady cooling process are studied from the standpoint of fracture mechanics. Figure 1 shows the cross section of a cracked two-phase solid (matrix: ARALDITE F, fibers: steel) containing three matrix cracks as well as two interface cracks in the fiber-matrix interfaces of two neighbouring fibers due to thermal loading after a special casting process. Thereby the thermal loading of the composite structure took place due to a cooling from the temperature $T_0 = 60$ deg C of the unstressed initial state to a loading temperature of $T_1 = -7.5$ deg C according to a cooling curve given in reference [12]. Further, the material properties of the two-phase composite structure as well as the geometrical parameters of the disk-like bimaterial specimens used in the investigations can also be taken from the paper just mentioned.

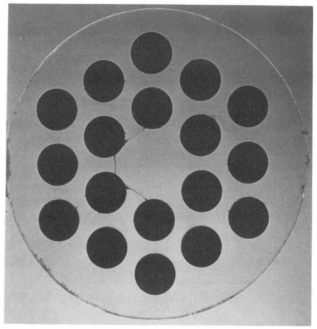

Figure   1. Elementary failure mechanisms in a thermally loaded two-phase composite structure.

In order to get the stress- and displacement states in the uncracked and cracked two-phase solids, respectively, as well as fracture mechanical data governing the propagation behaviour of the arising branched thermal crack systems the following boundary value problems of the plane thermoelasticity have to be solved.

Stress- and Displacement States in Thermally Loaded Two-Phase Composite Structures

By introducing a special nomenclature [13] according to Fig.2 of the different model geometries of the composite structures considered the following boundary value problems of the plane thermoelasticity have to be solved by assuming the existence of plane stress states in the self-stressed two-phase solids [11]

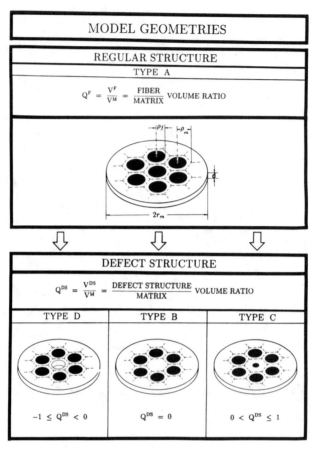

Figure 2. Nomenclature of different model geometries of self-stressed two-phase composite structures.

$$\sigma_{rr}^{m}(r_m,\phi) = \sigma_{r\phi}^{m}(r_m,\phi) = 0 \quad ; \quad (|\phi|\leqq\pi) \tag{1}$$

$$\sigma_{\rho\rho}^{j}(\rho_f,\psi) = \sigma_{\rho\psi}^{j}(\rho_f,\psi) = 0 \quad ; \quad (|\psi|<\psi_i) \quad , \quad (j=f,m) \tag{2}$$

$$[\sigma_{\rho\rho}(\rho,\psi)]_{\rho_i=\rho_f} = [\sigma_{\rho\psi}(\rho,\psi)]_{\rho_i=\rho_f} = 0 \quad ; \quad (\psi_i\leqq|\psi|\leqq\pi) \tag{3}$$

$$[u_{\rho}(\rho,\psi)]_{\rho_i=\rho_f} = [u_{\psi}(\rho,\psi)]_{\rho_i=\rho_f} = 0 \quad ; \quad (\psi_i\leqq|\psi|\leqq\pi) \tag{4}$$

Thereby in the boundary and continuity conditions (1)-(4) global plane polar coordinates r,ϕ with respect to the centers of the composite structures (with the outer radius r_m) as well as local coordinates ρ, ψ with the center in each fiber (with the radius ρ_f) have been introduced. By applying the basic equations of the stationary thermoelasticity for a plane stress state

$$\sigma_{ij} = \frac{E}{1+\nu} \left(\varepsilon_{ij} + \frac{\nu}{1-2\nu} \varepsilon_{kk}\delta_{ij} - \frac{1+\nu}{1-2\nu} \alpha\Theta\delta_{ij} \right) \tag{5}$$

$$\sigma_{ij,j} = 0 \tag{6}$$

$$\varepsilon_{ij} = \frac{1}{2} (u_{i,j} + u_{j,i}) \tag{7}$$

the associated boundary value problems mentioned above can be solved either by means of the experimental methods of photoelasticity or by using the finite element method. Thereby fields of principal stress trajectories in the cross-sections of uncracked and cracked self-stressed two-phase composite structures, respectively, as well as stress distributions along radial cuts between two fibers were already determined by means of the experimentally obtainable fields of isoclinics as well as by the aid of the shear stress difference procedure [11-13]. Further, the Figs.3-7 show the distributions of the stresses σ_x, σ_y and τ_{xy}, respectively, for an uncracked thermally loaded two-phase solid of type B as well as for different stages of the development of a matrix crack between two fibers of a self-stressed composite structure under the additional consideration of a partial debonding along two neighbouring fiber-matrix interfaces. Thereby Fig.3 gives the characteristic slopes of the stresses σ_x and σ_y, respectively, in agreement with earlier investigations performed by the shear stress difference procedure [13]. Moreover, because of the symmetry of the uncracked thermally loaded two-phase solid with respect to the x-axis the shear stresses τ_{xy} are vanishing along this line. But there arises a remarkable change in the graphs for the stresses σ_x and σ_y, respectively, after the development of a curvilinear matrix crack originating from one of two neighbouring fiber-matrix interfaces (cf. Figs. 4-6). Further, it can be seen from Fig. 6 that after the fully development of a matrix crack between two fibers the stresses σ_x and σ_y, respectively, show a very different behaviour. Namely, the stress σ_x is a continuous function along the x-axis with zero value at both crack surfaces whereas the stress σ_y jumps by traversing the crack. Moreover, Fig. 7 shows that this behaviour of the stresses remains valid also in case of a thermal crack system consisting of a matrix crack and two interface cracks, respectively, in the neighbouring fiber-matrix interfaces.

Figure 8 gives the corresponding crack surface displacements for such a branched thermal crack system in dependence on the crack tip position as well as of the simulation step in a quasi-threedimensional representation.

6

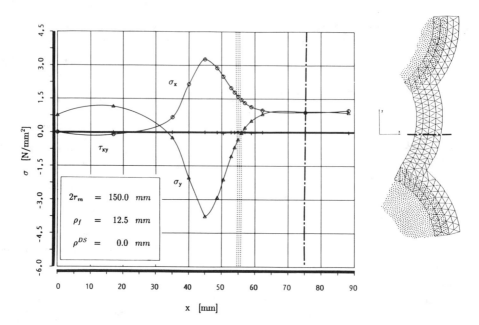

Figure 3.

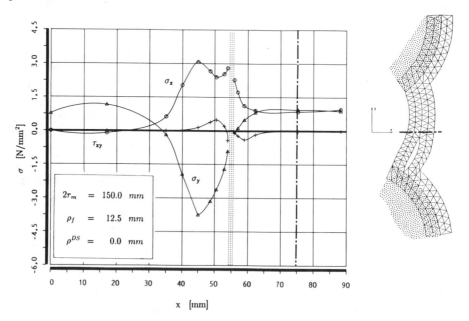

Figure 4.

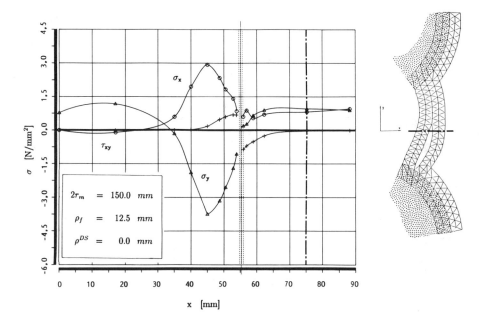

Figure 5.

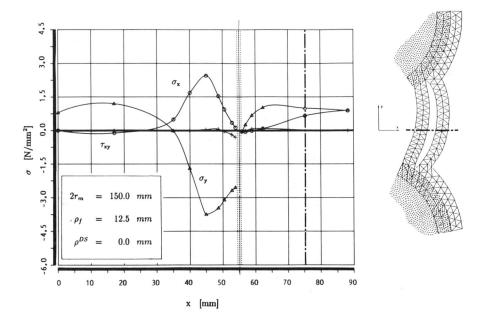

Figure 6.

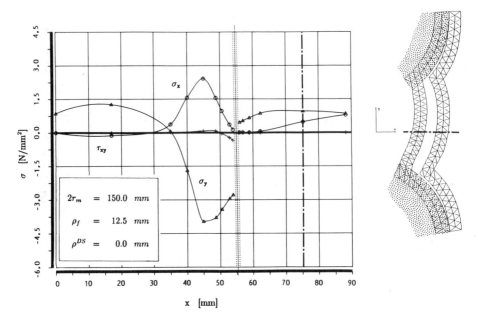

Figure 7.

Figs. 3-7. Distribution of normal and shear stresses between two fibers of thermally loaded uncracked and cracked two-phase composite structures, respectively.

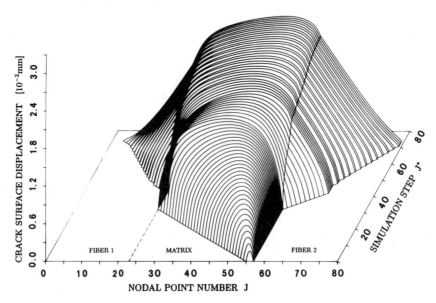

Figure 8. Crack surface displacements for a branched thermal crack system between two fibers of a self-stressed two-phase compound.

Crack Path Prediction and Fracture Mechanical Data

Moreover, finite element calculations were performed in order to predict the experimentally observed branching phenomenon of curvilinear thermal cracks in plane models of self-stressed fibrous composites [12]. Thereby, by an improvement of the method applied in reference [12] a crack growth criterion has been implemented in the finite element calculations in such a way that the two possible crack tip positions $a + \Delta a^{F1}$ and $a + \Delta a^{F2}$ (ref. Fig.9) could be obtained by an estimation of the variations of the elastic self-stress energy $\bar{U}$ stored in the thermally loaded two-phase composite structure according to the inequalities [14]

$$\frac{\bar{U}(a) - \bar{U}(a+\Delta a^{F1})}{\Delta a^{F1}} \underset{>}{\overset{\leq}{\underset{=}{}}} \frac{\bar{U}(a) - \bar{U}(a+\Delta a^{F2})}{\Delta a^{F2}} \tag{8}$$

Figure 9 shows the residual strain energy difference $\Delta U = \bar{U}(a+\Delta a^{F2}) - \bar{U}(a+\Delta a^{F1})$ in dependence on the simulated crack steps of a finite element calculation by using the so-called global energy method. Thereby the sign of the quantity ΔU decides over the crack extension into the matrix material (Δa^{F1}) or along the fiber-matrix interface (Δa^{F2}), respectively. Further, the graph shows a two-step development of the branched thermal crack system where at first the matrix crack has been initiated and also completed in the simulation ($\Delta U < 0$). Afterwards an interface crack is created due to a partwise debonding of the fiber 2 from the matrix material along the fiber-matrix interface. This interface crack is then arrested after a certain crack length. In a following step the same process takes place along the fiber-matrix interface of the fiber 1. Furthermore, according to Fig.9 an alternating extension of the interface cracks can be observed up to a certain crack length from which both interface cracks extend simultaneously along the corresponding fiber-matrix interfaces up to their respective arresting points [15]. Moreover, these numerical calculations show a good coincidence with the results of associated cooling experiments.

Figure 10 gives the total strain energy release rates G at the tips of matrix cracks for three different types of model geometries of self-stressed fibrous composite structures calculated by means of the global and local energy method, respectively.

The Figs.11-13 contain for the corresponding interface cracks at two neighbouring fibers the total strain energy release rate $G = \sum_{j=I}^{II} G_j$ as well as the single parts G_j (j=I,II) where the latter were determined by the aid of Irwin's modified crack closure integral [15].

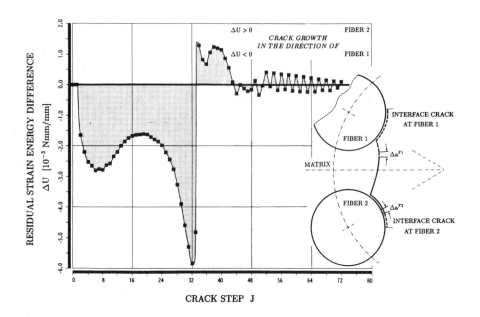

Figure 9. Residual strain energy difference in dependence on the simulated crack steps.

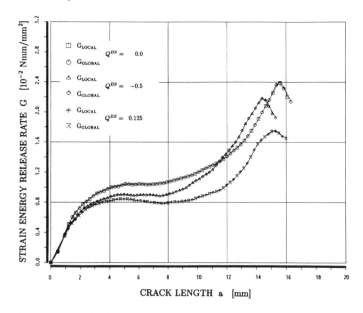

Figure 10. Total strain energy release rates in dependence on crack length at the tips of thermal matrix cracks for three different plane models of composite structures.

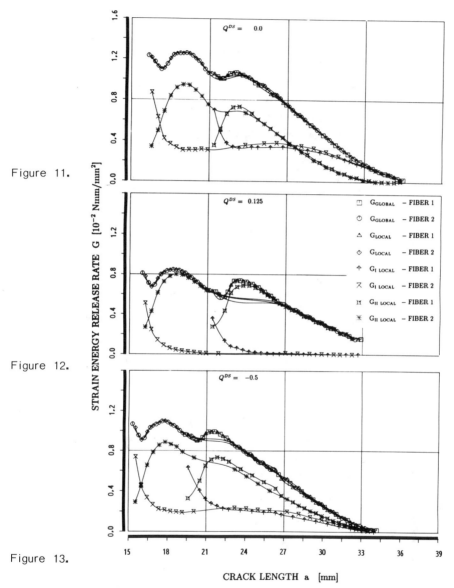

Figure 11.

Figure 12.

Figure 13.

STRAIN ENERGY RELEASE RATE G $[10^{-2}\,\mathrm{Nmm/mm^2}]$

CRACK LENGTH a [mm]

Figs. 11-13. Strain energy release rates in dependence on crack length at the tips of curvilinear interface cracks in the fiber-matrix interfaces of two neighbouring fibers.

Besides, Fig.14 shows a section of the refined finite element mesh in the neighbourhood of a branched thermal crack system used for the calculation of strain energy release rates at the tips of the matrix and interface cracks, respectively.

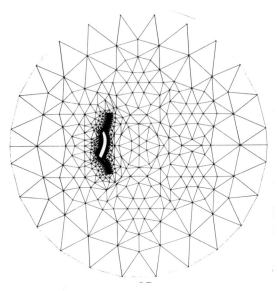

Figure 14. Section of a refined finite element mesh in the neighbourhood of a branched thermal crack system.

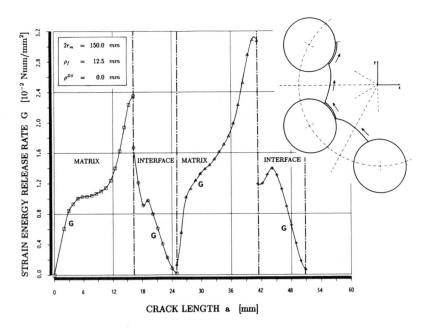

Figure 15. Total strain energy release rate G in dependence on crack length for a quasistatically extending curvilinear thermal crack system.

In addition, Fig.15 shows the total strain energy release rate G for a curvilinear thermal crack system consisting of two matrix and two interface cracks, respectively, in a self-stressed composite structure.

Crack Initiation

From a physical point of view the initiation of a first matrix crack durig the duration of a cooling experiment performed with different disk-like model geometries of fiber reinforced composite structures represents an important problem. By using disk-like specimens of type B it was shown experimentally that in all experiments for the same model geometry a first matrix crack arises in a reproduceable way in a small region at the fiber-matrix interface of fiber 2 (ref. Fig.17). By considering the principal stresses σ_i (i=1,2) around the fiber-matrix interface of fiber 2 the validity of the inequalities $\sigma_1 > O$ and $\sigma_2 < O$, respectively, around the discontinuity area could be demonstrated. Figure 16 gives the results of the numerical calculations where the principal tension stress σ_1 shows maximum values for the angles $\psi = \pm 22.5$ degrees. This result could be confirmed by consideration of the energy density variation around the fiber-matrix interface of fiber 2 in the direction orthogonal to the local principal tension stress σ_1.

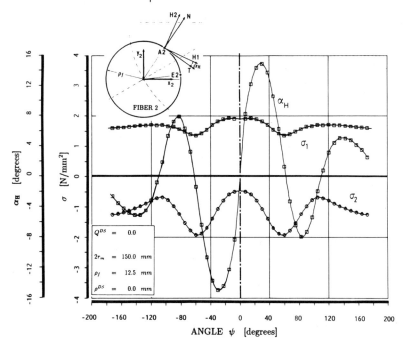

Figure 16. Principal stresses and angle of principal stress direction around a fiber-matrix interface.

Figure 17 gives the self-stress energy density variation w^* in dependence on the angle ψ of a local coordinate system located in the center of fiber 2. Thereby the quantity w^* which is defined as follows

$$w^* = \frac{w_A(\rho_f, \psi) - w_B(\rho_B, \psi_B)}{\Delta\rho_m \cos^{-1}[\alpha_H(\rho_f, \psi)]} \quad ; \quad |\psi| \leq \pi \tag{9}$$

with $\rho_B = \rho_f + \Delta\rho_m$ shows distinct maximum values for angles $\psi_I = \pm 23.1°$.

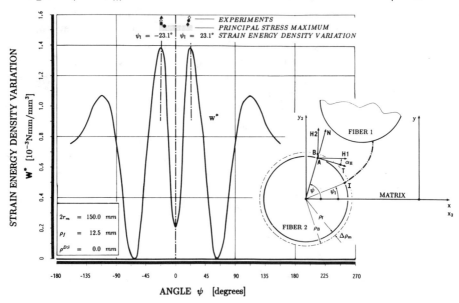

Figure 17. Strain energy density variation in dependence on the circumferential angle ψ.

EXPERIMENTAL STRESS ANALYSIS

Shadow Optical Method of Caustics

At present the shadow optical method represents a powerful tool in experimental fracture mechanics for the determination of stress intensity factors at the tips of quasistatically extending and fast running cracks, respectively. Further, the method of caustics can also be applied for mixed-mode crack problems as well as for the investigation of plastic zones around crack tips propagating in elasto-plastic materials. Moreover, due to the possibility of applying the method auf caustics in transmission as well as in reflection the treatment of cracked opaque model materials can also be performed. A comprehensive review about the shadow optical method of caustics has been given in reference [16].

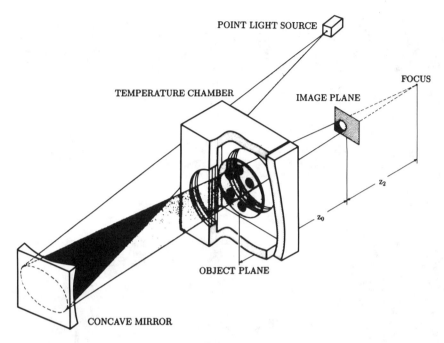

POINT LIGHT SOURCE

FOCUS

TEMPERATURE CHAMBER

IMAGE PLANE

z_2

z_0

OBJECT PLANE

CONCAVE MIRROR

Figure 18. Experimental set-up for the determination of fracture mechanical data by means of the shadow optical method in transmission.

Figure 18 shows the experimental set-up used for the determination of fracture mechanical data for a curvilinear thermal crack propagating in the matrix material (Araldite B) of a ceramics fiber reinforced composite structure. Because of the optical anisotropy of the matrix material the appearance of so-called double caustics around the extending crack tip of a curvilinear thermal matrix crack could be observed [13,17]. Figure 19 gives the different stages of a thermal matrix crack growth illustrated by the associated distributions of caustics and isochromatics, respectively, for a disk-like composite structure (model type B) consisting of a ceramics fiber reinforced Araldite B matrix.

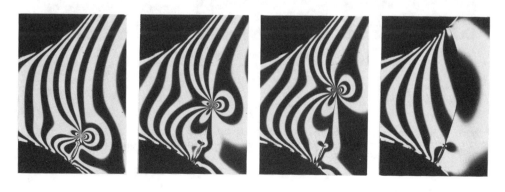

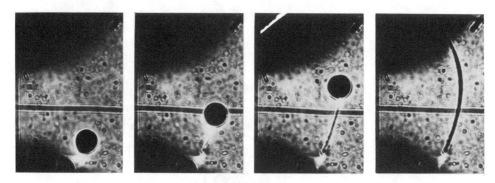

Figure 19. Distribution of isochromatics and caustics, respectively, for a thermal matrix crack extension.

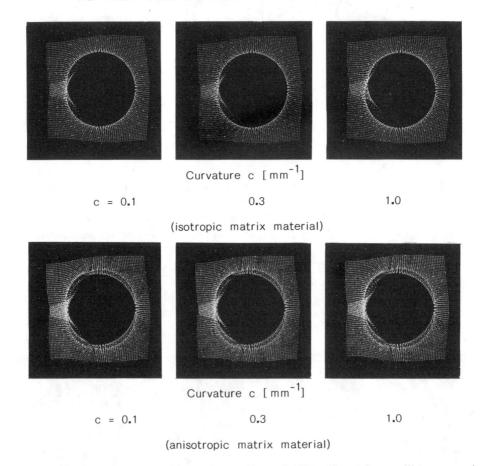

Curvature c [mm^{-1}]

c = 0.1   0.3   1.0

(isotropic matrix material)

Curvature c [mm^{-1}]

c = 0.1   0.3   1.0

(anisotropic matrix material)

Figure 20. Numerically calculated caustics at the tips of curvilinear cracks with different curvatures in isotropic and anisotropic matrix materials, respectively.

17

Moreover, Fig.20 shows numerically calculated caustics at the tips of matrix cracks with different curvatures under mode I-loading (K_I = 15 Nmm$^{-3/2}$).
The shadow spots exhibit deviations from the regular shape with increasing
values of curvature for both cases of the isotropic as well as of the anisotropic matrix material.

Finally, Fig.21 gives a summary of experimentally obtained results concerning the crack velocity and the opening-mode stress intensity factor K_I at
the tip of a curved thermal matrix crack propagating in a disk-like model of
a self-stressed composite structure. Further, the radius r_o of a small circle
(radius of the initial curve of the caustic) surrounding the crack tip is given
which has to be used for a determination of the diameter of the caustic according to the formula [17]

$$D_{a,i} = f_{a,i} r_o m \qquad (10)$$

with the dimensionless quantity $f_{a,i}$ calculable from the specific material constants of the corresponding anisotropic or isotropic materials, respectively, and
the representation scale m.

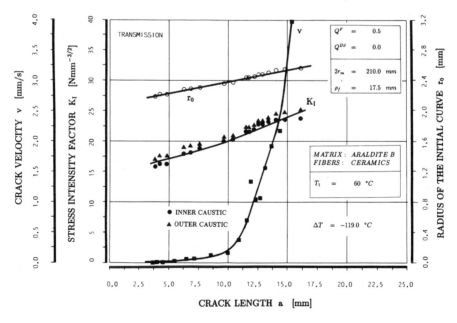

Figure 21. Crack velocity v, radius of the initial curve r_o, and stress intensity
factor K_I in dependence on crack length a at the tip of an extending curvilinear thermal matrix crack in a composite structure.

Moreover, it could be demonstrated that the values of the stress intensity factor K_I obtained experimentally by the method of caustics in transmission and reflection, respectively, for a curvilinear thermal matrix crack show a very good agreement with the corresponding numerical K_I-values gained by associated finite element calculations especially in the region of stable crack propagation [15].

CONCLUSIONS

Micromechanical modeling of thermal fracture effects in two-phase composite structures by means of numerical as well as experimental methods of fracture mechanics leads to a thorough understanding of the interaction of elementary failure mechanisms with the microstructure of the heterogeneous material. Thereby energetic considerations concerning the amount of elastic self-stress energy stored in a cracked two-phase solid as well as by implementation of an appropriate crack growth criterion based on a maximum strain energy release rate principle allow the numerical modeling of branched thermal crack systems along principal stress trajectories of thermal stress fields known from associated cooling experiments. Further, fracture mechanical data at the tips of curvilinear thermal matrix and interface cracks, respectively, were determined by means of finite element calculations showing in case of matrix cracks a very good coincidence with stress intensity factors experimentally gained by use of the shadow optical method of caustics.

REFERENCES

1. Kopyov, I.M., Ovchinsky, A.S. and Bilsagayev, N.K., Computer simulation of various fracture mechanisms in fibrous composite materials. In Fracture of Composite Materials, eds. G.C. Sih and V.P. Tamuzh, Martinus Nijhoff Publishers, The Hague/Boston/London, 1982, pp.45-52.

2. Goree, J.G. and Gross, S., Analysis of a unidirectional composite containing broken fibers and matrix damage. Engng. Fracture Mechanics, 1979, 13, 563-578.

3. Kanninen, M.F., Rybicki, E.F. and Griffith, W.J., Preliminary development of a fundamental analysis model for crack growth in a fiber reinforced composite material. In Composite Materials: Testing and Design, ASTM, 1977, STP 617, 53-69.

4. Reifsnider, K.L., Henneke, E.G., Stinchcomb, W.W. and Duke, J.C., Damage mechanics and NDE of composite laminates. In Mechanics of Composite Materials. Recent Advances, eds. Z. Hashin and C.T. Herakovich, Pergamon Press, New York/Oxford, 1983, 399-420.

5. Rosen, B.W., Failure of fiber composite laminates. In Mechanics of Composite Materials. Recent Advances, eds. Z. Hashin and C.T. Herakovich, Pergamon Press, New York/Oxford, 1983, 105-134.

6. Hashin, Z., Fatigue failure criteria for unidirectional fiber composites. J. Appl. Mech., 1981, **48**, 846-852.

7. Hashin, Z., Analysis of composite materials. J. Appl. Mech., 1983, **50**, 481-505.

8. Herrmann, K. and Braun, H., Quasistatic thermal crack growth in unidirectionally fiber reinforced composite materials. Engng. Fracture Mechanics, 1983, **18**, 975-996.

9. Herrmann, K., Curved thermal crack growth in the interfaces of a unidirectional Carbon-Aluminum composite. In Mechanics of Composite Materials. Recent Advances, eds. Z. Hashin and C.T. Herakovich, Pergamon Press, New York/Oxford, 1983, 383-397.

10. Buchholz, F.G. and Herrmann, K., Effects of micromechanical modelling on the fracture analysis of thermally loaded fibre reinforced composites. In Advances in Fracture Research, eds. S.R. Valluri et al., Pergamon Press, Oxford, Vol. **4**, 1984, 3005-3012.

11. Herrmann, K., Micromechanical analysis of basic fracture effects in self-stressed fibrous composites. In Fracture of Fibrous Composites, ed. C.T. Herakovich, AMD-Vol. **74**, 1985, 1-13.

12. Herrmann, K. and Ferber, F., Curved thermal crack growth in self-stressed models of fiber-reinforced materials with a brittle matrix. In Brittle Matrix Composites **1**, eds. A.M. Brandt and I.H. Marshall, Elsevier Applied Science Publishers, 1986, 49-68.

13. Ferber, F., Ph.D. Dissertation, Paderborn University 1986.

14. Herrmann, K.P. and Ferber, F., Failure mechanisms in thermally loaded models of fibrous composites with a brittle matrix. In Proc. Int. Symp. on Composite Materials and Structures, eds. T.T. Loo and C.T. Sun, Beijing, China, 1986, 483-488.

15. Herrmann, K.P. and Ferber, F., Numerical and experimental investigations of branched thermal crack systems in self-stressed models of unidirectionally reinforced fibrous composites. In Computational Mechanics '88, eds. S.N. Atluri and G. Yagawa, Springer Verlag, Berlin/Heidelberg/New York, 1988, 8.V.1-8.V.4.

16. Kalthoff, J.F., The shadow optical method of caustics. In Handbook on Experimental Mechanics, ed. A.S. Kobayashi, Prentice Hall, Englewood Cliffs, New Jersey, 1985.

17. Ferber, F. and Herrmann, K., Determination of stress intensity factors for curved self-stress cracks in thermally loaded models of fiber reinforced composites. (in German), VDI-Berichte **631**, 1987, 63-74.

STABLE MICROCRACKING IN CEMENTITIOUS MATERIALS

HERBERT KRENCHEL and HENRIK STANG
Department of Structural Engineering
Technical University of Denmark
ABK/DTH, 2800 Lyngby, Denmark

ABSTRACT

For centuries the practical use of cementitious materials has
been limited as a consequence of the brittleness of these materials. In the technical ceramics community, similar problems
have been overcome and it is now possible to make these highly
brittle materials ductile with an ultimate elongation at rupture which is several orders of magnitude higher than the
straining capacity of the traditional base material. This has
opened an immense new area for the practical use of technical
ceramics in the industry. A parallel development in the use of
cementitious materials with substantially improved ductility is
on its way.

Two different techniques are used for making these two
groups of materials ductile. In cement based materials the most
efficient and economical way is fibre reinforcement. In ceramics, on the other hand, the very sophisticated technique of
transformation toughening with two-phase crystal inclusions has
been introduced. In both cases the inclusions, the fibres as
well as the crystals, must be totally dispersed in the matrix
material to give full effect. The resulting mechanism by which
the two brittle materials are made ductile is the same in both
cases: a progressive internal rupture process by totally dispersed and stable microcracking in the matrix.

This dispersed, stable microcracking process has been investigated in detail in a series of tests performed on polypropylene fibre reinforced cement paste. The investigation included an accurate determination of the constitutive behaviour of
the fibres under controlled strain rate conditions followed by
a detailed investigation of the behaviour of the composite
material in uniaxial tension. Again the tests were performed
under controlled strain rate conditions. The stress-strain behaviour of the composite material was recorded and the detailed
knowledge of the constitutive behaviour of the fibres allowed
a determination of the average stress-strain relation in the
matrix during microcracking. The microcracking process was investigated indirectly by means of acoustic emission and direct-

ly by means of fluorescence analysis of thin sections representing different stages in the load history. In this way both quantitative and qualitative results describing the microcracking
were obtained as a function of the load history.

The experimental results obtained from the test series
thus contain information about the stress-strain behaviour of
the constituents along with information about the microstructure of the composite material. This type of information is
essential in order to understand the mechanism which govern
the microcracking process in cementitious composites.

The main conclusions drawn from the experiments are: The
microcracking process can be divided into two parts; the first
where initial cracks and flaws are totally stabilized and the
second where new microcracks are formed in a distributed manner.
The stress level at the second microcracking stage is highly
dependent on the fibre volume concentration. A uniaxial tension
stress up to 15 MPa has been observed in the matrix material at
the beginning of the second stage (with a fibre volume concentration of 0.13) while the strength of the unreinforced matrix
material is about 3.5 MPa. In the matrix material strain softening is observed on the average during second stage microcracking without any localization.

INTRODUCTION

The practical use of the cheaper types of building materials
such as glass, gypsum, ceramics, calcium silicate, cement, and
concrete is limited because of the notch sensitivity and brittleness of these materials. When loaded in tension the ultimate
straining capacity of such materials is only some 100-200
μ-strain (1 μ-strain = 10^{-6}) and the ductility and fracture
energy is negligible in comparison with most other materials.

Considering these facts it is surprising that cementitious
materials has any applicability at all since drying shrinkage
and thermal expansion strains alone can easily exceed the ultimate straining capacity.

The key to the understanding of how cementitious materials
can retain their continuity in aggressive environments after
all is their capability to form distributed microcracking
(cracks with lengths of 0.001-0.1 mm and widths of 1-5 μm) instead of one discrete crack when the ultimate tensile straining
capability is reached through drying, or thermal and/or mechanical loading. This ability to form microcrack patterns is obtained through addition of inhomogeneities or aggregate particles
which act as microcrack arrestors causing stable microcracks to
form in the hardened cement paste (hcp).

Stable microcracking is not harmful to the cementitious
material in the same drastic way as is the formation of discrete, visible cracks (cracks with widths from 5 μm and upwards).
The stable microcracks just make the hcp phase of the cementitious material slightly more porous and reduce its tensile
strength and elastic modulus to some degree, but this cracking
system is so fine that it will quite often heal again after
some time by further hydration of the cement.

However, it should be noted that this microcracking ability is limited in most cementitious materials. This means that the microcracks will eventually localize in certain regions forming a macrocrack pattern which in general depends on the total hardening history as well as the environmental and mechanical loading history.

STABILIZATION OF MICROCRACKS

It will be understood that the ductility and the durability of a cementitious material can be strongly improved if a totally stabilized microcracking system can be introduced in such a way that by straining the material, more and more microcracks are formed in the cementitious matrix, the crack length of each microcrack formed being kept so small that it never becomes unstable.

This technique for changing drastically the straining capacity and fracture energy of a brittle material has been used, in fact, for nearly a century now all over the world in the production of high quality asbestos cement. Here the matrix material is neat hcp being, in fact, as brittle as glass. By mixing in some 7 vol.-% of finely distributed asbestos fibres [1] with a high specific fibre surface giving a very low fibre spacing [2] a composite material has been made with highly improved ductility and excellent durability, as we all know.

In modern high performance ceramics quite another technique is used with the same aim and result. Here the microcracks are stabilized by so-called transformation toughening. Here finely distributed two-phase crystals (e.g. zirconia oxide) work as crack arrestors causing microcrack patterns to develop all through the material [3-5]. Thereby the straining capacity of the matrix material is improved with a factor of ten to one hundred, and these types of ceramics are being used today for technical purposes in parallel with metals and sustaining much higher temperatures and giving no corrosion problems at all.

FRC-MATERIALS

Considering the above it is obvious that the quality of a fibre reinforced brittle material is totally dependent on the cracking pattern developed in tensile loading.

If discrete cracks are formed in the matrix the material is not much better than ordinary reinforced concrete or ferrocement. If, on the other hand, the fibres are able to stop and stabilize all microcracks formed in the matrix material, as in e.g. the best types of asbestos cement, a new high quality composite material has been created with an optimal fibre effect.

Different parameters are responsible for this cracking behaviour, the dominant ones being, as far as we can see today, the fibre spacing, the specific fibre surface [SFS], the degree of fibre dispersion in the material and the fibre matrix bond.

We do not yet fully understand how these parameters are interrelated and how they should be evaluated, but it is obvious that a low SFS and a poor bond will always result in a low quality composite with a discrete crack pattern as observed in ordinary reinforced concrete.

In the following a series of uniaxial tensile tests with polypropylene roving fibre reinforced hcp material will be examined. With a high fibre loading (c_f = 0.04-0.13) and using a split fibre with a low Denier cross section (average fibre cross section: 35×150 μm ~ 55 Den.) and taking care during the manufacturing process of the test specimens that the cement paste is being smeared onto the entire fibre surface and that the fibres are being totally dispersed - a high quality FRC-material has been made with an apparent elastoplastic behaviour in tensile loading including apparent plastic yielding from a very well defined yield point followed by a strain hardening zone. Nothing but stable microcracking was observed in the matrix even at strains up to 3-5%.

TABLE 1
Matrix composition

Composite material	Epoxy Resin (Ciba-Geigy)	Hardener	Portland Cement (Aalborg)	Smearing agent (Dow Chem.)	Plasti-ziser (SKW-A/G)	Water
FRE gr/charge	BY 158 100	HY 2996 28				
FRC gr/charge			Rapid hardening PC 2,000	Methocel M 228 2.4	Melment L 10 48.0	650

All FRC-matrix ingredients were homogenized in a 5 liter ball mill, (50 minutes, 28 rev/min).

The FRE test specimens were hardened at room temperature (22° C) for 24 hours before demoulding.

The FRC test specimens were stored in water (20° C) for 7 days after demoulding, and hereafter in laboratory atmosphere (22° C, 45-55% RH) until time of testing (min. 30 days).
Air content in the FRC-specimens at time of demoulding was between 0.02 and 0.04.

TEST SPECIMENS

In order to determine not only the composite material response but also the response of each of the two phases (fibres and matrix), two sets of test specimens were manufactured. The first set consisted of KRENIT polypropylene fibre*) reinforced epoxy (FRE) specimens while the second set consisted of KRENIT polypropylene fibre reinforced cement paste (FRC) specimens.

*) Made by Danaklon A/S, Varde, Denmark.

The epoxy matrix used was a low viscous slow hardening epoxy type*) while the cement paste had a low w/c-ratio and contained super plastiziser and a smearing agent. The two matrix materials were composed according to Table 1.

In each set of test specimens uncut fibre rovings, i.e. sheets of fibrillated polypropylene film were used thus all specimens were reinforced with continuous fibres in a uniaxial arrangement.

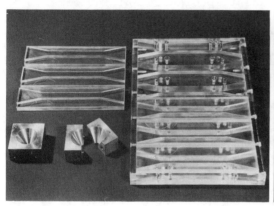

Figure 1. The plexiglass moulds and steel nozzles (for $c_f \sim 0.06$ and 0.13 respectively) used for casting the FRC-specimens (left). 6 FRC specimens after demoulding (right).

Both set of specimens were manufactured using basically the same technique consisting of: a) Impregnation of the fibre roving with matrix material. In order to impregnate the fibres with the Portland cement matrix it was essential to open the fibre mesh by stretching it perpendicular to the fibre direction. In this way all fibre surfaces were exposed to the cement matrix which was smeared out on the fibres by hand. b) Vacuum treatment of the impregnated roving to reduce air content. c) A reversed extrusion process (so-called pulltrusion) where the impregnated roving was pulled through a nozzle. This process was the main fibre volume concentration controlling process and by varying the diameter of the opening of the nozzle a wide range of volume concentrations were obtained. d) After the pulltrusion process the rovings were cut to a length corresponding to the specimen mould and a number of matrix impregnated rovings finally made up the specimen. Since the roving weight was determined before impregnation and since the density of the polypropylene was known the exact volume concentration for each specimen could be calculated when the number of rovings in each specimen was known.

The moulds were transparent plexiglass moulds with plexiglass lids (see Fig. 1) ensuring an optimum specimen finish on all four sides. The specimens were all dog bone shaped with rectangular cross section for handling convenience. The FRC specimens had a constant width section of $10 \times 15 \times 110$ mm, while the FRE specimens had a constant width section of

*) Viscosity $\eta \sim 100$ cP at 22° C.

5 × 10 × 100 mm. The fibre volume concentrations varied from 0 to 0.13 in the FRC specimens and from 0 to 0.33 in the FRE specimens.

TESTING PROCEDURE AND ANALYSIS

The basic idea using two sets of test specimens is that the fibre stress-strain response can be determined quite accurately by analyzing the stress-strain response of the FRE specimens with different fibre volume concentrations. Then the fibre stress-strain response can be used to determine the average matrix response in the FRC specimens even after microcracking has occurred, provided that the cracking is well distributed and stabilized.

Both the FRE and the FRC specimens were tested in uniaxial tension in an Instron 6022 closed loop testing machine. The tests were deformation controlled with a feed back signal provided by two axial extensometers placed on opposite sides of the specimens. In this way the overall strain rate over the gauge length (50 mm) was controlled during the experiments. The testing machine provided computerized data logging and was furthermore interfaced with a personal computer where the final analysis was carried out along with preparation of graphical output.

The analysis of the FRE specimen tests was based on the following general equation

$$\sigma_c^a(\varepsilon(t)) = c_f \sigma_f^a(\varepsilon(t), c_f) + (1-c_f)\sigma_m^a(\varepsilon(t), c_f) \qquad (1)$$

where σ^a means axial stress, subscript c, f, and m means composite, fibre, and matrix respectively, c_f denotes fibre volume concentration, and $\varepsilon(t)$ means strain history.

Assuming that σ_f^a and σ_m^a are independent of c_f and that only a load history with a constant strain rate is considered eq. (1) reduces to

$$\sigma_c^a(\varepsilon) = c_f \sigma_f^a(\varepsilon) + (1-c_f)\sigma_m^a(\varepsilon) \qquad (2)$$

which can be used to determine $\sigma_f^a(\varepsilon)$ (and $\sigma_m^a(\varepsilon)$) when test results with more than one value of c_f are known at each strain level.

In Fig. 2 it is shown that a linear description as proposed in eq. (2) is correct within the experimental accuracy. In Fig: 3 the uniaxial fibre stress-strain relation is shown for a constant strain rate of 1.67 10^{-4} Hz. Equation (1) can also be used to determine cyclic stress strain response (an example is shown in Fig. 3) or even an analytical description of the relaxation behaviour of the fibres, however time dependent phenomena will not be discussed here.

The advantages of the method described above for the determination of fibre stress-strain behaviour is, that handling of the fibres in the testing machine becomes extremely easy involving no other experimental equipment than the equipment required for the FRC specimen tests. Testing of the fibrillated film 'as is' is extremely difficult as reported by Hughes and

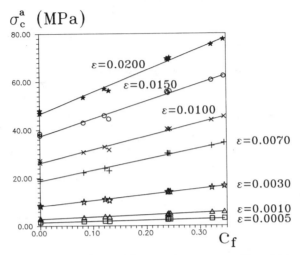

Figure 2. Stress in the FRE-specimens as function of fibre
volume concentration and strain level.

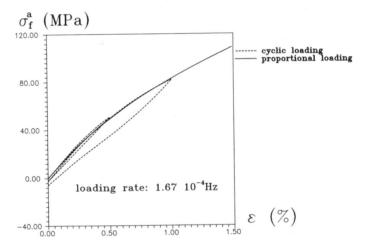

Figure 3. The stress-strain relationship for the pp. KRENIT
fibre for two different load histories.

Hannant [6]. With the present setup (see also [1] pp. 68-73)
the use of extensometers present no problem and the fibre material is tested in its final configuration rather than in the
shape of a film as reported by Hughes and Hannant [6] and Keer
[7]. Finally it should be noted that the method applies to all
fiber types which are available as 'infinitely' long fibres.
    The FRC specimens were tested using the same testing pro-
cedure as described above. Optimal composite material stress-
strain behaviour was obtained with the specimens referred to

in Fig. 4 showing stress-strain curves for 6 specimens with a fibre volume concentration of 0.125-0.131. Note that the concept of a stress-strain relation for the cracking composite material is only valid if the cracking is taking place as distributed microcracking with a characteristic length (crack spacing) which is small in comparison with the gauge length. As it will be shown later this is indeed the case here. Each of the test were stopped at different strain levels in order to investigate the microcrack patterns as discussed later.

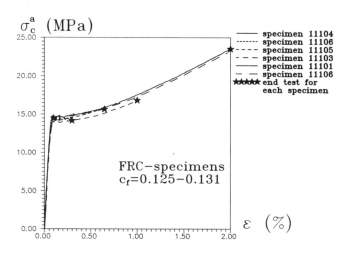

Figure 4. The stress-strain relationship for 6 FRC specimens with fibre volume concentrations of 0.125-0.131.

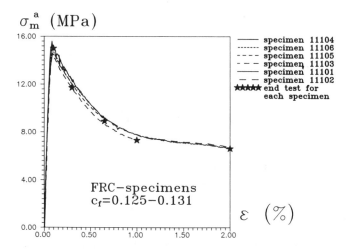

Figure 5. The average matrix stress-strain relationship for the specimens referred to in figure 4.

In order to analyze the FRC results we again refer to eq.
(1) which, again using the same argument, is reduced to eq. (2).
It is not evident, however, that the arguments to reduce eq. (1)
applies to the FRC specimens as well. An elastic analysis of a
uniaxial fibre reinforced material using e.g. Hashin's composite cylinder model (see [8] or [9]) shows that the stress
field in the fibres is uniaxial for a uniaxial composite material stress only when the matrix and the fibres have the same
Poisson's ratio. If not, the stress state in the fibres becomes
triaxial and the degree of triaxiality depends on the fibre volume concentration. However, in order to simplify the analysis
it was assumed here, that eq. (2) applies to the FRC specimens
as well. Using eq. (2) it is now easy to determine the average
matrix stress-strain response as shown in Fig. 5 showing the
stress-strain curves for the same specimens as shown in Fig. 4.
In order to further characterize the composite material behaviour the load carried by the fibres relative to the total composite material load was calculated on the basis of the fibre
stress-strain curve and compared to the average relative load
carried by the matrix. Three of the optimal relative load
curves obtained are shown in Fig. 6.

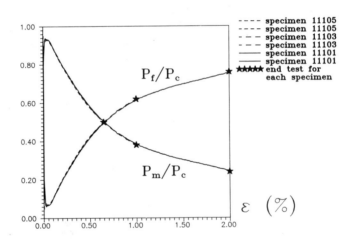

Figure 6. The relative load carried by the fibres and the
matrix, respectively, in 3 of the 6 specimens
referred to in figure 4.

Looking at Fig. 4 first of all the limited experimental scatter
should be noted. The specimens behave almost identically, a
rare sight when dealing with cementitious materials in tension.
Secondly we can characterize the stress-strain curve in the
following way; the curve starts as a nearly linear curve with
a tangential modulus of about 28 GPa at (0,0). At a well defined stress level (BOP) the stress-strain relationship suddenly changes; first a short stress plateau is reached (ε =
0.0005-0.003), and here after the slope is slowly increasing
over a long strain range (ε = 0.003-0.03). The specimens fail

at about ε = 0.05, however in this study primarily the first parts of the stress-strain relationship was investigated. The magnitude of the BOP stress is depending heavily on the fibre volume concentration as shown in Fig. 7, where the BOP stresses for 23 test specimens are reported along with the results of 3 tensile tests of the matrix material without fibres. A linear or 2. order polynomial relation between the BOP-stress and c_f seems to be a reasonable description which even includes a prediction of the matrix tensile strength without fibres.

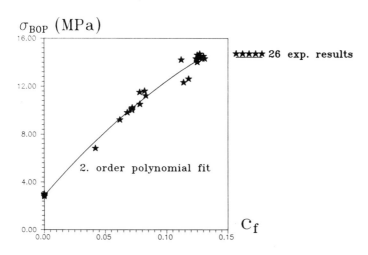

Figure 7. The BOP stress as function of fibre volume
          concentration.

Note that these results are somewhat contradictory to other experimental results found in the literature e.g. [6],[7], and [10] where constant or even decreasing stresses are reported in a very long strain interval (between 0.01 and 0.02 for the volume concentrations in the range reported here). Keer [7] reports no clear relationship between strain (or stress) at BOP and c_f also contradicting the findings reported here.
     The stress-strain behaviour in the matrix meterial shows a very smooth and well controlled strain softening after the BOP and again the behaviour of the 6 specimens are very similar. It is noteworthy that the average stresses in the matrix material reaches 15 MPa at the peak point (this is approximately 5 times the unreinforced matrix tensile strength) and that the matrix material continues to carry average stresses two times the tensile strength of the unreinforced material even at ε = 0.02.
     Looking at Fig. 6 it is evident that the matrix is an important load carrying component of the composite in the total range from ε = 0 to ε = 0.02.

CHARACTERIZATION OF MICROCRACKS

The microstructural changes which obviously takes place in the specimens during the loading history has been characterized in two ways.

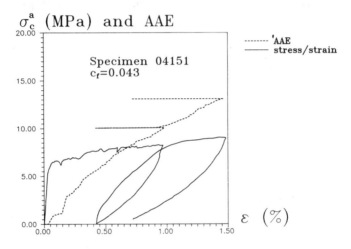

Figure 8. The stress and accumulated AE vs. strain relationship for a test specimen with fibre volume concentration of 0.043 subjected to a two cycle load history.

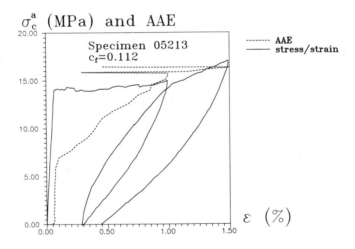

Figure 9. The stress and accumulated AE vs. strain relationship for a test specimen with fibre volume concentration of 0.112 subjected to a two cycle load history.

First acoustic emission (AE) recordings were performed during
loading using a single transducer connected to an amplifier and
an electronic counter which counted the number of acoustic emission pulses with an amplitude larger than a preset trigger level which was held constant during the experiments. In Figs. 8
and 9 typical results are reported the graphs showing stress
and accumulated acoustic emission (AAE) vs. strain. The figs.
are referring to two cycle tests on two specimens with a fibre
volume concentration of 0.043 and 0.112 respectively. The AE
typically starts immediately before the BOP is reached. The AE
activity then rises suddenly with a maximum count rate around
the strain corresponding to BOP (ε = 0.0002-0.0005). The count
rate then slowly decreases with increasing strain. It is shown
in Figs. 8 and 9 that with the trigger level used in the experiments unloading produces no AE, and AE activity is not recorded until the unloading strain level is reached again. The differences between the shape of the AE-curves at different fibre
concentrations is noteworthy: At the larger fibre volume concentration a very large part (approx. 40%) of the AE activity
recorded takes place in a very narrow strain band around BOP
while only a small part of the total activity takes place in
the same band when the fibre volume concentration is smaller.
This is consistent with observations in other specimens as well.
Looking at Figs. 8 and 9 the total AE activity seems to increase with increasing fibre volume concentration. However
looking at the whole range of c_f's investigated here this
observation does not seem to hold in general.

The results shown here indicate that larger fibre volume
concentrations introduces more microcracking in the early
stages of the load history, however, the underlying mechanisms
for this AE activity is not fully understood and further investigations using a more refined AE recording technique is obviously necessary.

The second way in which the microcracking process has been
characterized is by direct observation using fluorescence microscopy. Tensile tests with constant strain rate were performed
and stopped at different strain levels as shown in Fig. 4. The
strain in the test specimens was then kept constant while steel
blocks were glued to the specimen surface with a fluorescent
epoxy as shown in Fig. 10. After the epoxy adhesive was hardened the specimens were removed from the testing machine, the
steel blocks maintaining the strain level in the specimen.
Then the specimens were cut, vacuum impregnated with a low viscous fluorescent epoxy (the same epoxy used for gluing the
steel blocks to the specimen), and finally thin sectioned. The
microcracking pattern could now be investigated in a microscope
equipped for fluorescence microscopy. Using this technique very
fine cracks with an opening of 1 µm or less could easily be detected. Until now only qualitative analysis has been performed
though quantitative results obtained using a digital image
analysis system is under way.

The qualitative observations can be characterized in the
following way: In the specimen loaded to a maximum strain of
0.001 a distributed microcracking pattern perpendicular to the
fibres can be observed. No cracks penetrate the specimen from
one side to the other and many cracks are limited to a length
corresponding to the average distance between adjacent fibres.

Figure 10. (Left) FRC-specimens with steel blocks glued to the
surface in order to maintain the strain level at a
particular point in the load history. The specimens
have been loaded to a strain of 0.000 and 0.030 respectively. (Note the difference in length).

Figure 11. (Right) Microcracking patterns in FRC specimens
with a volume concentration of 0.13. The image
has been obtained using fluorescence microscopy,
(strain = 0.030. Average microcrack spacing: 0.3 mm).

With increasing strain levels the number of individual cracks
clearly increases, however, it is not clear to what extend the
crack length distribution changes. At large strain levels
(0.02 and 0.03) considerable crack opening can be observed in
the largest cracks but there is still no macrocrack separating
the specimen from one side to the other. Fibres can be observed
bridging the cracks but even at the points of largest crack
opening no fibre-matrix debonding can be observed, even though
it is clear that such a debonding has taken place. A typical
microcrack pattern at $\varepsilon = 0.030$ is shown in Fig. 11.

ACKNOWLEDGMENTS

Herbert Krenchel wishes to acknowledge the support of NATO
Double Jump Research Contract No: D.J-702/84: Brittleness of
High-Strength Cement-Based Composites. Henrik Stang wishes to
acknowledge the financial support from the Danish Council for
Scientific and Industrial Research. Grant No. 16-4239.B and
from "Grosserer Emil Hjort og Hustrus Legat". Both authors wish
to thank Professor S.P. Shah for valuable discussions during
the project.

REFERENCES

1. Krenchel, H., Fibre reinforcement, (Diss. Technical University of Denmark). Akademisk Forlag, Copenhagen, 1964, pp. 159.

2. Krenchel, H., Fibre spacing and specific fibre surface. RILEM Symposium: Fibre Reinforced Cement and Concrete, London 1975. The Construction Press, Horneby, Lancaster, 1976, pp. 69-79 and pp. 511-513.

3. Garvie, R.S., Hannik, R.H.J. and Pasco, R.T., Ceramic steel, Nature (London), 1975, **258**, pp. 703-705.

4. Advanced Ceramic Materials, Jan. 1986, Vol. 1, No. 1, News: Superplasticity of high-performance ceramics, p. 7.

5. Lambropoulos, J.C., Effect of nucleation on transformation toughening. American Ceramic Society, ACS. Journ., Vol. 69, No. 3, March 1986, pp. 218-222.

6. Hughes, D.C. and Hannant, D.J., Brittle matrices reinforced with polyalkene films of varying elastic moduli. Journal of Materials Science, 1982, 17, pp. 508-516.

7. Keer, J.G., Influence of fibre characteristics on polyolefin reinforced composites under limited cyclic loading. The International Journal of Cement Composites and Lightweight Concrete, 1987, 9, pp. 145-156.

8. Hashin, Z. and Rosen, B.W., The elastic moduli of fibre-reinforced materials. Journal of applied mechanics, 1964, 31, pp. 223-232.

9. Christensen, R.M., Mechanics of composite materials. A Wiley-Interscience Publication, 1979.

10. Hannant, D.J., Hughes, D.C., and Kelly, A., Toughening of cement and other brittle solids with fibres. Philosophical Transactions, Royal Society London, Series A, 1987, 310, pp. 175-190.

STRUCTURAL CHARACTERIZATION OF STEEL FIBRE REINFORCED CONCRETE

PIET STROEVEN
Stevin Laboratory, Delft University of Technology
Stevinweg 4, 2628 CN Delft, The Netherlands

ABSTRACT

Steel fibre reinforced concrete can be designed by a law of mixtures concept, in which efficiency parameters account for the fibre dispersion details. The model can also be employed for the boundary zone. The controlling micromechanical model is fibre pull-out. The paper outlines the stereological modelling of the structure and the characterization of mechanical behaviour, encompassing also the shearing of fibres over crack edges. Additionally, anchoring effects are included.
Experiments show the fibres mostly distributed according to a partially-planar system, with lower reinforcement ratio's in boundary layers. Placement and compaction further lead to anisometry and segregation in the fibre structure. This is found to be directly refleced in anisotropical behaviour in splitting tensile tests.
The estimates based on the outlined theoretical concepts are shown to satisfactorily agree with such experimental characteristics.

INTRODUCTION

Macroscopically heterogeneous materials, such as concrete, are even in the virgin state severely cracked. The size distribution of these cracks is directly governed by that of the particles in the material. Steel fibres dispersed in such a matrix have a length roughly twice the maximum grain size. A composite model of steel fibre concrete on mesolevel inevitably has to encompass particles, cracks and fibres. They mutually interact in a complicated way when the material body is stressed. Only somewhat simplified concepts are however accessible for design purposes.
It is well-known that the short fibres that are conventionally employed debond under loading, whereupon the transfer of stresses in the cracked region is accomplished by interfacial friction. Recent experiments using holographic interferometry have demonstrated the process of debonding to start under relatively low loadings. Complete debonding was found at ultimate pull-out loading [1]. At ultimate tensile strength (UTS) the fibres intersecting the main crack are therefore solely contributing by in-

terfacial friction. This allows to attribute a constant value for the coefficient of friction, τ_f , to the model underlying the estimate for UTS.
When larger global deformations are imposed on the material body one has to
account for a reduced value of τ_f due to a polishing effect.

Cracking at discontinuity in plain concrete is probably restricted to
isolated particle-matrix debonding [2]. Fibres can hardly influence this
degradation process. Hence, they will not significantly improve the "crack
initiation strength". This is confirmed by experimental data. A composite
SFRC model is therefore only relevant for UTS and post-ultimate loading
stages. Also fracture toughness estimates could be based on such a SFRC
composite model.

Two further approximations have to be considered for the development
of a mechanical model. Firstly, it is assumed that actual fibre
distributions can be treated globally as a linear combination of 3-D and
2-D portions, so that a partially planar system is modelled. Occasionally
it is necessary to add a 1-D portion, too. However, this can easily be
encompassed in the model concept [3]. 3-D and 2-D systems are considered to
be statistically uniform at random (UR). Secondly, the curvature of the macrocrack is assumed sufficiently smaller than that of the structuring elements (e.g. particles), so that micromechanical "building blocks" can be
defined consisting of a matrix pocket divided by a crack that is inter-
sected by steel fibres.

A stereological approach to describing the dispersion characteristics
of all fibres contributing to this stress transfer mechanism is developed.
A previously proposed concept was solely based on the fibre pull-out mechanism [4]. Brandt [5] has demonstrated however that shearing over the crack
edges and plastic deformation roughly contribute in equal terms to the
energy dissipation during mechanical loading. A more general concept will
therefore be outlined in ths paper.

STRUCTURAL CONCEPT

Estimates for "stiffness" can either be based on a continuum model
(elasticity theory for an aeolotropic solid) or on that of an intensily
cracked medium [6]. It has been demonstrated that differences are insignificant, seeing the small contributions of the fibres to Young s modulus,
Ec, of the composite. A similar concept can be used for UCS. Hence, we have
(E_f , E_m , V_f being Young's moduli of steel and mortar and volume fraction
of fibres, respectively)

$$E_c = E_m (1 - V_f) + V_f E_f \overline{\cos^4\theta} \qquad \text{with}$$

$$\overline{\cos^4\theta} = \int_0^{\pi/2}\cos^4\theta \sin\theta d\theta \Big/ \int_0^{\pi/2}\sin\theta \, d\theta = 1/5 \qquad (1a)$$

$$E_c = E_m (1 - V_f) + (\tfrac{1}{2} V_f) E_f \overline{\cos^2\theta} \qquad \text{with}$$

$$\overline{\cos^2\theta} = \int_0^{\pi/2}\cos^2\theta\cos\theta \sin\theta \, d\theta \Big/ \int_0^{\pi/2} \cos\theta \sin\theta \, d\theta = 1/2 \qquad (1b)$$

At UTS an arbitrary fibre intersecting the macrocrack under an angle
$(\pi/2-\theta)$ is transmitting a load perpendicular to the crack plane $P_f(\theta)\cos\theta$,
where $P_f(\theta)$ is the load acting in the fibre at the crack plane. A 3-D UR

fibre system obviously leads to a projection factor $\cos\theta$

$$\overline{\cos\theta} = \int_0^{\pi/2} \cos\theta \, \cos\theta \, \sin\theta \, d\theta \Big/ \int_0^{\pi/2} \cos\theta \, \sin\theta \, d\theta = 2/3 \tag{2}$$

We see that the sub set of fibres intersecting the crack plane is non-randomly distributed. The orientation distribution function $\Psi(\theta)$ is given by

$$\Psi(\theta) = \tfrac{1}{2}\sin\theta \, \cos\theta \tag{3}$$

where $\cos\theta$ stands for the relative probability that the crack intersects a fibre enclosing an angle $(\pi/2-\theta)$ with the crack plane. $\sin\theta$ represents the relative size of an infinitely small surface element of a sphere with unit radius.

For the 2-D UR system , the equivalent of eq (2) is

$$\overline{\cos\theta} = \int_0^{\pi/2} \cos\theta \, \cos\theta \, d\theta \Big/ \int_0^{\pi/2} \cos\theta \, d\theta = \pi/4 \tag{4}$$

The out-of-plane component is of course zero.

The stress transmitted in the macroplane also depends on the number of fibres per unit of area, N_A . Since $N_A = N_V \, \overline{H}$, we have to find a stereological expression for the average tangent height perpendicular to the crack plane, $\overline{H}$, for the 3-D and 2-D portions. $\overline{H}$ is a volume average respectively given by

(3-D)
$$\overline{H}_3 = 1 \int_0^{\pi/2} \cos\theta \, \sin\theta \, d\theta \Big/ \int_0^{\pi/2} \sin\theta \, d\theta = 1/2 \tag{5a}$$

(2-D)
$$\overline{H}_2 = 1 \int_0^{\pi/2} \cos\theta \, d\theta \Big/ \int_0^{\pi/2} d\theta = 21/\pi \tag{5b}$$

so that

(3-D)
$$N_{A_3} = \tfrac{1}{2}1 \, N_{V_3} = \tfrac{1}{2}L_{V_3} \tag{6c}$$

(2-D)
$$N_{A_2}(0) = \frac{2}{\pi}1 \, N_{V_2} = \frac{2}{\pi}L_{V_2} \; ; \quad N_{A_2}\left(\frac{\pi}{2}\right) = 0 \tag{6d}$$

and for the partially planar fibre structure

$$N_A(0) = N_{A_2} + N_{A_3} = \frac{2}{\pi}L_{V_2} + \tfrac{1}{2}L_{V_3} \; ; \quad N_A\left(\frac{\pi}{2}\right) = \tfrac{1}{2}L_{V_3} \tag{6c}$$

A good approximation to the boundary layer problem is presented in [6]. Two independent angles are required to define the spatial orientation of a fibre, as shown in fig. 1. The projection factor is as a consequence given by $\overline{\sin\phi \, \cos\beta}$. For a 3-D UR system $\sin\phi \, \cos\beta$ indeed yields the result of eq (2). The boundary zone is imaginary sliced. The contribution of all composite slices to the reinforcement ratio in the boundary layer – assumed to extend over half the fibre length- is formulated. To that end, only those fibres having their centre in an arbitrary slice and intersecting the crack plane are considered.

It must be obvious that a 3-D UR system would lead to intersections with the external surface. These fibres are removed from te model! When the slice is situated on a distance t from the external surface this implies all fibres intersecting the z-axis (perpendicular to the external surface) under an angle $\phi<\phi_0$, with $\cos\phi_0 = 2t/1$. Hence,

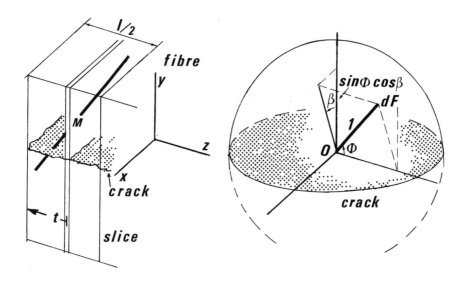

Fig. 1. Imaginary sliced boundary zone in SFRC (left) and sphere with unit
radius displaying the fibres having their centre in a slice and
intersecting the crack plane.

(3-D) $\overline{\cos\beta\ \sin\phi}$ $= \int_0^{\pi/2} \int_{\phi_0}^{\pi/2} \sin^3\phi\ \cos^2\beta d\phi\ d\beta\ /\ \int_0^{\pi/2}\int_0^{\pi/2}\sin^2\phi\ \cos\beta\ d\phi\ d\beta =$

$$= \cos\phi_0 - \cos^3\phi_0\ /3 \qquad (7)$$

Operating similarly for $\overline{H}$ we have

(3-D) $\overline{H} = 1 \int_0^{\pi/2}\int_{\phi_0}^{\pi/2}\sin^2\phi\ \cos\beta\ d\phi\ d\beta\ /\ \int_0^{\pi/2}\int_0^{\pi/2}\sin\phi\ d\phi\ d\beta$

$$= 1\ (1 - \frac{2}{\pi}\phi_0 + \frac{1}{\pi}\ \sin2\phi_0)/2 \qquad (8)$$

The relevant 2-D portion, the orientation plane being parallel to the
external surface, is not influenced by the external surface.

For $t \rightarrow 1/2$ we have $\phi_0 \rightarrow 0$ and eq (7) reduces to eq (2) and eq (8) to
eq (5a). N_A is in the boundary zone in case of a partially planar fibre
system as a consequence given by

$$N_A = \tfrac{1}{2}L_{V_3}\ (1 - \frac{2}{\pi}\phi_0 + \frac{1}{\pi}\sin2\phi_0) + \frac{2}{\pi}L_{V_2} \qquad (9)$$

A degree of partial orientation is defined by

$$\omega = L_{V_2}\ /\ L_{V_3} \qquad (10)$$

so that for 3-D systems $\omega=0$ and for 2-D ones $\omega=1$. Substitution in eqs (9) and (6c) yields for bulk and boundary values of the reinforcement ratio

$$N_A \ (t \leq 1/2) = \tfrac{1}{2}(L_V - \omega \ L_V)(1 - \tfrac{2}{\pi}\phi_0 + \tfrac{1}{\pi}\sin 2\phi_0) + \tfrac{2}{\pi}\omega L_V$$

$$= \tfrac{1}{2}L_V\gamma \ [1 + (\tfrac{4}{\pi\gamma} - 1)\omega] \tag{11a}$$

$$N_A \ (t \geq 1/2) = \tfrac{1}{2}(L_V - \omega \ L_V) + \tfrac{2}{\pi}\omega L_V$$

$$= \tfrac{1}{2}L_V \ [1 + (\tfrac{4}{\pi} - 1)\omega] \tag{11b}$$

with

$$\gamma = (1 - \tfrac{2}{\pi}\phi_0 + \tfrac{1}{\pi}\sin 2\phi_0) \tag{12}$$

The ratio of eqs (11a) and (11b) is

$$\frac{N_A(t \leq 1/2)}{N_A(t \geq 1/2)} = \gamma\frac{\{1+(\tfrac{4}{\pi\gamma} - 1)\omega\}}{\{1+(\tfrac{4}{\pi} - 1)\omega\}} \tag{13}$$

Since $\gamma \leq 1$ we see that the reinforcement ratio is predominantly (linearly) declining with γ as given by eq (12). This is only slightly counteracted by the more favourable orientation distribution of the fibres. The normalized contribution to the reinforcement ratio in the boundary zone of the slice at the exterior of the specimen is found by substitution of $\gamma=0$ in eq (13)

$$N_A \ (t=0)/N_A \ (t \geq 1/2) = \tfrac{4}{\pi} \ \omega \ /[1 + (\tfrac{4}{\pi} - 1)\omega] \tag{14}$$

whch is of course solely governed by the 2-D portion. As a result, eq (14) yields zero and unity for $\omega=0$ and $\omega=1$, respectively.

The average reduction of the reinforcement ratio in the boundary zone is obtained upon averaging over all possible values of t. A value of 0.58 is found for a 3-D UR fibre dispersion. Substitution in eq (13) for values of ω between roughly 0.2 and 0.4 demonstrates the reinforcement ratio of a partially planar fibre composite to decrease in the boundary layer to 70 to 80% of its bulk value!

In this approach it is assumed that slices outside the boundary zone contribute to the reinforcement in the boundary layer to the same degree as fibres inside the boundary zone contribute to the reinforcement outside this layer. The result can obviously be only slightly biased, which can be confirmed theoretically.

MECHANICAL CONCEPT

At the point where a fibre intersects the crack an axial load $P_f' + P_f$ is transmitted. Fibre and crack plane enclose an angle $\tfrac{\pi}{2} - \phi$ (see fig. 2). Because of preloading the material body is cracked and the fibre plastically deformed at the crack. Perpendicular to the crack plane a load $(P_f' + P_f)\cos\phi + f(P_f'+P_f)\sin\phi$ is transferred. P_f is due to interfacial friction along the shortest embedded fibre end, with a length of l_e. The second term of the transferred stress is coming from the shearing of the fibre over the crack edge. P_f' is the anchor force at the end of the fibre. For a large number of fibres the normal stress transferred by these fibres is obviously given by

39

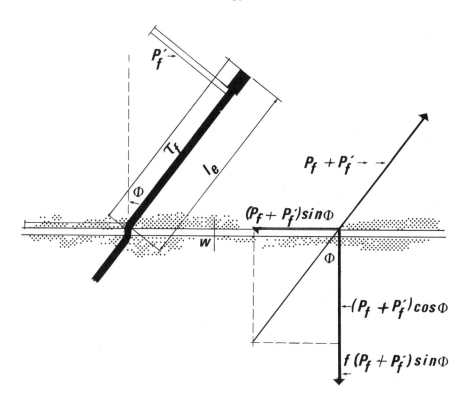

Fig. 2. Fibre intersecting the crack plane and transmitting a load over the
crack due to interfacial friction, τ_f , shearing over the crack
edge and an anchor load, P_f'.

$$\sigma_{nf} = (P_f' + \pi d \tfrac{1}{4}\tau_f)\, N_A\, (\overline{\cos\theta} + f\overline{\sin\theta}\,) \tag{15}$$

For a 3-D UR system $\overline{\cos\theta}$ and $\overline{\sin\theta}$ amount to 2/3 according to eq (2). Hence

$$\sigma_{nf} = \tfrac{2}{3}(P_f' + \tfrac{\pi}{4}d l\tau_f)(\tfrac{1}{2}L_V)(1 + f) \tag{15a}$$

With $V_f = \pi d^2 L_V /4$ and a=1/d eq (15a) can be written in the form

$$\sigma_{nf} = \frac{4\,V_f}{3\pi d^2}(1+f)\,P_f' + \tfrac{1}{3}a\,V_f\tau_f(1+f) \tag{15b}$$

When the crack significantly opens up l_e and N_A will be reduced linearly
with $2w/l$, w being the crack opening. The post ultimate strength is
therefore by good approximation given by

$$\sigma_{nf} = \frac{4V_f}{3\pi}(1+f)\frac{P_f'}{d^2}(1-\tfrac{2w}{l})+ \tfrac{1}{3}aV_f\tau_f(1+f)(1-\tfrac{4w}{l}) \tag{15c}$$

Assuming for simplicity that $P_f' = \tau_f' \pi d \overline{l}_e$ for $w \ll \overline{l}_e$, eq (15c) would reduce to the expression

$$\sigma_{nf} = \frac{1}{3} a\, V_f\, (\tau_f + \tau_f')(1+f)(1-\frac{4w}{l}) \tag{15d}$$

which includes the pull-out mechanism, the effect of shearing over the crack edges and the load transfer by anchoring facilities. Experimentally it would be impossible to distinguish between the various factors, so that a fictitious friction resistance τ_f^* is obtained, that is related to the parameters in eq (15d) by

$$\tau_f^* = (\tau_f + \tau_f')(1 + f) \tag{16}$$

We see that the "first order" approximation to formulating strength properties of SFRC, i.e.

$$\sigma_{nf} = \frac{1}{3} a \tau_f^* V_f \tag{17}$$

indeed correctly describes the material behaviour.

It should be noted that for larger crack openings the first term in eq (15c) tends to zero for $w=1/2$, whereas the second term tends to zero for $w=1/4$, a factor previously mentioned in the literature. Pull-out experiments have shown τ_f to deminish due to polishing. Experiments by Bien´ and Stroeven [1] indicate the drop to be considerable. The first part of the post ultimate curve will therefore tend to zero for values below $w=1/4$. This author has suggested a value of $w=1/6$, that is derived from eq (17) reorganized in the form

$$\sigma_{nf} = \tau_f^* (\tfrac{1}{2} L_V)(\pi d \tfrac{1}{6}) \tag{18}$$

demonstrating 50% of the fibres to be effective over a depth perpendicular to the crack plane of 1/6. Experimental values obtained on Bekaert and Thibo fibres reinforced concrete amounted to 0.183 l and 0.170 l, which are somewhat between 1/4 and 1/6 [7].

The model can easily be extended to cover the 2-D UR set as well. The only relevant component is the in-plane one. The component perpendicular to the orientation plane is zero. Substitution of eqs (4) and (6b) into eq (15) yields

$$\sigma_{nf} = (P_f' + \frac{\pi}{4} dl\tau_f)(\frac{2}{\pi} L_{V_2})(\frac{\pi}{4} + \frac{1}{2} f) \tag{19}$$

$$= \frac{1}{2} P_f' L_{V_2} (1+\frac{2}{\pi} f) + \frac{\pi}{8} d\tau_f L_{V_2} (1+\frac{2}{\pi} f) \tag{19a}$$

$$= \frac{4 V_{f_2}}{\pi d^2}(1+\frac{2}{\pi} f)P_f' + \frac{1}{2} a\, V_{f_2} \tau_f(1+\frac{2}{\pi} f) \tag{19b}$$

with the post ultimate equivalent

$$\sigma_{nf} = \frac{2 V_{f_2}}{\pi}(1+\frac{2}{\pi} f)\frac{P_f'}{d^2} (1-\frac{2w}{l}) + \frac{1}{2} a\, V_{f_2} \tau_f (1+\frac{2}{\pi} f)(1-\frac{4w}{l}) \tag{19c}$$

Assuming here again that $P_f' = \frac{\pi}{4} dl\tau_f'$ ($w \ll l_e$) eq (19c) can be written as

$$\sigma_{nf} = \tfrac{1}{2} a \, V_{f_2} \, (\tau_f + \tau_f')(1 + \tfrac{2}{\pi} f)(1 - \tfrac{4w}{l}) \tag{19d}$$

which finally leads with eq (16) to

$$\sigma_{nf} = \tfrac{1}{2} a \, V_{f_2} \, \tau_f^{*}(1 + \tfrac{2}{\pi} f)/(1 + f) \tag{20}$$

In the "first order" approximation we have developed the expression

$$\sigma_{nf} = \tfrac{1}{2} a \, V_{f_2} \, \tau_f^{*} \tag{20a}$$

that turns out to be only slightly biased, the error being $(1 + \tfrac{2}{\pi} f)/(1+f)$. Upon combining eqs (17) and (20) a strength concept for a partially planar steel fibre composite is obtained. Hence,

$$\sigma_{nf} = \tfrac{1}{3} a \, V_{f_3} \, \tau_f^{*} + \tfrac{1}{2} a \, V_{f_2} \, \tau_f^{*} \, (1 + \tfrac{2}{\pi} f)/(1+f)$$

Taking $C = (1 + \tfrac{2}{\pi} f)/(1+f)$ and introducing ω, this leads to

$$\sigma_{nf}(0) = \tfrac{1}{3} a \, V_f \, \tau_f^{*}[1 + (\tfrac{3}{2} C - 1)\omega] \tag{21a}$$

In the orthogonal direction we have of course

$$\sigma_{nf}(\tfrac{\pi}{2}) = \tfrac{1}{3} a \, V_f \, \tau_f^{*}(1 - \omega) \tag{21b}$$

With respect to the "first order" approximation outlined in [4] the changes are marginal. The gain is nevertheless that in addition to the pulling out mechanism, the shearing over the crack edges and the anchoring effect are encompassed.

The average strength of the boundary zone is found upon substitution of eqs (7) and (8) into the general strength expression. Hence for the 3-D UR system

$$\sigma_{nf}(t \leq l/2) = \tfrac{\pi}{4} d \, l \tau_f^{*} N_V \overline{H \, \cos\beta \, \sin\phi}$$

$$= \tfrac{\pi}{8} d l N_V \tau_f^{*} l \int_0^{\pi/2}(1 - \tfrac{2}{\pi}\phi_0 + \tfrac{2}{\pi}\sin\phi_0 \cos\phi_0)(\cos\phi_0 - \tfrac{1}{3}\cos^3\phi_0)/ \int_0^{\pi/2} d\phi_0$$

$$= \tfrac{1}{5} a \, V_f \, \tau_f^{*} \tag{22}$$

A strength reduction of 40% is found!

A partially planar system is in a more favourable situation. The strength can be calculated from eq (21a). Taking C=1 for simplicity reasons

$$\overline{\sigma_{nf}}(t \leq l/2) = \tfrac{1}{5} a \, V_{f_3} \, \tau_f^{*} + \tfrac{1}{2} a \, V_{f_2} \, \tau_f^{*} = \tfrac{1}{3} a \, V_f \, \tau_f^{*}(0.6 + 0.9 \, \omega) \tag{23}$$

For $\omega = 0.4$ the strength is reduced to 80% as compared to bulk property!

A method also presenting insight into structural and mechanical changes within the boundary zone will be presented elsewhere. The outcomes for this layer as a whole fit those presented.

DISCUSSION

A stereological concept for structural characterization of a partially planar fibre dispersion in SFRC is presented. In a law of mixtures concept for strength and stiffness it provides for the efficiency parameters. Additionally the boundary zone is modelled. The structural concept reveals anisometrical features and a declining reinforcement ratio towards the exterior of the specimen.

Experiments described in [8] have demonstrated this concept to be most relevant. The degree of orientation was found to increase roughly linearly with volume fraction of fibres. As a consequence, the reinforcement ratio's in orthogonal sections change parabolically with V_f as is obvious from eqs (6c) and (10). Upon introduction of $\omega = c\, L_V$ these expressions can be written in the following way

$$N_A(0) = \tfrac{1}{2} L_V \left(1 + (\tfrac{4}{\pi}-1)\,\omega\right) = \tfrac{1}{2} L_V \left(1 + (\tfrac{4}{\pi}-1)c\, L_V\right) \qquad (23a)$$

$$N_A(\tfrac{\pi}{2}) = \tfrac{1}{2} L_V (1 - \omega) = \tfrac{1}{2} L_V (1 - c\, L_V) \qquad (23b)$$

In the boundary zone –extending inwards over half the fibre length– the reinforcement ratio in sections perpendicular to the outer surface (controlling a crack that moves inwards from the outside) is predicted to decline to about 75% of its bulk value for average vales of ω. Experimental observations on the anisometry of the fibre structure in the bulk of the material body accurately matched the theoretical estimates according to eqs (23), as shown in [8]. Eq (23b) as well as the experimental data display a maximum within the considered range of V_f-values!

Mechanical behaviour of the fibre composite is partly governed by eqs (21). Completed with the mortar contribution we obtain

$$\sigma_{nf}(0) = \sigma_m(1 - V_f) + \tfrac{1}{3} a\, V_f\, \tau_f^{*}\, [1 + (\tfrac{3}{2}C - 1)\,\omega] \qquad (24a)$$

$$\sigma_{nf}(\tfrac{\pi}{2}) = \sigma_m(1 - V_f) + \tfrac{1}{3} a\, V_f\, \tau_f^{*}\, [1 - \omega] \qquad (24b)$$

$$C = (1 + \tfrac{2}{\pi} f)/(1 + f) \qquad (24c)$$

The splitting tensile data reported on in [8] quite accurately agree with the theoretical estimates according to eqs (24). So far we have taken C=1 in a "first order" approximation. A reduced value would make the correlation even better.

CONCLUSIONS

A complete stereological concept is nowadays available for quantitative image analysis of section or projection images of SFRC specimens. Actual reinforcement ratio's can be determined in this way. Generally, anisometry and segregation will also be detected by this approach; the fibre distribution will be found to deviate from the conventionally assumed uniform randomness.

The stereological modelling approach for mechanical characterization, presented in this contribution, encompasses the basic mechanisms for stress transfer in a pre-cracked and pre-loaded SFRC element. The developed formulas nevertheless turn out to be quite similar to the ones previously presented in a "first order" approximation solely based on fibre pull-out. A 3-D concept of the boundary zone is added. It completes the description of a SFRC material body. Estimates for the reinforcement ratio agree with experimental findings.

Strength estimates could be relevant for cracking studies and in case of size effect problems. A significant 20% reduction in strength is theoretically predicted for a boundary layer with a thickness of half the fibre length. It should be noted that this is an average value. The external layer of the boundary zone will theoretically have the lowest strength. The given value holds for a partially planar fibre structure. For a 3-D UR system a larger strength reduction can be expected, as shown in this paper. Moreover, the orientation plane of the planar portion is assumed to be parallel to the external surface. For an external surface perpendicular to the orientation plane of the 2-D portion, the distribution of those fibres will be affected as well in the boundary zone. Hence, the strength reduction will be larer than 20%. Derivation of this value follows the same lines as given in this paper for the 3-D portion. Instead of volumetric averages, planar ones are to be used, however. This is much simpler. The more complicated case is treated in this paper. Moreover, it is the more relevant one, since the experimental values concerned specimens sawn from larger plates. Hence, structural gradients at top and bottom surface can be expected to be governed by changes in the 3-D portion only.

REFERENCES

1. Bień, J. and Stroeven, P., Holographic interferometry study of debonding between steel and concrete. In Engineering applications of new composites, eds., S.A. Paipetis and G.C. Papanicolaou, Omega Scientific, Hartnolls Ltd, Bodmin, 1988, pp. 213-18

2. Stroeven, P., Characterization of microcracking in concrete. In Proc. Eur. Conf. Cracking in concrete and durability of constructions, Saint Rémy-lès-Chevreuse, August 31-Sept. 2, 1988, to be published

3. Stroeven, P. and Shah, S.P., Use of radiography-image analysis for steel fibre reinforced concrete, In Proc. RILEM Symp. Testing and test methods of fibre cement composites, ed. N. Swamy, Constr. Press, Lancaster, 1978, pp. 345-53

4. Stroeven, P., Morphometry of fibre reinforced cementitious materials. Part II. Mat. Constr., 1979, 12, 67, pp 9-20

5. Brandt, A.M., The optimization of fibre orientation in brittle matrix composite materials, Report Stevin Laboratory, Delft University of Technology, Delft, 1985

6. Stroeven, P., Structural and mechanical characterization of bond-based wire-reinforced cementitious composites. In Proc. Int. Conf. Advancing with composites, ed. I. Visconti, CUEN, Napoli, 1988, pp 159-66

7. Stroeven, P. Micro- and macromechanical behaviour of steel fibre reinforced mortar in tension. Heron, 1979, 24, 4, pp 7-40

8. Stroeven, P., Babut, R., Fracture mechanics and structural aspects of concrete. Heron, 1986, 31, 2

INFLUENCE OF FIBRE ORIENTATION ON MECHANICAL
STRENGTH OF FIBER REINFORCED CONCRETE

G. DEBICKI[*], J. RACLIN[**], P. HAMELIN[*]

*   INSA LYON Laboratoire Bétons et Structures
    Bât 304 - 69621 VILLEURBANNE CEDEX
** IUT de GRENOBLE
    B.P. 40 - 38403 SAINT MARTIN D'HERES

ABSTRACT

Previous investigations and experiences led to the assumption that fibres have not always been distributed uniformly in the concrete. In order to get futher informations about this problem, a particular fibre reinforced concrete is studied with two approaches : first the morphological features of fibre scatter is investigated and secondly an experiment consisting of directional axial compression or traction of oriented specimens is performed. The fibres used are metallic glass ribbons, the fresh concrete is placed by the pouring method. The results show that fifty percent of the fibers are disposed in planes perpendicular to the filling direction of the mould, the others being randomly distributed. In axial compression, strength, crack mode and strain are influenced by the position of the principal stress in relation to the natural axis of symmetry of material. Tensile strength is linked to a morphometric coefficient of the material. A strength criterion, giving failure stress and crack mode, is proposed.

INTRODUCTION

Design calculations for concrete work pieces consider that the material used is both homogeneous and isotropical where as it is in reality a heterogeneous material composed of aggregates, cement paste and voids. Will the inclusion of fibres in the concrete modifie this traditional out look ?

In trying to answer this question, we followed two complementary lines of research : first a study of the physical structure of the material, in order to determinate fibre layout in the concrete and secondly an analysis of the mechanical behaviour of the material under load. Our object is to assess the degree of heterogeneity of this material and it's influence on mechanical behaviour.

MATERIAL STUDIED

For the present study we used Saint Gobain's (French company) metallic glass ribbons, the dimensions of which are the following: length - 30 mm, breadth - 1 mm, thickness = 30 μm. Their strength is 2500 MPa, their elastic modulus is 130000 MPa. Due to their geometrical characteristics, we find a great number of fibres for a small volumetrical percentage consequently, there is a large contact surface between fibres and concrete.

For the composition of the mix used, we take into account the workability of the material, choosing a low fibre percentage (0.5 % of volume) and a plastic consistency (W/C - 0.5 with plasticizer, sand : Max Dia = 3.15 mm, C = 535 kg/m^3). Among the various techniques of placing fibre reinforced concrete, we used pouring method.

In order to study the natural layout of the fibers in the mass of the concrete it is necessary to make sufficiently large specimens to avoid the surface influence of the mould. Blocks (dimensions : 32 x 32 x 40 cm), were moulded on a flow table, the first series was filled in three times, followed each time by ten successive drops. The second series was filled in ten times, followed each time by ten drops.The quantity of energie received during moulding stage is much higher for series n° 2. The specimens were cured in the open air till the testing date.

LAYOUT STUDY OF THE FIBERS

It is during the pouring stage that the material organizes and structures itself. If we try pouring the fibres alone, we notice that, due to gravity and to their geometrically linear shape (thirty times wider than thick), they have a natural tendency to orient themselves horizontally, and so perpendicularly to the filling direction. In fresh concrete the fibers have the same tendency, but they are obstructed by the other materials composing the concrete. We can then say that favorable elements for the natural fibre orientation are : a good workability, fine gravel, high density of fibers, firm compaction. So, we conclude that several elements work together to orient the fibers perpedicular to the filling direction. Consequently a natural axis of symmetry linked to the fiber distribution mode appears. This allows us to recognize the material as being a cylindrical orthotropical material.

In the 32 x 32 x 40 cm blocks we sawed tile-like slabs (about 8 mm thick) as described in fig 1. Next, we projected, on a plane, the fibres included in the slabs by X-ray radiography. Then, we analysed the pictures obtained using principles of stereology. The number of intersections $P_L(\theta)$ per unity of length between an oriented test grid(θ is the angle between test grid and the axis of symmetry) and the fibers is used to characterise fiber distribution for a given direction [1].

Block

dimensions in cm.

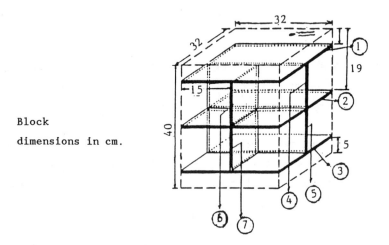

Figure 1. Positions of slabs in the block

Figure 2 shows the rose diagram of $P_L(\theta)$ for both series of specimen (measurements were made on a mean number of eight pictures per block) Partial orientation of fibres is made clear in this diagram.

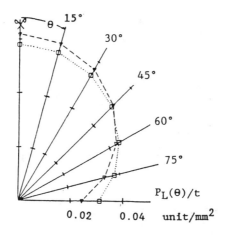

Figure 2. Rose diagram of $P_L(\theta)$; t is the thickness of the slabs.

For each series, an identical rose diagram may be obtained by superimposing a volumetric fraction of randomly distributed fibres, V_{Viso}, and a fraction of fibres randomly scattered in a plane perpendicular to the filling direction V_{Vor}. Table 1 gives the numerical results of such a comparison. Approximately 50 % of the fibres position themselves at 90° to the filling direction. We have to notice that the ribbons, of the oriented fraction V_{Vor}, lay horizontally on their wider breadth. So, our analysis brings out a "partial stratification" effect of the material caused by the fibres.

TABLE 1
Mean repartition of fibre pourcentage between 3D and 2D

Fibre pourcentage	Series 1	Series 2
V_{Viso} (3D)	0,33 %	0,26 %
V_{Vor} (2D)	0,17 %	0,24 %
V_V	0,5 %	0,5 %

In the case studied, tight compaction of concrete has little influence on fibre orientation, but decreases macroporosity of the matrix, by reducing air and water retained by the fibres, and ensuring a better bleeding.

STUDY OF ANISOTROPIC BEHAVIOUR

J. BONZEL and M. SCHMIDT [3] had already noticed the fibres' tendency to orient themselves perpendicular to the filling direction, but without quantifying it. They cut out test specimens at angles 0° and 90° to placing direction and then studied their mechanical behaviour. Their results show that tensile strength is increased when the fibres are parallel to stress, and compressive strength is increased when they are perpendicular. The influence of fibre orientation on tensile strength was studied in more detail by A.M. BRANDT [4].

To analyse anisotropic behaviour we core-drilled test specimens (ϕ = 80 mm, l = 160 mm) out of the block, at seven different angles, 0°, 15°, 30°, 45°, 60°, 75°, 90°, all measured from the block's natural axis of symmetry. On each specimen the preferencial fibre distribution plane is noted. θ is the angle between the natural axis of symmetry of material and the axis of load.

Compression

The core-drilled cylinders were loaded at a constant rate. For tests, we eliminated frictionnal restraint, on the specimen-testing device interfaces, with oilded aluminium and plastic sheets piled.

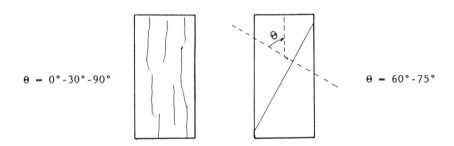

θ = 0°-30°-90°   θ = 60°-75°

Figure 3. Cracks in compressive specimens

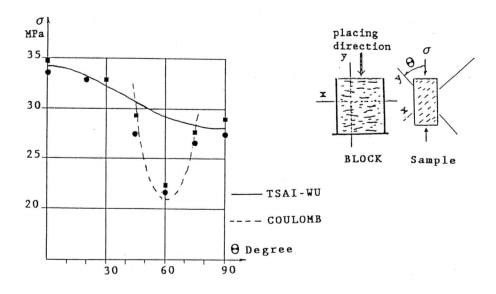

Figure 4. Compression strength in function of θ (angle between principal stress and natural axis of symmetry of material

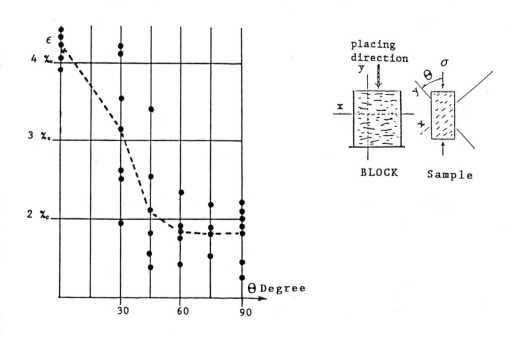

Figure 5. Strain at the pic load.

Under compression, on the specimens at 0°, 15°, 30° and 90°, vertical cracks aligned in the direction of loading appeared as traditionaly in concrete (crack mode I). On specimens at 60° and 75°, a single crack appeared parallel to the fibre stratification planes, afterwards the fracture process continued by slidding on this crack plane (crack mode II) figure 3. For specimen at 45°, the crack direction was uncertain. For each series, the mean yield values are noted in figure 4. The material is not mechanically isotropic. The behaviour of the material is different following θ - 0° and θ - 90°, but it must be noted that the lowest yield value is at θ - 60°. This phenomenon exists for rocks that exhibit a "bedding plane" type of anisotropy, also for composites made of short fibres.

Figure 5 presents the strain at the pic load. We remark a continuous variation of strain in function of θ.

Tensile strength

Strengths against unidirectional tensile load for core drilled test specimens are noted in figure 6. Rupture follows a plane perpendicular to the main stress direction, and this regardless of the angle at which the specimen was drilled. The role of the matrix is therefore essential.

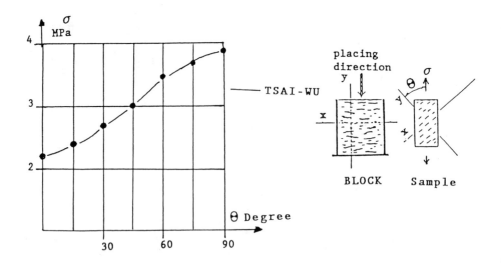

Figure 6. Tensile strength in function of θ (angle between principal stress and natural axis of symmetry of material)

Figure 7 shows the relationship between σ_T and the value of $\mu(\theta)$ which represents the total projected length of the fibres in the stress direction over the total length of included fibres. In this way we link tensile strength to a morphologic characteristic of the material.

$$\sigma_T - 1,8 + 6 \ [\mu(\theta)]^2$$

This equation has yet to be confirmed by further tests using different fibre percentages.

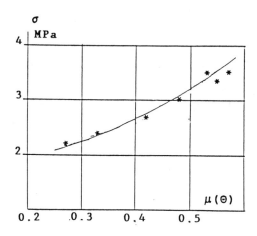

Figure 7. Tensile strength in function of μ (θ).

STRENGTH CRITERIA

We put forward a strength criterion which takes into account both the anisotropic strength of the material and the fact that rupture follows two different crack modes. The criterion has two branches, each associated with a rupture mode. Similar methods have been put forward by HASHIN [5] for unidirectional fibre composites and BOEHLER, RACLIN [6] for fibre-glass mat.

For criterion in mode I, we had no multiaxial test results, but we had different traction/compression results associated to given orientations.

Simplicity lead us to choose TSAI-WU's quadratic criterion [7]. For uniaxial load it is written as follows :

$$\sigma_n^2 \; [F_{11} \sin^4\theta + F_{33} \cos^4\theta + \cos^2\theta.\sin^2\theta(2F_{31} + F_{44})] +$$
$$\sigma_n \; (F_1 \sin\theta + F_3 \cos^2\theta) - 1$$

In order to determinate the five coefficients, it is necessary to know for example compression strength for orientations of 0°, 30° and 90°, and tensile strength for 0° and 90°.

For the criterion in mode II we chose Coulomb's criterion with ϕ_0 being the angle of friction and c_0 being the shear strength. The numerical application with this material gives tg ϕ_0 - 0,61 and c_0 - 5,85 MPa.

$$\sigma_n \cos\theta \; [\sin\theta - \cos\theta.\text{tg} \; \phi_0] - c_0$$

In figure 4 and figure 6, we can see that theoretical and experimental results match. The two criteria chosen allow us to predict both material strength and rupture mode.

CONCLUSION

The fibre reinforced concrete we studied presents a partially oriented fibre structure which gives it an anisotropic mechanical behaviour.

In uniaxial compression, strength, crack mode and strain are influenced by the position of the principal stress against the natural axis of symmetry of material (θ). Two modes of crack are observed depending which of the two the matrix or the "stratification" influences cracking. The lower strength is obtained for θ - 60° with a shear mode crack surface parallel to the preferential planes of fibre positioning.

Strength under tensile load is linked to a morphologic coefficient of the material. This coefficient caracterises fibre distribution.

The association of the two strength criteria chosen gives satisfactory results in predicting both directional strength and rupture mode.

When calculating structures it will be important to take into account, especially in multiaxial stress zones, the anisotropic characteristics confered to the material during the moulding stage.

REFERENCES

1. DE GUILLEBON B. SHOM J.M. : "Metallic glass ribbons - A new fibre for concrete reinforcement" - RILEM SYMPOSIUM, 13-17 July Sheffield 1986.

2. SALTYKOV S.A. : Stereometric metallography, second edition, Moscow : metallurgizdat 1958.

3. BONZEL J., SCHMIDT M. : "Distribution and orientation of stell fibres in concrete and their influence on the characteristics of steel fibre concrete" - RILEM SYMPOSIUM, 13-17 July Sheffield 1986.

4. BRANDT A.M. : "Influence of the fibre orientation on the mechanical properties of fibre reinforced cement (FRC specimens)" - RILEM INTERNATIONAL CONGRESS - Matériaux mixtes et Composites : élaborations et propriétés - Versailles FRANCE 7-11 septembre 1987.

5. HASHIN Z. : "Failure criteria for unidirectional fiber composites" - J. Applied Mechanics, Vol. 47, 1980, p. 329-334.

6. BOEHLER J.P., RACLIN J., DELAFIN M. : Rapport contrat DGRST, n° 78-70610/0611, 1981.

7. TSAI S.W., WU E.M. : "A general theory of strength for anisotropic materials" J. Comp. Mat., pp. 58-80, vol. 5, 1971.

OPTIMAL FIBRE ORIENTATION IN CONCRETE-LIKE COMPOSITES

MARIA MARKS
Institute of Fundamental Technological Research,
Polish Academy of Sciences,
Swietokrzyska 21, 00-049 Warsaw, Poland

ABSTRACT

The paper deals with composite materials in which brittle matrix is reinforced with two systems of parallel continuous fibres. The stress and deformation states are determined for a composite element subjected to normal and tangent loads. The optimal directions of the systems of fibres are obtained from the criterion of the minimum of strain energy. The directional angle θ_1 of the first system of fibres and the angle α between the two systems are considered as independent variables. The solution of the equations for the minimum strain energy is obtained for the parameters corresponding to a certain concrete-like composite.

INTRODUCTION

Previous papers on the optimization of composite materials reinforced with fibres concern mostly the high strength composites [1],[2],[3]. Paper [4] deals with optimization of the internal structure of different kind of a material composed of brittle matrix and ductile fibres. In [4] the optimum fibre orientation is determined using maximum fracture energy as a criterion. The present paper concerns the optimization of a composite element in which brittle matrix is reinforced with ductile fibres, but at another optimization criterion: the optimal directions of fibre systems are determined from the minimum strain energy of the element.

STRAIN IN A COMPOSITE ELEMENT

The composite element is in shape of a plate of thickness 2h, in which the matrix is reinforced with two families of parallel fibres inclined at angles θ_1 and $\theta_1+\alpha$ to axis x_1 - Fig.1.

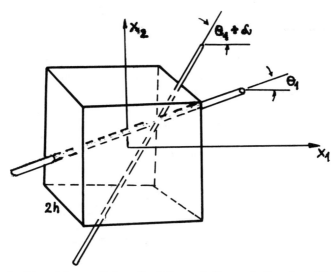

Figure 1. Directions of two families of the reinforcing fibres.

The composite element is in plane state of stress which can be described by three stress components, corresponding to certain mean values over the thickness of the element [5]. From the definition of a generalized plane stress state the following relation may be found for the mean composite stress $\sigma_{\alpha\beta}^{(c)}$, mean matrix stress $\sigma_{\alpha\beta}^{(m)}$ and stresses in both fibre systems $\sigma_{a}^{(s)}$ and $\sigma_{b}^{(s)}$:

$$\sigma_{\alpha\beta}^{(c)} = \frac{h_m}{h}\sigma_{\alpha\beta}^{(m)} + \frac{h_a}{h} \sigma_a^{(s)} a_\alpha a_\beta + \frac{h_b}{h} \sigma_b^{(s)} b_\alpha b_\beta, \qquad \alpha,\beta=1,2. \qquad (1)$$

here h_m defines the total thickness of the matrix, h_a and h_b define the thickness of both layers of fibres ($h=h_m+h_a+h_b$). The stresses $\sigma_a^{(s)}$ and $\sigma_b^{(s)}$ are oriented along the fibres, it means according to tangents of the both fibre systems − vectors $\underline{a}$ and $\underline{b}$.

Assuming,that the matrix and thin fibres are build of linear-elastic, isotropic and homogeneous materials, the corresponding constitutive equations have the form:

$$\sigma_{\alpha\beta}^{(m)} = \frac{E^{(m)}}{1 + \nu} \varepsilon_{\alpha\beta}^{(m)} + \frac{\nu E^{(m)}}{(1 + \nu)(1 - \nu)} \delta_{\alpha\beta}\varepsilon_{\delta\delta}^{(m)}$$

$$\sigma_a^{(s)} = E^{(s)} \varepsilon_a^{(s)} \quad , \quad \sigma_b^{(s)} = E^{(s)} \varepsilon_b^{(s)} \qquad (2)$$

here $E^{(m)}$ and $E^{(s)}$ are Young's moduli of matrix and fibres respectively and ν is the Poisson's ratio of the matrix.

Next, assumed is the continuity of strain of the matrix and fibres:

$$\varepsilon_a^{(s)} = \varepsilon_a^{(m)} \quad , \qquad \varepsilon_b^{(s)} = \varepsilon_b^{(m)} \quad . \tag{3}$$

Substituting equations (2) and (3) into equation (1) and using the following relations:

$$\varepsilon_a^{(m)} = \varepsilon_{\gamma\delta}^{(m)} a_\gamma a_\delta \quad ,$$

$$\varepsilon_b^{(m)} = \varepsilon_{\gamma\delta}^{(m)} b_\gamma b_\delta \quad ,$$

leads to an expression for the mean stress:

$$\sigma_{\alpha\beta}^{(c)} = \frac{E}{1+\nu} \left(\varepsilon_{\alpha\beta}^{(m)} + \frac{\nu}{1-\nu} \delta_{\alpha\beta} \varepsilon_{\delta\delta}^{(m)} \right) \frac{h_m}{h} +$$
$$+ E \frac{{}^{(s)}h_a}{h} \varepsilon_{\gamma\delta}^{(m)} a_\gamma a_\delta a_\alpha a_\beta + E \frac{{}^{(s)}h_b}{h} \varepsilon_{\gamma\delta}^{(m)} b_\gamma b_\delta b_\alpha b_\beta \tag{4}$$

From this equation the strain components $\varepsilon_{11}^{(m)}$, $\varepsilon_{12}^{(m)}$ and $\varepsilon_{22}^{(m)}$ can be uniquely derived.

Let us suppose that the element is subjected to two tensile stresses: p along the axis x_2 and q along the axis x_1 and to a shear stress τ. If the fibres of both systems are uniformly distributed, that is if $\dfrac{h_a}{h} = \dfrac{h_b}{h} = \dfrac{1}{2} \dfrac{h - h_m}{h}$, and if for compactness the substitutions are made $\gamma = E \dfrac{{}^{(m)}h_m}{h}$, $\beta = E \dfrac{{}^{(s)}h - h_m}{h}$, the strain components are given in the following form:

$$\varepsilon_{11}^{(m)} = \frac{1}{D} \left\{ q \left[\frac{\gamma^2}{(1+\nu)^2(1-\nu)} + \frac{\gamma\beta}{4(1+\nu)(1-\nu)} \left[1 - \cos2(2\theta_1+\alpha)\cos2\alpha \right] + \right. \right.$$

$$+ \frac{\gamma\beta}{2(1+\nu)} \left[\frac{3}{4} - \cos(2\theta_1+\alpha)\cos\alpha + \frac{1}{4}\cos2(2\theta_1+\alpha)\cos2\alpha \right] + \frac{1}{8}\beta^2 \left[\cos\alpha + \right.$$

$$\left. -\cos(2\theta_1+\alpha) \right]^2 \sin^2\alpha \Bigg] + p \left[\frac{-\nu\gamma^2}{(1+\nu)^2(1-\nu)} - \frac{\gamma\beta}{8(1-\nu)} \left[1 - \cos2(2\theta_1+\alpha)\cos2\alpha \right] + \right.$$

$$+ \frac{\beta^2}{16} \left[\cos 2\alpha - \cos 2(2\theta_1 + \alpha) \right] \sin^2 \alpha \right] + \tau \left[- \frac{\gamma \beta}{2(1+\nu)} \sin(2\theta_1 + \alpha) * \cos \alpha + \right.$$

$$\left. - \frac{\gamma \beta}{4(1-\nu)} \sin 2(2\theta_1 + \alpha) \cos 2\alpha - \frac{\beta^2}{4} \left[\cos \alpha - \cos(2\theta_1 + \alpha) \right] \sin(2\theta_1 + \alpha) \sin^2 \alpha \right] \right\} ,$$

$$\varepsilon_{22}^{(m)} = \frac{1}{D} \left\{ q \left[- \frac{\nu \gamma^2}{(1+\nu)^2 (1-\nu)} - \frac{\gamma \beta}{8(1-\nu)} \left[1 - \cos 2(2\theta_1 + \alpha) \cos 2\alpha \right] + \right. \right.$$

(5)

$$\left. + \frac{\beta^2}{16} \left[\cos 2\alpha - \cos 2(2\theta_1 + \alpha) \right] \sin^2 \alpha \right] + p \left[\frac{\gamma^2}{(1+\nu)^2 (1-\nu)} + \right.$$

$$+ \frac{\gamma \beta}{4(1+\nu)(1-\nu)} \left[1 - \cos 2(2\theta_1 + \alpha) \cos 2\alpha \right] + \frac{\gamma \beta}{8(1+\nu)} \left[\cos 2(2\theta_1 + \alpha) \cos 2\alpha + \right. \right.$$

$$\left. + 4 \cos(2\theta_1 + \alpha) \cos \alpha + 3 \right] + \frac{\beta^2}{8} \left[\cos(2\theta_1 + \alpha) + \cos \alpha \right]^2 \sin^2 \alpha \right] +$$

$$+ \tau \left[- \frac{\gamma \beta}{2(1+\nu)} \sin(2\theta_1 + \alpha) \cos \alpha + \frac{\gamma \beta}{4(1-\nu)} \sin 2(2\theta_1 + \alpha) \cos 2\alpha + \right.$$

$$\left. - \frac{\beta^2}{4} \left[\cos(2\theta_1 + \alpha) + \cos \alpha \right] \sin(2\theta_1 + \alpha) \sin^2 \alpha \right] \right\} ,$$

$$\varepsilon_{12}^{(m)} = \frac{1}{D} \left\{ q \left[- \frac{\gamma \beta}{4(1+\nu)} \sin(2\theta_1 + \alpha) \cos \alpha - \frac{\gamma \beta}{8(1-\nu)} \sin 2(2\theta_1 + \alpha) \cos 2\alpha + \right. \right.$$

$$\left. - \frac{\beta^2}{8} \left[\cos \alpha - \cos(2\theta_1 + \alpha) \right] \sin(2\theta_1 + \alpha) \sin^2 \alpha \right] +$$

$$+p\left[\frac{-\gamma\beta}{4(1+\nu)}\sin(2\theta_1+\alpha)\cos\alpha+\frac{\gamma\beta}{8(1-\nu)}\sin2(2\theta_1+\alpha)\cos2\alpha-\frac{\beta}{8}\left[\cos(2\theta_1+\alpha)+\right.\right.$$

$$\left.+\cos\alpha\right]\sin(2\theta_1+\alpha)\sin^2\alpha\right]+\tau\left[\frac{\gamma^2}{(1+\nu)(1-\nu)}+\frac{\gamma\beta}{(1+\nu)(1-\nu)}+\right.$$

$$\left.\left.-\frac{\gamma\beta}{4(1-\nu)}\left[1-\cos2(2\theta_1+\alpha)\cos2\alpha\right]+\frac{\beta^2}{4}\sin^2(2\theta_1+\alpha)\sin^2\alpha\right]\right\},$$

here:

$$D=\frac{\gamma}{(1+\nu)(1-\nu)}\left[\frac{\gamma^2}{1+\nu}+\frac{\gamma\beta}{1+\nu}+\frac{\beta^2}{4}\left[1-\cos^4\alpha-\nu\sin^4\alpha\right]\right].$$

OPTIMAL DIRECTION OF FIBRE SYSTEMS

The minimum of the strain energy is chosen as optimization criterion to determine the directions of the fibres. The energy may be expressed by a following relation:

$$U=\frac{1}{2}\iiint_V \overset{(c)}{\sigma}_{ij}\overset{(c)}{\varepsilon}_{ij}\,dV=h\iint_\Omega\frac{1}{2}\overset{(c)}{\sigma}_{\alpha\beta}\overset{(m)}{\varepsilon}_{\alpha\beta}\,d\Omega .$$

The angle θ_1 of the direction of the first system of fibres and the angle α between the two systems are independent variables.

Because the functional U does not depend on the derivatives of θ_1 and α , therefore the necessary conditions for the minimum are:

$$\frac{\partial W}{\partial\theta_1}=0 \quad\text{and}\quad \frac{\partial W}{\partial\alpha}=0 , \tag{6}$$

here:

$$W=\frac{1}{2}\overset{(c)}{\sigma}_{\alpha\beta}\overset{(m)}{\varepsilon}_{\alpha\beta}.$$

The sufficient conditions for the minimum of the functional have in this case the following form:

$$\frac{\partial^2 W}{\partial\theta_1^2}\frac{\partial^2 W}{\partial\alpha^2} - \left[\frac{\partial^2 W}{\partial\theta_1\partial\alpha}\right]^2 > 0 \qquad \text{and} \qquad \frac{\partial^2 W}{\partial\theta_1^2} > 0 \quad , \quad \frac{\partial^2 W}{\partial\alpha^2} > 0 \quad . \qquad (7)$$

The form of the function W can be obtained after substituting the strain components from (5) and it is as follows:

$$W = \frac{1}{2D}\left\{K_1 - \left[\left[p-q\right]^2 - 2\tau^2\right]\left[\frac{\gamma\beta}{8(1-\nu)}\cos2(2\theta_1+\alpha)\cos2\alpha + \right.\right.$$

$$-\frac{\beta^2}{8}\cos^2(2\theta_1+\alpha)\sin^2\alpha\right] - 3\tau\left[p-q\right]\left[\left[\frac{\gamma\beta}{4(1-\nu)}+\frac{\beta^2}{16}\right]\sin^2\alpha +\right.$$

$$\left. - \frac{\gamma\beta}{8(1-\nu)}\right]\sin2(2\theta_1+\alpha) + \left[p^2-q^2\right]\left[\frac{\gamma\beta}{2(1+\nu)}+\frac{\beta^2}{4}\sin^2\alpha\right]\cos(2\theta_1+\alpha)\cos\alpha +$$

$$- 3\tau\left[p+q\right]\left[\frac{\gamma\beta}{4(1+\nu)}+\frac{\beta^2}{8}\sin^2\alpha\right]\sin(2\theta_1+\alpha)\cos\alpha + \tau^2\frac{\beta^2}{4}\sin^2\alpha +$$

$$(8)$$

$$\left. + \left[p+q\right]^2\frac{\beta^2}{8}\sin^2\alpha\cos^2\alpha\right\} .$$

here:

$$K_1 = \left[q^2 - 2\nu pq + p^2 + (1+\nu)\tau^2\right]\frac{\gamma^2}{(1+\nu)^2(1-\nu)} + \left[q^2 - (1+\nu)qp + p^2 + \right.$$

$$\left. + 4\tau^2 - (1+\nu)\tau^2\right]\frac{\gamma\beta}{4(1+\nu)(1-\nu)} + \left[p^2+q^2\right]\frac{3\gamma\beta}{8(1+\nu)} .$$

The necessary conditions (6) for the minimum of the strain energy W – eq.(8) have the form of nonlinear equations and only an approximate minimum of the function W may be looked for. Approximate solution will be obtained for a material which is composed of concrete matrix and of steel fibres of volumetric content 2%.

The values of the function W will be calculated for different ratios $k=\dfrac{q}{p}$ and $l=\dfrac{\tau}{p}$ (Fig.2), and for angles $\theta_1 = 0°, 10°, \ldots, 90°$ and $\alpha = 0°, 10°, \ldots, 90°$

using the following data: $E^{(m)} = 30000$ MPa , $E^{(s)} = 210000$ MPa , $\nu = \dfrac{1}{6}$, $\dfrac{h_m}{h} = 0.98$.

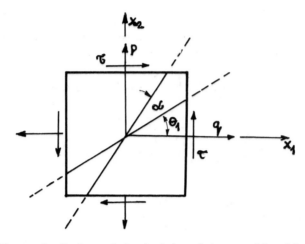

Figure 2. System of loads taken into consideration.

The calculations lead to a conclusion, that the strain energy for different ratios of loads k and l reaches the minimum for different values of the angle θ_1 and the angle α equal 0° or 90°. We may conclude, that for the concrete-like composites the optimal angle between the fibres of both systems is either 0° or 90°, it means that the solution has the form of a single fibre system or of two orthogonal systems. When the angle α may be fixed, then it is much simpler to get the analytical solutions of the equations which yield the angles θ_1 corresponding to the minimum of the strain energy. The necessary condition for the minimum of the strain energy has the form:

$$
\frac{\partial W}{\partial \theta_1} = -\frac{p}{2D} \left\{ \left[2l^2 - (1-k)^2 \right] \left[\frac{\gamma\beta}{2(1-\nu)} \cos 2\alpha - \frac{\beta^2}{4} \sin^2\alpha \right] \sin 2(2\theta_1 + \alpha) + \right.
$$

$$
+ 3l(1-k)\left[\left[\frac{\gamma\beta}{1-\nu} + \frac{\beta^2}{4} \right] \sin^2\alpha - \frac{\gamma\beta}{2(1-\nu)} \right] \cos 2(2\theta_1 + \alpha) +
$$

$$
+ (1-k^2)\left[\frac{\beta^2}{2} \sin^2\alpha + \frac{\gamma\beta}{1+\nu} \right] \sin(2\theta_1 + \alpha)\cos\alpha + 3l(1+k)\left[\frac{\beta^2}{4} \sin^2\alpha + \right.
$$

$$
\left. + \frac{\gamma\beta}{2(1+\nu)} \right] \cos(2\theta_1 + \alpha)\cos\alpha \left. \right\} = 0 \ ,
$$

(9)

here:

$$D = \frac{\gamma}{(1+\nu)(1-\nu)}\left[\frac{\gamma^2}{1+\nu} + \frac{\gamma\beta}{1+\nu} + \frac{\beta^2}{4}\left[1-\cos^4\alpha-\nu\sin^4\alpha\right]\right].$$

The sufficient condition (7) for the minimum is expressed in the form:

$$\frac{\partial^2 W}{\partial\theta_1^2} = -\frac{p^2}{D}\left\{\left[2l^2-(1-k)^2\right]\left(\frac{\gamma\beta}{1-\nu}\cos2\alpha - \frac{\beta^2}{2}\sin^2\alpha\right)\cos2(\theta_1+\alpha) + \right.$$

$$- 3l(1-k)\left[\left(\frac{2\gamma\beta}{1-\nu} + \frac{\beta^2}{2}\right)\sin^2\alpha - \frac{\gamma\beta}{1-\nu}\right]\sin2(\theta_1+\alpha) + (1-k^2)\left(\frac{\gamma\beta}{1+\nu} + \right.$$

$$+ \frac{\beta^2}{2}\sin^2\alpha\right)\cos(2\theta_1+\alpha)\cos\alpha - \frac{3}{2}l(1+k)\left[\frac{\beta^2}{2}\sin^2\alpha + \right.$$

$$\left. + \frac{\gamma\beta}{1+\nu}\right]\sin(2\theta_1+\alpha)\cos\alpha\left.\right\} > 0 . \tag{10}$$

In the case of a single fibre system when $\alpha=0$, condition (9) is expressed by the following equation:

$$\left[2l^2-(1-k)^2\right]\sin4\theta_1 - 3l(1-k)\cos4\theta_1 + (1+k)\frac{1-\nu}{1+\nu}\left[2(1-k)\sin2\theta_1 + \right.$$

$$\left. + 3l\cos2\theta_1\right] = 0 , \tag{11}$$

and the condition (10) has the form:

$$- 2\left[2l^2-(1-k)^2\right]\cos4\theta_1 - 6l(1-k)\sin4\theta_1 - (1+k)\frac{1-\nu}{1+\nu}\left[2(1-k)\cos2\theta_1 + \right.$$

$$\left. - 3l\sin2\theta_1\right] > 0 . \tag{12}$$

In the case of two orthogonal fibre systems, the necessary condition for the minimum of strain energy corresponds to the equation:

$$\mathrm{tg}4\theta_1 = \frac{3l(1-k)}{2l^2-(1-k)^2} , \tag{13}$$

and the sufficient condition to the inequality:

$$\left[2l^2-(1-k^2) \right] \cos4\theta_1 + 3l(1-k)\sin4\theta_1 < 0 \ . \tag{14}$$

If the angle θ_1^o satisfies the equation (11) and the inequality (12), then for $\theta_1=\theta_1^o$ the strain energy minimum is obtained with a single fibre system.

If the angle θ_1^* satisfies the relations (13) and (14), then for $\theta_1=\theta_1^*$ the reinforcement with two orthogonal systems is required. Next we calculate value of the strain energy $W(\theta_1^o)$ and $W(\theta_1^*)$. The smaller value of the strain energy corresponds to optimal orientation of the fibres.

Let us consider the particular cases of load to illustrate the calculations. For k=-1, l=1 - it means for q=-p and τ=p - the equation (11) is identical to the equation (13), and has a following form:

$$\text{tg } 4\theta_1 = -3 \ .$$

This necessary condition is satisfied at $\theta_1=27.11^o$ and $\theta_1=72.11^o$. The sufficient conditions (12) and (14) are identical and may be expressed by the following inequality:

$$-2\cos4\theta_1+6\sin4\theta_1 < 0 \ .$$

This inequality is satisfied at $\theta_1=72.11^o$. If we compare the values of the strain energy for $\theta_1=72.11^o$, $\alpha=0^o$ and for $\theta_1=72.11^o$, $\alpha=90^o$, we obtain that for the concrete-like composites: $W(\theta_1=72.11^o;\alpha=0) > W(\theta_1=72.11^o;\alpha=90^o)$. If the composite element is subjected to loads q=-p and τ=p, then the optimal orientation of fibres corresponds to two orthogonal systems of fibres which are inclined at angle 72.11^o.

Taking into account the necessary and sufficient conditions (11)-(14) for the minimum strain energy, we may find the optimal directions of fibre systems for different ratios of the normal and shear loads. In the Table 1 shown are the optimal angles of fibres for several ratios of the loads: k=q/p=-2,-1,0,1,2 and l=τ/p=0,1,2.

TABLE 1

Optimal angles of orientation of fibres

	k =-2	k =-1	k =0	k =1	k =2
l=0	$\theta_1=0$ $\alpha =0^O$	$\theta_1=0$ $\alpha =90^O$	$\theta_1=90$ $\alpha =0^O$	θ_1 $\alpha =90^O$	$\theta_1=0$ $\alpha =0^O$
l=1	$\theta_1=166.93^O$ $\alpha =0^O$	$\theta_1=72.11^O$ $\alpha =90^O$	$\theta_1=62.59^O$ $\alpha =0^O$	$\theta_1=45^O$ $\alpha =0^O$	$\theta_1=27.69^O$ $\alpha =0^O$
l=2	$\theta_1=158.18^O$ $\alpha =0^O$	$\theta_1=62.89^O$ $\alpha =90^O$	$\theta_1=54.97^O$ $\alpha =0^O$	$\theta_1=45^O$ $\alpha =0^O$	$\theta_1=35.25^O$ $\alpha =0^O$

The optimal angles θ_1 and α correspond to minimum values of the strain energy. Figure 3 shows the dependence of a certain quantity W^* corresponding to optimal angles θ_1 and α as shown in Table 1, depending on the ratio of the loads k and l. The quantity W^* is defined as:

$$W = \frac{(1+\nu)(1-\nu)p^2}{2\gamma} * W^* .$$

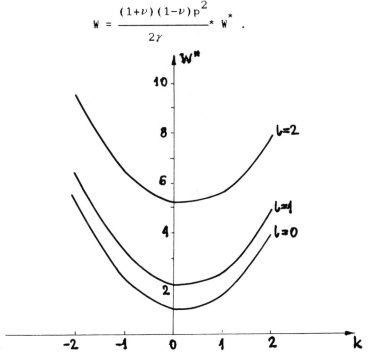

Figure 3. Effect of k and n on the strain energy.

From a study of the results, we may conclude that if the composite element is subjected to normal loads only: p=q and $\tau=0$ (l=0, k=1), the strain energy attains minimum at two orthogonal systems of fibres, which are inclined at any angle θ_1. If p=q and $\tau=0$ then the optimal orientation of fibres corresponds to a single fibre system, which is inclined at angle 45° to axis x_1.

If the composite element is subjected to the compressive load q=-p along the axis x_1 (k=-1), and the tensile load p along the axis x_2, then the strain energy attains minimum for two orthogonal families of fibres, inclined at angle θ_1, which depend on the shear load. In the remaining cases of loads minimum strain energy corresponds to a single fibre system inclined at angle θ_1 to axis x_1, which is dependent on the ratio of the normal and shear loads q/p and τ/p.

REFERENCES

1. Brandmaier, H.E., Optimum filament orientation criteria. J. Comp. Mater., 1970, 4, pp. 422-5.

2. Cox, H.L., The general principles governing the stress analysis of composites. Proc. Conf. Fibre Reinforced Materials: design and engineering applications, Inst. of Civil Engineers pap. No.2, 9-13 London March 1977.

3. Kelly, A., Davies, G.J., The principle of the fibre reinforcement of metals. Metallurgical Reviews, 1965, 10, 37, pp. 1-77.

4. Brandt, A.M., On the optimal direction of short metal fibres in brittle matrix composites. J. Materials Science, 20, 1985, pp. 3831-41.5.

5. Marks, M., Composite elements of minimum deformability reinforced with two families of fibres (in Polish). Rozprawy Inzynierskie - Engineering Transactions, 1988 (in print).

FINITE ELEMENT MODEL FOR SHORT FIBER REINFORCED CONCRETE

S.Majumdar, I.Banerjee and A.K.Mohan
Indian Institute of Technology, Kharagpur

ABSTRACT

Finite element models for short fiber reinforced composites
in plane stress and bending have been developed. Random distribution and orientation of fibers are generated by monte
carlo simulation. The entries of the stiffness matrix are
treated as means of associated random variables. Two numerical
examples demonstrate their behaviour and limitations. Limitations of the present model have also been discussed.

INTRODUCTION

A quest for better strength to weight ratio, efficient
force transmission, easy fabrication, dependability and quality control has steered the growth of different types of composites. If we consider the volume of any one structural material which is used in the industry that of concrete by far
exceeds any other. Although concrete as a building material
has many advantages its use is somewhat restricted because of
its poor performance in tension. This drawback has been coupled with its susceptibility to size effect which compels the
cross-sections to have a minimum dimension. To overcome these
difficulties ferro concrete and short fiber reinforced concrete have been introduced. Their particular capacity to arrest
microcracks is also making them an attractive choice. Growing
interest in this field has led to a number of researches in
recent years (1-3). To best of our knowledge finite element
formulation for short fibers has not been adopted so far.
The main obstacle to the application of finite element model
is the randomness in distribution and orientation of the
fibers. An analytical approach is hopelessly complex. Hence a
direct monte carlo approach has been adopted to simulate the
distribution and orientation of fibers. In this way it is also
possible to generate any distribution of our choice. In the

present paper we have considered a rectangular plane-stress
element and a triangular plate bending element reinforced with
short fibers. There are three major limitations to the present
approach. Firstly, a smooth deformation field is assumed for
both the matrix and the fibers. In an actual situation there
will be local interaction between the two. However in the
present application we are interested in the macroscopic behaviour and hence have neglected the local effect. Secondly,
only the mean of the components of the element stiffness matrix has been considered and thus a very important information,
that is the standard deviation has not been utilised in the
analysis. Thirdly, we have assumed that the ratio of the fiber
length to the characteristic dimension of the element is small.
For this reason we are unable to appeal to the arguments of
convergence as mesh size is decreased. However for practical
applications element dimensions are invariably larger compared
to fiber length and hence for engineering considerations the
last observation is not so critical.

Break up of the paper is as follows. In the next two
sections we have described the development of the inplane
stress element and the plate bending element. In the fourth
some numerical experiments are presented.

RECTANGULAR PLANE STRESS ELEMENT

In many of the applications panels of concrete are used
to transfer inplane stresses. Since these panels are most of
the time cast on the horizontal plane the orientations of the
fibers are restricted in the plane of the panel. In a particular element there are two random variabels, the position of
the center of gravity of the fiber and the orientation of the
fiber with respect to one of the axes. Fig.1 shows such an
element with one single fiber.

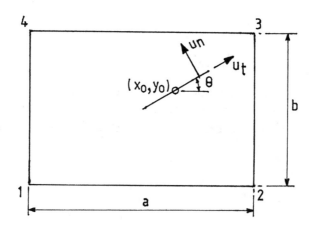

Fig.1 A single fiber in a rectangular element

Since fibers are assumed to deform in conformity with the matrix, strain energy in these elements may be split into two components.

$$\pi = \pi_M + \pi_F \tag{1}$$

where π_M and π_F stand for the energy stored in the matrix and the fibers respectively.

For the base material plane stress rectangular element with serendepity interpolation functions are chosen with local variables r and s.

$$\underline{U} = \left[N\right]^T \underline{U}_o \tag{2}$$

Since the derivation of the stiffness matrix for such elements are fairly standard (4,5) it has not been shown in details.

$$\pi = 0.5 \int \left[B\right]\left[D_M\right]\left[B^T\right] da = 0.5 \underline{U}_o^T K_M \underline{U}_o \tag{3}$$

The fiber strain energy is stored in terms of three components associated with deformations due to axial force, bending moment and shear.

$$\pi_F = \pi_{FA} + \pi_{FB} + \pi_{FS} \tag{4}$$

where

$$\pi_{FA} = 0.5 \int AE \, e \, dl$$

$$\pi_{FB} = 0.5 \int EI \, x \, dl$$

$$\pi_{FS} = 0.5 \int \propto AG \, dl$$

where A is the cross-sectional area, I is the moment of inertia, E is modulus of elasticity, G is shear modulus and $\propto$ is a factor to account for the shape of the cross-section.

If we write the generalised strain vector as

$$\underline{e}_F = \left[\frac{\partial u_t}{\partial l} \quad \frac{\partial^2 u_n}{\partial l^2} \quad \frac{\partial^3 u_n}{\partial l^3} \right] \tag{5}$$

where u_t and u_n are the axial and transverse displacements at a point in the fiber (Fig.1) and may be expressed in terms of nodal degrees of freedom.

$$\underline{U}_F = \left\{ \begin{array}{c} u_t \\ u_n \end{array} \right\} = \left[\begin{array}{cc} \cos \theta & \sin \theta \\ -\sin \theta & \cos \theta \end{array} \right] \left\{ \begin{array}{c} u \\ v \end{array} \right\} \tag{6}$$

in the form

$$\underline{e}_F = \left[B_F \right] \underline{u}_F \tag{7}$$

The constitutive equation relating generalised strain and stress is given by the diagonal matrix.

$$D_F = \left[\begin{array}{ccc} AE & 0 & 0 \\ 0 & EI & 0 \\ 0 & 0 & AG \end{array} \right] \tag{8}$$

The strain energy in the fiber may be written in the form

$$\pi_F = 0.5 \int \left[B_F \right] \left[D_F \right] \left[B_F \right]^T dv = 0.5 \, \underline{U}_0 \, K_F \, \underline{U}_0 \tag{9}$$

The total element stiffness matrix is then written as

$$k = k_M + k_F \tag{10}$$

co-ordinates x_o, y_o, z_o and θ. Orientation of the fiber within the triangle is as shown in Fig.2 .

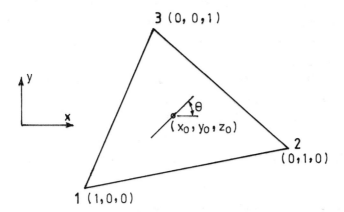

Fig.2 A fiber within a triangular element

For a straight fiber the following linear relation holds

$$m\,(\xi_1 - \xi_{o1}) = (\xi_2 - \xi_{o2}) \qquad (13)$$

where m is a constant of proportionality defined below. The two other natural co-ordinates may be written in terms of m as

$$\xi_2 = m\,(\xi_1 - \xi_{o1}) + \xi_{o2}$$
$$\xi_3 = 1 - (1 + m)\,\xi_1 + m\,\xi_{o1} - \xi_{o2} \qquad (14)$$

The slope is given as

$$q = \tan\theta = \frac{dy}{dx} = \frac{y_1 + m\,y_2 - (1 + m)\,y_3}{x_1 + m\,x_2 - (1 + m)\,x_3} \qquad (15)$$

where (x_i, y_i) are the co-ordinates of vertices.

Hence the value of m is expressed as

$$m = \frac{q\,(x_1 - x_3) - (y_1 - y_3)}{(y_2 - y_3) - q\,(x_2 - x_3)} \qquad (16)$$

The distance s of a general point on the fiber from its mid-point may be expressed in the form

In order to generate the stiffness matrix on a digital
computer at first the stiffness contribution of the base
material is computed. Then the distribution of a specified
number of fibers are generated and their stiffnesses are
added to the stiffness matrix of the base material. In the
present application attention is restricted to uniform distribution of the location of the center of gravity and the orientation of the fibers. The main advantage in this process is
that any type of distribution having preferential distribution
and orientation may be generated by this approach.

TRIANGULAR PLATE BENDING ELEMENT FOR COMPOSITES

The plate bending element for composites differs from
the plane stress element in two aspects. The random distribution of the fibers across the thickness of the plate should
also be considered. The axial force in the fiber is mainly
due to their location away from the neutral axis. The computation of energy proceeds in the same manner as the plane stress
element. In order to be able to discretise any arbitrary poly-
gonal domain a triangular plate bending element has been developed here. The following interpolation function in natural
co-ordinate is chosen to represent the transverse displacement
field.

$$\underline{w} = \underline{N}^T \underline{w}_o \qquad (11)$$

where

$$\underline{N}^T = \begin{bmatrix} 1 & \varepsilon_1 & \varepsilon_2 & (\varepsilon_1\varepsilon_2) & (\varepsilon_2\varepsilon_3) & (\varepsilon_3\varepsilon_1) & (\varepsilon_1\varepsilon_2 - \varepsilon_2\varepsilon_1^2) \\ & (\varepsilon_2\varepsilon_3^2 - \varepsilon_3\varepsilon_2^2) & (\varepsilon_2\varepsilon_1^2 - \varepsilon_1\varepsilon_3^2) \end{bmatrix} \qquad (12)$$

Degrees of freedom for each node include transverse
deflection and rotations about the two co-ordinate axes. The
stiffness matrix for the base material is derived in the
standard manner (5). To this the stiffness contributions of
each of the fibers are added.

Four random variables are generated for each fiber

$$s = c(\varepsilon_1 - \varepsilon_{o1}) \tag{17}$$

where c is given as

$$c = \left[(x_1 + mx_2 - (1+m)x_3)^2 + (y_1^2 + my_2 - (1+m)y_3)^2 \right]^{1/2} \tag{18}$$

Curvature of the fiber in the vertical plane is then

$$X = \frac{1}{c^2} \frac{d^2u}{d\varepsilon_1^2} \tag{19}$$

If the fiber is at a distance z from the neutral plane then the strain is expressed as

$$\underline{e}_F = -\frac{z}{c^2} \frac{d^2u}{d\varepsilon_1^2} \tag{20}$$

Associated strain-displacement matrix B is given as

$$B_F = \frac{z}{c^2} \left[0 \quad 0 \quad 0 \quad 2m \quad -2m(1+m) \quad -2(1+m) \right.$$
$$\left. (c_1\varepsilon_1 + d_1) \quad (c_2\varepsilon_1 + d_2) \quad (c_3\varepsilon_1 + d_3) \right] \tag{21}$$

where

$$c_1 = 6m^2 - 6m$$

$$d_1 = (-m\,\varepsilon_{o1} + \varepsilon_{o2})(4m - 2)$$

$$c_2 = 12m^3 + 14m^1 + 2m$$

$$d_2 = (-m\varepsilon_{o1} + \varepsilon_{o2})(12m^2 + 12m + 2) - 6m^2 - 4m$$

$$c_3 = -12 - 18m - 6m^2$$

$$d_3 = (6 + 4m)(1 + m\varepsilon_{o1} + \varepsilon_{o2})$$

Hence the fiber stiffness matrix is defined as

$$k_F = \int_{-L/2}^{L/2} \left[B_F \right] \left[D_F \right] \left[B_F \right] dl \tag{22}$$

where $\left[D_F \right]$ = EI of fiber

NUMERICAL RESULTS

A numerical example is given below. Fig.3 shows a square panel subjected to a concentrated tranverse load at the center. Particulars of the plate and the fibers are given below.

MORTAR   COPPER

Plate width   =   .136 m   Fiber diameter = 0.00019 m

Plate thickness =   .045 m   Fiber length   = 0.01 m

E for Matrix   = $0.26(10)^9$ E for Fiber   = $0.120(10)^{11}$
                     N/m^2 N/m^2

Poisson's ratio =   .35   Poissons ratio = 0.30

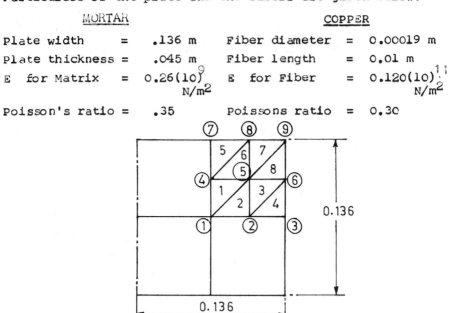

Fig.3 A simply supported square plate subjected to transverse
       load.

several numerical exercises are carried out with different numbers of fiber content. It shows a steady increase in the stiffness of the plate. In order to compare the stiffness for plain matrix to that after addition of 200 fibers per element some of the stiffness coefficient are tabulated below.

TABLE-1   Comparison of the coefficients of the stiffness
          matrix (SI)

Component $x(10)^{-3}$	k_{11}	k_{22}	k_{33}	k_{12}
Plain Matrix	0.233598	0.275897	0.275897	-0.123598
With 200 fibers	0.305214	0.416106	0.351855	-0.156030

Even though elements are of same type the stiffness
matrices are generated independently. It is interesting to
compare the variation in the values of a particular stiffness
coefficient, say k_{11}, for different elements of same type
(Table-2).

TABLE-2 Variation in the values of the coefficients k_{11}.

Element No.	2	4	6	8
Values	0.305214	0.24070	0.234465	0.325293

Deflections and rotations at different locations of the
plate are shown in Fig.4.

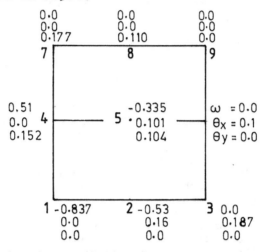

Fig.4 Deflections and rotations at nodal points

For space costraint it is not possible to show the
results associated with the plane stress problem. However it
is interesting to note that when a test example is run with
fiber length comparable to the element size the deviation in
the stiffness components are found to vary over a wide range.
As a consequence nodal deflections tend to loose exact
symmetry. This may be due to the fact that for most of the
fibers part of them are lying outside the element. A comparison with experiment carried out with steel fiber reinforced
concrete may be found in reference (2).

CONCLUSIONS

An attempt has been made to construct the stiffness properties of inplane and plate bending elements for short fiber reinforce materials. The elements are suitable for calculation of macroscopic behavior of thin flat composite elements. It is possible to generate composites having preferential fiber orientation. However, it has several limitations at this stage. i) It does not take into account local interaction between the matrix and the reinforcement, ii) Only the means of the stiffness coefficients have been considered and information related to variance is ignored, iii) The question of convergence has not been dealt with, which may be an intiriguing aspect. It has been assumed in the course of the derivation that the ratio of the fiber length to the element dimensions is small.

REFERENCES

1. Ghalib, M.A., Moment capacities of steel fiber reinforced small concrete slabs, Tech. Paper, May and June 1981, ACI Journal, Title No. 77-27 , pp. 247-257.

2. Kumar, Anupam., Strength of fiber reinforced concrete slabs under uniformly distributed load for variable fiber concentration, M.Tech. Thesis, Dept. of Civil Engineering, IIT., Kharagpur, 1984.

3. Brandt, A.M., Influence of the fiber orientation on the mechanical properties of fiber reinforced cement (FRC) specimen. Proc. of Ist international RILEM congress 1987, (edit, J.C. Maso) pp. 651-658.

4. Bathe, K., and Wilson, E.L., Numerical Methods in Finite Element Analysis, Prentice Hall of India, 1978.

5. Holand, I. and Bell, K. (eds) Finite Element Methods in stress analysis, Tapir, Norway, 1972.

BENDING OF CONCRETE AND REINFORCED CONCRETE ELEMENTS
PARTIALLY FILLED WITH STEEL FIBRES

KONSTANTIN YAMBOLIEV
Central Laboratory of Physico-Chemical Mechanics
Bulgarian Academy of Sciences
Acad. G. Bonchev Str., Block 1, 1113 Sofia
Bulgaria

ABSTRACT

Herein are contained the experimental results of an investigation
into the influence of short steel fibres on the bearing capacity and
deformability of concrete and reinforced concrete elements subject to
bending. The fibres are either dispersed at random along the entire
cross-section or in a part of the tensile zone. The beams were loaded
until fracture using two concentrated forces. The following parameters
were measured and observed: ultimate flexural strength, deflections in
the mid-point of the span, flexural stiffness and the strain on the
lateral faces of the beams.
The effectiveness of reinforcing fibres in increasing the bearing
capacity and deformability is clearly established.

INTRODUCTION

Steel fibres considerably increase the bearing capacity and
substantially improve the deformability and cracking resistance of
elements subject to bending. In [1] the results are shown for testing
of concrete elements containing fibres along the entire cross-section.
The fibre influence on bearing capacity and deformability of reinforced
concrete elements subjected to bending is considered in [2]. These
investigations were aimed at establishing the bearing capacity and
deformability of concrete and reinforced concrete elements with the
tensile zone partly or entirely cast using steel fibre reinforced
concrete (FRC).

MATERIALS AND INVESTIGATION PROGRAMME

Seven concrete elements numbered from 0 to 6 and seven reinforced concrete elements numbered from 7 to 13 were tested (Fig. 1). The elements 0 and 13 did not contain fibres and those numbers 1 and 7 were reinforced with fibres to full depth. The concrete composition for all the elements was as follows : cement - 350 kg/m^3, water-cement ratio 0.6 and maximum grain diameter 10 mm. At the time of testing, after 6 months ageing, the concrete cube strength was 37 MPa. The fibres were smooth, straight, without curved ends, 20 mm long and with 0.2 mm cross-section diameter. The volume contents of fibres was 1.5%. The bar reinforcement was composed of 3 bars of diameter 3 mm, i.e. $\mu\% = 0.237$ in the tensile zone and two constructional bars in the compression zone. One of the bars was bent due to structural reinforcement and every element had 4 stirrups. The elements were loaded by two forces as shown (Fig. 1).

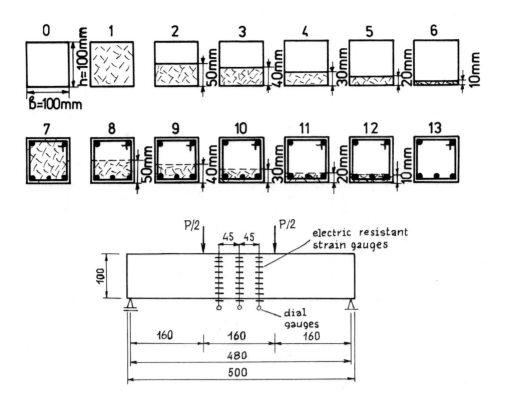

FIGURE 1   Size of elements, dislocation of FRC, measuring devices.

The deflections if the element were measured by dial gauges at three cross-sections. The strain deformations along the depth of the beam were measured by electrical resistance strain gauges. In each case a loading was applied in 1 kN increments until fracture occurred. A detailed description of the test programme and procedure is given in [3].

CONCRETE ELEMENTS

ULTIMATE FLEXURAL STRENGTH

Evaluation of the ultimate flexural strength is given by the load at fracture P_u and the calculated corresponding bending moment M_u and stress σ_u (Table 1). In element 1, made wholly of FRC, the stress σ_u was 2.6 times greater than in element 0. An increase of σ_u of 74% was observed in element 6 with only 10% of FRC. In element 2, the stress $_u$ was only 5.5% smaller than element 1. These comparisons show that reinforcement with fibres should only be introduced in the tensile region and thus the compressive strength of non-reinforced concrete can more effectively be used. The results show that FRC increases the ultimate flexural strength.

TABLE 1

E1 elem. No	FRC depth (mm)	Load at 1st crack				At ultimate load				k_m	k_c
		P_{cr} (kN)	σ_{cr} (MPa)	f (mm)	EJ (kNm^2)	P_u (kN)	σ_u (MPa)	f (mm)	EJ (kNm^2)		
0	0	2.5	1.2	0.100	167	7.84	3.70	0.28	165	-	-
1	100	9.5	5.7	0.125	447	20.00	9.60	0.47	250	2.59	1.68
2	50	7.5	4.5	0.115	383	19.00	9.12	0.42	266	2.46	2.00
3	40	6.0	3.6	0.100	353	16.00	7.68	0.36	261	2.07	2.13
4	30	5.1	3.0	0.080	316	18.00	8.64	0.39	271	2.33	2.88
5	20	5.1	3.0	0.080	316	15.80	7.58	0.33	258	2.05	2.52
6	10	4.0	2.4	0.090	261	12.80	6.44	0.32	235	1.74	2.68

DEFORMABILITY OF ELEMENTS

The deflections f of the beams 0-6 are plotted against loads P in Fig. 2. The curves show that all the elements containing FRC had smaller deflections than element 0. Even element 6 with only a 10 mm deep FRC layer had deflections smaller than those of element 0.

All the elements with FRC layer had a longer linear branch of the P-f diagram. After P_{cr} the elements continued to steadily bear increasing load up to complete pull-out of the fibres from the matrix. In table 1 is given the bending stiffness values EJ at crack appearance P_{cr} and fracture P_u. In the region up to P_{cr} element 1 has considerably greater bending stiffness than the other elements. But in the region up to P_u the bending stiffness of element 2 is practically equal to that of element 1. This shows that FRC in the compressive region does not have a substantial effect on the stiffness of the element provided that the resistant moment is taken for the entire cross-section.

The stiffness of element 6 is the smallest of all the elements containing fibre reinforcement but greater than that of element 0. The values of bending stiffness of sections working without cracks are on average 34.5% greater than those at maximum deflection. The intensifying role of the fibres, or their effectiveness, can be determined by the co-efficient for matrix intensification $k_m = \delta_u^c / \delta_u^m$. Here δ_u^c is the stress corresponding to destruction of the element and m is the stress corresponding to the destruction of the matrix. The k_m values are given in table 1. The intensification is the greatest in element 1 and the smallest in element 6, k_m indicates the FRC effectiveness when put into the entire section or alternatively only into part of the tensile region. The coefficient $k_c = \delta_u / \delta_{cr}$ shows the correlation between part of the P-f diagram in which the section works with cracks and its linear part. This shows that element 4 works with crakcs under mostly prolonged loading and element 1 has the smallest k_c due to the long linear branch of this diagram.

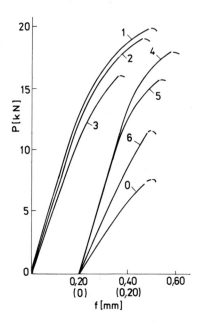

Figure 2   Deflections of Concrete elements.

The unit deformations measured along the depth of elements give an opportunity to determine the position of the neutral axis. The results of strains measured on the lateral faces of the beam under ultimate load P_u are given; similar diagrams due to load P_{cr} at first crack are shown in Fig. 4. On the base of the measured strains ε_c - due to compression and ε_t - due to tension, the depth x of the compressive zone is calculated. At P_{cr} element 1 is the smallest x. For all other elements x is between 51 and 56 mm. Characteristic of the stress zone at P_u is shown in element 1 to be the most shallow. Close to it are elements 2, 3 and 4 while in elements 5 and 6 x is the greatest. At ε_c the diagrams show weak slight curving but still near to the triangle ones, while at ε_t they are near to a trapezium according to the data shown by the inner indicators. Generally, after Fig. 3 it may be concluded that the position of the neutral axis varies slightly from one element to another, while their deformations depend considerably on the depth of the FRC layer.

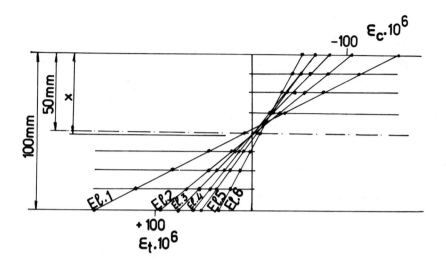

Figure 3.   Results of strain measurements and variation of
the neutral axis position in concrete elements
at P_{cr}.

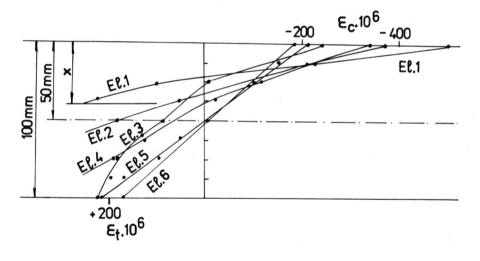

Figure 4.   Results of strain measurements and variation of
the neutral axis position in concrete elements
at P_u.

REINFORCED CONCRETE ELEMENTS

ULTIMATE FLEXURAL STRENGTH

Evaluation of the bearing capacity of reinforced concrete elements numbered 7 to 13 can be made on the basis of the load P_u and the bending moment M_u (Table 2). The influence of fibres on the bearing capacity is given by the ratio of M_u of an element to the $M_{u,13}$ of element 13 which was cast without fibres.

TABLE 2

Ele m. No	FRC depth (mm)	Load Load			At Fracture			k_1	k_u
		$(EJ)_u$ (kNm)	f (mm)	$(EJ)_1$ (kNm^2)	P_u (kN)	Mu (kNm)	f (mm)		
7	100	248	0.126	324	25.0	2.00	0.59	3.6	2.0
8	50	279	0.103	365	22.0	1.76	0.43	3.2	1.8
9	40	231	0.120	339	20.6	1.65	0.52	2.9	1.7
10	30	224	0.149	273	18.0	1.44	0.47	2.6	1.5
11	20	219	0.160	254	15.0	1.20	0.40	2.1	1.2
12	10	227	0.158	257	14.0	1.12	0.36	2.0	1.1
13	0	190	0.188	216	12.0	0.96	0.37	1.7	-

The values calculated as $k_u = M_u/M_{u,13}$ are given in Table 2. The greatest k_u is that of element 7 having more than twice the bearing capacity, however, the difference between k_u of elements 7 and 8 is only 25%.

The life load of element 13 calculated according to [4] is P_1 = 6.958 kN and the bending moment is M_1 = 0.553 kNm. The coefficient k_1 = M_u/M_1 indicates that the safety margin every element works with. The value of k_1 is smallest for element 13 while for element 7 it is twofold greater. The coefficient k_1 shows the extent to which the fibres contribute to increasing the difference betwen P_1 and the experimentally obtained P_u.

DEFORMABILITY OF ELEMENTS

The influence of the fibres for improving the element deformability is investigated by taking into consideration the deflections and strains measured on the lateral face of each beam. In figure 6 deflections f are plotted against loads P for all elements. The values of P_{cr}, M_{cr} and f are given in table 3. Under the same load all elements containing FRC had smaller f than that of element 13 and a considerably longer linear branch of the P-f diagram, even elements 11 and 12. An evaluation of the quasi-plastic properties of the elements is possible using the curve-linear branch of P-f diagram. It is the longest for elements 7, 8 and 9 containing deeper FRC layers. The bending stiffnesses EJ best characterise the deformed state of each element. The values of $(EJ)_{cr}$

at the first crack sections and $(EJ)_u$ at ultimate load are given in the aforementioned tables.   The EJ_{cr} amd EJ_u stiffnesses of element 8 are respectively greater by 26% and 12.5% as compared to those of element 7. It is evident that FRC located in the compressive zone does not contribute to an appreciable increase in bending stiffness.   Such a conclusion was also reached for the concrete elements.   The FRC increase in the elements leads to increased values for $(EJ)_1$ and $(EJ)_u$.

TABLE 3

Element No.	FRC depth (mm)	At first crack			
		P_{cr} (kN)	M_{cr} (kNm)	f (mm)	EJ^{cr} (kNm^2)
7	100	12.0	0.96	0.220	319
8	50	11.0	0.88	0.160	402
9	40	9.0	0.72	0.152	346
10	30	7.5	0.60	0.151	290
11	20	6.2	0.50	0.136	266
12	10	6.0	0.48	0.135	260
13	0	5.5	0.44	0.147	219

The deflections f and the stiffnesses $(EJ)_1$ at design life load P_1 are given in Table 2.   There, as well as in Fig. 5, can be seen that for elements 7, 8, 9 and 10 which have considerably prolonged linear branches of the P - f diagram their P_1 is in this branch, while for the remaining elements P_1 is in the curve-linear branch.   These comparisons are important for the design of reinforced concrete elements containing FRC by the ultimate state method [4].

In Fig. 6 is indicated the compressive and tensile areas of all the elements at ultimate load P_u.   The neutral axis of elements 7, 8, 9 and 10 is strongly displaced upwards and the compressed area is triangle like.

It can be concluded that the presence of FRC in the tensile area or any part of it leads to the appearance of larger compressive stresses. Thus, the considerable compressive strength of the concrete matrix will be used more effectively.   This is valid for concrete as well as reinforced concrete elements.

CONCLUSION

The experimental results shown herein, their analysis and conclusions show the influence of the FRC layers on the behaviour of concrete and reinforced concrete elements under bending.   The beams with FRC layers are more rigid and have longer linear behaviour, and these properties increase with depth of FRC.   This is particularly important for reinforced concrete elements and their design.   The prolonged linear and curvilinear branches of ε_c graphs and the greater FRC tensile strength lead to substantial changes when checking for opening and

propagation of cracks in reinforced concrete elements. The introduction
of approximate modifications to the ultimate state method on which design
is based [4] will help the design of reinforced concrete elements with
better deformability and enhanced cracking resistance. However, an
accumulation of sufficient experimental data is a necessary prerequisite
for this goal to be achieved.

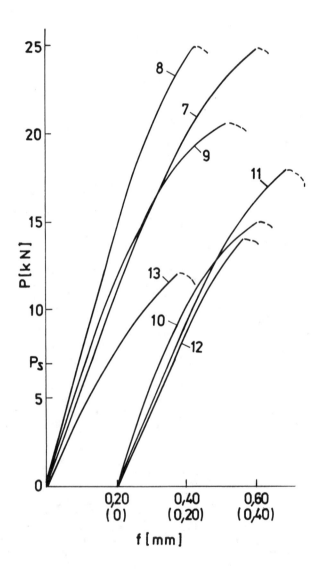

Figure 5.   Deflections of reinforced concrete elements

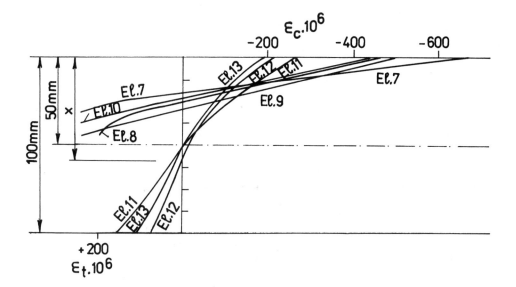

Figure 6.   Results of strain measurements and variation of the
             neutral axis position in reinforced concrete elements
             at P_u.

REFERENCES

1.   Babut, R., The Bearing Capacity and the Deformability of Fibre
      Reinforced Concrete Elements Under Bending (in Polish), ITFR
      Reports, No. 32, Warsaw, 1979.

2.   Swamy, R.N., Sa'ad A. Al-Ta-Ta'an, Deformation and Ultimate
      Strength in Flexure of Reinforced Concrete Beams Made with Steel
      Fibre Concrete. ACI Journal, 1981, Sept/Oct. pp 395-405.

3.   Yamboliev, K.A., Concrete Elements Partially Reinforced with Steel
      Fibres Under Bending (in Bulgarian). Technicheska misal, BAN,
      Sofia, 1988, 2, pp 73-80.

4.   Norms for Design of Concrete and Reinforced Concrete Structures
      (in Bulgarian). Ministry of Construction and Structural
      Materials, Bulgaria, 1981.

INDENTATION TESTING OF SiC-WHISKER REINFORCED Al$_2$O$_3$ COMPOSITES

INGA-LILL EKBERG, ROBERT LUNDBERG, RICHARD WARREN* and ROGER CARLSSON
Swedish Institute for Silicate Research, Göteborg, Sweden
*Chalmers University of Technology, Göteborg, Sweden

ABSTRACT

Indentation testing as a method of measuring fracture toughness of SiC-whisker reinforced Al$_2$O$_3$ composites was evaluated. Polishing was found to influence the measured K_{1c} significantly. The different equations that exist to correlate the crack length with K_{1c} were critically compared. A large variation in the level of K_{1c} calculated with the different equations was observed. The load dependence over a large range of indentation loads was determined for the equations. The least load dependent equation was chosen to characterize the material. A "true" K_{1c} level was obtained after calibrating this equation against a traditional fracture toughness method. K_{1c} values consistent with other reported values were obtained indicating the usefulness of indentation testing for whisker reinforced ceramics.

INTRODUCTION

SiC-whisker-reinforced Al$_2$O$_3$ ceramics have recently been introduced as a candidate ceramic for cutting tool applications, their advantages being a higher toughness than the existing SiAlON or Si$_3$N$_4$ cutting tools coupled with the high wear resistance of Al$_2$O$_3$ ceramics. Fracture toughness and hardness are important mechanical properties influencing the cutting behaviour of a ceramic cutting tool material. The purpose of the present investigation was to investigate the possibility of using indentation testing, which is convenient for small samples such as cutting tools, to determine the fracture toughness of SiC-whisker/Al$_2$O$_3$ composites. As an introduction, the possible mechanisms of whisker toughening, as well as the development of equations to correlate crack length of indentation cracks with the fracture toughness will be reviewed.

Whisker Toughening Toughening mechanisms for ceramics (1) and whisker reinforced ceramics (2) have been reviewed recently by other authors. Among the possible toughening mechanisms are: load transfer to the whiskers, pre-stressing of the matrix, crack-deflection, crack bowing, crack twisting, micro-cracking, crack branching, crack bridging (wake toughening), debonding and pull-out.

Since the whiskers are extremely strong and stiff <u>load transfer</u> to the whiskers may reduce the stress concentration of an advancing crack by redistributing the load over a larger volume of material. A strong fibre/matrix interface is required for effective load transfer. If the whiskers have a higher thermal expansion (α_{fibre}) than the matrix (α_{matrix}) which is <u>not</u> the case for the system SiC-whiskers/Al_2O_3, the matrix is pre-compressed during cooling from sintering temperature. Such <u>pre-stressing</u> also requires load transfer and a strong interface. Two mechanisms relevant to whisker composites, namely <u>crack bowing</u> and <u>crack deflection have</u> been discussed extensively for particulate composites and several theories and models exist to describe and explain the toughening observed. In the mid 60's Hasselman and Fulrath proposed a model explaining the strengthening of glass by Al_2O_3 particulates based on the idea that the critical flaw size, and thus strength, was determined by the interparticle spacing (3). Lange later tried to explain the same phenomenon by a line tension mechanism (4) (5) (6) claiming that flaw size was not limited but rather that the inclusions pinned the advancing crack causing it to bow. The model considered, however, only in-plane bowing. A more complete approach considering in-plane and out-of-plane deflection was presented by Faber and Evans (7) (8). If residual thermal mismatch stresses exist in the composite this may cause crack deflection. When $\alpha_{matrix} > \alpha_{fibre}$, the fibre is in compression and the matrix near the fibre is in radial compression and circumferential tension. This attracts the crack to the fibre. If, on the other hand, $\alpha_{matrix} < \alpha_{fibre}$ the crack is deflected away from the fibre (which is in tension while the matrix is in radial tension and circumferential compression). If, in the case where the crack is drawn towards the fibres, which is the case in SiC-whisker/Al_2O_3 composites ($\alpha_{Al2O3} > \alpha_{SiC}$), a weak interface exists, the crack is further deflected along the fibre/matrix interface. This has been observed for SiC-whisker/Al_2O_3 material (2). Whiskers are shown to have an effective geometry for crack deflection. The crack is pinned, in plane, and deflected by twisting or tilting, out of plane, thus effectively increasing the fracture surface area. Evans and Faber's analysis shows that since the crack plane is no longer normal to the applied stress an increase in the apparent mode I crack tip stress intensity is necessary for further crack growth. Thus, crack deflection can result in significant toughening.

An important mechanism operating in addition to crack deflection but after the crack has passed is wake toughening or <u>crack bridging</u> (9). Fibres, whiskers and even grains may bridge the crack faces in the wake region of the extending crack. Each bridge acts to shield the crack tip from the applied stress so that the K_1 at the crack tip is less than the apparently applied K_1. Since the magnitude of the wake toughening will depend on the number of bridging whiskers the resistance to crack extension will initially increase with crack size until a constant bridge length is established, i.e. R-curve behaviour should in principle be observed (1), although this would require a significant proportion of oriented fibres.

<u>De-bonding</u> can be thought of as an extreme form of crack deflection occuring at the fibre/matrix interface such that the whiskers are left without any bonding to the matrix. <u>Pull-out</u> of the whiskers may then occur in the wake region of the crack with the frictional sliding of the whiskers adding to the fracture energy. De-bonding and pull-out are important mechanisms for the toughening of continuous fibre composites but probably play only minor roles in randomly oriented whisker reinforced ceramics.

With sufficiently large mismatch in thermal expansion (α_{matrix} > α_{fibre}) <u>microcracking</u> can occur in the matrix during cooling after preparation of the composite (10). Such microcracks could contribute to the toughening by encouraging crack deflection and branching. However since the microstructure isweakened by microcracking it is questionable whether a significant net gain in toughness can be achieved. A potentially more effective form of microcrack toughening is achieved if the micocracking occurs only in association with the loading of a crack tip. In this case the microcracked zone dilates and provides a crack tip shielding effect. It has yet to be demonstrated that the necessary density of microcracks can be generated in whisker reinforced ceramics to provide significant toughening by this means.

Whisker toughening of ceramics can be explained by a combination of the above mentioned mechanisms. The most dominant one is probably different for different whisker/matrix combinations. A recent indentation study, for example, of the influence of whisker orientation on the fracture toughness of SiC-whisker/Si_3N_4 composites (11) showed a maximum K_{1c} perpendicular to the whisker direction with almost no evidence of crack deflection. In SiC_w/Al_2O_3 material, however, both crack deflection and crack bridging has been shown to occur (1) (2). Even though efforts have been made to model the most probable mechanism for SiC-whisker toughening of Al_2O_3 (12) much remains before a full understanding of these phenomena is attained. Hopefully in the future, it will be possible to predict, and tailor, the whisker size and volume fraction, the degree of whisker orientation, and perhaps most important, the strength of the whisker/matrix bond in order to maximize toughness and strength.

Indentation Testing – Fracture Toughness A large number of test procedures to measure K_{1c} exist today (1) (13). Most of them are fairly complicated to perform and require specially fabricated test specimens. Indentation testing requires only a polished surface large enough to place one, or several, indents, and may be performed on virtually any specimen or component geometry. Furthermore, standard hardness testing equipment may be used to make the indents. These advantages make indentation fracture toughness measurement a very interesting alternative test method, especially for rapid quality control in production.

When ceramics are indented with a Vickers or Knoop diamond, cracks are seen at the corners of the impression . Sven Palmqvist performed pioneering work in trying to empirically relate the length of these cracks to the fracture toughness of the indented material (14). However, it was not until recently that models, based on stress analysis, have been developed to derive equations valid for all types of brittle materials. The formation of indentation cracks can be explained by the elastic and permanent deformation developed during indentation and unloading (15) (16). Cracks are first formed below the surface during loading; they then grow to the surface during unloading since the elastic compressive stress at the surface becomes zero, while the tensile residual stress remains. This tensile stress arises from the mismatch between the elastic deformation of the bulk and the permanent deformation due to the indentation. Depending on the magnitude of the load, either median cracks or Palmqvist cracks are formed , as shown in figure 1.

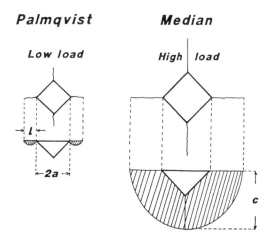

Figure 1. Comparison of geometries of cracks formed at different load levels.

Several equations exist, especially for median cracks, relating the load and measured crack length (c or l in figure 1) with K_{1c}. The hardness (H) as calculated from the impression diagonal (a in figure 1) and the load (P), Young's modulus, (E) and some form of calibration constant are found in most equations. The earliest work was done by Lawn and Fuller, on glass, showing that $P/C^{3/2}$ was constant (17). Based on their measurement an equation for K_{1c}, independent of Young's modulus, was proposed by Evans and Charles (18) in 1976:

$$K_{1c} \cdot \emptyset / (H \cdot \sqrt{a}) = 0.15 \cdot K \cdot (c/a)^{-3/2}$$

The left side of this equation was obtained by stress analysis, and the right by curve fitting. With $\emptyset = 3$ and $K = 3.2$ for large values of c/a ie high load and median cracks, the following equation is obtained:

$$K_{1c} = 0.16 \cdot H \cdot \sqrt{a} \cdot (c/a)^{-3/2}$$

EVANS & CHARLES (1976)
(MEDIAN CRACKS)

The equation was claimed to deviate by less than 30% from the true K_{1c} value. The equation was later modified (19) to contain Young's modulus, and is said to differ less than 10% from the true K_{1c}

$$K_{1c} = (E/H)^{0.4} \cdot H \cdot \sqrt{a} \cdot 10^y$$

EVANS (1979)
(MEDIAN CRACKS)

$(y = f(c/a)$ see ref (19))

Lawn, Evans and Marshall (20) tried a more analytic approach than with the earlier semi empirical equations also taking into account the residual components of the stress field. What is unique about their analysis is that an expression was derived specifically for the surface trace of the median crack, i.e. the measurable crack in opaque material. The analysis resulted in the following equation:

$$K_{1c} = 0.028 \cdot H \cdot \sqrt{a} \cdot (E/H)^{1/2} \cdot (c/a)^{-3/2}$$

LAWN, EVANS, MARSHALL (1980)
(MEDIAN CRACKS)

Niihara (21) proposed an equation based on a similar approach, stress analysis and curve fitting, as Evans & Charles:

$$(K_{1c} \cdot \emptyset / (H \cdot a)) \cdot (H/(E \cdot \emptyset))^{0.4} = 0.129 \cdot (c/a)^{-3/2}$$

With the expression for hardness and $\emptyset = 3$ the following equation is obtained.

$$K_{1c} = 0.0309 \cdot (E/H)^{0.4} \cdot (P/c^{3/2})$$

NIIHARA ET AL (1981)
(MEDIAN CRACKS)

Anstis et al (22) employed dimension analysis and proposed the equation below, in which the calibration constant has an upper and lower limit.

$$K_{1c} = (0.016 \pm 0.004) \cdot (E/H)^{1/2} \cdot P/c^{3/2}$$ ANSTIS ET AL (1981)
(MEDIAN CRACKS)

The factor E/H was studied further by Laugier (23) who found the exponent 2/3 better than the 1/2 in the equation of Anstis et al.

$$K_{1c} = 0.010 \cdot (E/H)^{2/3} \cdot P/c^{3/2}$$ LAUGIER (1985)
(MEDIAN CRACKS)

These are all equations valid for median cracks. A fairly high load is required to obtained median cracks, especially in high toughness technical ceramics. Relations between the crack length of Palmqvist cracks, which are formed at a lower load, and K_{1c} may therefore be of importance. Most commercially available hardness testing equipments are limited to a maximum loading of 100N, which for many ceramics is not sufficiently high to create median cracks. An equation relating Palmqvist cracks with K_{1c}, has been proposed by Niihara et al (21). Palmqvist cracks are considered to be predominant when c/a < 3 i.e. at low loads, then:

$$(K_{1c} \cdot \emptyset/(H \cdot \sqrt{a})) \cdot (H/(E \cdot \emptyset))^{0.4} = 0.035 \cdot (1/a)^{-1/2}$$

This equation yields, with the expression for hardness, and $\emptyset = 3$, the following:

$$K_{1c} = 0.0123 \cdot E^{0.4} \cdot H^{0.1} \cdot (P/1)^{1/2}$$ NIIHARA ET AL (1981)
(PALMQIVST CRACKS)

For heterogeneous materials, such as particulate or whisker reinforced ceramics it is probable that a fairly large crack will be necessary in order to get reproducible K_{1c} values. The crack has to be larger than several interparticle (or interwhisker) spacings. There is, however, nothing in the fundametal theory of indentation fracture toughness testing that, in view of the toughening mechanisms operating in whisker composites, prevents its application on such materials.

EXPERIMENTAL PROCEDURE

Material The materials used in this study were obtained by glass encapsulated HIPing (as developed by ASEA Cerama, Robertsfors, Sweden) of slip cast or pressed green bodies. Japanese SiC-whiskers (Tokawhisker, Tokai Carbon, Japan) and commercial Al_2O_3 powder (A16SG, Alcoa, USA) were used. The processing (24) (25) and some mechanical properties (25) of this material has been described previously . The processing involves whisker "cleaning" by sedimentation, dispersion and wet mixing of whiskers and powder prior to forming, encapsulation and HIPing. Composites with 5, 10, 15, 20, 25, 30, 35, 40 and 45 vol% SiC whiskers were fabricated.

Machining The HIPed samples were machined according to the following procedure:

1. Machining, 65 μm diamond wheel, 10 min

2. Coarse polishing, 6 μm diamond spray on Petrodisc, 10 min

3. Polishing, 6 μm diamond spray on cloth, 4 min

4. Polishing, 3 μm diamond spray on cloth, 15 min

This resulted in an optically perfect surface finish. To investigate the influence of further polishing on K_{1c} the 3 μm diamond spray polishing was further extended and K_{1c} measured at 3 minute intervals. The polishing depth was measured by the reduction in diameter of indents placed on the polished surface before the extended polishing. A Planopol (Struers) equipment was used for all machining and polishing.

Indentation All indents were made with a Vickers diamond (136°). For loads below 100N a commercial hardness tester (Zwick 3212) was used, whereas for higher loads a specially constructed indenting equipment was utilized. With this equipment the diamond is loaded at a constant loading rate. All indents were made with a loading time of 5 s hold at full load. The crack length was measured with a light optical microscope at 500 x magnification.

Evaluation of K_{1c}-Equations As a criterion for choosing the best equation to calculate K_{1c} a minimum of load dependence was selected. (It is assumed that R-curve behavior is not sigificant in these materials). Indents were made at ten different loads from 20 to 700 N on slip cast material with 10 and 30 vol-% SiC whiskers. 3-4 indents were made at each load. K_{1c} was calculated using all the previously described equations and plotted against the load.

Multiple-Indentation Bend Tests (MIBT) To compare the indentation fracture toughness with a fracture mechanical method, indentation bend tests were performed. Bars 7 mm wide by 3 mm high by 50 mm long were machined as described above. Three indents were made as described by Cook and Lawn (26) and the bars were broken in 4-p bending, 40/20 mm spans, at a loading rate of 0.5 mm/min with a Zwick 1464 testing machine. A pressed material with 25 vol% SiC whiskers was used for this experiment.

K_{1c} vs. whisker content The equation exhibiting the least load
dependence was chosen to compare slip cast, HIPed materials with whisker
contents from 5 to 45 vol%. Four indents were made at two loads above
the load where the equation became loadindependent.

RESULTS AND DISCUSSION

Machining The polishing of the samples was found to strongly influence
the K_{1c} as measured by indentation. As can be seen in figure 2 the K_{1c}
value drops considerably before reaching a constant value after 3-4 μm
polishing. This is explained by the compressive surface layer created
during the coarse machining. The 3-4 μm compressive layer that has to be
removed corresponds to about 15 min further polishing with 3 μm diamond
spray after an optically perfect surface has been obtained.

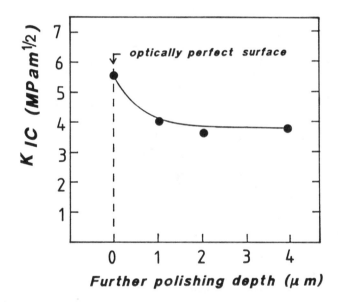

Figure 2. K_{1c} (Niihara (Palmqvist) 100N) versus further polishing depth
   for slip-cast, HIPed 10 vol% SiC whisker/Al_2O_3 composites.

The difference before and after the extra polishing was around 2
$MPa \cdot m^{1/2}$ (5.6-3.6), which clearly demonstrates the importance of
removing the compressive layer as well as the error that can arise if
the polishing is not reproducible and performed in a controlled manner.

Evaluation of K_{1c}-equations Since whisker composites are heterogeneous
materials, the generation of large median cracks was believed to reduce
the variation in K_{1c} caused by the inhomogeneous microstructure. The
material has a high toughness which probably increases the load
necessary to create median cracks. The load dependence for all equations
for median cracks was shown to be largest at low loads and to diminish
at loads over 300N. As an example two equations are compared in figure
3. Median cracks were assumed to be formed at loads exceeding 300N since
$c/a > 3$. To compare the equations a linear regression was performed over
300N, as shown in figure 3.

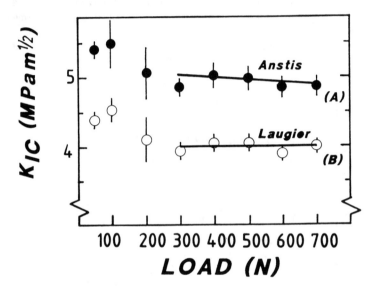

Figure 3. Load dependence of K_{1c} using two equations for median cracks.
(A) Anstis (0.016+0.004) (B) Laugier, (for slip cast HIPed 30 vol%
SiCwhisker/Al$_2$O$_3$ composites)

The regression lines were compared for all equations and, as can be seen
in figure 4 the level of K_{1c} varied over a surprisingly large range. The
Anstis equation contains a variable constant (0.016+0.004). With
(+0.004), values around 5 were obtained as compared to around 3 with
(-0.004). It is probable that the level of all the equations has to be
adjusted with a calibration constant, obtained by calibration against a
more conventional toughness method. The Laugier equation and the Evans,
Lawn, Marshall equation were found to be the best equations showing
almost no load dependence over 300N and the Laugier equation was
selected for further study of this material. The Niihara (Palmqvist)
equation which is extensively used for ceramics showed a remarkably high
load dependence and should be avoided, at least for whisker reinforced
Al$_2$O$_3$.

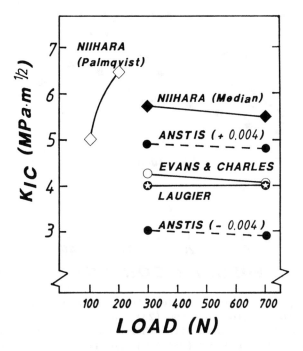

Figure 4. K_{1c} v.s. load as calculated with six different equations for 30 vol% SiCwhisker/Al$_2$O$_3$ composites.

Multiple-Indentation Bend Test To compare the results obtained with the Laugier equation, multipleindentation bend tests (MIBT) were performed with the following result: For a pressed, HIPed 25 vol% SiC whisker/Al$_2$O$_3$ composite the MIBT yielded a K_{1c} of 6.0 MP m$^{1/2}$ as compared to the 3.6 MPa m$^{1/2}$ obtained with the Laugier equation. 6.0/3.6 could be used as a calibration factor to convert the values calculated with the Laugier equation to a more true level.

K_{1c} VS. Whisker Content Composites with a whisker content ranging from 5 to 45 vol-% were tested, to determine the influence of whisker content on K_{1c} as compared to other reported values for SiC-whisker reinforced Al$_2$O$_3$.

The Laugier equation was chosen to calculate K_{1c} since this equation was found to yield a constant K_{1c} at all loads above 300N. In figure 5 the results are summarized and compared with other data from the literature.

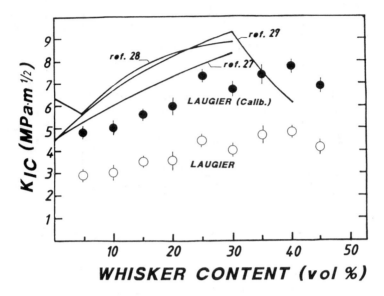

Figure 5. K_{1c} v.s. whisker content for slip cast, HIPed composites as calculated with the Laugier equation (23), and a calibrated Laugier equation, i.e. x 6/3.6, compared with other SiCw/Al$_2$O$_3$ data from the literature, measured with AMDCB (28) (29) and short rod (27) techniques.

The Laugier equation, however, yielded values substantially lower than those reported in the literature for other SiCw/Al$_2$O$_3$ composites. With the calibration factor 6/3.6, obtained as described in the previous section, values in the right range are obtained. Since a different manufacture of whiskers was used here from that in reference 27 and 28 identical results are not necessarilty expected.

In view of this, the K_{1c} obtained with the calibrated Laugier equation appears to be near the probable value for the material. The increase in K_{1c} with whisker content from between 4 and 5 to between 7 and 8 also agrees well with that reported in the literature. Indentation testing thus seems to be a useful method, accurately characterizing the toughening behaviour even of materials with a complex microstructure such as SiC whisker reinforced Al$_2$O$_3$.

CONCLUSIONS

The purpose of this study was to determine whether indentation testing could be used to measure fracture toughness of SiC whisker reinforced Al_2O_3. It was concluded that in order to do so, the influence of polishing, the load dependence and the relation to a traditional fracture toughness method have to be established. Once this is done indentation testing is an excellent, and simple, method for fracture toughness measurements. The polishing was found to lower the K_{1c} by about 2 MPa $m^{1/2}$ when a 3-4 μm compressive surface layer was removed. After that the K_{1c} remained constant with further polishing. All equations tested were more or less load dependent, the Laugier (23) and Evans, Lawn, Marshall equations being the most constant for this material. Loads higher than 300N have to be used, however, to create median cracks and obtain load independent results. The level obtained with the different equations varied from 3 to 7 MPa $m^{1/2}$ for the same material. After calibration against a more standard method realistic K_{1c} values were obtained. These values agreed well with other reported data for $SiCw/Al_2O_3$ composites. It is possible that a different calibration constant has to be established for each material type to be tested with indentation testing.

ACKNOWLEDGEMENTS

The authors wish to thank AB SANDVIK and SECO TOOLS AB for financially supporting this study, as well as Mrs U.-B. Jigholm for preparing the manuscript for publication.

REFERENCES

1. Freiman, S.W., Brittle Fracture Behavior of Ceramics, <u>Am. Ceram. Soc.Bull.</u>, 1988, 67 (2) 392-402

2. Homeny, J., Vaughn, W.L., Ferber, M.K., Processing and Mechanical Properties of SiC-Whisker-Al_2O_3-Matrix Composites, <u>Am. Ceram. Soc. Bull.</u>, 1987, 67 (2) 333-338

3. D.P.H. Hasselman, R.M. Fulrath, Proposed Fracture Theory of a Dispersion Strengthened Glass Matrix, <u>. Am. Ceram. Soc.</u>, 1966, 49 (2) 68-

4. Lange, F.F., The Interaction of a Crack Front with a Second-phase Dispersion, <u>Phil. Mag.</u>, 1970, 22 (179) 983-992

5. Lange, F.F., Fracture Energy and Strength Behaviour of a Sodium Borosilicate Glass - Al_2O_3 Composite System, <u>J. Am. Ceram. Soc.</u>, 1971, 54 (12) 614-620

6. Evans, A.G., The Strength of Brittle Materials Containing Second Phase Dispersions, Phil. Mag., 1972, 26 1327-1344

7. Faber, K.T., Evans A.G., Crack Deflection Processes – I. Theory, Acta Metall., 1983, 31 (4) 565-576

8. Faber, K.T., Evans, A.G., Crack Deflection Processes – II. Experiment, Acta Metall., 1983, 31 (4) 577-584 9.   Swanson, P.L.,

9. Fairbanks, C.J., Lawn, B.R., Mai, Y.-W., Hockey, B.J., Crack-Interface Grain Bridging as a Fracture Resistance Mechanism in Ceramics: I. Experimental Study on Alumina, J. Am. Ceram. Soc., 1987, 70 (4) 279-289

10. Evans, A.G., Faber, K.T., Crack-Growth Resistance of Microcracking Brittle Materials, J. Am. Ceram. Soc., 1984, 67 (4) 255-260

11. Corbin, N.D., Willkens, C.A., Yeckley R.L., Quantification of Microstructural Interrelationships in Si_3N_4 Based Materials, Presented at the 12th Annual Conference on Composites and Advanced Ceramics, Cocoa Beach, USA, jan 17-20, 1988 (to be published in Cer. Eng. Sci. Proc.)

12. Kageyama, K., Chou, T-W., Modelling and Analysis of Fracture Toughness of Short Fibre Reinforced Ceramic Matrix Composites, Proc. ICCM VI & ECCM 2, Elsevier, Ed. Matthews, F.L., et al., 1987, pp 2.60-2.69

13. Evans, A.G., Fracture Mechanics Determinations, Fract. Mech. Ceram. 1, Ed. Bradt, R.C., Hasselman D.P.H., Lange, F.F., Plenum Press, New York, 1974, 17-49

14. Palmqvist S, Rissbildungsarbeit bei Vickers-Eindrücken als Mass für die Zähigkeit von Hartmetallen, Arch. Eisenhutenwes., 1962, 33:629

15. Smith, S.S., Pletka, B.J., Indentation Fracture of Single Crystal and Polycrystalline Aluminum Oxide, Fract. Mech. Ceram. 6, Ed. Bradt, R.C., Hasselman D.P.H., Lange, F.F., Plenum Press, New York, 1983, 189-209

16. Moussa, R., Coppolani, I., Osterstock, F., Indentation Techniques Applied to Silicon Carbides, Brit. Ceram. Soc., 1982, 32 237-247

17. Lawn, B.R., Fuller, E.R., Equilibrium Penny-lika cracks in indentation fracture, J. Mater. Sci., 1975 10, 2016-2024

18. Evans, A.G., Charles, E.A., Fracture Toughness Determinations by Indentation, J. Am. Ceram. Soc., 1976 59 (7-8) 371-372

19. Evans, A.G., The Role of Indentation Techniques, ASTM STP 678, 1979 112-135

20. Lawn, B.R., Evans A.G., Marshall, D.B., Elastic/Plastic Indentation Damage in Ceramics: The Median/Radial Crack System, J. Am. Ceram. Soc., 1980, 63 (9-10) 574-581

21. Niihara, K., Morena, R., Hasselman, D.P.H., Indentation Fracture Toughness of Brittle Materials for Palmqvist Cracks, Fract. Mech. Ceram. 5, Ed. Bradt, R.C., Hasselman, D.P.H., Lange, F.F., Plenum Press, New York, 1983, 97-105

22. Anstis, G.R., Chantikul, P., Lawn, B.R., Marshall, D.B., A Critical Evaluation of Indentation Techniques for Measuring Fracture Toughness: Direct Crack Measurements, J. Am. Ceram. Soc., 1981 64 (9) 533-538

23. Laugier, M.T., The Elastic/Plastic Indentation of Ceramics, J. Mater. Sci. Lett., 1985, 4 (12) 1539-1541

24. Lundberg, R., Nyberg. B., Williander, K., Persson, M., Carlsson R., Processing of Whisker-Reinforced Ceramics, Composites 1987, 18 (2) 125-127

25. Lundberg R., Nyberg, B., Williander, K., Persson, M., Carlsson R., Glass Encapsulated HIP-ing of SiC Whisker Reinforced Ceramic Composites, Proc. First Int. HIP Conf., Luleå, Sweden, June 15-16, 1987

26. Cook, R.F., Lawn, B.R., A Modified Indentation Toughness Technique, J. Am. Ceram. Soc., 1983, 63 (11) C-200-C-201

27. Wei G.C, Becher P.F., Development of SiC-whisker reinforced ceramics, Am. Ceram. soc. Bull. 1985 64 (2) 298-304

28. Becher , P.F., Tiegs, T.N., Ogle, J.C, Warwick, W.H., Toughening of ceramics by whisker reinforcement, Fract. Mech. Ceram. 7, Ed. Bradt, R.C., Hasselman, D.P.H., Lange, F.F., Plenum Press, New York 1986, 61-63

29. Tiegs, T.N., Becher, P.F., Whisker Reinforced Ceramic Composites, Proc. Int. Conf. Ceram. Mater. & Components for Engines, Lübeck-Travemünde
Apr. 14-17 1986, DKG, Ed. by Bunk, W., Hausner, H., 193-200

THE INFLUENCES OF Al_2O_3 AND $CaSO_4 \cdot 2H_2O$ ON HYDROTHERMAL SYNTHESIS AND CHARACTERIZATION OF COMPOUNDS IN SYSTEM $CaO-SiO_2-H_2O$

CHUICHI TASHIRO and HIROSHI NISHIOKA

Faculty of Engineering, Yamaguchi University, Ube, Japan

ABSTRACT

The reaction diagram and characterization of the hydrothermal products obtained from the 0.5, 1.0 and 2.0 CaO/SiO_2 mixtures with an Al_2O_3 and $CaSO_4 \cdot 2H_2O$ addition were investigated. The reaction temperatures were 160, 200 and 250°C. The Al_2O_3 added was 5 to 30 mol% and the $CaSO_4 \cdot 2H_2O$ added was 5 mol%. The mixtures were press-molded into a frame mold of a cylindrical column of Ø8x6 mm under a pressure of 30 kg/cm^2 and were autoclaved. Then, the reaction products were examined by SEM observation, XRD and mercury pressured polosimeter. Also, the compressive strengths of the products were tested.

With a CaO/SiO_2 ratio of 0.5, the products were a tobermorite gel, hydrogarnet solid solution, γ-AlO(OH) and anhydrite. When 5% Al_2O_3 was added, the compressive strength of the product was 235 kg/cm^2 at 200°C; the pore were distributed from 37.5 Å to 270 Å, the total pore volume was 0.699 cm^3/g. When $CaSO_4 \cdot 2H_2O$ was added, the area of 200 kg/cm^2 or above regarding strength increased in temperature and Al_2O_3 content. In general, the strengths of the products obtained from 0.5 CaO/SiO_2 were larger than that of 1.0 or 2.0 CaO/SiO_2.

With a CaO/SiO_2 ratio of 1.0, the products were a tobermorite gel, CSH(I), tobermorite solid solution, xonotlite, hydrogarnet solid solution, γ-AlO(OH) and anhydrite. The tobermorite substituted about 15 mol% Al_2O_3 and the crystal had a plate-like shape. When 10% Al_2O_3 was added, the compressive strength of the products was 210 kg/cm^2 at 200°C.

With a CaO/SiO_2 ratio of 2.0, the products were portlandite, γ-C_2S hydrate, α-C_2S hydrate, tobermorite solid solution, xonotlite, hydrogarnet solid solution and anhydrite. The strength of the products was smaller than that of 0.5 or 1.0 CaO/SiO_2. As a rule, the porosity was large, and it was largest when more than 20% of Al_2O_3 was added.

INTRODUCTION

CSH, tobermorite, xonotlite and related materials are being widely utilized in recent years. It is produced on autoclaved lightweight concrete, autoclaved concrete, sand-lime blocks, hydrous calcium silicate heat-insulation materials, hydration products of cement and other building or construction materials.

The crystal structure, phase relations and mineralogical description of autoclaved calcium silicate materials have been investigated by many researchers in the past, but few reports (1-5) exist on the effect of the Al_2O_3 component or $CaSO_4 \cdot 2H_2O$ in raw materials or cement regarding the reaction diagram and strength of autoclaved products.

The present work reports on the reaction diagram of products obtained from 0.5, 1.0 and 2.0 CaO/SiO_2 mixture with Al_2O_3 and $CaSO_4 \cdot 2H_2O$ addition at 150, 200 or 250°C, and also discribes the reaction between the microstructure and compressive strength of the products.

EXPERIMENTAL PROCEDURE

The starting materials used were the following chemicals : $Ca(OH)_2$ as CaO, silica gel as SiO_2, γ-AlO(OH) as Al_2O_3 and $CaSO_4 \cdot 2H_2O$. The loss of ignition of silica gel contained 0.1% HF nonvolatite was 0.5 to 7.0 wt%. The silica gel was ground into powder below 50 μm. CaO/SiO_2 molar ratio used was 0.5, 1.0 or 2.0. The amounts of Al_2O_3 replaced in these preparations were 5, 10, 15, 20, 25 and 30 mol%, and the $CaSO_4 \cdot 2H_2O$ added was 5 mol%. These chemicals were mixed in a mortar for 1 hr. The mixtures were press molded into a frame mold of a cylindrical column Ø 8x6 mm under pressure of about 30 kg/cm^2. Then, the mixtures were immediately sealed after the preparation and autoclaved. After processing, the autoclaved products was quenched in air. The products was then dried at 100°C for 2 hrs. The processing was 160, 200 or 250°C for 100 hrs, respectively.

The autoclaved products were examined using a scanning electron microscope (SEM), X-ray powder diffraction (XRD) and mercury pressure porosimeter. Also, the compressive strengths of the products were tested.

RESULTS AND DISCUSSION

CaO/SiO_2: 0.5

Figure 1 shows the influence of Al_2O_3 or Al_2O_3 and $CaSO_4 \cdot 2H_2O$ addition on the synthesized phases and the strength of autoclaved 0.5 CaO/SiO_2 mixture. Irrespective of temperature, the products were always tobermorite gel when Al_2O_3 and $CaSO_4 \cdot 2H_2O$ were not added. With the Al_2O_3 addition, the tobermorite gel(ill-crystallized tobermorite), γ-AlO(OH) and hydrogarnet solid solution were obtained, and also anhydrite appeared with the $CaSO_4 \cdot 2H_2O$

addition.   It seems that the tobermorite gel substitutes were
about 15% Al_2O_3, which agreed with the data from Kalousek(1).

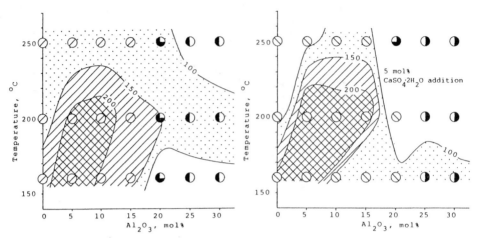

Figure 1. Influences of Al_2O_3 or Al_2O_3 and $CaSO_4 \cdot 2H_2O$ on the
synthesized phases and the strength of autoclaved 0.5
CaO/SiO_2 mixtures.

⊘ :Tobermorite gel
◑ :Tobermorite gel+γ-AlO(OH)
◐ :Tobermorite gel+Hydrogarnet solid solution+γ-AlO(OH)
⊘ :Tobermorite gel+Anhydrite
◑ :Tobermorite gel+γ-AlO(OH)+Anhydrite
◑ :Tobermorite gel+Hydrogarnet solid solution+γ-AlO(OH)+Anhydrite
✕✕✕ Over 200 kg/cm^2
⁄⁄⁄ From 150 to 200 kg/cm^2
∴∴∴ From 100 to 150 kg/cm^2

Figure 2 shows the XRD of the products.   Figure 3 shows the
SEM and Figure 4 shows the pore size distribution of the products.
The strengths of the products obtained from a 5 to 10% Al_2O_3

Without Al_2O_3 10 mol% Al_2O_3 addition

Figure 3. SEM of autoclaved 0.5 CaO/SiO_2 mixtures with or without 10 mol% Al_2O_3 addition at 200°C.

addition at 160 to 200°C were larger than those of the others.
The products obtained from a 5% Al_2O_3 addition at 200°C exhibited the following properties: the compressive strength was
235 kg/cm^2, most of the pore were distributed from 37.5 Å to
270 Å, total pore volume was 0.699 cm^3/g and the crystals were
probably foil-like habits.

When $CaSO_4 \cdot 2H_2O$ was added, the area of strength of 200 kg/cm^2
or above spreads out toward the high Al_2O_3 composition and high
temperature.

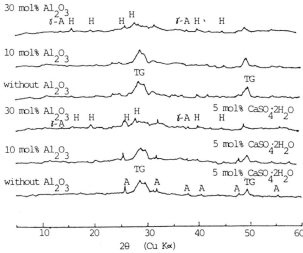

Figure 2. XRD of autoclaved 0.5 CaO/SiO$_2$ mixtures with or without Al_2O_3 and $CaSO_4 \cdot 2H_2O$ additions at 200°C.

TG:Tobermorite gel   H:Hydrogarnet solid solution
ɣ-A:ɣ-AlO(OH)   A:Anhydrite

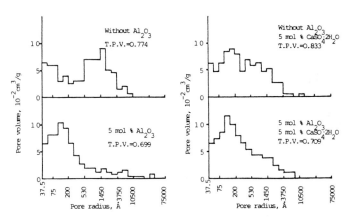

Figure 4. Pore size distributions of autoclaved 0.5 CaO/SiO$_2$
mixtures with or without Al_2O_3 and $CaSO_4 \cdot 2H_2O$
additions at 200°C.

CaO/SiO$_2$: 1.0

Figure 5 shows the influence of the Al$_2$O$_3$ or Al$_2$O$_3$ and CaSO$_4$·2H$_2$O addition on the synthesized phases and the strength of auto- claved 1.0 CaO/SiO$_2$ mixture, respectively. The products were tobermorite gel, well-crystallized tobermorite and xonotlite when Al$_2$O$_3$ and CaSO$_4$·2H$_2$O were not added. With the Al$_2$O$_3$ additions, the tobermorite gel, CSH(I), xonotlite, hydrogarnet and δ-AlO(OH) were obtained, and anhydrite appeared with the CaSO$_4$·2H$_2$O ad- dition. It seems that the Al$_2$O$_3$ added allowed the process to continue under conditions lower than those in the no addition. Namely, the tobermorite solid solution and xonotlite appeared even at 250°C when 5 or 10% Al$_2$O$_3$ was added.

When CaSO$_4$·2H$_2$O was added to the mixture with 5 or 10% Al$_2$O$_3$, the tobermorite solid solution and anhydrite appeared at 160, 200 and 250°C, with hydrogarnet not appearing. It seems that

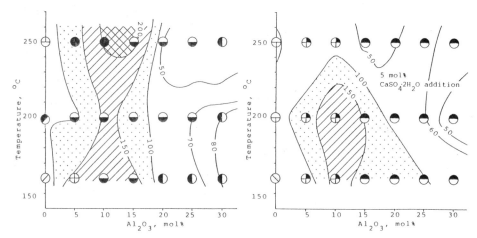

Figure 5. Influences of Al$_2$O$_3$ or Al$_2$O$_3$ and CaSO$_4$·2H$_2$O on the synthesized phases and the strength of autoclaved 1.0 CaO/SiO$_2$ mixtures.

⊘:Tobermorite gel
◖:Tobermorite
⊕:CSH(I)+Tobermorite solid solution
⊖:Tobermorite solid solution+Hydrogarnet solid solution
⊕:Xonotlite
●:Tobermorite solid solution+Xonotlite
◑:Tobermorite solid solution+Hydrogarnet solid solution+ δ-AlO(OH)
⊘:Tobermorite gel+Anhydrite
⊕:Xonotlite+Anhydrite
⊕:Tobermorite solid solution+Anhydrite
●:Tobermorite solid solution+Hydrogarnet solid solution+ Anhydrite
▩ Over 200 kg/cm^2
/// From 150 to 200 kg/cm^2
░ From 100 to 150 kg/cm^2

the $CaSO_4 \cdot 2H_2O$ added helped the substitution of Al_2O_3 in the tobermorite. The $CaSO_4 \cdot 2H_2O$ addition interferred with the formation of xonotlite at the 5 to 10% Al_2O_3 addition. It seems that the SO_3^- ion interferred with the transformation from tobermorite to xonotlite. This agreed with Sakiyama's data (6).

Figure 6 shows the XRD of the products and Figure 7 shows the SEM and Figure 8 shows the pore size distribution of the products. The products obtained from the 10 to 15% Al_2O_3 addition at 200°C had large strength and exhibited the following microstructure: reaction products consisted of tobermorite solid and hydrogarnet solid solution, and showed a compact structure; total pore volume of the product with 10% Al_2O_3 was 0.743 cm³/g. As a rule, the strength of the tobermorite substituted for 10% Al_2O_3 was larger than that of other products.

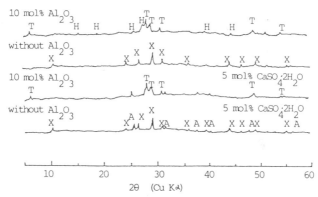

Figure 6. XRD of autoclaved 1.0 CaO/SiO_2 mixtures with or without Al_2O_3 and $CaSO_4 \cdot 2H_2O$ additions at 200°C.

X:Xonotlite   T:Tobermorite solid solution
H:Hydrogarnet solid solution   A:Anhydrite

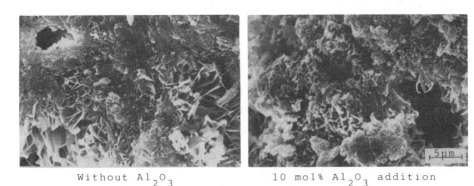

Without Al_2O_3 10 mol% Al_2O_3 addition

Figure 7. SEM of autoclaved 1.0 CaO/SiO_2 mixtures with or without 10 mol% Al_2O_3 addition at 200°C.

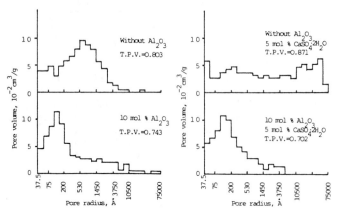

Figure 8. Pore size distributions of autoclaved 1.0 CaO/SiO$_2$
mixtures with or without Al$_2$O$_3$ and CaSO$_4$·2H$_2$O
additions at 200°C.

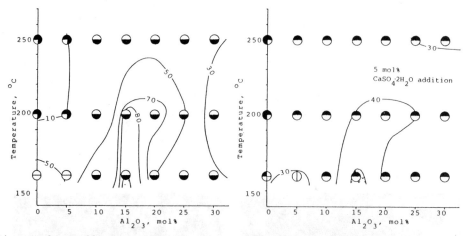

Figure 9. Influences of Al$_2$O$_3$ or Al$_2$O$_3$ and CaSO$_4$·2H$_2$O on the
synthesized phases and the strength of autoclaved 2.0
CaO/SiO$_2$ mixtures.

⊖ : Portlandite
◖ : γ-C$_2$S hydrate+Xonotlite
◑ : Tobermorite solid solution+Hydrogarnet solid solution
⊕ : α-C$_2$S hydrate+Anhydrite
⊘ : α-C$_2$S hydrate+Tobermorite solid solution+Anhydrite
◕ : γ-C$_2$S hydrate+Xonotlite+Anhydrite
◔ : Tobermorite solid solution+Hydrogarnet solid solution+
     Anhydrite
░░░ : Over 100 kg/cm^2

CaO/SiO_2: 2.0

Figure 9 shows the influence of the Al_2O_3 or Al_2O_3 and $CaSO_4 \cdot 2H_2O$ addition on the synthesized phases and strength of autoclaved 2.0 CaO/SiO_2 mixture. The portlandite, α or γ-C_2S hydrate, xonotlite, tobermorite solid solution and hydrogarnet solid solution were obtained, and also anhydrite appeared with the $CaSO_4 \cdot 2H_2O$ addition. An addition of $CaSO_4 \cdot 2H_2O$ should have negligible influence on the synthesized phases of the system $CaO-SiO_2-H_2O$.

Figure 10 shows the XRD of the products. Figure 11 shows the SEM of the products obtained from the mixture with or without 10% Al_2O_3 and 5% $CaSO_4 \cdot 2H_2O$ addition. Figure 12 shows the pore size distribution of the products. It seems that the addition of Al_2O_3 of 10% or above interferred with the forma-

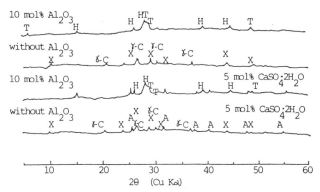

Figure 10. XRD of autoclaved 2.0 CaO/SiO_2 mixtures with or without Al_2O_3 and $CaSO_4 \cdot 2H_2O$ additions at 200°C.

X:Xonotlite γ-C:γ-C_2S hydrate T:Tobermorite solid solution H:Hydrogarnet solid solution A:Anhydrite

Without Al_2O_3 and $CaSO_4 \cdot 2H_2O$ 20 mol% Al_2O_3 and 5 mol% $CaSO_4 \cdot 2H_2O$ addition

Figure 11. SEM of autoclaved 2.0 CaO/SiO_2 mixtures with or without 20 mol% Al_2O_3 and 5 mol% $CaSO_4 \cdot 2H_2O$ additions at 200°C.

tion of $\gamma-C_2S$ hydrate and xonotlite, and helped the formation of the tobermorite solid and hydrogarnet solid solutions.

As a general rule, the strengths of the product were less than that of the products having a 0.5 or 1.0 CaO/SiO$_2$ molar ratio. The product obtained from the 15% Al$_2$O$_3$ addition at 160 °C exhibited the following properties: the strength was 101 kg/cm^2, the total pore volume was 1.190 cm^3/g, reaction phases obtained were tobermorite solid solution and hydrogarnet solid solution, crystal of tobermorite was mostly plate granular habits.

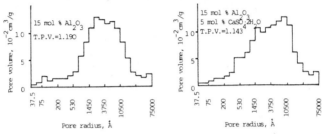

Figure 12. Pore size distributions of autoclaved 2.0 CaO/SiO$_2$ mixtures with Al$_2$O$_3$ or with Al$_2$O$_3$ and CaSO$_4$·2H$_2$O additions at 200°C.

CONCLUSION

The influence of Al$_2$O$_3$ and CaSO$_4$·2H$_2$O on the phase and compressive strength of reaction products obtained from Ca(OH)$_2$ and silica gel at 160, 200 and 250°C were investigated and the following conclusions were obtained:

1. With a CaO/SiO$_2$ ratio of 0.5, the products were tobermorite gel, hydrogarnet and γ-AlO(OH), and also anhydrite appeared with the CaSO$_4$·2H$_2$O addition. With a CaO/SiO$_2$ ratio of 1.0, the products were tobermorite gel, tobermorite solid solution, CSH(I), xonotlite, hydrogarnet and γ-AlO(OH). With a CaO/SiO$_2$ ratio of 2.0, portlandite, α or γ-C$_2$Shydrate, xonotlite and tobermorite solid solution were synthesized.

2. Al$_2$O$_3$ added allowed the process to continue under a condition lower than that with no addition. In the case of 1.0 CaO/SiO$_2$, a mixture with 5 to 10% Al$_2$O$_3$, the tobermorite solid solution appeared even at 250°C.

3. With the tobermorite substituted for about 15 mol% Al$_2$O$_3$, the crystal were small plate habits.

4. Additions of 5 to 10% Al$_2$O$_3$ and 5% CaSO$_4$·2H$_2$O to the 0.5 CaO/SiO$_2$ mixture helped the formation of a compact structure made up of tobermorite gel, and contributed to strength development.

5. CaSO$_4$·2H$_2$O added interferred with the transformation from tobermorite to xonotlite.

6. The strength of the products was directly proportional to

107

the total pore volume in the each product with the same CaO/SiO$_2$ ratio.

REFERENCES

1. Kaulousek,G.L., Crystal chemistry of hydrous calcium silicates;I, Substitution of aluminum in lattice of tobermorite, J.Am.Ceram.Soc.,1958,**41**,pp.124-132.

2. Diamond,S.,White,J.L. and Dolch,W.L., Effects of isomorphous substitution in hydrothermal-synthesized tobermorite, Amer. Mineral.,1966,**51**,pp.388-401.

3. Sakiyama,M. and Mitsuda,T., Effect of Al on the formation of tobermorite(in Japanese),Semento Gijutsu Nenpo,**31**,pp.46-49.

4. Mitsuda,T. and Taylar,H.F.W., Influence of aluminum on the conversion of calcium silicate hydrate gels into 11Å tobermorite at 60 and 120°C, Cem. Con. Res.,1975,**5**,pp.203-209.

5. Takahashi,T.,Hayashi,H. and Yamakita,J., Effect of Al$_2$O$_3$ on the formation of xonotlite(in Japanese),Semento Gijutsu Nenpo,1973,**27**,pp.41-44.

6. Sakiyama,M and Mitsuda,T., The effect of gypsum on the formation of tobermorite(in Japanese),1977,**31**,pp.49-52.

A NEW MODEL TO DESCRIBE THE DAMAGE IN A LAMINATED CARBON FIBRE REINFORCED CARBON COMPOSITE

C. Rief, K. Kromp
Max Planck Institut für Metallforschung
Institut für Werkstoffwissenschaften
Seestr. 92, D-7000 Stuttgart 1, FRG

ABSTRACT

Specimens of carbon-fibre reinforced carbon in a bidirectional layup were loaded in 3-point bending at various span lengths. The results are compared to previous bending tests, performed with "thin" specimens, smaller in breadth and width. A "Stack Model" succeded to explain the results, deviating from classical bending theory in the former investigation. This model predicts a change in the failure mode for "thick" specimens, larger in breadth and width used here. This change in failure mode is pointed out experimentally and it is proved that the "Stack Model" interprets the results exactly.

INTRODUCTION

CFRC is a kind of high performance composite material with a brittle matrix and a non metal fibre. It exhibits high strength up to very high temperatures, combined with low density and and good workability. One application for that material, with a SiC-coating to prevent oxidation, is in the Shuttle Orbiter Thermal Protection System /1/. CFRC is used in applications where very high thermal and mechanical resistance are required.

With specimens of a cross section of breadth B = 9 mm and a width W = 4.3 mm, it was shown in /2/ that with the help of a "Stack Model", the shear failure near the neutral axis could be explained, even at high span/width ratios, S/W, where failure in pure tension/compression was expected.This model considered the relative ply thickness $\Delta w/W$ in bending

and related it to the shear strength and the tensile strength of the material. It predicted further that, below a critical $\Delta w/W$-value, the mode of
failure is described by the classical bending theory /3/, i. e. S/W will
control the failure mode. The present paper investigates the validity of
the "Stack Model".

MATERIAL

The CFRC-Material was kindly made available by SIGRI GmbH Meitingen, FRG.
The material is processed from woven prepregs out of yarns of 3000 PAN
based high modulus fibers of 7 µm diameter. The stacking sequence is
(0°/90°). Each ply has an average thickness of $\Delta w = 0.3$ mm. The laminate
consists of 36 plies. It is pyrolysed, reimpregnated and repyrolysed, until
the desired density is reached. The final product consists of pure carbon.
Because of the several pyrolysations, which result in weigth loss, the
material is highly porous and contains presumable high residual stresses.
The high density of pores and cracks is visible in Figure 1. Figure 1 also
shows that the matrix between the plies has a negligible thickness.

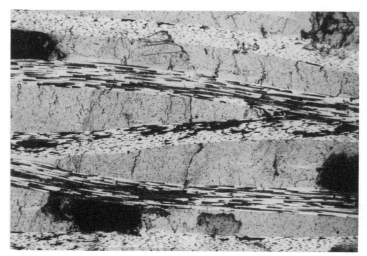

Fig. 1. Polished surface of reinforced carbon-carbon in a bidirectional
layup normal to the plies.

These imperfections are the reason for the ductile behaviour of the material, as already described for LAS/SIC-laminates /4/. The difference of the elastic moduli in warp and weft direction in the CFRC material is less than 15 %. Specimens with an average heigth of W = 9.3 mm and average breadth of B = 18.3 mm were prepared out of plates of the material described above.

EXPERIMENTAL PROCEDURE

Displacement controlled bending tests were performed with span lengths S = 60, 90, 120, 150, and 180 mm at room temperature by means of a hydraulic testing machine. S/W varied from 6.5 to 20. The specimens were loaded flatwise on stainless steel rods of 20 mm diameter with a 5 mm radius of curvature in order to minnimize support effects. The displacement was measured directly under the middle load bearing on the lower surface of the specimen. With the help of a function generator and a closed loop circuit different constant displacement rates could be achieved.

In order to compare the results of this investigation with previous results, the stress rates $\dot{\sigma}$ of both investigations should be the same,

$$\dot{\sigma} = \frac{3 \; F \; S}{2 \; B \; W^2 \; t} \quad (1) \qquad \dot{\delta} = \frac{F \; S^3}{4 \; E \; B \; W^3 \; t} \quad (2)$$

where $\dot{\delta}$ must be chosen such that the thick specimens are loaded with a stress rate comparable to that of the thin specimens. Combining (1) and (2), one obtains:

$$\dot{\delta} = \frac{\dot{\sigma} \; S^2}{6 \; E \; W} \quad (3)$$

Assuming an effective Young's modulus E=60 GPa and a stress rate $\dot{\sigma}$ of 0.2 N/mm^{-2}s^{-1} one obtains the following displacement rates:

S [mm]	180	150	120	90	60
$\dot{\delta}$ [µms^{-1}]	2.5	1.75	1.2	0.6	0.3

To determine the mode of failure, the specimen was photographed in the range of the central loading rod in the as received state. Afterwards the test was started. The test was stopped at the first observed deviation from linearity in the load/displacement diagram, F/δ-curve. Fig. 2 shows an original F/δ-curve for the tested specimens. It is assumed, that the first damage in this material will occur at F^*, the first deviation from linearity. This final state was photographed again. The damage could now be detected by comparing the as received state with the final state.

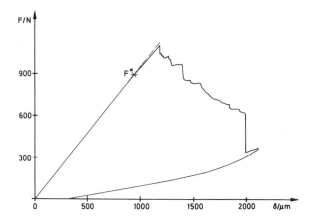

Fig. 2. Original load/displacement curve, span 150 mm, displacement rate 1.75 μm/s.

THE "STACK MODEL"

Previous experiments were performed on the same material but with cross sections 4.3 × 8.6 mm² /2/. S varied between 30 - 120 mm. The mode of failure could not be described by the classical bending theory. Even at S = 120mm (corresponding to S/W ≈ 30) failure occurred by delamination near the neutral axis in the range of the central load bearing. In /2/ a "Stack Model" was proposed, which succeeded to describe the experiments.

In the "Stack Model", two laminates are regarded, loaded in pure bending, both having the same radius of curvature (Figure 3). In the first case, the matrix transmits no shear force. In the second case, the matrix

Pure Bending

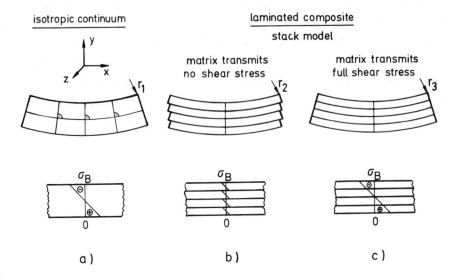

a)   b)   c)

Fig. 3. "Stack Model", the behaviour of an idealized laminate under load,
   (b) compared to a continuum, (a) and a strong bonded laminate, (c).

transmits the full shear force. By comparing these two cases, it can be
seen that the matrix must transmit a shear force to satisfy the Bernoulli
hypothesis and to obtain a continuous stress distribution (all cross
sections are plane and normal to the neutral axis). The shear stress $\tau_b{}^{\Delta w}$,
which could not be derived by classical bending theory, was postulated in
/2/ to be:

$$\tau_b{}^{\Delta w} = \frac{3}{2} \frac{F}{B} \frac{S}{w^3} \Delta w \quad (4)$$

The shear stress $\tau_b{}^{\Delta w}$, acting between two neighbouring plies, is
superimposed on to the shear stress τ_{xy} resulting from the shear force.
Figure 4 shows all stresses and stress distributions acting in a laminate
according to the "Stack Model".

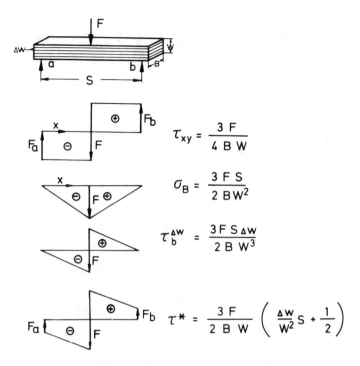

Fig. 4: Stress distributions in a laminate loaded in three-point bending according to the "Stack Model", schematically.

By knowing the shear strengths and the bending strengths, it was possible to describe the mode of failure even at S/W ≈ 30 with the "Stack Model". One could obtain a critical total shear stress $\tau^* = \tau^{xy} + \tau_b^{\Delta w}$, which at failure was the same for all specimens, i.e. the total shear stress was independent from S. At all S/W values failure by delamination, i.e. in shear, was observed. The stress τ^* described the results and thus the mode of failure /2/.

RESULTS AND DISCUSSION

The investigation for the "thick" specimens was started at the longest span of S = 180mm. Figure 5a shows the as received state in the range below the central load bearing. The specimen was now loaded until the first deviation from linearity was visible in the F/δ diagram. The test was stopped at the final state shown in figure 5b.

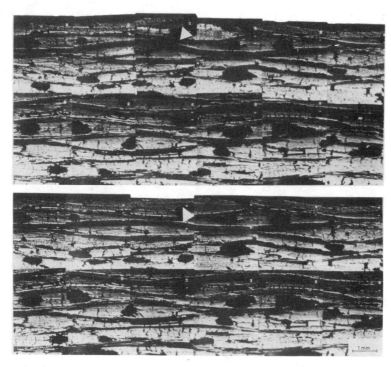

a)

b)

Fig 5a,b. Initiation of damage at S = 180 mm by buckling near the middle
    load bearing (arrows); as received state, (b) and final state, (a).

    Comparing the two figures one observes a buckling near the middle load
bearing as the mode of failure (arrows). All five specimen, tested at
S=180 mm failed in that mode.

    At S = 150 mm, the same mode of failure was observed. At S = 120 mm
and S = 90 mm, the mode of failure was buckling near the middle load bearing as well as delamination near the neutral axis.

    At S = 60 mm, all five specimens tested failed in delamination near
the neutral axis in the range of the central load bearing (figure 6a,b).
The arrows show the damage.

    The experiments show that the "Stack Model" can describe the mode of
failure:

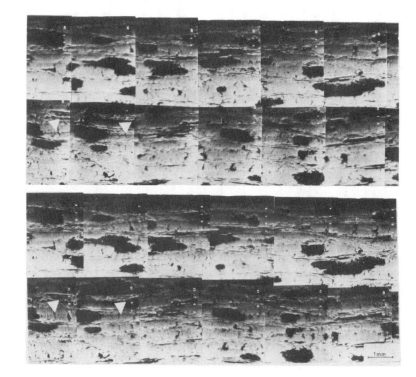

a)

b)

Fig 6a,b. Initiation of damage by delamination near the neutral axis at
S = 60 mm in the range of the central load bearing (arrows); as received
state, (b) and final state, (a).

- Fig. 7 shows the dependence of all stresses on S, for the "thin"
  specimens of the previous results (a,b) /2/ and the "thick" specimens of
  this investigation (c,d). Fig 7a shows the applied bending stress at
  failure for the "thin" specimens over the span length S. It is obvious,
  that in the area investigated this bending stress is increasing with S,
  so it can never be the bending strength, even not at S =120 mm
  (S/W = 30). Fig. 7b shows that the total shear stress τ^* at failure for
  these specimens is independent from S, which means always shear failure
  in the area investigated. This mode of failure was proved experimentally
  in /2/.

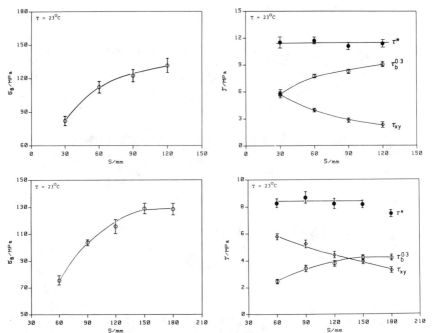

Fig. 7. All stresses in dependence on the span S of the "thin", (a,b) and the "thick", (c,d) specimens.

- In fig. 7c, for the "thick" specimens however, the applied bending stress at failure becomes independent from S at S = 150 mm (S/W > 16). So it can be assumed that the bending strength of this material for the tested specimen geometry is reached. According to that, fig. 7d shows, that τ^* decreases for S > 150 mm. τ^* decreasing means that the applied shear stress at failure decreases. According to that, the mode of failure for this specimen must change, i. e. failure in tension/compression must now occur. This is proved by the experimental results, at S = 150 mm and S = 180 mm failure by buckling near the central loading rod was observed.

So it is allowed to say that the total shear stress τ^* is the failure criterion, not τ_{xy}.

The shear strength of the "thick" specimens is only about 70 % of that of the "thin" specimens. So it can be supposed that the residual stresses are much higher in the thick specimens than in the thin specimens. One

method to estimate these residual stresses is proposed in /4/. The investigation of these stresses will be the subject of future experiments.

SUMMARY

The level of the total shear stress τ^*, postulated in the "Stack Model",
describes the mode of failure in the "thick" specimens of a CFRC - laminate, consisting of 36 plies as it did for "thin" specimens of 16 plies in
an earlier investigation. It was shown that the span to heigth ratio, S/W,
for the thick specimens indeed gives a criterion to determine the mode of
failure At S/W $\approx$ 6, pure failure in shear, at S/W $\approx$ 20 pure failure in
tension/compression was observed. The bending strength of the material was
found to be 129 MPa, the shear strength 8.5 MPa. The latter is only about
2/3 of the shear strength obtained by the same test method with the same
material, but consisting of only 16 plies. It is assumed that this decrease
in strength is caused by the residual stresses, which are thus presumed to
be higher in specimens consisting of 36 plies than in the specimens consisting of only 16 plies. This investigation shows clearly that for the
determination of the mode of failure in a CFRC-laminate not only the
S/W-ratio is decisive, but also the width W of the specimen. That means, it
is not arbitrous, in which way S/W is varied: by the span length S alone or
by S and W combined. Due to the results presented, a combined variation
should be avoided.

ACKNOWLEDGMENT

The authors gratefully acknowledge the support of the Deutsche Forschungs-
gemeinschaft under contract No Kr 970/1.

REFERENCES

/1/ Korb, L.J., Morant, C.A., Calland, R.M., and Thatcher, C.S., The
    Shuttle Orbiter Thermal Protection System, Ceramic Bulletin, Vol 60,
    Nr.11, 1981, pp 1188- 1193.

/2/ Rief, C., Kromp, K., Experimental Investigations and Model Proposal on
    Damage Mechanism in a RRC Material, to be published in ASTM STP.

/3/ Din Norm Nr 29971, German Institute of Standardization.

/4/ Sbaizero, O., Evans, A.G., Tensile and Shear Properties of Laminated
    Ceramic Matrix Composites, J. Am. Ceram. Soc., 69,[6] 481-86 (1986)

UTILIZATION OF CERAMIC ENGINEERING MATERIALS UNDER EXTREME CONDITIONS OF GAS ABRASIVE WEAR

Belousov V.Ya., Philipenko V.M.
Ivano-Frankovsk Oil and Gas Institute
Ivano-Frankovsk, Karpatskaya St.15, USSR

ABSTRACT

Face seals used in centrifugal type pumps of natural gas and general purpose pumps are subjected to most rapid wear. Locking fixture installed on gas wells operates under extreme conditions of gas abrasive wear. Austenite-martensite steels, oxide ceramics and hard alloys used to manufacture working organs of these units do not always satisfy the appropriate level of service characteristics. Ceramic engineering materials on Si_3N_4 and SiC base possess a valuable set of properties, namely, high values of strength, hardness, heat condition and crack resistance which makes them candidate materials to replace the conventional antifrictional materials and also the materials resisting gas abrasive wear. Authors found that ss-carbide and hp-Si_3N_4 can be successfully used to manufacture friction elements of face seals. It was also shown that Si_3N_4 material can reliably withstand extreme conditions of gas abrasive wear.

INTRODUCTION

Drilling equipment life is determined by the level of physical, mechanical and chemical properties of materials used in working organs. Rational and effective utilization of material properties is possible provided certain service conditions influencing wear peculiarities are considered.

Locking fixture installed in gas wells undergoes extreme gas abrasive wear.

Austenite-martensite steels used to manufacture it are nearing their limit as further increase in physico-mechanical properties of alloys does not essentially lead to longer life of parts,(1).

Face seals used in centrifugal-type natural gas pumps and general purpose pumps represent working units subjected to most severe stresses and rapid wear. On trunk lines in use today gas pressure reaches 7.5 MPa while surface sliding friction rate in seals is 50-60 m/sec. This presupposes extremely high requirements for face seals materials, the main of them being resistance to corrosive media and temperature gradient, absence of catching and sticking and enough resistance to thermal friction.

MATERIALS USED

Graphite-steel friction couple is currently used for ЦБН (centrifugal-type pumps) seals. The above materials possess favourable antifrictional, heat conducting and running-in characteristics.

Their disadvantage is small graphite strength and hardness causing considerable wear and lessening sealing characteristics. Such materials are used only under conditions of friction couple cooling with liquid when pressure does not exceed 1 MPa and sliding rate is no more than 10 m/sec.

Motor potential of such friction couple does not exceed 3,000-4,000 hours which is not only due to the lack of wear resisting characteristics of graphitized materials but also due to abrasive wear caused by transported gas contamination. Proper gas refining being rather difficult face seal operation will not be enough reliable,(2).

With considerable sliding rates (>10 m/sec) and friction couple pressure (>1 MPa) chromium steels and all kinds of surfacing are liable to thermal cracking effected by temperature gradients. Under these conditions hard alloys BK3, BK4, BK6 and BK8 give satisfactory results. The specific feature of these materials is that under conditions of liquid contact both elements of the couple are manufactured of hard alloy and working surfaces of these elements show no signs of catching or sticking.

The disadvantage of hard alloys is that in some corrosive media bond etching occurs resulting in working surface deterioration by protruding tungsten carbide grains. Despite possessing high wear resistance values hard alloys working at sliding rates 20 m/sec so far cannot complete with graphites when used as antifrictional materials for seals at high sliding rates (50-100 m/sec). The reason lies primarily in low heat conducting factors adherent in these materials which causes superheating of friction surface and hence its distortion and the appearance of microcracks, (3).

Face seal rings of ЦМ-322 material (sintered Al_2O_3) proved to work satisfactorily with all antifrictional materials subjected to liquid and half dry friction. However this material has no fracture resisting factor ($K_{1c} < 3$ MPa m$^{1/2}$ which leads to rings damage in the course of mounting and transporting), it is liable to thermal cracking at temperature difference 250-20°C and also shows inadequate operational characteristics under dry friction conditions which manifests itself at the point of starting the equipment.

Of all industrial materials used for face seals preference should be given to siliconized graphites (СГ-П,Т) due to their having better physico-mechanical characteristics. These materials are composed of silicon carbide, graphite and free silicon.

The optimum ratio of their phases chosen ensured high antifrictional characteristics and corrosion resistance. High heat conductivity of the above material and its strength characteristics resist rings thermocracking at V = 25 m/sec and friction couple load < 10 MPa.

But extreme brittleness of these materials presents a serious limitation to their use as fracture resisting factor of these materials is < 3 MPa. Rings failure occurs due to substantial difference in linear expansion coefficient as compared to steels and othermaterials and their low elasticity which causes preliminary stresses accompanying the assembly with metal drum. The use of conventional rings fastening techniques (glueing, brazing, welding) complicates the process of manufacturing face seals and increases percentage of rejected items.

Thus the possible way of solving the problem is the development of heat conducting materials possessing the following desirable properties: wear resistance, low linear expansion and friction factors, high levels of strength and hardness. The appropriate set of valuable characteristics is adherent in ceramic engineering materials developed on silicon carbide and silicon nitride base. Promising is the use of these materials in machines subjected to intense stationary and dynamic loading in a wide range of opetational temperatures 20°-$1,3oo^{\circ}$ combined with corrosive and errosive effect of gas flows, solid particles and other corrosive media. Supplies of necessary raw materials available for producing engineering ceramics on Si_3N_4 and SiC base are in fact inexaustible and the cost of ceramic items under conditions of batch production is not too high.

The use of these ceramic engineering materials instead of metallic alloys will lead to saving critical costly raw materials. Thus the earth's crust contains 0.01 : 0.02 and 0.003 per cent of Ni, Cr and Co respectively and their concentration in ores amounts to 1%, 35% and 1% while the corresponding values for silicon are 27% and 45%. The amount of waste products in the process of producing Si_3N_4 powder is considerably less as compared to that of nickel alloys manufacturing which gives Si_3N_4 a substantial ecological advantage. Heat power consumption necessary to obtain 1 ton of Ni alloys is 36,000 kw/hr while for Si_3N_4 it amounts to 19,000 kw/hr, i.e. twice as lower,(4).

In this connection in USA expenditures on the development of new ceramic materials for engineering chemical, optical, magnetic, nuclear and electrical uses are expected to go up to 6 billion in 2000 as against 0,6 billion dollars in 1980,(5).

With the view of obtaining information concerning

utilization trends of engineering ceramics in industry Kema
Nord (Sweden) made a poll among industrial firms, universities
and research institutions of Europe, USA and Japan. The
analysis made showed that the priority was given to Si_3N_4,
the second place went to SiC and the third - to zirconium
dioxide.

The necessity of creating commercial sources of this raw
material was also pointed out with the biggest efforts being
centered on securing high quality of powders obtained as
regards their chemical and physical characteristics,(6).

Ceramic engineering materials currently find their major
applications in new machinery while their use for general
engineering purposes is still limited.

The materials developed on Si_3N_4 base with MgO, Al_2O_3 and
also SiC additives are intended for high temperature
conditions,(7).

Our study is aimed at developing and using ceramic
engineering materials in oil and gas industry with the view
of obtaining antifrictional materials resisting gas abrasive
wear, possessing high values of strength, heat conduction,
hardness and the minimum value of linear expansion
coefficient. Such materials can be created on silicon nitride
base by hot pressing. Additional borum nitride increases
antifrictional characteristics of these materials. The
materials on SiC base obtained by hot forming and sintering
techniques possess an ideal combination of strength, heat
conductivity and hardness. Crack resistance factor of these
materials K_{1c} = 5-6 MPa·m$^{1/2}$, is enough for operational
conditions of face seals and under gas abrasive wear.

Comparing the characteristics of materials on Si_3N_4 base
with that of СГ and БСГ materials a considerable increase in
strength and hardness of silicon nitride and carbide can be
noted (table 1). This presupposes their higher wear
resistance with the increase in loading force. Low linear
expansion coefficient favourably distinguishes silicon
nitride from hard alooy BK8 and oxide ceramic material ЦМ-322
and serves as a guarantee to prevent from microcracking on

working surfaces. Linear expansion coefficient of SiC is $4.5 \cdot 10^{-6} \mathrm{K}^{-1}$ which is 50 per cent higher than that of Si_3N_4. Therefore under conditions of equal temperature gradient inner stresses are lower in parts of Si_3N_4 than in those of SiC. On the other hand, higher heat conductivity of silicon carbide as compared to silicon nitride accounts for lower temperature gradient in parts of SiC as against Si_3N_4 parts influenced by one and the same temperature gradient.

Thus the level of characteristics of Si_3N_4 and SiC makes them candidate materials to substitute for the conventional one used for face seal manufacturing. High level of their hardness (HRA 90) ensures their workability under conditions of gas abrasive wear.

TABLE 1

CHARACTERISTICS OF MATERIALS USED FOR FACE SEAL
COUPLES MANUFACTURING

Materials used	Packed Density gr/cm^3	Heat condition factor Wt/mk	Linear expansion Coefficient $10^6 c^{-1}$	Elasticity modular, HPa	Compression strength limit, MPa	Hardness
СГ-П	2,4-2,6	130	4,2	127	430-450	HRC 65-75
СГ-Т	2,5-2,8	85	4,6	95	300-320	HRC 65-78
БСГ-30	2,2-2,6	50	5,7	60	150	HRC 65-80
БСГ-80	2,0-2,4	35	6,0	89	500	HRC 70-80
Hard alloy ВК8 (92%WC, 8%Co)	14,5	50,2	5,1	600	4130	HRA 80-90
Ceramics -332 (99% Al_2O_3)	3,8	25	7,0	350	4500	HRA 90
Si_3N_4-hotformed	3,2	30	3,0	310	2000-3500	HRA 94-95
SiC-sintered	3,0-3,1	85-150	4,5	400	1500-2000	HV 19-25

DATA AND DISCUSSION

We have manufactured face seals with a friction couple of hot formed silicon designed for pumping liquid media and operating at 20 m/sec sliding rate and 2.5 MPa specific load. Tests showed that such friction couples tend to scratches formed on the friction track resulting in poor sealing characteristics. This occurs due to the peculiarities of hot formed Si_3N_4 microstructure consisting of B-crystals with 2-5 mcm cross-section and silicate phase of Si - Me - O -N. Their corresponding microhardness values are 34-45 HPa and 10-15 HPa. We suppose that deterioration initially occurs in silicate phase at $B-Si_3N_4$ grain interfaces. Further interaction between spinal grains leads to their deterioration migration of deteriorated particles with 2 mcm cross-section into friction plane which is primarily responsible for intensive abrasive wear of face couple.

The replacement of one of the Si_3N_4 rings by that of СГ-II material under similar tests gave encouraging results. Sealing bonds showed no signs of wear, the initial non-parallel arrangement has been practically preserved and friction paths had "mirror-like" surface. Wear resistance of such friction couple was secured both by high hardness level of the materials used (Si_3N_4 - HRA 94-95, СГ-II- HRC 65-75) and also by the presence of graphite in siliconized graphite structure which forms antifrictional layer between friction surfaces.

The best results were obtained due to the application of face seals designed for pumping liquid media with temperature gradient of 40-250°C whose friction couple was manufactured of ss-carbide. Face seals were tested at 25 m/sec friction rate and 2.5 MPa specific load during 4,000 hours. In this case uniform SiC grain displacement with microhardness of 20-28 HPa causes their gradual wear with no solid particles appearing in the friction area which prevents abrasive wear.

Si_3N_4 and SiC materials wear resisting characteristics withstanding intensive wear under various conditions combined with corrosive deterioration caused by gas flows and abrasive

wear by solid particles presupposes their utilization under
extreme service conditions, namely in working organs of gas
transporting equipment.

This study deals with the investigation of wear
resistance of hot pressed Si_3N_4 with MgO and ss-carbide
additives under conditions of gas abrasive wear.

Tests were conducted in a specially designed centrifugal-
type accelerating unit at normal temperature creating
abrasive particles flow rate of 100 m/sec, with attack angles
of 15°,45°,90°. Used as abrasive was quartz sand with 0.6 -
0.8 mm grain size. Standard patterns were made of steel grade
45 in tempered state.

Si_3N_4 material showed the highest abrasive wear
resistance. With attack angles of 45°-90° it appeared 4-7
times higher than that of the standard pattern. The reduction
of attack angle to 15° led to 2-5 times decrease in the above
value. Evaluating wear resistance reduction with the decrease
in attack angle we can suggest that materials deterioration
initiates and progresses at the expence of intergranular
silicate phase cutting whose microhardness is close to that
of abrasive material and is 10-12 HPa. The increase of attack
angle to 90° leads to higher wear resistance as the main
impact load is distributed amongst spinal B-crystals of Si_3N_4
phase whose microhardness greatly exceeds that of abrasive
materials and is 35-45 HPa Si_3N_4 material surface morphology
subjected to gas abrasive wear has a uniform relief (Fig.1a).

Low wear resistance charscteristics of SiC which at
various attack angles was 0.5-0.8 that of the standard
pattern can be accounted for easy deterioration of residual
silicon interfaced between grains. The centre of Fig.1b shows
single SiC grain surrounded by furrows filled with 5-10 mcm
abrasive particles inclusions. We suppose that deterioration
initiates in residual silicon and 15 mcm SiC particles
simultaneously and then as the furrows become deeper bigger
particles (40-50 mcm) are damaged and forced away.

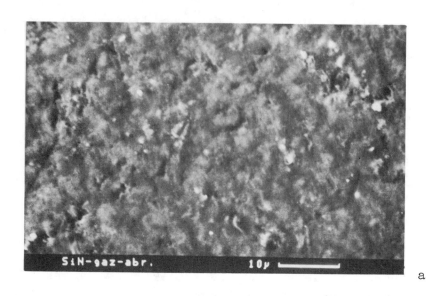

a

b

Figure 1. Surface morphology of materials subjected to gas
abrasive wear at sand particles flow rate 100 m/sec
(grain size - 0.6-0.8 mm) and attack angle - 45°-
90°; a) Si_3N_4 with 10% MgO additive; b) ss-carbide.

CONCLUSIONS

Potential use of Si_3N_4 and SiC materials for manufacturing friction couple elements in face seals has been established.

Silicon nitride was found promising for manufacturing working parts of locking fixture installed on gas wells.

Workability of Si_3N_4 and SiC materials is ensured by high levels of strength, hardness and heat conduction with low linear expansion coefficient. It should be noted that microstructure morphology of the above materials as well as microhardness of the main and intergranular phase essentially effect ceramic material resistance to gas abrasive wear.

REFERENCES

1. Belousov V.Ya. Wear resistance characteristics of machine parts made of composite materials. Lvov, Lvov University Publishing House, 1984 - 180 p.

2. Muzalevsky V.I., Sidorenko V.V. Selection of antifrictional materials for friction couple seals of centrifugal pumping units of natural gas. Vol.7, Moscow: Vnienergoprom, 1982 - 26 p.

3. .Contact seals of rotating shafts, G.A.Golubyev, G.M.Kukin, G.Ye.Lazarev, Moscow, Mashinostroyeniye, 1976, -264 p.

4. Davidge R.W. Economic and energetic considerations for nitrogen ceramics. Nitrogen Ceram. Prac. NATO Adv. Study Inst., Canterburu, 1976 . - Noordhof-Leuden.- 1977.- p.653-657

5. Skalny J. Ceramics in the 1990s: Will they still be brittle? Mater. and Soc.- 1985.-9, No.2 - p.73-184

6. Raw materials supply to the engineering ceramics industry. Ci News . - 1984. - 4, No.3. - p.3

7. Gnesin G.G. Oxygen-free ceramic materials. Kiev, Tehknika, 1987. - p.152

THE MULTI-LEVEL STUDY OF CONSTITUENT-STRUCTURE-INTERFACE-MECHANICAL BEHAVIOUR RELATIONSHIPS OF CONCRETE MATERIALS

HUANG YIUN-YUAN
Department of Materials Science and Engineering
Tongji University, Shanghai 200092, China

ABSTRACT

Concrete materials are investigated on four structural levels, namely: atom-molecular, micro, meso and macro. The relationships between the constituent, structure, interface on the first three levels and the macro mechanical behaviour of the materials are studied inter-disciplinarily with surface chemistry, micro fracture mechanics, solid statistic mechanics and computer simulation of materials systems. Fundamental conceptions and methods of approach are abstracted and new materials obtained with special mechanical properties are reported. The comprehensive design of mechanical behaviour of concrete materials according to their specific service conditions is seen to be possible and its prosperity is prosperous.

To predict the mechanical behaviour from constituent, structure and interface on multi-levels, to predict the constituent, structure and interface on multi-levels from raw materials and technical processing and to predict the service life under specific conditions are proposed to be three tasks of concrete materials science and engineering. The last two predictions are foresighted on the base of development of the first.

INTRODUCTION

Concrete materials as one category of composites are characterized by their aggregate-binder nature. Such materials reach their new stage of achievement in the last 20 years under the influence of development of materials science and engineering. The tasks of concrete materials science and engineering briefly are three, namely:

1. To design the materials according to their specific service requirements,
2. To design the technological process to produce them from the raw materials to have the required constituent, structure and interface on their different structural levels,
3. To predict or forecast the service life of them under the specified service conditions.

These three tasks, in fact, can be stated as three predictions:

1. To predict the mechanical, physical or other behaviour of materials according to the constituent, structure and interface on their different structural levels,
2. To predict the constituent, structure and interface on different structural levels of the formed materials according to the raw materials used and technological process adopted,
3. To predict the service life of the materials according to their service conditions, environment and physical and chemical changes from themselves.

The common base of these predictions is the constituent-structure-interface-behaviour relationships on different levels of the materials.

In our study, the concrete materials are investigated on four structural levels, namely: atom-molecular, micro (from 10 A to 1 mm), meso (from 1 mm to several cm) and macro (construction element size). Of caurse, there are no definite borders between them. Interface, in fact, is not a physical surface, but a transition zone of ions, molecules or micro-crysticals.

For all solid materials the mechanical behaviour is always very important, of caurse, for structural materials it is dominant. Mechanical behaviour of a material denotes its whole process from deformation to fracture. It is determined by a series of mechanical property parameters of the material and the applied stress condition. In our study performed in the past 10 years, the following series of parameters were used:

Series 1. (basic, simple stress condition)
Modulus of Elasticity, Strength, Fracture Energy, Brittleness Index (the ratio of elastic to plastic energy measured from the area under the $P - \delta$ curve of a loaded specimen at peak load or some other selected point on the decending branch of it).

Series 2. (complex stress condition)
Abrasive Resistance, Shocking Resistance, Fatigue Limit etc.

Series 3. (long term property and failure process)
Reological and Creep parameters and Ageing parameters etc.

The disciplines closely related to our study are The Physical Chemistry of Interface, The Micro Fracture Mechanics, The Solid Statistical Mechanics and The Computer Simulation of Materials Systems.

FUNDAMENTAL CONCEPTIONS ABSTRACTED

In our multi-level study of concrete materials, the following fundamental conceptions are abstracted and found to be very helpful(1):

1. The constituent, structure and interface on atom-molecular, micro and meso levels of a material influence its macro mechanical property in different aspects and to different extent. The resultant influence determines the whole macro mechanical behaviour.
2. So called 'Structure', in fact, is the collection of different bonds and structural units, mainly is the collection of different bonds and interfaces; and the 'Interface', in fact, is a transition zone of ions, molecules or micro-crystalines etc. In the whole history from structure formation to structure disintegration of a material, throughout occur disappearance, transition and formation of interfaces. For concrete, also occur those of electrical double layers.
3. Energy is a common physical quantity which can be used to run through all the structural levels. It is the main media to establish constituent-structure-interface-behaviour relationships. The tensile strength of a

material is a function of interfacial energy per unit volume of it.

4. For a real material, the collection of bonds and structural units always is statistical, thus the general property of it is only should be statistical.

5. If the strain energy of a material under stress can not transform into heat energy and then be dissipated in time, or it is not consumed in morphological transformation of crystals during deformation under stress or in other way of energy consumption, cracks are going to form to transform the strain energy into surface energy, edge energy and corner energy. Among them, surface energy is dominant and it still stores in whole material system to influence its mechanical behaviour.

6. If the constituent, structure and interface on different levels of materials could be changed to more favourable condition automatically under the influence of applied stress, its resistance to deformation and strength could be increased.

7. Porosity effects the mechanical behaviour of a material not only on the meso-level, but also on the micro-level(2). The size and shape of pores play role in such influence(3). A pore can not be considered only as a hole in the material without mass in it; in the case of very small pore or narrow seam, the van der Waals' and other long range forces between the walls of pore or seam is going to play role to effect the mechanical behaviour.

8. In solving the durability problem, the forecasting results obtained from even very satisfied fitting in some chosen function by method of least square may not be valid. For prediction of long term strength variation, special method of forecasting fitting should be developed.

   Of caurse, in the cause of further study, we may make supplement or adjustment to above abstracted conceptions upon the results of theoretical and experimental investigations.

NEW CONCRETE MATERIALS FORMED AND THE POSSIBILITY OF COMPREHENSIVE DESIGN OF THEIR MECHANICAL BEHAVIOUR

We obtained a series of new concrete materials with special mechanical properties from change of the constituent, structure and interface on different levels. Some of them are introduced as follows to see their typical properties as well as the possibility of comprehensive design of mechanical behaviour of such materials.

1. By increase of the degree of silicate polymerization of HCP, i.e. to change the c-s-f (constituent, structure and interface) on atom-molecular level, we obtained PSC (polymerized silicate cement) of very high compressive strength of 344 MPa with a brittleness index of 0.63 (as brittle as a cement mortar of about mark 200)(4).

2. Change of c-s-f on atom-molecular level can be also accomplished by change of the nature of bond between interfaces. Organic additives of appropriate dose in ordinary cement change the nature of bond between the aggregates and binder, and thus effect the strength and brittleness of the resulting concrete (polymer cement concrete PCC) considerably. Upon appropriate selection of additive and its dose, a high flexural/compressive strength ratio (1/4) concrete can be obtained(5).

3. To change the c-s-f on micro level, the micro cracks and pores in the dried cement mortar are filled by both polymer impregnation and polymerization under high pressure up to 200 atm.. The resulting PIC behaves superhigh compressive strength of 240 MPa with superbrittleness of

bursting nature during failure(6).

4. PIC formed from partial drying technology is another interesting case of change of c–s–f on micro level, in which the micro cracks are partially filled by water then their exits are blocked by polymer impregnation. Two sets of filler net work of micro thikness exist in the mortar, this makes the material have super–gastightness. Pipe line of 15 atm. working pressure ($\emptyset$ = 330 mm) was successfully established(7).

5. By changing the c–s–f on both atom–molecular and micro levels, PI(PCC) is obtained. Change of c–s–f on atom–molecular level is done by nature of bond and that on micro level is done by polymer impregnation. Such material with a compressive strength over 100 MPa is as brittle as an ordinary concrete mortar with strength of 30 – 40 MPa(5).

6. With constant total porosity of a HCP, the c–s–f on micro level can be changed by changing the specific surface area of the pores. We not only got the clear view of the role of size and shape of pores on the micro level and that on meso level to effect the strength of HCP, but also obtained the critical specific surface area, above which the total porosity would have no effect on the strength(2).

7. To change the c–s–f on meso level of an ordinary cement concrete, we change the nature of surface of the coarse aggregates so as to change that of the interface between them. Concrete of medium compressive strength with low brittleness can be obtained(8).

8. More detailed investigation is performed for ordinary cement concrete, in which the dynamic modulus of elasticity, compressive strength, fracture energy and brittleness index are chosen to be the parameters for mechanical behaviour, and coarse aggregates and aggregate–mortar interfaces are taken as the structural parameters on meso level; the big pores in mortar, sand–cement paste interfaces and the cracks arrested by intrusions characterize the features of structure on meso–micro level; on micro level the porosity of HCP, crystal/gel ratio or hard/soft base ratio ((crystal + unhydrated cement particles)/C–S–H gel), together with factors such as the size of crystals, the chain length of C–S–H gel and the interfaces between different phases are taken into consideration.

    Computerized tomography (CT), different kinds of porosimeter and point counting microscope are used in our experiments.

    The fundamental relationships between the c–s–f on these structural levels and four mechanical behaviour parameters are obtained and thus this shows the possibility of designing a cement concrete according to specific requirements of those behaviour parameters upon the accumulation of supplementary enough experimental data. Thus the comprehensive design of mechanical behaviour of concrete materials is prosperious(9).

9. In our study, it is also seen that the possibility of designing a concrete by method of computer simulation of materials systems according to required σ – ε curves under uniaxial compression was presented(10).

10. We also perform the c–s–f–behaviour relationships for concrete materials under abrasive, shock and vibration. Very interesting results are obtained, but here has no space to inform them separately even very briefly(11, 12, 13).

TO RULE THE RESEARCH ON TECHNOLOGICAL PROCESSING
OF MATERIALS ON THE TRACK OF MATERIAL SCIENCE

The tasks to rule the research on technological processing of concrete materials on the track of materials science are very imperative. The raw materials selected undergo a series of intermediate stages in the processing line and finally the designed materials with specified c-s-f on different structural levels and with required mechanical behaviour are thus formed. At certain stage the inert constituents of various size find their final relative position in micro-meso structural level and the general shape of the element is moulded. At certain stage, like curing or polymerization, chemical reaction and phase transformation occur, and the c-s-f become the designed condition on the atom-molecular level. Obviously, the technological process should be ruled on the track of materials science.

The reological parameters of concrete materials in the whole history of (concrete mix) - (flesh concrete) - (hardened concrete) - (aged concrete) - (failure) reflect the variation of stage of c-s-f on different levels in the cause of time. This is another important approach to establish the relationship between the c-s-f and mechanical behaviour which could run through from technology of processing and service performance to durability.

THE FORESIGHT OF PREDICTION OF SERVICE LIFE

Service life prediction is a very complex, difficult but non-avoidable problem. Up to now, usually, some indirect parameters obtained by tentative accelerated methods are used for comparison only. We may foresight that the c-s-f-behaviour relationship on multi-levels together with general principle of basic science, computer optimization, development of new forecast fitting method, necessary accelerated tests and long-term observations and tests would be considerably promoted. In this field, we have forecast the service life of concrete road pavement and concrete crane girder under fatigue damage process with good agreements(13), and have proposed a new method to forecast the residue strength of fly-ash silicate wall blocks in natural outside environment for 26 years from its 3 years strength data; with result far better than that from forecast fitting by method of least square(14).

CONCLUSIONS

The multi-level study of constituent, structure and interface of concrete materials is promoting the development of comprehensive design of mechanical behaviour of them according to their service requirements even including the service life. The task to rule the research on technological processing of concrete materials on the track of materials science and engineering is very imperative, otherwise, we would have no way to produce the material with specific service behaviour in high precision requirements and on optimum economical base.

The birth of very high compressive strength concrete with low brittleness from appropriate simple technology would open the new era of constructional engineering, in which the prestressed concrete structure may be completitive with steel structure in strength/weight ratio.

The prosperity of the three tasks of predictions in concrete materials science and engineering is prosperious.

133

REFERENCES

1. Huang Yiun-yuan, On the advance in the research on science of concrete materials (part I and II), J. Concrete and R. C. (in Chinese), 1982, 5, pp. 2-7, 6, pp. 16-26.

2. Huang Yiun-yuan, Ding Wei and Lu Ping, The influence of pore-structure on the compressive strength of hardened cement paste, MRS Symp. Proc., 1984, 42, Very High Strength Cement-based Materials, pp. 123-131.

3. Wang Jien-guo, The theoretical analysis of pore structure of hardened cement paste by fractal and critical percolation theory, Ph.D. Thesis, Tongji University, Shanghai, China, 1988, June.

4. Lu Ping, The research on the intrinsic properties of micro structure of hardened cement paste and its properties modification, Ph.D. Thesis, Tongji University, Shanghai, China, 1985, June.

5. Huang Yiun-yuan, Wu Keru, Tan mohua and Shen Zheng-jia, The mechanism of modification in mechanical behaviour of PC, PCC and PI(PCC), Proc. of 4th Intl. Congress on Polymer in Concrete, 1984, pp. 429-433.

6. Yu Fei-xiong and Huang Yiun-yuan, Mechanical properties of PIC impregnated and polymerized under high pressure, Polymer Concrete, Uses, Materials and Properties, SP-89 ACI Committee 548, 1983, pp. 161-176.

7. Zhuang Ju, Lei Yong-ming, Chen Xing, Tan Mohua and Huang Yiun-yuan, Partial drying impregnation technology in the manufacture of 15 atm. PIC gas-supply pipe, Proc. of 3rd Intl. Congr. on Polymer in Concrete, 1981, pp. 974-986.

8. Zhou Jien-hua, The influence of the interfacial bond between coarse aggregates and hardened cement paste on the mechanical properties of concrete and its computer analysis, Ph.D. Thesis, Tongji Univ. Shanghai, China, 1988, June.

9. Zhang Xinhua, The study on the relationships between structure of concrete on different levels and mechanical behaviour, Ph.D. Thesis, Tongji University, Shanghai, China, 1988, June.

10. Xu Zhiqing and Huang Yiun-yuan, The mechanical behaviour of concrete under uniaxial compression, Proc. of Intl. Conf. on Fracture of Concrete and Rock, Houston, Texas USA, 1987.

11. Zhang Xinhua and Huang Yiun-yuan, Fundamental study of the abrasive behaviour of concrete, Fracture Toughness and Fracture Energy of Concr. Proc. of Intl. Conf. on Frac. Mech. of Concr., Laussanne, Switherland, 1985.

12. Zhang Tie-wei, The mechanical behaviour of concrete under shock load, M.S. Thesis, Tongji Univ., Shanghai, China, 1984, Oct..

13. Zhang Bin-sheng, Study on the fatigue damage characteristics of concrete, Ph.D. Thesis, Tongji University, Shanghai, China, 1988, June.

14. Min Hua-ling, A new method of forecasting fitting, J. of Applied Probability and Statistics, 1988, 4, No.2, pp. 163-170, (in Chinese).

NUMERICAL ANALYSIS OF THE STRESS FIELD PARAMETERS IN THE FRACTURE PROCESS ZONES IN CONCRETE

DARIUSZ STYŚ
Institute of Building Engineering
Technical University of Wrocław
Plac Grunwaldzki 11, 50-370 Wrocław, Poland

ABSTRACT

Two physical models were applied for the material characterization in the fracture process zone at the crack tip in concrete: "a"-an isotropic, elastic-brittle body in which increasing strain is acompanied by elastic degradation of the material, "b"-an elastic, cylindrically-anisotropic body. Theoretical considerations were based on results obtained by the complementary application of the photoelastic coating method and the strain measurement using electric resistant strain gauges. The experiments were performed on concrete beams with initial notches, loaded by two symmetrical forces. In both "a" and "b" models, the problem was solved numerically, applying the Newton-Raphson's procedure and the least-squares minimization process. The desired parameters occur as non-linear parameters in the fringe equations. The set of data consisted of 30 points located on isochromatic fringes. It was possible to obtain "theoretical" fringe orders numerically and to compare them with the experimental results. It is remarkable, that in a cylindrically-anisotropic body, the rate of singularity λ depends on the constants of anisotropy. The singularity λ for concrete decreases with its elastic degradation.

INTRODUCTION

Since a long time attention was focussed on, and much work devoted to the technical application of fracture mechanics in concrete composites. Stress concentration areas in the vicinity of notches and cracks' zone [1] had been taken into special consideration. In these areas in concrete elements, there exists physical non-linearity which corresponds to the fracture process zone. Fracture process appears as the zone of intensive microcracking. One of the aims to deal with fracture mechanics

in concrete is to determine concrete mix proportions taking
into account the relations between its physical structure and
mechanical properties. Another important goal is to develop the
parameters describing the fracture process in concrete under
various conditions. So far the existing theories attempting to
describe physical pecularities of concrete in the stress
concentration areas are often insufficient.

MATERIALS AND METHODS

Material Modelling in the Crack Tip Area in the Composites with Cement Matrix

The phenomena associated with the microcracking process in the
stress concentration area at the crack tip have been described
in terms of two models:
    (a) an isotropic, elastic-brittle body,
    (b) a cylindrically-anisotropic body.
Material discontinuity proceeding in this area changes its
physical characteristics. A microcracking, common for materials
like rock and concrete, results in elastic degradation of a
body. The model of an elastic-brittle body seems to be adequate
in this case (Figure 1).

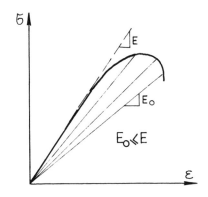

Figure 1. Elastic-brittle model for concrete.

It was assumed that Young's modulus E_b as well as Poisson's
ratio ν_b attained their fictitious values in the microcracking
region. The fictitious Young's modulus for concrete E_b^* can be
derived analytically provided the strain loci in the notch
cross-section is known (Figure 2). The function $\sigma(x)$ in the
equilibrum equations is given as a product of the functions
$\varepsilon(x)$ and $E_b(x)$:

$$\sigma(x) = \varepsilon(x) \ E_b(x) \qquad (1)$$

Equilibrum equations and restrictions imposed on the range
of variability of $E_b(x)$ function allow to define $E_b(x)$ as

a second degree polynomial. In the next step one can calculate the mean value E_b^* ,characterizing material properties in the microcracking zone:

$$E_b^* = \frac{1}{d} \int_0^d E_b(x)\, dx \qquad (2)$$

where: d is the extent of microcracking zone (Figure 2).

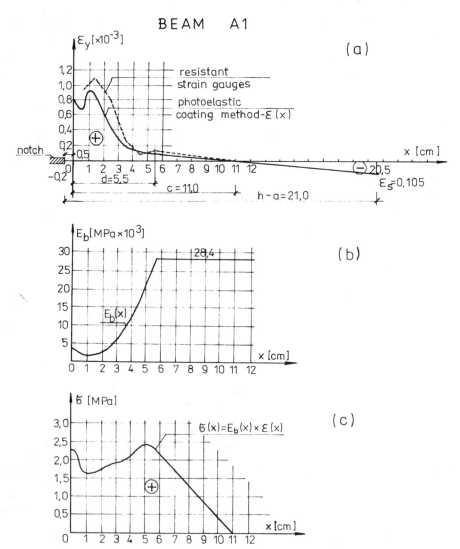

Figure 2. Strain loci $\varepsilon(x)$ in the notch cross-section (a) , $E_b(x)$ loci in microcracking region (b) , stress loci $\sigma(x)$ (c) .

The same fictitious value ν_b^*, taking into account microcracks formation, can be found for Poisson's ratio [2].

The experiments performed on initially notched beams made of photoelastic material, exhibit the radial and circumferential isostatic lines pattern (Figure 3).

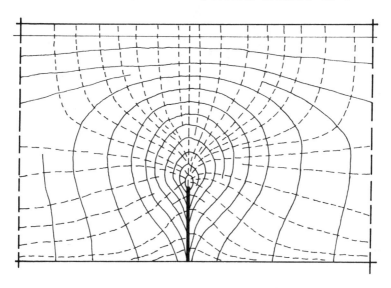

Figure 3. The appearance of the isostatic lines pattern at the vicinity of a notch for a beam in pure bending.

It seems that in concrete composites with brittle matrix, in the crack tip area, the most probable directions of microcracks formation are those which are perpendicular to the isostatic lines. Principal stress trajectories in the vicinity of a notch are connected with the polar coordinate system — microcracks initiate radially and circumferentially, so the body may be considered as cylindrically-anisotropic.

Mathematical Description of the Models

Mathematical formulation was based on the solution of a cylindrically-anisotropic wedge problem [3]. Having assumed a polar coordinate system, as in Figure 4, the stress-strain relations are given as:

$$\varepsilon_r = \beta_{11}\,\sigma_r + \beta_{12}\,\sigma_\theta \tag{3}$$

$$\varepsilon_\theta = \beta_{21}\,\sigma_r + \beta_{22}\,\sigma_\theta \tag{4}$$

$$\varepsilon_r = \beta_{66}\,\tau_{r\theta} \tag{5}$$

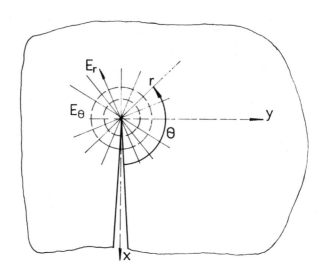

Figure 4. Location of a polar coordinate system.

The strains can then be computed in terms of $f(\theta,\lambda)$ function:

$$\varepsilon_r = -\beta_{11} r^{\lambda-2}[\lambda f(\theta,\lambda) + f''(\theta,\lambda)] + \beta_{12}\lambda(\lambda-1)r^{\lambda-2}f(\theta,\lambda) \quad (6)$$

$$\varepsilon_\theta = \beta_{21} r^{\lambda-2}[\lambda f(\theta,\lambda) + f''(\theta,\lambda)] + \beta_{22}\lambda(\lambda-1)r^{\lambda-2}f(\theta,\lambda) \quad (7)$$

$$\varepsilon_{r\theta} = -\beta_{66}(\lambda-1)r^{\lambda-2}f'(\theta,\lambda) \quad (8)$$

and:

$$f(\theta,\lambda) = Ae^{t_1\theta} + Be^{t_2\theta} + Ce^{-t_1\theta} + De^{-t_2\theta} \quad (9)$$

$$t_{1,2} = \{0.5[-m \pm (m^2-4n)^{0.5}]\}^{0.5} \quad (8)$$

$$m = (2\beta_{12}+\beta_{66})\,\beta_{11}^{-1}(\lambda-1)^2+2 \quad (11)$$

$$n = \beta_{22}\beta_{11}^{-1}\lambda(\lambda-1)^2(\lambda-2)-\lambda(\lambda-2) \quad (12)$$

The constants B,C,D can be eliminated applying boundary conditions. The necessary boundary conditions at the stress-free edges can be expressed in a matrix form:

$$[W]\,[C] = 0 \quad (13)$$

$$[W] = \begin{bmatrix} 1 & 1 & 1 & 1 \\ t_1 & t_2 & -t_1 & -t_2 \\ e^{2\pi t_1} & e^{2\pi t_2} & e^{-2\pi t_1} & e^{-2\pi t_2} \\ t_1 e^{2\pi t_1} & t_2 e^{2\pi t_2} & -t_1 e^{-2\pi t_1} & -t_2 e^{-2\pi t_2} \end{bmatrix} \; ; \; [C] = \begin{bmatrix} A \\ B \\ C \\ D \end{bmatrix} \qquad (14)$$

Determinant of matrix [W] is its characteristic function:

$$\det[W] = -8t_1 t_2 - (t_1 - t_2)^2 [e^{(t_1 + t_2) 2\pi} + e^{-(t_1 + t_2) 2\pi}]$$

$$+ (t_1 + t_2)^2 [e^{(t_1 + t_2) 2\pi} + e^{-(t_1 - t_2) 2\pi}] = 0 \qquad (15)$$

For the fixed coefficients of anisotropy $\beta_{11}, \beta_{22}, \beta_{12}, \beta_{66}$,the roots of equation (15) are values of the λ_i (i=1,2,...,n) parameter. Eigenvector of [W] matrix corresponds to every value of λ_i :

$$[T] = A_i [1 \quad B_i (A_i)^{-1} \quad C_i (A_i)^{-1} \quad D_i (A_i)^{-1}]^T \qquad (16)$$

where: $A_i \neq 0$ is an arbitrary constant. For i=1, parameter $\text{Re}(\lambda_1)$ attains the smallest value in the range (1,2). It is worth to point out that in contrary to the existing solutions of the plane state of stresses in orthotropic bodies, the rate of singularity λ in cylindrically-anisotropic model depends on coefficients of anisotropy. It comes directly from equation (15).

Distortions of the isochromatic fringes caused by irregularities of concrete and boundary effects, in an isotropic, elastic model, were described by two Westergaard's type stress functions [4]. Functions ϕ_I ,ϕ_{II} are connected to the modes I and II of a crack propagation.

$$\phi_I (r, \theta) = K_I (2\pi r)^{-0.5} (e^{-0.5i\theta} + \gamma_1 r e^{0.5i\theta}) \qquad (17)$$

$$\phi_{II} (r, \theta) = K_{II} (2\pi r)^{-0.5} (e^{-0.5i\theta} + \gamma_2 r e^{0.5i\theta}) \qquad (18)$$

The stress tensor is determinated in terms of these functions with addition of an uniform stress field σ_{ox} , parallel to the crack line :

$$\sigma_x = \text{Re}\phi_I + 2\text{Im}\phi_{II} - y(\text{Im}\phi_I' - \text{Re}\phi_{II}') + \sigma_{ox} \qquad (19)$$

$$\sigma_y = \text{Re}\phi_I + y(\text{Im}\phi_I{}' - \text{Re}\phi_{II}{}') \qquad (20)$$

$$\tau_{xy} = \text{Re}\phi_{II} - y(\text{Im}\phi_{II}{}' + \text{Re}\phi_I{}') \qquad (21)$$

RESULTS

Experimental Procedure

The experiments were carried out on concrete beams of two types (A,B). Their dimensions and loading scheme are shown in Figure 5.

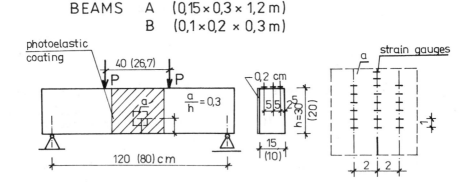

BEAMS A (0,15 × 0,3 × 1,2 m)
B (0,1 × 0,2 × 0,3 m)

Figure 5. Test arrangement and beams' dimensions.

Each series, A and B, consisted of 3 elements. In the midsection of beams, the artificial notch of relative length 0.3 was moulded. Notches were formed by thin steel plate 0.001m thick, covered with a silicon parting agent, inserted while casting. To decrease the influence of shrinkage, specimens were cured in water for 90 days. After drying of concrete beams, the photoelastic coating 0.002m thick was glued on one lateral surface in the midsection of each beam,and a set of strain gauges was fixed on the other one. The strain gauges indicated the strains in the compression area of the beams,and at the position of the neutral axis. Isochromatic and isoclinic fringe patterns were recorded at the loading levels of 0.5 Pn and 0.9 Pn (Pn-ultimate load).

Numerical Analysis

In both cases mathematical models were verified applying numerical procedures. The set of data consisted of a polar or cartesian coordinates of 30 points located on the isochromatic fringes of orders N_i (i=1.0; 1.5; 1.75; 2.0). The Newton-Raphson's procedure and the least-squares minimization process were involved in numerical calculations [5]. The basic equations of photoelastic coating method can be determined in terms of strains or stresses. They are found to be of the form f_k and ω_k functions which describe isochromatic fringes:

$$f_k = (\sigma_x - \sigma_y)^2 + (2\tau_{xy})^2 - [Nf_\sigma E_b^*(1+\nu_w)]^2 [2E_w(1+\nu_b^*)]^{-2} \qquad (22)$$

$$\omega_k = (\varepsilon_r - \varepsilon_\theta)^2 + \varepsilon_{r\theta}^2 - [Nf_\sigma(1+\nu_w)]^2 (2E_w)^{-2} \qquad (23)$$

where: f_σ ,ν_w ,E_w characterize properties of a photoelastic coating. Combining equations (6,7,8,19,20,21) and (22,23) gives the stress or strain formulation with respect to the parameters of a cylindrical-anisotropy or Westergaard's type stress functions and material constants (E_b^* ,ν_b^*). The desired parameters occur as non-linear parameters in the fringe equations. In matrix formulation, the procedure for determining the best-fit values of the unknown parameters consists of the following steps:

(a) assume initial values of the parameters— for example K_I , K_{II} , γ_1 , γ_2 , σ_{ox} for isotropic, elastic-brittle model,
(b) compute the elements of the matrix $[f_k]$ for each point of the data set,
(c) compute the elements of the matrix of derivatives. The column elements of this matrix are the derivatives of equation f_k with respect to each of the unknowns: K_I , K_{II} , γ_1 , γ_2 , σ_{ox} evaluated at each point k,
(d) compute the matrix of correction factors $[\Delta]$,
(e) improve the estimates of the unknowns,
(f) repeat steps (b),(c),(d),(e) until $[\Delta]$ becomes acceptably small.

It was possible to obtain "theoretical" fringe orders numerically and to compare them with the experimental results. One example of this comparison is shown in Figure 6. The range of a disturbance at the vicinity of a crack tip entailed by 3-D state of stresses and crack tip geometry had been evaluated as 0.005m (Figure 6).

CONCLUSIONS

Taking into account a heterogeneity of concrete composite, a relatively good agreement can be observed between the theoretical and experimental isochromatic fringe patterns. In isotropic body, described by fracture mechanics parameters, the influence of the stress intensity factor (SIF) K_{II} was strongly marked. In investigated elements, its value mounts from 32% to 130% of the SIF K_I . The values of K_I oscillate in the range 0.49-1.19 MNm$^{-3/2}$. This finding is of great importance while determining the critical values Kc of SIF for cement matrix composites. A model of cylindrically-anisotropic body operates with a new material characteristics connected to the physical structure of a body and to the stress field. The interesting feature of this model is the relation between the rate of singularity λ and coefficients of anisotropy. Singularity λ

is weaker in concrete, than in isotropic, elastic bodies (−0.5) and its value in investigated elements oscillates from −0.44 to −0.24 . The singularity λ for concrete decreases with its elastic degradation. The parameter A can be considered as the stress intensity factor analog in a cylindrically-anisotropic body. Attention should be also paid to the dependance of the physical dimension of A-parameter on the singularity λ [MNm $^{-\lambda}$] .

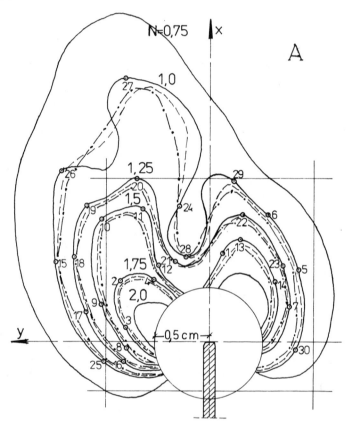

Figure 6. Isochromatic fringe patterns: experimental (solid-line), theoretical for isotropic body (dashed-line) and for cylindrically-anisotropic body (dot-and-dashed-line).

REFERENCES

1. Wittman, F.H., <u>Fracture Mechanics of Concrete</u> , Elsevier Applied Science Publishers, Amsterdam, 1983.
2. Kato, K., Microcracks, deformation and physical properties of plain concrete. Proc. of <u>the 5 International Conference on Fracture</u>, Cannes 1981, vol.5, pp.2275−80.

3.Erdogan, F., Boduroglu, H. and Delale, F., Stress singularities at the vertex of a cylindrically-anisotropic wedge. Int . J. Fracture,1982,19,pp.247–56.
4.Rossmanith, H.P., Analysis of mixed-mode isochromatic crack-tip fringe patterns. Acta Mech., 1979,34,pp.1–38.
5. Sanford, R.J. and Dally, J.W., A general method for determining mixed-mode stress intensity factor from isochromatic fringe patterns. Engng Fracture Mech., 1979, 11, pp. 621–33.

VARIATIONAL INEQUALITIES IN BRITTLE FRACTURE MECHANICS

LE KHANH CHAU

National Center for Scientific Research of Viet nam
Institute of Mechanics
208 D Doi Can Road, Ha noi, Viet nam

ABSTRACT

An energy criterion of equilibrium for a nonlinear elastic body with a slit is formulated, from which statical equations and boundary conditions are obtained . In dynamics an evolution variational inequality is postulated by generalizing the energy criterion. From this inequality dynamical equations and boundary conditions of crack propagation are derived.

INTRODUCTION

Let us consider an elastic body in its natural state with a surface of discontinuity ,which is called a slit. This slit is assumed to settle on a smooth surface Ω , being bounded by a contour $\partial\Omega$. A natural configuration, occupying a region $V_\Omega = V$

$(\Omega \cup \partial\Omega)$ of Euclidean three space is choosen to be a reference configuration . The Cartesian coordinates of a particle are denoted by X_a , $a=1,2,3$. In a deformed state this particle will have the Cartesian coordinates x_i given by

$$x_i = x_i(X_1, X_2, X_3), \quad i=1,2,3, \quad x_i \in v \qquad (1)$$

If the deformed state is equilibrium, then the functions $x_i(X_a)$ perform a one-to-one continuously differentiable transformation from V to v with the condition that $\det|\partial x_i / \partial X_a| > 0$, $X_a \in V$. Tracks of x_i on two sides of Ω draw surfaces of the slit $\omega^\pm$ in the deformed configuration. The first problem is to find a criterion of equilibrium for the configuration $x_i(X_a)$. For this purpose we formulate a variational principle: a necessary and sufficient condition for equilibrium of a configuration of a body with a slit is that the variation of its energy at this configuration should be more than or equal to zero in the class of admissible configurations. A configuration will be called admissible if its surface of discontinuity contains or coincides with Ω. The above stated principle is a generalization of the classical variational principles of nonlinear elasticity [1-3] and linear fracture mechanics [4-5]. It is important to note that the restriction for the admissible configurations forbids the body to be back in the state with the "healed" slit. Therefore the theory has evident irreversibility.

In dynamics with the inclusion of the force of inertia, the above stated principle can be extended to the principle of virtual work, expressed in the form of an evolution inequality . But in contrast to the classical theory, in the evolution inequality of fracture mechanics the virtual flux of kinetic energy entering into a tip of a slit (or a propagating crack) must be taken into account. It is resulted from analysis of an energy balance equation.

DERIVATION OF BASIC EQUATIONS

In order to derive equilibrium equations, according to the energy criterion one must define an energy functional of a body on its arbitrary admissible configuration. Let a configuration $x_i(X_a)$ have a surface of discontinuity Σ. By analogy with Griffith's theory [4,5] we postulate the following expression for the energy functional

$$\mathcal{E}[x_i(X_a)] = \int_{V_\Sigma} U(x_{i,a}, K_B) \, dX + \int_\Sigma 2\gamma dA + \int_{V_\Sigma} \rho_0 \Phi dX - \int_{\partial V_T} T_i x_i dA \qquad (2)$$

Here $V_\Sigma = V \setminus (\Sigma \cup \partial \Sigma)$, $\partial V = \partial V_x \cup \partial V_T$, dX and dA denote the element of volume and surface area respectively, ρ_0 is the mass density of the material in natural state, U and γ are the internal energy per unit volume and the surface energy per unit area. The tensor $x_{i,a} = \partial x_i / \partial X_a$ corresponds to the gradient of deformation, while $K_B(X_a)$ (B=1,2,...,N), the characteristics of the material. The potential of the body force is denoted by $\Phi(x_i)$ and T_i is the dead traction, prescribed on ∂V_T. On the rest of the boundary ∂V_x the values of x_i are given: $x_i = r_i(X_a)$. Here and afterwards the comma is employed to denote partial differentiation and the repeated suffix to denote summation.

From the formulated energy criterion, the configuration $x_i(X_a)$ with the surface of discontinuity Ω will be equilibrium if for all admissible configurations $y_i = y_i(X_a, \varepsilon)$ with surfaces of discontinuity Ω^ε, satisfying the restrictions $\Omega^\varepsilon \supseteq \Omega$, $y_i(X_a, 0)$ $=x_i(X_a)$ and $y_i(X_a, \varepsilon) = r_i(X_a)$ when $X_a \in \partial V_x$, we have the inequality

$$\delta\mathcal{E} = \frac{d}{d\varepsilon}\Big|_{\varepsilon=0} \mathcal{E}[y_i(X_a, \varepsilon)] \geqslant 0 \tag{3}$$

For obtaining consequences from (3) one must calculate $\delta\mathcal{E}$. Since the surface of discontinuity Ω^ε of the admissible configuration $y_i(X_a, \varepsilon)$ can differ from Ω when $\varepsilon=0$, it is convenient to introduce a one-to-one differentiable transformation from V to V according to a rule $Y_a = Y_a(X_a, \varepsilon)$ so that here the surface Ω is transformed to Ω^ε, moreover $Y_a(X_a, \varepsilon) = X_a$ when $\varepsilon=0$ or $X_a \in \partial V$. By this transformation one can show that

$$\delta\mathcal{E} = \int_{V\Omega} [-(T_{ai,a} + \rho_o F_i)\delta y_i + (-\mu_{ab,b} + \frac{\partial U}{\partial K_B} K_{B,a} + \rho_o F_i x_{i,a})\delta Y_a] dX$$

$$+ \int_\Omega \{(-T^+_{ai}\delta y^+_i + T^-_{ai}\delta y^-_i)N_a + [(-\mu^+_{ab} + \mu^-_{ab})N_b - 4\gamma H N_a$$

$$-\rho_o(\Phi^+ - \Phi^-)N_a]\delta Y_a\} dA + \int_{\partial\Omega}(2\gamma v_a - J_a)\delta Y_a dS + \int_{\partial V_T}(T_{ai}N_a - T_i)\delta y_i dA \tag{4}$$

In this formula a symbol δ is used to denote partial differentiation with respect to ε when $\varepsilon=0$ and $X_a=$const. Thus

$$\delta y_i = \frac{\partial}{\partial\varepsilon}\Big|_{\varepsilon=0, X_a=const} y_i(Y_a(X_a,\varepsilon),\varepsilon)$$

$$\delta Y_a = \frac{\partial}{\partial\varepsilon}\Big|_{\varepsilon=0, X_a=const} Y_a(X_a,\varepsilon)$$

The Piola-Kirchhof stress tensor T_{ai} is given by the formula $T_{ai} = \partial U/\partial x_{i,a}$, $F_i = -\partial\Phi/\partial x_i$ is the body force per unit mass, $\mu_{ab} = -T_{bi}x_{ia} + U\delta_{ab}$ is an analogy of a chemical potential in the theory of phase transition [1-3]. In the surface integral the indexes $+,-$ indicate the limit values of quantities on two sides of Ω, N_a is the outward unit normal of surfaces (on Ω it is in the direction of $+$), H denotes the average curvature of Ω. In the contour integral dS denotes the element of length, v_a is the outward unit normal on $\partial\Omega$. If the virtual surfaces Ω^ε smoothly continue Ω, then v_a is the tangential vector on Ω. But in general case it is not obligatory, and v_a only indicates the normal direction of $\partial\Omega^\varepsilon$. At last J_a is the vector of the energy flux entering into the tip of the slit [6,7] to be calculated by

$$J_a = \lim_{|\Gamma| \to 0} \int_\Gamma \mu_{ab} \kappa_b \, dS = \lim_{|\Gamma| \to 0} \int_\Gamma (-T_{bi} x_{i,a} \kappa_b + U \kappa_a) \, dS \qquad (5)$$

where the contour Γ, settling on the transversal to $\partial\Omega$ plane surface, surrounds the point X_a on $\partial\Omega$, κ_a is the outward unit normal on Γ. The formulae (4),(5) are deduced on supposing the following asymptotic behaviour of the stress field near $\partial\Omega$

$$\lim_{|\Gamma| \to 0} \int_\Gamma T_{ai} \kappa_a \, dS = 0.$$

It is obvious that δy_i and δY_a can have arbitrary values in V_Ω as well as δy_i does on ∂V_T. Let us find out the restrictions, which must be satisfied by δy_i and δY_a on Ω and $\partial\Omega$. If the banks of the slit are not in contact with each other in the deformed state, then δy_i can have any values on Ω. Let the surface Ω be referred to a curvilinear two-coordinates system. Let us denote sub-areas of Ω by Ω^+ and Ω^-, whose points after deformation will be in contact with each other

$$x^+_i(\eta_\alpha) = x^-_i(\theta_\alpha), \quad \eta_\alpha \in \Omega^+, \quad \theta_\alpha \in \Omega^-, \quad \alpha = 1,2$$

For those points one can show that

$$[\delta y^+_i(\eta_\alpha) - \delta y^-_i(\theta_\alpha)] n_i \geqslant 0$$

where n_i is the common unit normal on the deformed contact surfaces $\omega^\pm$ in the direction of $+$. Simultaneously, $\delta y_i \, x_{i,\alpha}$ can have any values on $\Omega^\pm$, where $x_{i,\alpha} = \partial x_i / \partial \eta_\alpha$. The friction force between ω^+ and ω^- is to be omitted. On supposing $\Omega^\varepsilon \supseteq \Omega$ for δY_a we have the restrictions $\delta Y_a N_a = 0$ on Ω; $\delta Y_a v_a \geqslant 0$, $\delta Y_a \pi_a = 0$, $\delta Y_a \tau_a = 0$ on $\partial\Omega$, where τ_α and π_α are the tangential and binormal   vectors on $\partial\Omega$ respectively.

Taking into account all above restrictions, one can show that (3) and (4) lead to

$$T_{ai,a} + \rho_o F_i = 0, \quad T_{ai} = \partial U/\partial x_{i,a} \quad \text{in} \quad V_\Omega$$

$$x_i = r_i(X_a) \quad \text{on} \quad \partial V_x \ , \quad T_{ai} N_a = T_i \quad \text{on} \quad \partial V_T$$

$$T^\pm_{ai} N_a = 0 \quad \text{on} \quad \Omega \setminus \Omega^\pm, \; T^\pm_{ai} N_a x_{i,\alpha} = 0 \quad \text{on} \quad \Omega^\pm \qquad (6)$$

$$T^+_{ai} N_a n_i \sqrt{a}\, |_{\eta_\alpha} = T^-_{ai} N_a n_i \sqrt{a}\, |_{\theta_\alpha} = -p \leqslant 0 \quad \text{on} \quad \Omega^+$$

$$|J_\alpha| = \sqrt{J_a J_a - J_3^2} \leqslant 2\gamma \quad \text{on} \quad \partial\Omega$$

where a = det$|a_{\alpha\beta}|$, $a_{\alpha\beta} = X_{a,\alpha} \cdot X_{a,\beta}$, $X_{a,\alpha} = \partial X_a/\partial\eta_\alpha$, $J_3 = J_a\tau_a$.
The relations   (6)   are the equilibrium conditions, which must be satisfied by equilibrium configuration.
If there is no configuration satisfying   (6), then the slit becomes the propagating crack. In order to find a real motion of the body $x_i = x_i(X_a,t)$ with the surface of discontinuity Ω_t , we introduce vitual motions $y_i(X_a,t,\varepsilon)$ with the surfaces of discontinuity Ω^ε_t ,   satisfying   the restriction $\Omega^\varepsilon_t \supseteq \Omega_t$. Since Ω^ε_t can differ from Ω_t ,   as before a transformation from V to V is constructed by a rule $Y_a = Y_a(X_a,t,\varepsilon)$ so that here the surface Ω_t is transformed to Ω^ε_t ,   moreover $Y_a(X_a,t,\varepsilon) = X_a$ when $\varepsilon=0$ or $X_a \in \partial V$. Now we postulate the principle of virtual work :   at the real motion of the body the following variational inequality

$$\delta\mathcal{E} + \int_{V_t} \rho_o \ddot{x}_i(\delta y_i - x_{i,a} \delta Y_a) \, dX - \int_{\partial\Omega_t} Q_a \delta Y_a \, dS \geqslant 0 \qquad (7)$$

takes place for all t and variations δy_i , δY_a of virtual motions. Moreover the postulate demands the absolute equality from (7) as an energy balance equation, if δy_i and δy_a will be replaced by $\dot{x}_i$ and $\dot{X}'_a$. Here for the real motion the functions $X'_a(X_a,t')$ are defined as a transformation $X'_a(X_a,t')$: $V \to V$ so that $\Omega_t \xrightarrow{X'_a} \Omega_{t'}$,$t' > t$. Thus $\dot{X}'_a = (\partial X_a/\partial t')_{t'=t,X = \text{const}}$·

In the inequality (7) , $\dot{x}_i = (\partial x_i/\partial t)_{X_a} = $ const ,
$\ddot{x}_i = (\partial^2 x/\partial t^2)_{X_a} = $ const , which are vectors of velocity and acceleration of a particle. An energy and a flux of kinetic energy Q_a are given by

$$\mathcal{E} = \int_{V_t} U(x_{i,a}, K_B)\, dX + \int_{\Omega_t} 2\gamma\, dA + \int_{V_t} \rho_0 \Phi\, dX - \int_{\partial V_T} T_i x_i\, dA \tag{8}$$

$$Q_a = \lim_{|\Gamma_t|\ 0} \int_{\Gamma_t} \frac{1}{2} \rho_0\, \dot{x}_i \dot{x}_i\, \kappa_a\, dS$$

In the formulae (8) $V_t = V \backslash (\Omega_t\ \cup\ \partial\Omega_t)$, Γ_t is a contour, surrounding X_a on $\partial\Omega_t$, the other symbols are explained from above. As before, from (7) we have

$$\int_{V_t}[\,(\rho_0\ddot{x}_i - T_{ai,a} - \rho_0 F_i)\,\delta y_i + (-\mu_{ab,b} + \frac{\partial U}{\partial K_B} K_{B,a} + \rho_0 F_i x_{i,a} - \rho_0\ddot{x}_i x_{i,a})\,\delta Y_a]\,dX$$

$$+ \int_{\Omega_t}[\,(-T^+_{ai}\,\delta y^+_i + T^-_{ai}\,\delta y^-_i)N_a + (-\mu^+_{ab} + \mu^-_{ab})\,N_b\delta Y_a]\,dA$$

$$+ \int_{\partial\Omega_t}(2\gamma v_a - I_a)\,\delta Y_a\, dS + \int_{\partial V_T}(T_{ai}N_a - T_i)\,\delta y_i\, dA \ \geqslant\ 0 \tag{9}$$

where

$$I_a = \lim_{|\Gamma_t|\ 0} \int_{\Gamma_t}[-T_{bi}\, x_{i,a}\,\kappa_b + (U + 1\ \rho_0 x_i\, x_i)\,\kappa_a]\, dS$$

When deriving (9) we assume $\delta Y_a N_a = 0$ on Ω_t . Inequality (9) results in

$$T_{ai,a} + \rho_0 F_i = \rho_0 \ddot{x}_i\ ,\quad T_{ai} = \partial U/\partial x_{i,a}\qquad \text{in } V_t$$

$$x_i = r_i(X_a)\ \text{ on } \partial V_x\ ,\quad T_{ai}\, N_a = T_i\qquad \text{on } \partial V_T$$

$$T^\pm_{ai}\, N_a = 0\ \text{ on } \Omega_t\backslash\Omega^\pm_t\ ,\quad T^\pm_{ai}\, N_a\, x_{i,\alpha} = 0\qquad \text{on } \Omega^\pm_t \tag{10}$$

$$T^+_{ai}\, N_a\, n_i\, \sqrt{a}\,|_\eta = T^-_{ai}\, N_a\, n_i\, \sqrt{a}\,|_\theta = -p \leqslant 0\qquad \text{on } \Omega^+_t$$

$$|I_\alpha| = \sqrt{I_a I_a - I_3^2}\ \leqslant\ 2\gamma\qquad \text{on } \partial\Omega_t$$

where $I_3 = I_a \tau_a$. In order to obtain more detailed information we use the second part of the formulated postulate. On replacing in (9) δy_i and δy_a by $\dot{x}_i$ and $\dot{x}'_a$ and taking (10) into account, we reduce (9) to

$$\int_{\Omega_t} (-T^+_{ai} \dot{x}^+_i + T^-_{ai} \dot{x}^-_i) N_a \, dA + \int_{\partial\Omega_t} (2\gamma v_a - I_a) \dot{x}'_a \, dS = 0 \qquad (11)$$

Here $\dot{x}'_a v_a$ defines the normal velocity of the crack tip. One can show that (11) is equivalent to the equation of energy balance

$$\frac{d}{dt}(\mathcal{E} + \mathcal{T}) = 0 , \quad \mathcal{T} = \int_{V_t} \frac{1}{2} \rho_0 \dot{x}_i \dot{x}_i \, dX$$

Thus, the equality (11) leads to the relations

$$p > 0 \;\Rightarrow\; [\dot{x}^+_i(\eta_\alpha) - \dot{x}^-_i(\theta_\alpha)] n_i = 0$$
$$\text{on} \quad \Omega^+_t$$
$$p = 0 \;\Rightarrow\; [\dot{x}^+_i(\eta_\alpha) - \dot{x}^-_i(\theta_\alpha)] n_i \geqslant 0$$
$$\qquad (12)$$
$$|I_\alpha| < 2\gamma \;\Rightarrow\; \dot{x}'_a = 0 \qquad \text{(no propagation)}$$
$$|I_\alpha| = 2\gamma \;\Rightarrow\; \dot{x}'_a v_a \geqslant 0, \; 2\gamma v_a = I_a - I_3 \tau_a$$

The relations (10), (12) compose the system of dynamical equations and boundary conditions, which must be used to find the motion of the body with the propagating crack.

CONCLUSION

Variational inequalitties (3),(7) are the compact formulations of the nonlinear statistical and dynamical problems of a body with its crack. The criterion of no propagation (or propagation) of the crack is expressed by (6) (or (12)). In general case the derived criterion is different from the well-known criterions [4-7]

REFERENCES

1. Gibbs J.W. On the equilibrium of heterogeneous substances, Trans.Connect.Acad., 1876,3, pp. 108-248, 1878,3,pp.343-524.

2. Sedov L.I. Variational methods of constructing models of continuous media. In Irreversible aspects of continuum

<u>mechanics</u>,Springer Verlag,Wien and New York, 1968,pp.17-40.

3. Berdichevski V.L. <u>Variational principles of continuum</u> <u>mechanics</u>, Nauka, Moscow, 1983, pp.180-4 (In Russian).

4. Griffith A.A. The phenomenon of rupture and flow in solids. <u>Phil. Trans. Roy.Soc</u>., 1920,ser.A, **221**, 163-98.

5. Griffith A.A. The theory of rupture. In <u>Proc. 1st Int.Congr.</u> <u>Appl. Mech.</u>, Delft, 1924, pp.55-63.

6. Cherepanov G.P. The propagation of cracks in continuum media. <u>Prikl. Math. Mech.</u> , 1967, **31**, No 3,476-88 (In Russian).

7. Rice J.R. A path independent integral and the approximate analysis of strain conentration by notches and cracks. <u>J. Appl. Mech.</u>, 1968,**35**, 379-88

NUMERICAL SIMULATION OF 3D CRACK FORMATIONS IN CONCRETE STRUCTURES

H. SCHORN and U. RODE
University of Bochum, FRG

ABSTRACT

"Numerical Concrete" as a keyword means a theoretical method for calculating material behaviour under external or internal loading including fracture process by a finite element method. At Bochum University a numerical simulation method of that kind has been developed. The "Bochum Method" delineates parameters determing concrete structure to a mechanical model, a framework model, and calculates the model behaviour including its fracture process up to the totally collapsed system. Using mechanical models of different size and geometry the necessity of using three-dimensional models instead of plane models or meshes, respectively, becomes obvious.

INTRODUCTION

Concrete is a microscopical inhomogeneous material with a very complicated damage process under external load which is determined by microcrack opening, microcrack growing, and by microcrack accumulation zones. This damage process leads to macrocrack opening and fracture of the concrete body. The characteristic fracture mechanism of concrete is very different from that type of single cracks which are assumed in using fracture mechanics. Developing fracture mechanic theories for composite materials as concrete demands knowledge on characteristic microcrack opening, microcrack propagation, and microcrack accumulation until a fracture determing macrocrack has been formed.

Delineating a real material structure to a mechanical model crack

opening processes can be calculated - if you are able to solve the complex
problems in using finite element method up to a totally collapsed
mechanical model, representing the totally collapsed material structure due
to external uniaxial or multiaxial load. The Bochum Method of numerical
simulation allows to study damage and fracture processes theoretically.

MODEL CONCEPTION AND NUMERICAL REALISATION

3D-models for the simulation of crack growth in concrete can only work
effectively, if the real complex material structure is presented in a
sufficient approximation. The Bochum-Model conception is based on the idea
of modeling stress trajectories between equidistant lattice points /1/.
Regular cubes with edge struts, surface diagonal struts and space diagonal
struts are gathered to a framework model, see fig. 1, which degenerates
under load. Due to the choosen material structure the struts are endowed
with different characteristics. The struts behave linear elastic up to
given strain rates. On exceeding strengh the affected struts are
irreversible removed from the system representing cracks. During the
loading or deformation process crack zones and bands are forming, which
finally mark the failure mechanism.

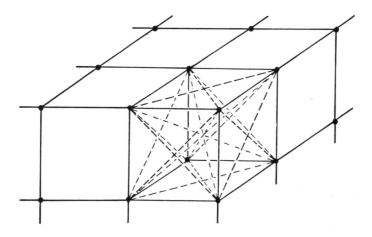

Figure 1: 3D framework model

The simulation of growing cracks with a degenerating framework requires a full geometrical and physical nonlinear static analysis of the system. A vector computer CDC Cyber 205 provides the computing power required here to solve problems also with large system sizes. All FEM algorithms have been adapted for the vector computer to utilize its computing speed in an optimal way /2/.

The basic program system has been developed in 1984 /3/. The following program variations are possible:

- 2D / 3D simulation analyses
- uniaxial / multiaxial loading
- compression / tension tests
- load / displacement control
- rigid load / "weak" load induction
- with / without influence of viscosity

Additionally the program system allows an implementation of any material laws. Actually, applications of nonlinear softening are studied.

SIMULATED CONCRETE STRUCTURES

Exact geometrical modeling of the material structure is not required for simulation analyses on the MACRO-level. The strut parameters are endowed with values by given distributions (see next chapter). Due to the model conception the Bochum-Model allows also simulations on MICRO- and MESO-levels without altering the number or arrangement of elements. Only the strut parameters have to be changed due to the alteration of concrete structure. Fig. 2 shows an example of a normal concrete structure, which has been numerical simulated in 3D. Not all aggregate particles can be represented in this way. The influence of small aggregate particles has to be added to the cement matrix.

Fig. 3 shows examples for simulations on the MESO-level, which have been performed to study the influence of the arrangement of four aggregate particles in the structure /4/.

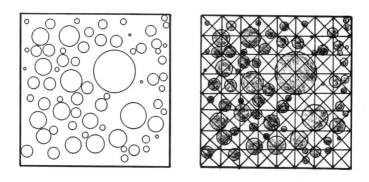

Figure 2: Computer simulated structure of normal concrete

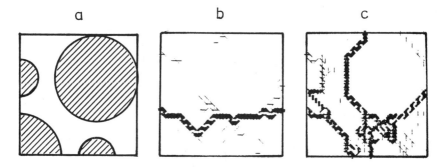

Figure 3: MESO level model (a) and its failure mechanism under
tension (b) and compression (c) loading

SIMULATION STUDIES

A 3D body consisting of 9x9x5 nodes or 8x8x4 macro-elements is shown in
fig. 4. The bearing conditions guarantee an ideal uniaxial tensile stress
state in the body; all nodes can move freely in the upper and in the lower
load transducing plane. Thus, transverse strain will not be affected.

The behaviour of the rods is determined by a macroscopical scale of
delineating material structure to the framework model. "Macroscopical
scale" means rod lengths greater than maximum grain diameter. The values of
stiffness of the rods are distributed stochastically as shown in fig. 5.

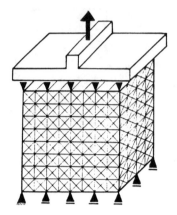

Figure 4: 3D model with 9x9x5 nodes

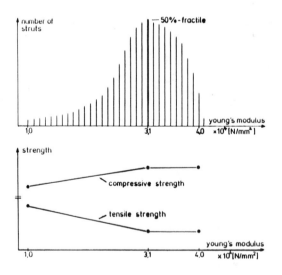

Figure 5: Input parameters: stiffness, tensile strengh, compressive strengh

For investigating the influence of the tensile strength of the rods to the behaviour of the model three values of tensile strength are studied: $\alpha = 1.0$; $\alpha = 3.0$; $\alpha = 0.333$. All other parameters are unaffected. The results of the calculations are shown in fig. 6 for those points on stress strain curve marked in fig. 7.

157

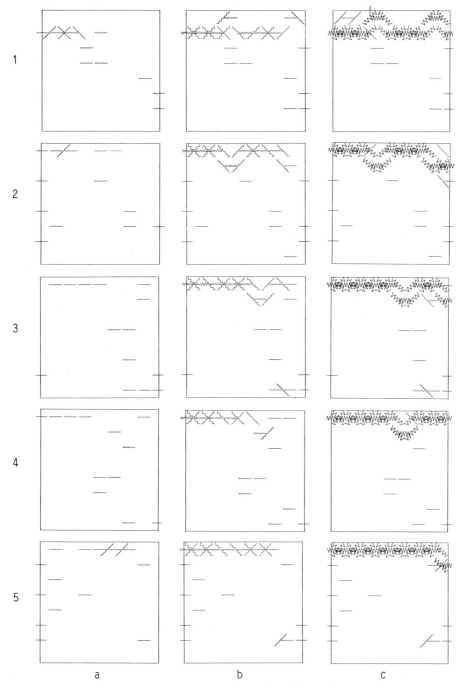

Figure 6: Crack formations in 5 sections (1..5) of the body for 3 load
levels (a,b,c)

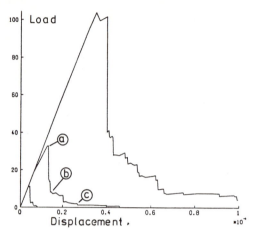

Figure 7: Results of the simulation study 1

In all cases (α = 1.0 / 3.0 / 0.333) the shapes of the curves as well as the crack patterns are very similar to each other. In the increasing branch of the curve some microcracks open and accumulate near the strength or at the point of strength. But there is not yet given a beginning macrocrack. Macrocracks only occur in the descending branch of the stress strain curve. The cross sections shown in fig. 6 give an imagination of the irregular shape of the plane formed by macrocrack opening. As exspected, the results are independent of the choosen value of the tensile strength of the rods.

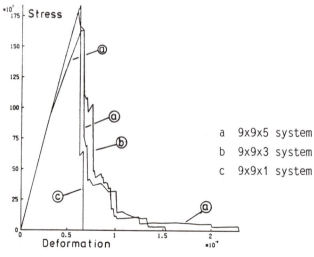

a 9x9x5 system

b 9x9x3 system

c 9x9x1 system

Figure 8: Results of the simulation study 2

159

Another simulation series has been carried out, to study the influence of model size under uniaxial tensile loading. Three simulation frameworks with 9x9x5 (a), 9x9x3 (b) and 9x9x1 (c) nodes have been used with the same parameter distribution shown in fig. 5. The results are presented in fig. 8. Not only the shapes of the descending branches of the load-displacement curves, but also the fracture energy of the three models are different from each other due to the model size.

CONCLUSIONS

1. The numerical simulation method is a helpful tool for studying damage and fracture processes in concrete structure, but it supplies experimental works - it does not substitute tests on real material.

2. The "Bochum Method" has been developed for delineating material structure to a 3D-framework model, calculated in its behaviour up to the totally collapsed system by a vector computer CDC Cyber 205.

3. Using a plane mechanical model or mesh, respectively, is not sufficient for investigating crack states and crack propagation in concrete by a numerical simulation method. A 3D-model is necessary.

REFERENCES

/1/ Schorn, H.: Zur Einführung numerischer Berechnungsverfahren in die Ermittlung strukturorientierter Stoffgesetze, Strukturmechanik und numerische Verfahren, K.-H. Schrader zum 60. Geburtstag, Hrsg. R. Diekkämper, H.-J. Niemann, Köln, 1982

/2/ Diekkämper, R.: Vectorized Finite Element Analysis of Nonlinear Problems in Structural Mechanics, International Conference on Parallel Computing 83, Berlin, 1983

/3/ Diekkämper, R.: Ein Verfahren zur numerischen Simulation des Bruch- und Verformungsverhaltens spröder Werkstoffe, Ruhr-Universität Bochum, TWM 84-7, Bochum, 1984

/4/ Schorn, H.; Rode, U.; Lu, J.P.: Microscopical Studies of Crack Growth in Concrete by Numerical Simulation, International Conference on Advanced Experimental Mechanics (ICAEM), Tianjin, China, May 1988

PLASTIC DEFORMATION UNDER MODE I AND MODE II LOADING IN WC-Co

SIEGFRIED SCHMAUDER
Max-Planck-Institut für Metallforschung
Institut für Werkstoffwissenschaften
Seestraße 92, D-7000 Stuttgart 1, FRG

ABSTRACT

A Finite Element (FE) analysis is carried out to examine plastic deformation of cobalt binder ligaments bridging the crack faces in the wake of a tungsten carbide matrix crack in a WC-Co hardmetal. Under mode I loading plasticity is confined to a narrow band between the carbide crack tips at the ligaments in accordance with experimentally found deformation patterns. The development of the plastic zone is examined in detail. The calculations demonstrate that ligaments are simultaneously deformed to varying degrees. The present work emphasizes the similarities and principal differences of plastic deformation between mode I and mode II loading. The model yields detailed information about void formation which is known to control failure of the ligaments.

INTRODUCTION

A review on the most important models applicable to the mode I fracture of WC-Co hardmetals has been given in [1]. It was concluded that tremendous disagreement exists between different authors. Most of the models (e.g. [2]) rely on a plastic zone size r_y comparable to the mean intercept length of the Co-binder L_{Co} due to a linear relation between L_{Co} and the toughness K_{Ic} of WC-Co. Recent FE-calculations even yielded a much larger size of the plastic zone [3] while in [4] a smaller size was assumed:

$$r_y < L_{Co} \qquad (1)$$

This latter relationship was experimentally confirmed in [1] but still remained in discussion. However, a more sophisticated FE-calculation [5] proved eqn. (1) to be correct.

In this paper simulations of the crack tip region under mode II loading are presented. Of central interest are common features with previously obtained results under mode I loading [5] concerning plastic deformation and the ductile failure of Co-ligaments as well as differences in the deformation of the crack tip region.

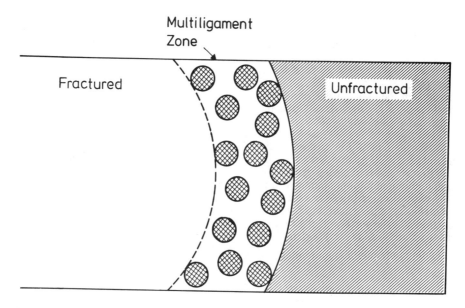

Figure 1. Schematic illustration of the crack plane in WC-Co with a multiligament zone at the crack tip. These ligaments fail according to the surrounding crack in the carbide frame in the course of the loading history.

Figure 2. MLZ in WC-Co with 10 wt.% Co (Courtesy L. Sigl).

MATERIAL

WC-Co is a two-phase material consisting of hard WC-grains interconnected with each other in a three-dimensional frame embedded in a Co-binder phase. Qualitatively, the relation between toughness and microstructure is well developed [6], [7]. Although the fracture path primarily depends on the crack advancing in the carbide phase, toughness is essentially a function of the deformability of the softer Co-binder in the multiligament zone (MLZ) [8] at the crack tip (Fig. 1). The crack resistance of the alloy mainly depends on the plastic work performed in the ductile binder [1]. It has been mentioned in the literature [7] that actual plastic deformation is exclusively restricted to the MLZ under mode I loading conditions.

For technically interesting materials the carbide grain size d_{WC} (0.3 μm - 3 μm) is much smaller compared with the binder grain size (0.1 mm - 1 mm) [1]. Depending on the microstructure the MLZ extends to approximately 4 d_{WC} (about 3.5 μm - 11 μm) and comprises typically between two to four binder ligaments. A typical MLZ in a WC-Co alloy with a coarse microstructure is shown in Fig. 2 after extensive loading under mode I. Obviously, the formation of voids may not be observed due to a limited deformation constraint at the surface of the material. However, plastic deformation is restricted to the fracture path and does not extend into the wedges of the ligaments.

METHOD AND MODEL

To adopt the FE-method (FEM) the program system ASKA was used [9]. Several

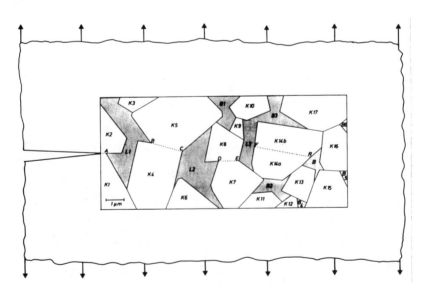

Figure 3.   Contour plot of the model representing the MLZ of Fig. 2 embedded in a WC-Co-surroundings under mode I loading (L=ligament, B=binder region, K=carbide grain).

reasons required an improved FE-model for the simulation of a WC-Co hardmetal: (i) unrealistic Westergaard boundary conditions were applied to the model for the two-phase alloy in the previous calculations [3], (ii) the influence of the MLZ was also neglected, (iii) the subdivision of Co-regions into a few elements provided a too stiff behaviour for Co [3]. Therefore, the two-dimensional representation of a MLZ shown in Fig. 2 was modelled with about 50 finite elements per grain, ligament or binder phase region and the surrounding alloy was also modelled (Fig. 3). (The mesh of the MLZ can be seen in Fig. 11.)

A comparison between the MLZ of Fig. 2 and Fig. 3 indicates only small differences with respect to the microstructure. The size of the surrounding is chosen to be 300 x 200 μm^2. The whole size of the model comprises more than 2000 triangular elements. The elastic properties E_{WC}=714 GPa, ν_{WC}=0.194; E_{Co}=211 GPa, ν_{Co}=0.31; E_{alloy}=595 GPa and ν_{alloy}=0.216 are adopted from [6]. Hardening of the binder is described through the Voce-law

$$\sigma = \sigma_y + (\sigma_s-\sigma_y) (1-\exp(-\phi/\phi_c)) \qquad (2)$$

with strain ϕ, stress σ, normalization $\phi_c = (\sigma_s-\sigma_y)/\Theta_o$, initial hardening rate $\Theta_o = E_{Co}/12$, saturation stress $\sigma_s = 12.19$ GPa and yield stress $\sigma_y = 2.19$ GPa. The high yield stress was recently justified by comparison with the hardness of the Co regions through microhardness indentation tests [10]. Mode I toughness is given in [1] as $K_{Ic} = 16.73$ MN/$\sqrt{m^3}$ while a mode II toughness value is not available from the literature.

RESULTS

Mode I Loading

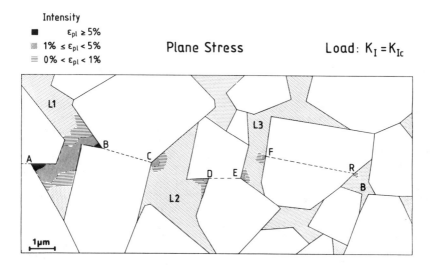

Figure 4. Plastic deformation at the surface under mode I loading (K_{Ic} = K_{Ic} (plane stress)).

Results from mode I loading have recently been reported in some detail [5], [11], [12], [13]. For brevity, two important findings will be repeated here: Only the first ligament L1 ahead of the fully broken alloy is plastically deformed to a large degree while other ligaments feel the carbide crack tips in a strongly reduced manner (Fig. 4 and Fig. 5). Under fracture load L1 is plastically deformed along the expected fracture path. Ligament wedges and other binder regions remain undeformed in agreement with experiment.

A somehow different behaviour is obtained in the bulk material (Fig. 5): Plastic deformation spreads primarily along phase boundaries parallel to the load direction leading to high stress triaxialities ahead of crack tips A and B in ligament L1. The combination of high stress triaxiality and large equivalent plastic strains may lead to void formation [14] and concomitant ductile failure of the binder. However, in both cases, large plastic strains are concentrated at the local crack tips.

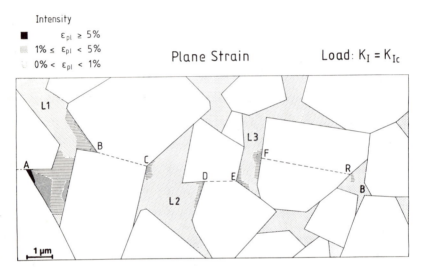

Figure 5.   Plastic deformation in the bulk under mode I loading.

Mode II loading
The results in case of mode II loading are shown in Figs. 6 - 11 without representing the ligaments in dark to distinguish from mode I loading. Similar to the results in mode I loading [5], the first ligament is most strongly plastically deformed with a kind of horizontal direction in the development of the plastically deformed region (Fig. 6). As in mode I loading, the intensity distribution of plastic deformation follows the development of the border between the elastic/plastic region (Fig. 7). Due to numerical instabilities the load could not be increased beyond 98.8 MN/m^2 in contrast to the plane strain calculation where the load was increased up to 280.5 MN/m^2. At this load level ligament L1 is nearly fully deformed and also large areas of ligaments L2 and L3 are plastified.

Remarkably non-reversible plastic deformations arise also away from the crack tips at some corners and edges in ligament L3 and other binder regions B, B2 and B4 (Fig. 8) which are not connected with the plastic zones at the crack tips. Additionally, all ligaments have been plastified through their cross sections. Fig. 9 shows that the intensity distribution of equivalent plastic strains follows again this deformation pattern. The differences between plane strain and plane stress for· a loading level comparable to

$K_{II} \approx 0.57\ K_{Ic}$ are depicted in Fig. 10. A comparison with the crack opening under mode I loading [5] shows that the rigid WC–Co material provides the same displacement pattern under mode II loading (Fig. 11) with their magnitudes in accordance to the ratio of stress intensities of both simulations.

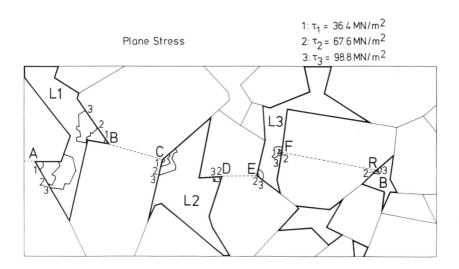

Figure 6.   Plastic deformation at the surface under mode II loading.

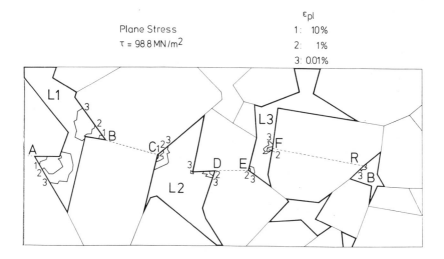

Figure 7.   Intensity distribution of plastic deformation at the surface under mode II loading.

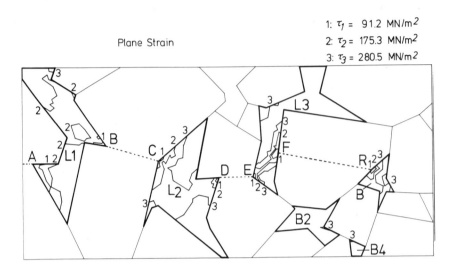

Figure 8. Plastic deformation in the bulk under mode II loading.

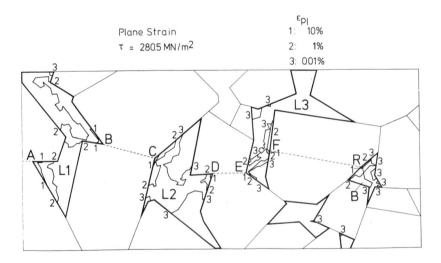

Figure 9. Intensity distribution of plastic deformation in the bulk under mode II loading.

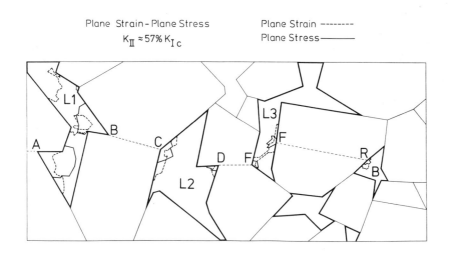

Figure 10.   Comparison of plastic deformation at $K_{II} \approx 0.57\ K_{Ic}$.

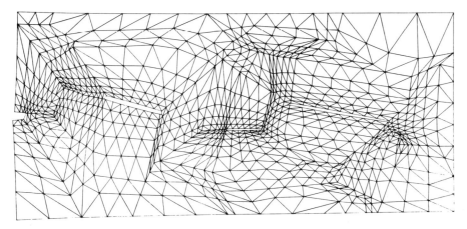

Figure 11.   Crack opening and displacements (magnified ten times) under mode II
loading at the surface.

DISCUSSION

In contrast to mode I loading, plastically deformed regions are larger in mode II loading
and do not arise only between the local carbide crack tips but also at carbide facets in

other ligament regions as well as in binder regions apart from the crack plane. This may be attributed to the presence of stress concentrations at phase boundaries which are already present under internal load [15].

In mode I and in mode II loading the intensity distribution of plastic deformation is in close agreement with the calculated deformation pattern, as expected. In both loading modes ligaments directly ahead of the fully broken material are much more plastically deformed compared to the other ligaments.

Concentrations of hydrostatic stresses ahead of and equivalent plastic strains at the local carbide crack tips arise in both mode I and mode II loadings as shown in [5] and [11] – [13]. They give rise to the assumption of localized void nucleation and growth in both failure modes (Fig. 12). These processes are known from experiment to control the failure process under mode I loading [1]. The results presented suggest a crack propagation by void nucleation, growth and coalescence as shown in Fig. 13.

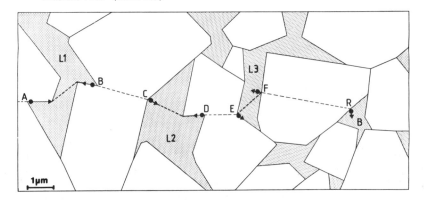

Figure 12.   Ductile fracture of the ligaments induced by preferred void growth at local crack tips. Damage occurs at first in ligament L1 at position A.

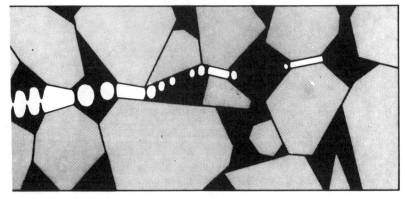

Figure 13.   New model of crack propagation in the MLZ.

CONCLUSION

The fine FE mesh used in our calculations is a necessary prerequisite to give agreement with experimental observations and to resolve the history of plastic deformation and intensity distribution in ligaments of the MLZ in detail. Results presented in this paper provide an idea about the details during ligament failure in MLZs of metal reinforced ceramic composites under mode I and mode II loading. Especially, progress is made in understanding the heterogeneously distributed nucleation of voids ahead of local carbide crack tips which is in strong contrast to former speculations about homogeneous void nucleation along the probable crack path in the ligaments.

REFERENCES

1. Sigl, L., Das Zähigkeitsverhalten von WC-Co-Legierungen, Fortschrittsberichte VDI, Reihe 5: Grund- und Werkstoffe Nr. 104, VDI-Verlag, Düsseldorf, 1986.
2. Almond, E.A., Microstructural Basis of Strength and Toughness in Hardmetals. In Speciality Steels and Hard Materials, eds., N.R. Comins and I.B. Clark, Pergamon Press, Oxford, 1982, pp. 353-60.
3. Ljungberg, A.B., Chatfield, C., Hehenberger, M. and Sundström, B., Estimation of the Plastic Zone Size Associated with Cracks in Cemented Carbides. In Inst. Phys. Conf. Ser. No. 75, 1988, pp. 619-30.
4. Krstic, V.A. and Komac, M., Toughening in WC-Co Composites. Phil. Mag., 1985, A51, 191-203.
5. Schmauder, S., Sigl, L., König, M., Fischmeister, H., Elasto-Plastic Modelling of a Multiligament Zone in Tungsten Carbid-Cobalt Hardmaterials. In Proc. 15th Int. FEM-Congress, IKOSS, Stuttgart, 1986, pp. 213-36.
6. Sigl, L.S. and Fischmeister, H., On the Fracture Toughness of Cemented Carbides. Acta Metall., 1988, 36, 887-97.
7. Sigl, L.S. and Exner, H.E., Experimental Study of the Mechanics of Fracture in WC-Co Alloys. Met. Trans. A, 1987, 18A, 1299-1308.

8. Evans, A.G., Heuer, A.H. and Porter, D.L., The Fracture Toughness of Ceramics. In Proc. 4th Int. Conf. on Fracture, ed., D.M.R. Taplin, Pergamon Press, New York, 1977, pp. 529-56.

9. Balmer, H.A., Doltsinis, J.St. and König, M., Elastoplastic and Creep Analysis with the ASKA Program System. Comp. Meth. Appl. Mech. Engng., 1974, 3, pp. 87-104.
10. Sigl, L.S., Exner, H.E., unpublished research.
11. Schmauder, S., Finite Element Studies of Crack Growth in a WC-Co Multiligament Zone. In Proc. 4th Int. Conf. Num. Meth. in Fracture Mech., eds., A.R. Luxmoore, D.R.J. Owen, Y.P.S. Rajapakse, M.F. Kanninen, Pineridge Press, Swansea, 1987, pp. 387-400.
12. Fischmeister, H.F., Schmauder, S., Sigl, L., Finite Element Modelling of Crack Propagation in WC-Co Hardmetals. Mat. Sci. Engng. (in press).
13. Sigl, L.S. and Schmauder, S., A Finite Element Study of Crack Growth in WC-Co. Int. J. Frac., 1988, 36, 305-17.
14. Barnby, J.T., Shi, Y.W. and Nadkarni, A.S., On the Void Growth in C-Mn Structural Steel during Plastic Deformation. Int. J. Frac., 1984, 25, 273-83.
15. Chiu, Y.P., On the Stress Field due to Initial Strains in Cuboid Surrounded by an Infinite Elastic Space. J. Appl. Mech., 1977, 45, 587-90.

STRENGTH CHARACTERISTICS OF WC-Co COMPOSITES

DUSZA JÁN, LEITNER GERT[*], RUŠČÁK MARTIN,
PARILÁK ĽUDOVÍT, ŠLESÁR MILAN
Institute of Experimental Metallurgy SAS Košice, ČSSR
[*]Zentralinstitut für Festkörperphysik und Werkstofforschung
DAdW, Dresden, DDR

ABSTRACT

This work was designed to investigate the strength characteristics of four WC-Co composites with different cobalt content and mean grain size of tungsten carbide. The sintering
and cooling regime of these materials were studied on the
basis of a complex thermoanalytical investigation and the
microstructure of final product was analysed by quantitative
metallography. The strength properties were measured in three
and four-point bend mode and the achieved values were analysed by Weibull's statistical analysis. An attempt has been
made to determine the subcritical crack growth parameters by
the help of the different strain rate test and from the life
time measurements at static bending tests. The fracture origins characteristic for the individual grades of studied materials have been determined by fractographic analysis.

INTRODUCTION

Classic WC-Co hard metals are applicable in a wide range of
material processing when besides high hardness, wear resistance and strength an adequate fracture toughness is required,
too. These characteristics are determined by the technological regime, from which the sintering is the most important
part. During the heating up some effects which influence the
quality of final products, their microstructure and fracture-
mechanical properties must be optimized. In the past a number
of papers dealt with the influence of microstructure and defects on the fracture and mechanical properties of WC-Co com-

posites /1-2/. The main attention has been devoted to the
bending strength and fracture toughness. It was found that
strength values are often degraded by the present technological defects (pores, inclusions, groups of grains, etc.) or
surface roughness and the microstructure influence on strength
values is often repressed. On the other hand the fracture
toughness of these materials show a good correlation with
microstructural parameters (mean free path in binder -L_{Co} and
contiguity -C_{WC}). In some cases it is very important to know
the time dependence of strength of these materials at the
room or elevated temperatures. For ceramics is elaborated this
problem exactly, but the subcritical crack growth in hardmetals was studied up to now only in a few papers /3,4/.

The aim of this paper is to study the strength characteristics of WC-Co composites in connection with thermoanalytical investigation and metallographic-fractographic examination,
to compare the possibilities of three and four-point bend
tests in hardmetal research and to study the subcritical crack
growth parameters of these materials.

MATERIALS AND METHODS

The work was performed on four grades of WC-Co hardmetals
with mean grain size of WC from 0.95 to 1.94 /um and with the
cobalt content from 6 to 20 wt.%, Tab. 1. The processes taking
place during the heating up and sintering were studied on
basis of a complex thermoanalytical investigation (DTA, ther-
mogravimetry, dilatometry). For parameters (dynamic heating-
up regime, argon atmosphere, etc.), see / 5 /. The microstructure and substructure analyses were carried out by help of
light, scanning and transmission electron microscopy, Figs.1,
2. The microstructural parameters were established by point
and linear analyses, strength values for grades B,C,D have
been measured in three and four-point bend mode (span of 30
mm and 16/32 mm, resp.) with a cross-head speed 50 /um/mm.
The distribution of strength values was analysed by Weibull's
statistical analysis based on the equation:

$$P_f = 1 - \exp\left[-\left(\frac{\sigma - \sigma_u}{\sigma_o}\right)^m\right] \qquad (1)$$

where σ_u, σ_o and $\underline{m}$ are Weibull parameters.

TABLE 1
Materials and their properties

Material	A			B		C		D	
wt.%of Co	6			6		9.5		20	
f$_{Co}$ /%/	9.2			9.4		19.5		29.7	
D$_{WC}$ //um/	0.95			1.73		1.94		1.33	
L$_{Co}$ //um/	0.28			0.55		0.84		0.96	
C$_{WC}$	0.77			0.66		0.56		0.4	
Hardness	1750			1395		1260		1007	
Bend mode	3/1	3/2	3/3	3	4	3	4	3	4
σ_B /MPa/	1714	1596	1617	1781	1417	2046	1885	3018	2310
σ_o /MPa/	1807	1685	1683	1870	1469	2125	1953	3175	2452
m	9.1	9	12.6	10.6	17.5	13.8	16.4	9.8	23
Samples with fract. origin/%/	80	75	75	15	40	10	20	15	40
Type of fracture origin	sharp pores Fig. 9			second phases groups of WC grains Fig. 10				inclusions spheric. pores	
Mean size of def. //um/	80			45		45		60	

3/1, 3/2, 3/3 are different strain rates, see text
σ_B - the mean bend strength

Figure 1. Microstructure of material grade C, light microscopy

Figure 2. Microstructure of material grade A, TEM

Assuming that a_o is the depth of the surface microcrack and that σ is the tensile strength at the crack location, the stress intensity factor $\underline{K}$ characterizing the stress condition ahead of the crack tip is given according to:

$$K = \sigma \cdot Y \cdot (a)^{1/2} \tag{2}$$

where y is the geometrical correction factor.
Failure of a mechanically stressed component occurs when $\underline{K}$ reaches the critical value K_c. It means that for each particular stress level there is a corresponding critical crack length

$$a_c = (K_{IC}/\sigma \cdot Y)^2 \tag{3}$$

The original defects (a_o a_c) often arise from a_o to a_c by a subcritical crack growth which causes limited life-time t_f:

$$t_f = \frac{2}{A \cdot Y^2 \cdot (n-2) \cdot K_C^{n-2}} \cdot \sigma_i^{n-2} \cdot \sigma^{-n} \tag{4}$$

where $\underline{A}$ and $\underline{n}$ are parameters of the subcritical crack growth:

$$v = da/dt = A \cdot K^n \tag{5}$$

For determination of $\underline{n}$ and $\underline{A}$ several methods have been developed: a) double torsion method, b) different strain-rate method, c) static bending test method, d) determination of $\underline{v}$ vs. $\underline{K}$ relation from life-time measurements.

In this work we used methods a), c) and d) (only for grade A). The different strain rate test has been carried out at the cross-head speeds $\mathcal{E}_1 = 10$ μm/min, $\mathcal{E}_2 = 500$ μm/min and $\mathcal{E}_3 = 10$ mm/min, in three-point bend mode. The mean strength values and Weibull modulus were established. For calculation of $\underline{n}$ it is possible to use the following equations:

$$\mathcal{E}_1/\mathcal{E}_2 = (\sigma_1/\sigma_2)^{n+1} \quad \text{and} \quad m_s = m_i (n+1/n-2) \tag{6,7}$$

where $\mathcal{E}_1$, σ_1 and $\mathcal{E}_2$, σ_2 are the corresponding strain rates and mean strength values, m_i and m_s are Weibull parameters for innert strength (without the subcritical crack growth) and for bend strength under the condition of subcri-

tical crack growth, resp. Static bending tests were carried
out in three-point bend mode on samples with natural and
artificial defects (Vicker's crack created at a load of 1000 N).
The time to failure had been measured at different stress
levels. The n value was calculated on the basis of following:

$$t_1/t_2 = (\sigma_2/\sigma_1)^n \quad \text{and} \quad m_t = m/(n-2) \qquad (8,9)$$

where m_t and m_c are the Weibull moduli for the life-time and
innert strength distributions. The v vs. K relation was determined from the life-time measurements of stressed samples
with artificial defects according to the equation /3/:

$$v(K_{Ii}/K_{IC}) = - \frac{2 \cdot K_{IC}^2}{Y^2 \cdot \sigma_b^2 \cdot t_f} \cdot \frac{d(\ln K_{Ii}/K_{IC})}{d(\ln t_f)} \qquad (10)$$

with Y=1.5 and K_{IC}= 9 MPa.m$^{1/2}$.

After the bending tests a detailed fractographic analysis
of the fractured surfaces has been made to determine the fracture origins of studied materials.

RESULTS AND DISCUSSION

The characteristic thermal effects within the temperature-
time regime for the production of hardmetals are described in
/ 5 /. In the range of 600 - 900 $^\circ$C the reduction of the oxide
layers on the hard material particles leads to a remarkable
mass loss, Fig. 3. The resulting gas products must be removed
from the samples before closing the pores. In the range of
1200 - 1350 $^\circ$C (in the solid state) the greatest part of
shrinkage takes place by rearrangement of the whole particles,
Figs. 3,4. At the formation of the liquid eutectic the open
porosity disappears and the distribution of the binder is
substantially improved. At higher temperatures, especially
during isothermal treatment at sintering temperature, a significant change and accomodation of the hard metal particle
shape takes place, connected with grain growth (D_{WC}) and the
development of the final binder distribution (L_{Co}), the formation of the hard material skeleton (contiguity C_{WC}).

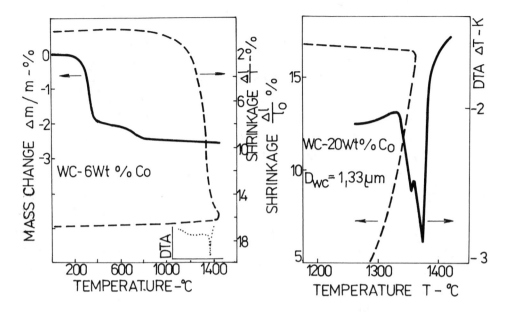

Figure 3. Mass change and shrinkage vs. temperature for grade A

Figure 4. Shrinkage and DTA vs.
temperature for mat. grade D

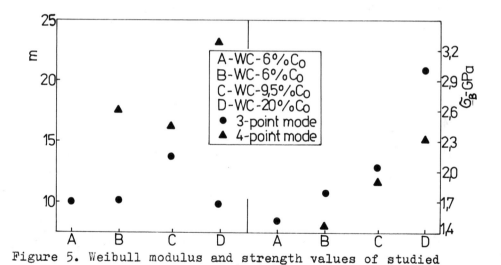

Figure 5. Weibull modulus and strength values of studied
materials

The results of microstructural analyses are illustrated
in Table 1. A relatively high porosity level in grades A,B and
free carbon precipitates in the grade B were found. Only sing-

le defects were found by metallographic analysis of grades C and D.

Three and four-point bend tests values and Weibull modulus for the studied materials are illustrated in Table 1 and on Fig. 5. It is evident that Weibull's moduli for four-point bend strength values are higher than those for three-point bend strength in all cases. On the other hand, the main strength values are higher in the case of three-point mode. These results are in agreement with the fact that at four-point bend test configuration the volume of material stressed to an equal level is larger than for the 3-point mode. The possibility that for this stress level there exists a defect with critical size in this volume is higher, too. Fig. 6 shows an example for Weibull's modulus and the mean strength values in the case of three and four-point mode. At the grade D we found out that the Weibull's line for four-point bend strength distribution consists of two lines with different slopes, Fig. 7. Detailed fractographic analysis confirmed that the reason of this phenomenon is the occurence of defects with different types, sizes and sites, Fig. 8 a,b. The characteristic features of fracture origins for all studied materials are in Table 1. The application of four parametric Weibull statistics

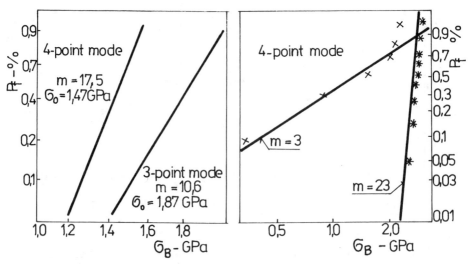

Figure 6. Weibull's lines achieved for material grade B

Figure 7. Weibull's lines achieved for material grade D

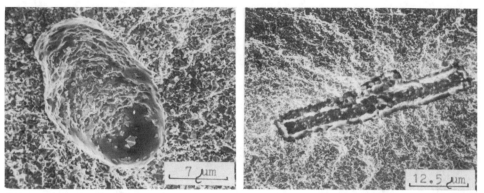

Figure 8. a,b Fracture origins ror material grade D

Figure 9. Fracture origins for   Figure 10. Fracture origins for
     material grade A   material grade B

on the strength values shows that a remarkable value for σ_u
in the case of grade B was achieved only, Fig. 11.

Results of subcritical crack growth analysis show that
for hard metals it is not easy to find out the parameters $\underline{n}$
and $\underline{A}$. The different strain rate method is probably not applicable for these materials because of a large scatter of
strength values, Fig. 12. An approximate calculation based on
these results gives a value of $\underline{n}$ cca 200. From results achieved in static bending tests on specimens with artificial cracks
(Fig. 13) using equations (8) and (9) we have achieved for $\underline{n}$
a value of 180 and 60, resp. According to our opinion for
determination of $\underline{n}$ and $\underline{A}$ the most suitable method is the life-
time measurement and the determination of the $\underline{v}$ vs. $\underline{K}$ relation

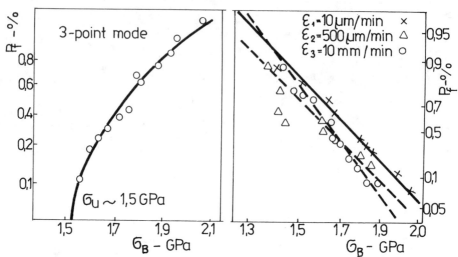

Figure 11. Three parametric Weibull's line for grade B

Figure 12. Results of different strain rates for grade A

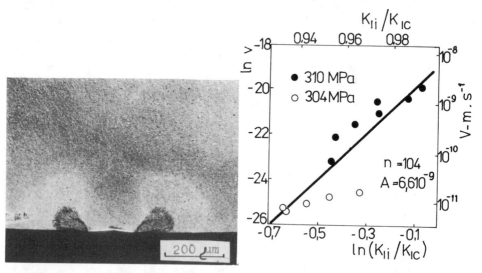

Figure 13. Fractured surface with the artificial crack

Figure 14. $\underline{v}$ vs. $\underline{K}$ relation for the material grade A

according to the equation (11), see Fig. 14. Fig. 14 shows that the subcritical crack growth starts at the value of K_{Ii} very close to the K_{IC} and the crack velocity is about 10^{-10} m.s^{-1}. These results explain why it is not possible to use

the different strain rate method for measuring n; the crack
velocity is 3 or 4 order of magnitude lower than the lowest
strain rate during the different strain rate test and what is
more, the crack growth starts at a very high value of K_{Ii}/K_{IC}.

CONCLUSION

1. Due to the differences in configuration between three and
   four-point bend mode the strength values are higher in case
   of three, and Weibull's moduli in case of four-point test
   mode. It can be said that the four point bend mode is more
   suitable for the determination of present technological defects. These cause that the threshold stresses for WC-Co
   materials are usually very low.
2. Results of subcritical crack growth measurements showed
   that there exists a difference between hard metals and ceramics, caused mainly by different apriori defects (type,
   shape, distribution) and by the presence of binder in WC-Co.
   Both, the achieved value of n and the threshold value, were
   high. Improvement of measuring methods and extending the
   knowledge about the subcritical crack growth on further
   hardmetals and at high temperatures requires more experiments.
3. Fractographic analysis carried out after bend tests showed
   that for strength and Weibull moduli degradation, the present defects are responsible. These arise during the pre-
   sintering operations (mixing, forming) and the sintering
   process is not able to remove these.

REFERENCES

1. Almond,E.A and Roebuck,B.,Defect-initiated fracture and
   the bend strength of WC-Co hard metals.Met.Sci.,1977,11,58.
2. Dusza,J.,Parilák,L. and Slesár,M.,Fract.characteristics of
   ceramic and cermet cutting tools,Ceram.Int.,1987,13,133-7.
3. Fett,T. and Munz,D., Prediction of time to failure of alumina under constant loading as deduced from bending strength
   tests,Cfi/Ber.DKG 4/5-84, 190-197.
4. Braiden,P.M.,Failure criteria for WC-Co materials under
   multiaxial stress states or long term static stresses,In:
   Conf.on Recent Advances in Hardmetal Production,Loughbo-
   rough, 17-19 Sept.1979, 28-1.
5. Leitner,G.,Schultrich,B.,Kubsch,H.,Sint.behaviour of TiC-
   Mo_2C-Ni and WC-Co.7th Int.Conf.on PM,CSSR,1987,Part 1,155.

THREE-DIMENSIONAL FAILURE MODES IN DOUBLE-PUNCH
INDENTATION OF BRITTLE MATERIALS

ANDRZEJ BUCZYŃSKI
Warsaw Technical University; Warsaw; Poland

ABSTRACT

An approximate approach to double-punch indentation into ceramic or rock materials is presented using the upper-bound theorem of limit analysis based on kinematic approach. It is assumed that a modified Coulomb-Mohr yield condition is valid with
an associated flow rule. Several kinematically admissible
three-dimensional failure modes are presented for different
geometric parameters of the punch and material. The modes corresponding to the lowest splitting force are selected for specific geometry parameters. The analysis results are compared
with experimental data obtained by the author. The method presented may prove useful in studying technological problems
associated with cutting, crushing, excavating, narrow-blade cutting, etc. of rocks or ceramics.

INTRODUCTION

Solutions of boundary-value problems in solid mechanics require
constitutive equations of material behaviour under applied loading. For problems concerned with failure modes and assessment
of critical load factors, the rigid-plastic model of material
is usually assumed. The analysis is then associated with a limit equilibrium state and an incipient failure mode [1,2].
A large class of solutions was obtained for plane strain, plane stress and axisymmetric problems [3,4,5], using the method
of slip-lines or providing lower and upper bounds to limit load
factors. However, experimental studies indicate that in many
technical problems such as coaxial plate cutting (Fig.1a),punch
indentation into a quarter-space (Fig.1b), etc., the failure
modes are three-dimensional and their form is much effected by
both geometrical and material parameters. Numerous papers were
devoted to the problem of plate cutting for ductile or brittle
materials [6,7,8]. Some experimental verification was provided
in [9,10] and the agreement between theory and experiment was
obtained for cases where plane strain condition was predominant.

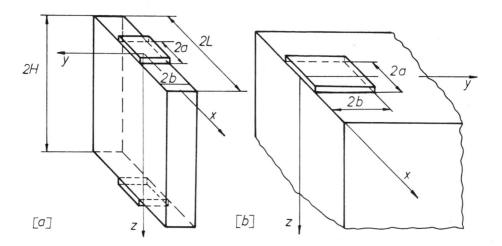

Figure 1. Examples of 3-d deformation modes
a) indentation of a plate by to opposite punches;
b) indentation of a "quarter-space" by a punch.

This paper provides an extension of previous studies by considering also three-dimensional failure modes. The experiments carried out by the present author provide some information on possible failure modes. Their description is provided next and the uppar bound to limit load factor is obtained by applying the limit analysis theorem. As there are many kinematically admissible modes, those associated with the lowest value of limit load are selected as best approximations to actual failure modes.

The rigid-plastic model assumptions imply the failure mechanism to operate instantaneusly. In actuality, in view of elastic compliance such failure mechanism will be developing progressively producing different degrees of damage in particular domains. The resulting failure load may then be considerably lower from that predicted by the rigid-plastic model. The theoretical predictions should therefore be backed up by experimental evidence. This is especially important in the case of brittle materials, for which cracking mechanisms in tensile zones are accompanied by plastic deformation in zones under compressive stresses, with different rates of softening in these zones. The rigid-plastic analysis provides only a much simplified insight into failure mechanism. The sequence of progressive failure could be traced by an elasto-plastic analysis with account for crack growth and development of plastic flow.

Experimental data

Rectangular samples made of brittle ceramic material were indented by two opposite punches (all dimensions being noted in
Fig.1a). Keeping two sample dimensions 2H and 2L as fixed,
the plate thickness 2b was changed,namely 2L=150 mm, 2H=40 mm,
2b=10,15 and 20 mm.The lateral punch width was equal to 2b and
its dimension 2a took values 5,10,15 and 20 mm. Introducing two
non-dimensional parameters:

$$m = \frac{2a}{2b}, \quad n = \frac{2H}{2b},$$

their values for respective tests are presented in Table 1.

TABLE 1

j \ i	n[i]	m[i,j]			
		1	2	3	4
1	2	0.25	0.50	0.75	1.00
2	2.66	0.33	0.66	1.00	1.33
3	4	0.50	1.00	1.50	2.00

TABLE 2

j \ i	$P_c^E[i,j]$ [N]			
	1	2	3	4
1	740	896	1345	1655
2	545	877	967	1145
3	350	545	702	742

Here i and j representing particular tests correspond to rows
and columns of all tables of this paper. Each test was repeated six times.Mean value of maximal forces P_c^E are given in
Table 2. Maximum upper and lower deviation values of measured
forces from their mean values reached the range specified by
its limits +11.3% and -7.9%. The two limits appeared for plate
thickness 2b=20 mm and punch width 2a=10 mm.
In all experiments, the three-dimensional failure modes were
observed with the two characteristic features:
    i) local modes with four rigid blocks moving laterally in
directions specified by vectors $\bar{V}yz$, cf. Fig.2a(for clarity,
only two visible blocks are shown in Fig.2a)
    ii) global modes with blocks moving in the direction speci-
fied by vector $\bar{V}y$ in Fig.2b(one block is shown in Fig.2b).
Depending on the value of parameter m, the occured blocks in
the form of pyramid or prism(further called x-prism and y-prism
respectively), cf. Fig.3. Some actual failure modes for different dimensions of sample and punch are shown in Fig.4.
The method described in[12,13]was used to determine values of
material parameters: cohesion C, angle of internal friction
and tensile rupture stress St'. The following values were
obtained: C=1.437[MPa], φ=12.5°, St=0.849[MPa].

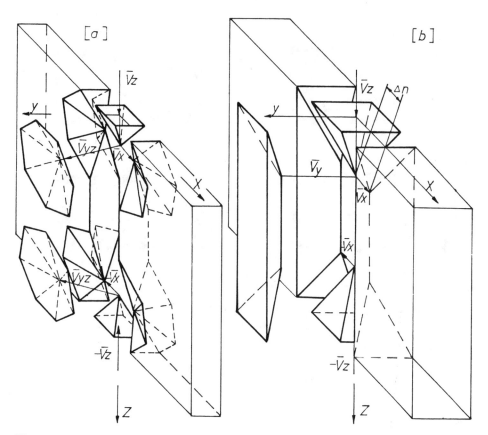

Fig.2 Examples of fracture modes of a plate and two punches:
a)mode with local blocks; b)mode with global blocks.

$$m = \frac{2a}{2b}$$

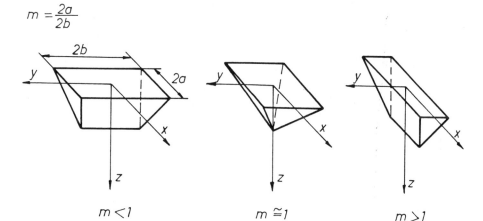

$m < 1$ $m \cong 1$ $m > 1$

Fig.3 Shapes of blocks appearing beneath punches.

a)

b)

c)

d)

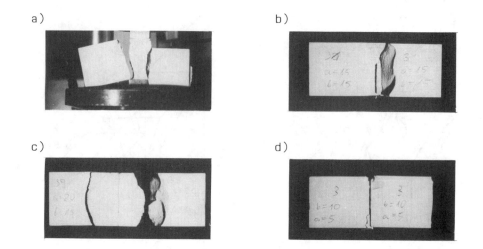

Fig.4 Fotographs of fracture modes for different plate
                and punch dimensions:
    a) 2b*2a=10*20 [mm] ; 2L*2H*2b=150*40*10 [mm] ; n=4,   m=2
    b) 2b*2a=15*15   ; 2L*2H*2b=   *15   ; n=2.66,m=1
    c) 2b*2a=15*20   ; 2L*2H*2b=   *15   ; n=2.66,m=1.33
    d) 2b*2a=10*5   ; 2L*2H*2b=   *10   ; n=4   ,m=0.5

Yield criterion, flow rule and dissipation rate.
The Mohr-Coulomb failure criterion with tension cutt-off
as shown in Fig.5, is adopted in this paper.

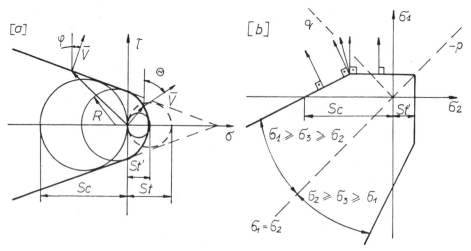

Fig.5 The modified Mohr-Coulomb yield criterion with
      tension cutt-off:
    a) in Mohr plane; b)in principal stress plane.

Parameters S_c and S_t' denote the unconfined compressive and tensile strengths, respectively. Compressive strength S_c and tensile strength S_t' can be expressed in terms of cohesion(C) and internal friction angle (φ), namely:

$$S_c = \frac{2*C*\cos\varphi}{1-\sin\varphi} \quad , \quad (1) \qquad\qquad 0 < S_t' < \frac{2*C*\cos\varphi}{1+\sin\varphi} \quad , \quad (2)$$

It is further assumed that the material is isotropic and initially homogeneous. The limit analysis method is based upon the idea of perfectly plastic behaviour of solids in which the strain rate tensor $\dot{\varepsilon}_{ij}$ is related to the stress tensor σ_{ij} by the potential flow rule associated with the yield condition $F(\sigma_{ij})=0$

$$\dot{\varepsilon}_{ij} = \lambda \frac{\partial F(\sigma_{kl})}{\partial \sigma_{ij}} , \quad \lambda > 0 \quad (3)$$

where λ is a non-negative scalar function dependent strain rate Using the kinematic approach providing the upper bound to limit load, the kinematically admissible failure mechanism should be selected. Following experimental observation, this mechnism is assumed to be composed of several rigid blocks translating with respect to each other along contact planes. Velocities of block translation are denoted by $\vec{V}x, \vec{V}z, \vec{V}yz$ in Fig.2a, and by $\vec{V}x, \vec{V}z, \vec{V}y$ in Fig.2b. In view of symmetry of failure modes with respect to three planes, only blocks on one side of these planes need to be considered.
The plastic dissipation occurs along thin interface planes which are velocity discontinuity lines. Both tangential and normal velocity discontinuity occur for shear zones corresponding to Coulomb yield condition. Only normal velocity discontinuity occurs for planes where tension failure condition is satisfied. Using the associated flow rule (3), the rate of dissipation D per unit area of discontinuity plane can be expressed as follows, cf.Izbicki and Mróz[3],

$$D = |\vec{V}|*(S_c*\frac{1-\sin\theta}{2} + S_t*\frac{\sin\theta - \sin\varphi}{1-\sin\varphi}) = k*|\vec{V}| \quad , \quad (4)$$

where $|\vec{V}|$ is the absolute value of velocity discontinuity across the discontinuity plane, θ is the dilatancy angle, shown in Fig.5a. Calculating the total dissipation rate Dt for an assumed failure mode and equating it with the rate of work of applied loads De,

$$Dt = De \quad , \quad (5)$$

the upper bound to limit load factor can be calculated.

Indentation of a plate by two opposite punches.
Consider the following boundary-value problem. A plate of dimensions $2H*2L*2b$ is indented by two opposite punches of dimensions $2b*2a$, Fig.1a.
Our aim is to derive formula specifying a value of the critical penetration force associated with the failure mode. Six three-dimensional modes presented in Figs.6,7 and 8 were considered.

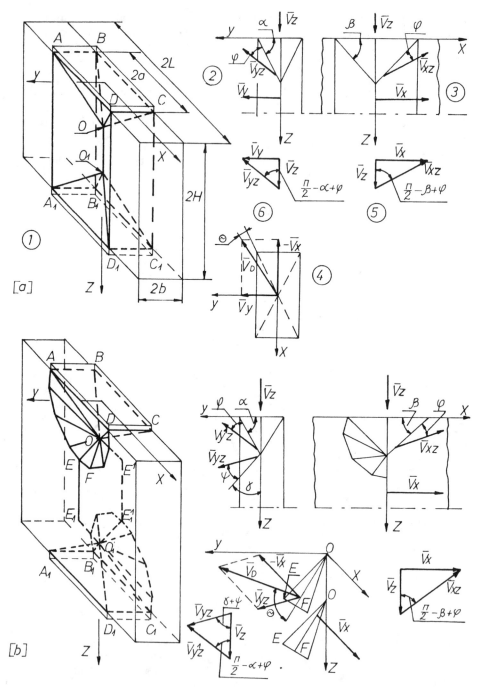

Fig.6 Theoretically assumed fracture modes with pyramid
    blocks beneath the punches:
    a)with global blocks; b)with local blocks.

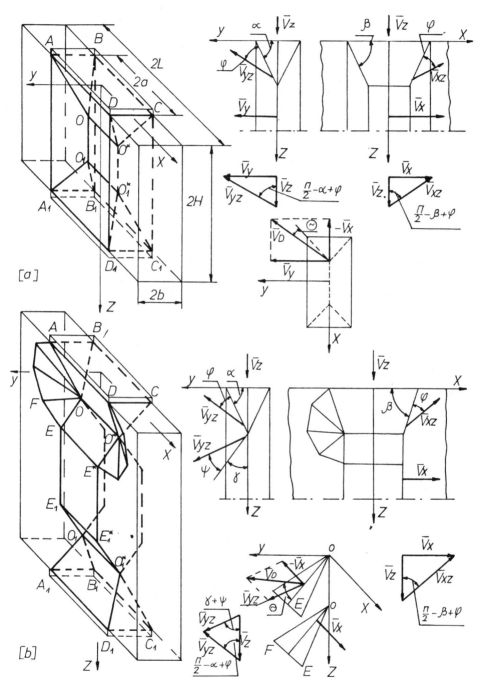

Fig.7 Theoretically assumed fracture mechanism with X-prism
blocks beneath the punches:
a)with global blocks; b)with local blocks

188

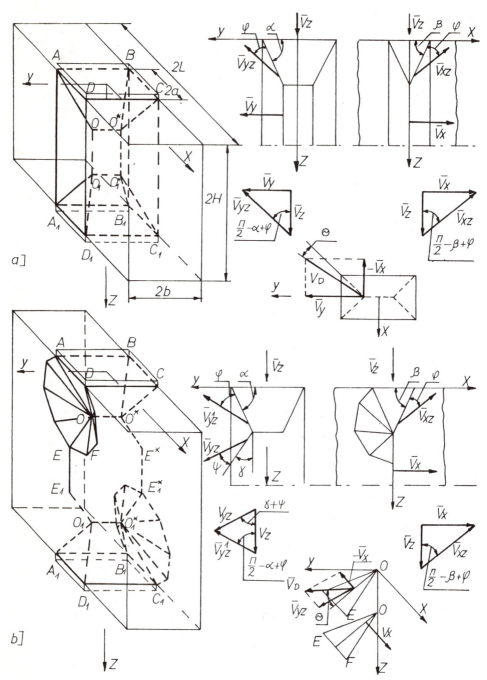

Fig.8 Theoretically assumed fracture modes with Y-prism
      blocks beneath the punches:
   a)with global blocks; b)with local blocks.

Depending on geometric parameters, the most plausible mode corresponding the lowest value of critcal force is selected. The procedure of determining the critical force is illustrated by considering the mode of Fig.6a. Denote by α and β the angles of walls ADO and ABO with the pyramid base ABCD. Note that the angles and are not independent $\alpha = \alpha(\beta)$, so the limit load can be expressed as follows:

$$P_C^T = p(\beta) \quad , \quad (6)$$

Two pyramidal blocks placed under punches move along z-axis with the velocity $\vec{V}z$. Two prismatical blocks move horizontally along the y-axis with the velocity $\vec{V}y$ and two blocks move along the x-axis with the velocity $\vec{V}x$. Relative velocities of blocks are $\vec{V}yz$, $\vec{V}xz$, $\vec{V}d$. These vectors can be regarded as velocity discontinuities occuring on the plane areas ABO, ADO and AOO_1A_1, respectively. The upper bound to the penetration force can be expressed as follows:

$$P_C = 2/|\vec{V}z| * \left[|\vec{V}yz| * k * S_{ABO} + |\vec{V}xz| * k * S_{ADO} + |\vec{V}d| * k * S_{AOO_1A_1} \right] \quad , \quad (7)$$

where k is specified by eq.(4) and S denotes the areas of discontinuity planes.
For Mohr-Coulomb yield condition the dilatation angle θ satisfies the inequality:

$$\varphi \leqslant \theta \leqslant \Pi/2 \quad , \quad (8)$$

In our case, we have $\theta < \Pi/2$ since there is no tensile fracture regime. It is assumed that the dilatation angle on ABO and ADO equals φ , $\theta = \varphi$, on the plane AOO_1A_1, the angle θ is specified by calculating $\vec{V}d$ and its orientation with respect to OA, cf. Fig.6a. We have:

$$\cos = (a * V_x + b * V_y)/(|\vec{a}| * |\vec{V}d|) \quad , \quad (9)$$

where a,b denote components of OA along x and y-axis, V_x and V_y are the components of $\vec{V}d$. Since

$$\vec{V}z + \vec{V}xz = \vec{V}x \quad , \quad (10)$$

then from the velocity hodograph it follows that

$$V_x = ctg(\beta - \varphi) \quad , \quad (11) \qquad V_{xz} = 1/\sin(\beta - \varphi) \quad , \quad (12)$$

where it was assumed that $|\vec{V}z| = 1$.
Similary, there is:

$$\vec{V}z + \vec{V}yz = \vec{V}y \quad , \quad (13)$$

and

$$V_y = ctg(\alpha - \varphi) \quad , \quad (14) \qquad V_{yz} = 1/\sin(\alpha - \varphi) \quad , \quad (15)$$

$$|\vec{V}d| = [(V_y)^2 + (V_x)^2]^{1/2} \quad , \quad (16)$$

The areas of ABO, ADO and AOO_1A_1 are:

$$S_{ABO} = b^2 * m/\cos\beta \quad , \quad (17)$$

$$S_{ADO} = b^2 * m * (1 + m^2 * tg^2\beta) \quad , \quad (18)$$

$$S_{A00_1A_1} = 2*b^2*m*(1+m^2)^{1/2}*(n-1/2*m*tg\beta) \quad , \quad (19)$$

Finally, using (12),(15),(16),(17),(18),(19), after some transformations the upper bound to the critical penetration force is expressed in the form:

$$P_C^T = 2*b^2*\left[m*C*\cos\varphi*(\frac{1}{\sin(\beta-\varphi)*\cos\beta} + \frac{1+m^2*tg^2\beta}{m*tg\beta*\cos\varphi-\sin\varphi}) + \right.$$

$$+ (S_C*\frac{1-\sin\theta}{2} + S_t*\frac{\sin\theta-\sin\varphi}{1-\sin\varphi})*(1+m^2)^{1/2}*(n-1/2*m*tg\beta) *$$

$$\left. * (\frac{1+m^2*tg^2\beta}{(m*tg\beta*\cos\varphi-\sin\varphi)^2} + \frac{\cos2(\beta-\varphi)}{\sin^2(\beta-\varphi)}) \right] \quad , \quad (20)$$

The minimum value of P_C^T and the angle is specified from the condition:

$$\frac{\partial P_C^T}{\partial\beta} = 0 \quad , \quad (21)$$

Similar procedure is applied to remaining fracture modes. The only difference is that there are more independent geometrical parameters. Thus, we have $P_C=p(\beta,\gamma)$ for modes of Figs.7a and 8a, $P_C^T=p(\alpha,\beta)$ for mode of Fig.6b and $P_C^T=p(\alpha,\beta,\gamma)$ for modes of Figs.7b and 8b. Note that for the local modes of Figs.6b,7b,8b it is assumed that on the plane EOF there is $\theta=\varphi$, the vector $\vec{V}yz$ is parallel to y-z plane and

$$n^y*V_{yz}^y + n^z*V_{yz}^z = |\vec{n}| * |\vec{V}_{yz}| * \cos(\Pi/2-\varphi) \quad , \quad (22)$$

where $\underline{n}$ is normal to EOF, and $\underline{V}yz$ is the velocity vector of local block motion. Its components along y and z-axis can be calculated from (22). Finally, the angle ψ is obtained from the relation:

$$\cos\psi = (V_{yz}^y*OE^y + V_{yz}^z*OE^z)/(|\vec{V}yz|*|\vec{OE}|) \quad , \quad (23)$$

where OE^y, OE^z are the components of $\vec{OE}$ along y and z. Table 3 presents the calculation results for particular values of geometric parameters.

TABLE 3

j\i	$P_C^T[i,j]$ [N]			
i	1	2	3	4
1	839	1059	1668	1993
2	619	924	1232	1536
3	421	675	871	948

TABLE 4

j\i	$P[i,j]$ [%]			
i	1	2	3	4
1	11.8	15.4	19.4	16.9
2	11.9	5.1	21.5	25.4
3	16.9	19.2	19.4	21.7

Comparison of theoretical and experimental results.
Differences between theoretical and experimental results are presented in the form:

$$P = \frac{P_C^T - P_C^E}{P_C^T} * 100\% \quad , \quad (24)$$

as shown in Table 4. The results of calculating and of experimental measurements for changing values of n and m are presented in Fig.9.

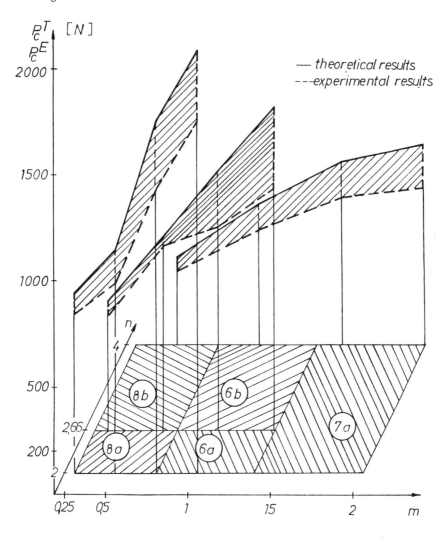

Fig.9 Diagrams of theoretical and experimental values of P_C^T and P_C^T .

The marked areas on the plane m,n in Fig.9 correspond to
theoretical failure modes corresponding to the lowest value
of limit load P_C^T with respect to other modes. For instance,
at the point n=4,m=0.5 , the selected failure mode is that
of Fig.8b. This is in close agreement with exsperimentally
observed fracture mode shown in Fig.4d. The difference between
theoretical and experimental results equals P=16.9%.
The agreement is less satisfactory for modes of Figs. 4a,b,c.
Table 4 shows the respective values P reaching 21.7%,21.5%
and 25.4%. These observed modes are closer to the modes
schematically presented in Fig.10.

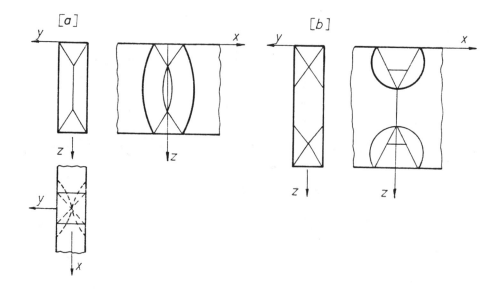

Fig.10 Schemes of other 3-d fracture modes:
    a) global mode with curvi-linear fracture trace on
       the plate wall;
    b) local mode

For n=4,m=2 , cf. Fig.4a, the best theoretical mode is that
of Fig.7a. However, the mode of Fig.7b shows better resemblance
to the experimental mode of Fig.4a.
Some new modes are therefore to be studied.Fig.10 presents
such modes where curvi- linear traces of fracture surfaces are
assumed. Refering to Fig.7b, assume that the points E* and E_1^*
coincide with E and E_1. In this way, we otain the mode of
Fig.10b. The mode of Fig.10a may be related to the experimental
mode of Fig.4a which does not show close resemblance to its
theoretical scheme of Fig.6a.

Conclusions.

Using the upper bound limit analysis theorem, the theoretical study of three-dimensionat failure modes can be provided for the case of plate cutting. In general, there is a satisfactory agreement between theoretical predictions and experimental measurements. However, the indentification of proper failure modes may be more difficult since theoretical modes are too simplified. The modes of Fig.10 provide better predictions and are now studied. Though, applicability of limit analysis to study of failure modes in brittle materials is questionable, it nevertheless provides an useful information on values of limit loads and also provides a tool for indentifiction of proper failure modes.

REFERENCES

1. Chen, W.F., Limit analysis and soil plasticity. Elsevier Sci. Pub. Comp., New York, 1975.

2. Mróz, Z., Drescher, A., Podstawy teorii plastyczności ośrodków rozdrobnionych. Ossolineum, Wrocław, 1972.

3. Mróz, Z., Izbicki, R., Metody nośności granicznej w mechanice gruntów i skał. PWN, Warszawa, 1976.

4. Hill, R., The mathematical theory of plasticity. Clarendon Press, Oxford, 1950.

5. Sokołowski, W.W., Teoria plastyczności. PWN, Warszawa, 1957.

6. Prandtl, L., Anwendungsbeispiele zu einen Henckyschen Satz über das plastische Gelichgewicht. Zeitschr. Math. Mech. 3, 1923, pp. 401-420.

7. Szczepiński, W., Indentation of a plastic block by two opposite narrow punches. Bul. Acad. Polon. Sci., Ser. Sci.Tech. 14, 1964, pp.671-676.

8. Chen, W.F., Extensibility of concrete and theorems of limit analysis. J. Eng. Mech. Div. Proceed. ASAE, vol.96, 1976, pp. 341-352.

9. Jaeger, J.C., Failure of rock under tensile conditions. Int. J.Rock Mech. Min. Sci., 1967, pp. 219-227.

10. Jaeger, J.C. and Cook, N.G.W., Fundamentals of rock mechanics. Methuen and Co LTD, London, 1969.

11. Chen, W.F. and Drucker, D.C., Bearing capacity of concrete blocks or rocks. J. Eng. Mech. Div. Am. Soc. Civ. Engrs. 95, 1969, pp. 955-978.

12. Kwaszczyńska K., Mróz, Z. and Drescher, A., Analiza ściskania krótkich walców z materiału Coulomba. Prace IPPT PAN,29, Warszawa 1968.

13. Kwaszczyńska, K., Mróz, Z. and Drescher, A., Analysis of compression of short cylinders of Coulomb material. Int. J. Mech. Sci., 11, 1969, pp. 145-158.

FATIGUE LIFE OF DIAMOND/HARD ALLOY COMPOSITE MATERIALS

Novikov N.V., Chepovetskiy G.I.,
Kulakovskiy V.N.
Institute for Superhard Materials,
the Academy of Sciences of the UkrSSR,
Kiev, USSR

Diamond-containing composite material (DCM) with WC-Co hard
alloy matrix is currently widely used for setting drilling
bits. High wear resistance and excellent cutting properties
of diamond single crystals are the main advantages of this
material which enabled production of highly efficient bits of
crushing and microcutting types. The use of bits, however,
showed that one of the factors limiting the bit durability is
the low strength of DCM under fatigue load characteristic for
drilling and particularly for drilling deep rocks of high
strength and hardness.

At the Institute for Superhard Materials of the Ukraini-
an Academy of Sciences DCM were tested for strength under
repeated-variable loads. Four lots of specimens were tested
having the similar diamond grit size, 315/250 μm (as the most
widely met in practice). The WC-6Co alloy was used as a matrix of the material. This paper presents an experimental investigation of the fatigue life of DCM as a function of diamond strength measure and diamond concentration in the volume
of the material. The measure of diamond strength is evaluated
from the failure load per a grain in compression. Diamond
concentration is taken from the volume of the whole material.
The specimens of the first lot were found to have the diamonds
with measure of strength 32 H and the concentration 25%,

those of the second, third and fourth lots had the measure of strength 82 H and the concentration I8.7%, 25% and 37.5%, respectively.

The test specimens were made in the form of prismatic rods and measured 4.6x5.2xI7 mm. They were tested on a fatigue machine based on an electrodynamic vibrator, in three-point bending. The frequency of load alternations was 400 Hz, the cycle shape was sinusoidal. The load was applied with the cycle asymmetry close to zero one R=0.05. The maximum stress in the specimen attained 200 MPa, which is appr. 0.7 of the G_{bend} for the weakest material under study, i.e. for the composite of the fourth lot (G_{bend} is the bending strength of the material).

TEST RESULTS AND DISCUSSION

Test results (Fig.I) show that the second lot composite with diamond strength measure 82 H and diamond concentration I8.7% exhibits the highest fatigue life, the lowest fatigue life is exhibited by the first lot composite with diamond strength measure 32 H and diamond concentration 25%, the ratio with respect to durability (in the case of 50% risk of rupture) being about 65:I.

Changing the diamond grits strength measure from 32 H to 82 H at constant diamond concentration results in 20-fold increase in fatigue life of the composite. At the same time the decrease in synthetic diamond concentration from 37.5% down to I8.7%, the other parameters being constant, results in 8-fold increase in a number of fatigue cycles to failure of DCM specimens.

In mechanical tests fracture surface of diamond-containing materials was examined under the CamScan-4DV scanning electron microscope.

In the zone of the crack initiation brittle step-wise fracture of diamond grains was observed with cleavage micro-steps on them which are similar to fatigue grooves but are

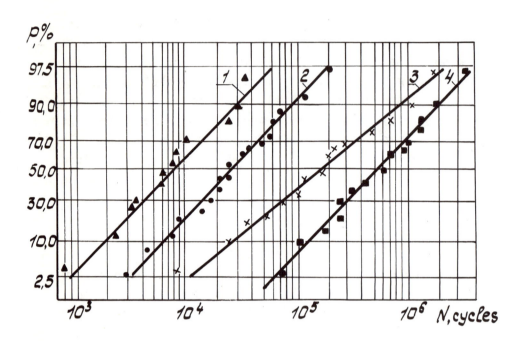

Figure I. Fatigue life distribution curve for DCM samples:
I – the measure of strength (MS) – 32 H (25%); 2 – MS – 82 H
(37.5%); 3 – MS – 82 H (25%); 4 – MS – 82 H (I8.7%)

vaguely oriented. The composite fracture surface has the appearance characteristic for the case of multiple development
of failure when the cyclic cracks grown from the various
fracture initiators tend to coalesce into the main crack in a
bit. Fracture initiation is more commonly caused by diamond
grains damaged either during manufacturing of composite materials or due to the presence of solvent–metal inclusions
therein which are revealed on the crystal fracture surface by
the use of X–ray microanalyzer. Such damages when accumulated
give rise to the breaking down of diamond grains. The percentage of damaged grains in composites reinforced with diamonds
of higher strength is much lower and as a rule the crack

detours around the diamond grains. In the zone of the specimen postfracture the cleavages of diamond grains with perfect
orientation in the direction of the main crack front displacement are observed.

Considering the general fracture growth pattern for DCM
one should classify the material resistance to loads as the
strength of the material which is heterogeneous in structure
and reinforced with dispersed particles with predetermined
defects (which include both cracks in diamond and decohesion
at the diamond-hard alloy interface). The strength of DCM is
influenced greatly by the level of the tensile stress applied.
This is attributed to the fact that after sintering the diamond grains are in the state of uniform compression which re-
tards the extension of their defects.

The test results offer the possibility to draw up recom-
mendations as to the use of tools set with diamond-containing
composite materials.

A STATISTICAL THEORY OF BRITTLE FRACTURE

Ruifeng Wu   Chun-hang Qiu
Professors/Dept. of Engineering Mechanics
Dalian University of Technology, Dalian, China

ABSTRACT

The Weibull's chain model of strength of materials, as today's foundation of statistical theory of brittle fracture, been widely applied and advanced due to its simplicity in mathematical calculation has in the last deca des. However, it has an obvious insufficiency. It does not take into account the interaction between various elements of calculation and in many cases, especially in the case of nonuniform stress field, it is the main cause which leads to errors. The present paper, based on a large amount of experimental data, attempts to put forward a hypothesis dealing with the mutual actions of the adjacent elements along one direction and to correct the aforesaid insufficiency. The formulas obtained seem somewhat more complicated than Weibull's, but they are more accurate when verified by the experimental data.

INTRODUCTION

Brittle fracture is much dependent on local strength of a body and the strength is related to the dimensions of the body. Because of many influencing factors existing, the scatter of strength in the brittle fracture is larger than that in the ductile fracture and the brittle strength is a random variable. Under the situation of not very clear to know the fracture mechanism, a precise calculation of the strength of a body is difficult to get. For this reason the macro theory of strength is unable to deal with and the strict determination of the strength by micro theory is also an unsolved problem. An actually possible idea is that we give up the strict micro structural consideration and use some simplified suppositions which can take account of some major influenced strength factors. By using the theory of probability and statistics,, we study the synthetic effects of a large number of micro structures and evaluate the macro physical values. this is so called the statistic theory of strength.

A famous chain model of strength for brittle fracture was proposed by Weibull in 1939[1] . He proposed that any body may be ideally divided into numerous small elements and a fracture is occurred if real stresses in any one of the elements is equal to or greater than the defecting strength of the same element. This creative work had laid the foundation of theory of brittle fracture.

In [2, 3] a note on [1] was made. It noted that the main unsatisfactory point of Weibull's hypothesis is not taking into account the interaction between various elements of the given body. The Weibull's hypothesis is rather reasonable for the case of uniformly distributed stress state, but is not for the non-uniformly distributed stress state, especially in the case of existing a large stress gradient. When the fracture of one element occured, the transferred stress may not introduce the fracture to the adjacent elements, on other words, the neighboring elements can give support to the fractured element and maintain the whole body in working state. This is the basic point of view of present authours.

The present paper, based on a large amount of experimental data, attempts to put forward a hypothesis dealing with the mutual actions between adjacent elements along one direction and to correct the aforesaid defect of Weibull's hypothesis. The obtained formulas seem somewhat more complicated than Weibull's and two more parameters are added.

Relation to the Average Strength
With Dimensional Effect

Herein the chain model is also accepted, but the hypothesis of mutual actions between the adjacent elements along one direction is adopted.

Let the body consists of m elementary strips, and the center line of those strips is the possible fracture line L*. Every center line of elementary strip is L_i (Fig.1) and any elementary strip can be decomposed into n_i small elements, the strength of those are represented by $\xi_1^i, \xi_2^i, \ldots, \xi_n^i$, and $\{\xi^i\}$ is a set of random variables.

$$\eta_i = \min\left(\frac{\xi_1^i}{f(x_1^i, y_1^i, z_1^i)}, \frac{\xi_2^i}{f(x_2^i, y_2^i, z_2^i)}, \ldots, \frac{\xi_n^i}{f(x_n^i, y_n^i, z_n^i)}\right) \tag{1}$$

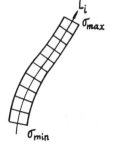

Fig.1 Elementary strip

where $f(x,y,z)$ is a dimensionless function of coordinates, and is determined as follows

$$\sigma = \max \sigma \cdot f(x, y, z) = R \cdot f(x, y, z)$$

in which σ is the stresses of point (x,y,z), R is the maximum stress. The $(x_k^i\ y_k^i\ z_k^i)$ represent the point coordinates in the k-th element of the i-th elementary strip.

The event of "$\eta_i < R$" equivalent to the fracture of a small elementary strip and the random variable of fracture of the body is

$$\eta = \min(\eta_1, \ \eta_2, \ldots, \ \eta_n).\tag{2}$$

Obviously, event "$\eta < R$" equivalent to at least a event of "$\eta_i < R$" is occured, but "$\eta_i < R$" represents at least any one of the elements is fractured. Therefore the η represents the strength of the whole body. Now, let us introduce the hypothesis of mutual action of the adjacent elements along one direction as follows:

(1) $\eta_1, \eta_2, \ldots, \eta_m$ are of the m independent random variables, that means no interaction between elementary strips, or interaction exists only in one direction.

(2) For the i-th given elementary strip the strength of its k-th element depends on ξ_{k-1}^i and ξ_{k+1}^i, but does not depend on any other element in i-th elementary strip, that means the adjacent relations are existing only.

Under these two hypotheses the problem now becomes to seek out the relation between the distribution function η and ξ. For this reason, (in the first) we find the relation between η_i and ξ_i at first.

The distribution law of η_i Let the strength of element obeys the negative exponential distribution, i.e.

$$P(\xi_j^i < \sigma) = F_0(\sigma) = \begin{cases} 1 - \exp\left[-\dfrac{\Delta V_j^i}{V_0} \ \dfrac{\sigma - \sigma_0}{\sigma_e}^{\alpha}\right], & (\sigma \geq \sigma_0) \\ \\ 0, & (\sigma < \sigma_0) \end{cases}\tag{3}$$

where α and σ_e are two positive constants; σ_0 is the non-negative constant representing the minimum strength of elements; V_0 is a standard volume; ΔV_j^i is the volume of j-th element. then the fracture probability of i-th elementary strip is

$$F_{\eta_i}(R) = P(\eta_j < R) = 1 - P(\eta_i \geq R) = 1 - P\left(\frac{\xi_1^i}{f(x_1^i y_1^i z_1^i)} \geq R, \right.$$

$$\left. \frac{\xi_2^i}{f(x_2^i, y_2^i, z_2^i)} \geq R, \ldots, \frac{\xi_n^i}{f(x_{n_i}^i, y_{n_i}^i, z_{n_i}^i)} \geq R\right)\tag{4}$$

Let

$$A_j^i = \frac{\xi_j^i}{f(x_j^i, y_j^i, z_j^i)} \geq R, \qquad (1 \leq j \leq n)$$

By the hypothesis of adjacent relationship, formula (4) may be simplified to

$$F_{\eta_i}(R) = 1 - P(A_{n_i}^i | A_{n_i-1}^i) \cdot P(A_{n_i-1}^i | A_{n_i-2}^i) \cdots P(A_2^i | A_1^i) \cdot P(A_1^i)\tag{5}$$

Now let's see the property of the conditional probability $P(A_k^i | A_{k-1}^i)$

Introduce following symbols:

$$P\left(\frac{\xi_k^i}{f(x_k^i, y_k^i, z_k^i)} \geq R, \frac{\xi_{k-1}^i}{f(x_{k-1}^i, y_{k-1}^i, z_{k-1}^i)} \geq R\right) = F_{k, k-1}^i(R, R),$$

$$P\left(\frac{\xi_{k-1}^i}{f(x_{k-1}^i, y_{k-1}^i, z_{k-1}^i)} \geq R\right) = F_{k-1}^i(R), \tag{6}$$

$$P(A_k^i | A_{k-1}^i) = \phi_{k, k-1}^i(R).$$

Then we have

$$\phi_{k, k-1}^i(R) = \frac{F_{k, k-1}^i(R, R)}{F_{k-1}^i(R)} \tag{7}$$

(1) From the physical meaning there are

$$\lim_{R \to 0} \phi_{k, k-1}^i(R) = 0, \qquad \lim_{R \to \infty} \phi_{k, k-1}^i(R) = 1$$

and $\phi_{k, k-1}^i(R)$ is a non-increased function, then

$$1 - \phi_{k, k-1}^i(R) = P\left(\frac{\xi_k^i}{f(x_k^i, y_k^i, z_k^i)} < R \,\middle|\, \frac{\xi_{k-1}^i}{f(x_{k-1}^i, y_{k-1}^i, z_{k-1}^i)} \geq R\right) \tag{8}$$

is a distribution function. Based on the theorem in the theory of probability there exists a random variable ζ_k^i, cooresponding the distribution function (8), (i.e.) we have

$$P(\zeta_k^i < R) = P\left(\frac{\xi_k^i}{f(x_k^i, y_k^i, z_k^i)} < R \,\middle|\, \frac{\xi_{k-1}^i}{f(x_{k-1}^i, y_{k-1}^i, z_{k-1}^i)} \geq R\right). \tag{9}$$

(2) $P(A_k^i | A_{k-1}^i)$ represents the non-fracture probability of the k-th element under the condition of non-fracture of the k-1-th element.

Along the fracture line L the larger stress gradient shows itself more different in stress values of the adjacent elements. In this case, the non-fracture condition of k-1-th element, which sustanins less stress, can provides Less insurance of non-fracture of k-th element, which sustains more stress than k-1-th element, i.e. conditional probability $P(A_k | A_{k-1}^i)$ decreases with increasing of $\partial f / \partial l$. Noted that $P(\zeta_k^i < R) = 1 - P(A_k^i | A_{k-1}^i)$, so $P(\zeta_k^i < R)$ increases with increasing of $\partial f / \partial L$. Therefore we introduce the $\partial f / \partial L$ as a parameter between random variables ζ_k^i and original random variables

$$\frac{\xi_k^i}{f(x_k^i, y_k^i, z_k^i)} \qquad \text{i.e.}$$

$$\zeta_k^i = \psi\left(\frac{\partial f}{\partial L}, \frac{\xi_k^i}{f(x_k^i, y_k^i, z_k^i)}\right). \tag{10}$$

As a first degree of approximation, let ζ_k is linearly dependent on

$$\frac{\xi_k}{f(x,y,z)}$$

and based on the second property we may suppose that

$$\zeta_k^i = \frac{1}{C\left(\dfrac{\partial f(x,y,z)}{\partial L}\right)^{\delta}\Big|_{x,y,z=(x_k^i,y_k^i,z_k^i)}} \cdot \frac{\xi_k^i}{f(x_k^i,y_k^i,z_k^i)} \quad , \tag{11}$$

where C and δ are two material constants, which can be determined experimentally. So we have

$$F_{\eta_i}(R) = 1 - P(\eta_i \geq R) = 1 - \{[1 - P(\xi_{n_i}^i < RCf(x_{n_i}^i, y_{n_i}^i, z_{n_i}^i)$$

$$\left(\frac{\partial f(x_{n_i}^i, y_{n_i}^i, z_{n_i}^i)}{\partial L}\right)^{\delta})] \cdots$$

$$[1 - P(\xi_2^i < RCf(x_2^i, y_2^i, z_2^i)(\frac{\partial f(x_2^i, y_2^i, z_2^i)}{\partial L})^{\delta})] \cdot$$

$$P(\xi_1^i \geq RCf(x_1^i, y_1^i, z_1^i))\} \tag{12}$$

Substitute the negative exponential distribution in (11) and let ΔV_j^i is sufficient small the fracture probability can be written as follows

$$F_{\eta_i}(R) = 1 - \exp\left[-\frac{1}{V_0}\int_{RCf(\partial f/\partial L)^{\delta} > \sigma_0}(\frac{RCf(\partial f/\partial L)^{\delta} - \sigma_0}{\sigma_e})^{\alpha}dV^i\right]. \tag{13}$$

where σ_0 is the minimum strength and σ_e is a constant.
From (2) and $\eta_1, \eta_2, \ldots, \eta_n$ are independent random variables, we have

$$F_{\eta}(R) = 1 - \exp\left[-\frac{1}{V_0}\int_{RCf(\partial f/\partial L)^{\delta} > \sigma_0}(\frac{RCf(\partial f/\partial L)^{\delta} - \sigma_0}{\sigma_e})^{\alpha}dV\right]. \tag{14}$$

the symbol under integration shows that the integrated region is of $RCf(\partial f/\partial L)^{\delta} > \sigma_0$ only.
The mathematical expectation of fracture strength η is

$$M\eta = \sigma_0 + \int_{\sigma_0}^{\infty} \exp\left[-\frac{g(R)}{V_0}\right]dR \tag{15}$$

where

$$g(R) = \int_{RCf(\partial f/\partial L)^{\delta} > \sigma_0} \left\{\frac{RCf(x,y,z)\left[\dfrac{\partial f(x,y,z)}{\partial L}\right]^{\delta} - \sigma_0}{\sigma_e}\right\}^{\alpha}dV \tag{16}$$

If put the minimum strength $\sigma_o = 0$, and let

$$V_{**} = \int_v \left[\ Cf(x,y,z) \ (\frac{\partial f \ (x,y,z)}{\partial L} \ )^\delta \ \right]^a dV \tag{17}$$

where V_{**} is so called transferring volume, (15) can be rewritten as

$$M\eta = \sigma_e \ (\frac{V_o}{V_{**}})^{1/a} \ \ \Gamma(1 + \frac{1}{a}) \tag{18}$$

where Γ is the gamma function.

Obviously, if two different complex stresses states are compared, the proportion is obtained

$$\frac{M\xi_1}{M\xi_2} = (\frac{V_{**2}}{V_{**1}})^{1/a} \tag{19}$$

Formula (19) is similar to Weibull's one, but is different from transferring volume V_{**}. There are two more material constants than Weibull's which can be determined by one set of specimens in uniform stress state and two sets of non-uniform stress state.

Experimental Verification

All specimens are made of two kinds of gypsum A and B with water-gypsum ratio of 1:1, $a = 9.4$ for kind A $a = 13$ for B.

1. Uniform stress state
    (1) Simple tension test. The sketch of specimens is shown in Fig.2.
    (2) Thin walled tube test under uniform inner pressure with open ends. The length of tubes are 30 cm, thickness of the tube is 0.5 cm, and outer radius is 7.5 cm.
    The experimental data are listed in Table 1 and 2, from which it can see that the Weibull's formula has sufficient precision for uniform stress state.

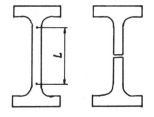

Fig.2 Specimen of tension test.

2. Complex stress state
    (1) Bending test of beams with rectangular cross setions.
    (2) thick walled tubes under uniform inner pressure (Fig.3).
    From Table 3 and 4 it may be seen that formula (19) proposed by present paper has more precision than Weibull's.
    Recently in the test for determining the stress intensity factor K_{1c} of concrete beams with artifitial preparated crack, the dimensional effects of concrete are evidently exhibited in all of those tests. How to determine the K_{1c} of concrete from small specimens to the structure? Experimental research work in River-Ocean University [4] shows that the formula (19) has a satisfactoried precision. The experimental data are listed in Table 5.

Table 1.   Tension Test

Gypsum	Set No.	Dimensions (cm)	Number of specimens	Fracture strength		Errors %
				Test	Theory	
A a =9.4	1	1x1x10	37	19.32	–	–
	2	2x2x10	60	17.17	16.67	2.91
	3	2x4x10	26	15.05	14.83	1.46
	4	2x4x15	24	14.05	14.21	1.36
	5	3x4x15	24	13.50	13.46	0.30
B a =13	1	1x1x10	77	20.65	–	–
	2	1x2x10	22	19.64	19.52	0.61
	3	2x2x10	20	18.00	18.56	3.11
	4	2x4x15	14	16.99	17.06	0.41

Table 2.   Thin Walled Tube

Number of specimens	Volume cm³	Inner pressure kg/cm²	Fracture strength		Errors %
			Test	Theory	
35	683	0.8633	12.52	12.33	1.5

Table 3.   Benging Test

Set	Width x height x length cm³	Number of specimens	Fracture strength kg/cm²						Type of Loading
			Test	Weibull's	Errors %	Formula (19)	Errors %		
1	2x2x20	12	27.36	27.44	0.29	–	–	
2	2x3x30	78	26.32	25.14	4.37	–	–	
3	3x5x50	30	23.55	21.63	8.15	24.02	2	
4	4x6x60	16	23.02	20.18	12.3	22.88	0.6	
5	6x10x100	31	20.41	17.33	15.1	20.86	2.2	
6	2x3x30	13	22.55	20.48	9.18	21.48	4.75	
7	3x5x30	17	19.55	18.58	4.96	20.49	4.81	
8	4x6x60	8	20.84	17.68	15.3	20.10	3.55	

α = 9.4; C = 1.014; δ = 0.117

Table 4.   Thick Wall Tube with Open Ends

Set	Number of specimens	Outer dia.	Inner dia.	Fracture strength kg/cm²					Thickness	Wei.'s Trans. Volume V_*	V_{**}
				Test	Wei.	Error %	Fomu. (19)	Error %			
1	31	7.5	6	15.99	12.38	22.5	15.05	5.88	1.5	657.9	1048
2	13	7.5	2.5	17.30	15.47	10.6	16.18	6.47	5.0	80.8	52.9

CONCLUSIONS

1. The strength of brittle fracture has dimensional effects evidently.
2. Weibull's chain model can be used to predict the strength of brittle fracture in the case of uniformly distributed stress state.

Table 5   Stress Intensity Factor $(kg/cm^{3/2})$

Speci-mens	Dimensions	$\dfrac{a}{w}$	Test k	Formula (19)				Weibull's	
				α	δ	K_{Ic}	Error %	k_{Ic}	Error %
9-1	10x10x40		53.70			–	–	–	–
9-2	15x15x60	0.3	63.50	6.65	0.366	63.54	0.06	54.77	13.7
9-3	20x20x80		70.78			71.59	1.14	55.54	21.5
11-1	10x10x40		50.19			–	–	–	–
11-2	15x15x60	0.5	58.22	6.65	0.318	58.24	0.03	51.19	12.1
11-3	20x20x80		63.05			64.72	2.65	51.91	17.1

3. The mutual action between elements of a body must be considered on the statistical theory of brittle fracture.

4. The improved formulas in the present paper which are based on the proposal of mutual action between the adjacent elements along one direction can be convinently used in the case of nonumiformly distributed stress state for predicting the strength of brittle fracture.

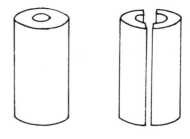

Fig.3 Specimen of thick walled tube

5. The proposed formula (19) can be used to predict k_{Ic} for concrete structure by using the experimental data from small specimen.

REFERENCES

1. Weibull, W., Proc. Rog. Swedish Inst. Eng. Res., Stockholm, 1939, 151.
2. Wu, R.F., The State-of-art of Theory of Brittle Fracture and Its Development, Journal of Dalian Institute of Technology, 1964, 1, 88-99.
3. Wu, R.F., C.H.Qiu, A Statistical Theory of Brittle Fracture, Journal of Dalian Institute of Technology, 1979, 3, 49-60.
4. Xu, D.Y., Feng, P.L., Fu, X.L., The Fracture Behaviour of Concrete and The Measurement of Stress Intensity Factors, The Research of Engineering Mechanics, 1987, 2, 1-16.

PREDICTION OF FATIGUE TENSILE CRACK GROWTH USING THE THEORY OF DAMAGE

Jean-François DESTREBECQ
Laboratoire de Génie Civil - Université Blaise Pascal
Clermont-Ferrand - France

ABSTRACT

Using the theory of damage, a model is elaborated which is likely to account for the propagation of tensile cracks in RC structure members subjected to fatigue. According to this model, the crack growth results from a redistribution of the internal forces due to softening of the concrete behaviour, along with strength decrease of the concrete due to compressive-tensile fatigue in the vicinity of the crack tip. The model is applied to plain concrete prisms tested under eccentric cyclic loading, and to RC-beams subjected to bending fatigue. The computed values of the tensile damage factor depict the propagation of the tensile crack. The predicted modes of failure prove in fair agreement with available experiments results.

INTRODUCTION

Cracks growth affecting structure members under service loading is a subject of worry for engineers dealing with concrete constructions. For structures subjected to repeated loading, the fatigue of the constitutive materials cannot be ignored as a prime cause of cracks initiating and growing. To assure a fair prediction of cracks propagation, any analytical model has to account for the behaviour of the material up to the failure, and refer to a failure criterion. In this paper, we discuss the applicability of the damage theory to concrete subjected to repeated tensile loading or to repeated tensile-compressive loading, in order to represent the behaviour of the concrete in the vicinity of a tension crack tip. Then, whe apply our model to plain concrete prisms and to RC flexural members subjected to fatigue loading.

MODEL FOR PLAIN CONCRETE SUBJECTED TO FATIGUE

When subjected to repeated compressive loading, the concrete exhibits a progressive degradation. Internal micro-cracks appear and grow steadily with increasing number of applied cycles. A decrease in stiffness is observed

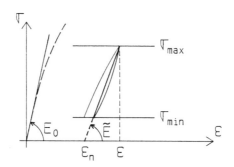

FIGURE 1 : Stress-strain relationship
for fatigue damaged plain concrete (1).

along with an increasing instant strain at unloading (Fig.1). Provided the
creep strain is negligible, the stress-strain relationship can be expressed
as follows :

$$\sigma = \tilde{E} \ (\varepsilon - \varepsilon_n) \qquad (1)$$

$\tilde{E}$ represents the actual value of the secant modulus of elasticity,
measured on the descending branch of the loading curve, whereas ε_n represents the instant remaining strain after cycle n. Using the theory of damage
$\tilde{E}$ and ε_n can be related to a damage factor D increasing from zero for virgin material to a critical value at failure (see (1) for more details) :

$$\tilde{E} = (1 - D) \ E_0$$
$$0 \leq D \leq D_{cr} = 1 - \frac{E_f}{E_0} \qquad (2)$$
$$\varepsilon_n = D \ \varepsilon_{nf}$$

E_0 is the tangent modulus of elasticity measured on untouched material.
E_f and ε_{nf} are the values gained by $\tilde{E}$ and ε_n just before the failure occurs.

In a previous work (1), we established that the damage factor evolution
is independent of load level and number of cycles to failure, provided the
tests duration remains short enough (pure fatigue tests). Under these conditions, the damage factor gains continuously increasing values that depend
only on the cycle ratio to failure (Fig.2) :

$$D = f(\beta) = \frac{-0.0174}{\beta+0.044} - \frac{0.0174}{\beta+1.044} + 0.245\beta + 0.378 \qquad (3)$$

with $\beta = n/N_f$, n being the number of applied cycles. For fatigue under identically compressive repeated loading, the number of cycles to failure N_f is
estimated using Aas-Jakobsen linear formula :

$$S_{max} = \frac{\sigma_{max}}{f_c} = 1 - 0.0685 \ (1 - \frac{\sigma_{min}}{\sigma_{max}}) \ Log \ N_f \qquad (4)$$

σ_{max} and σ_{min} being the maximum and the minimum stress levels of the loading cycle. f_c is the compressive strength of the concrete determinated

208

from cylinder tests. Eq.4 has proved to be valid for tensile repeated loading as well (2). S_{max} refers then to f_t instead of f_c, f_t being the static tensile strength.

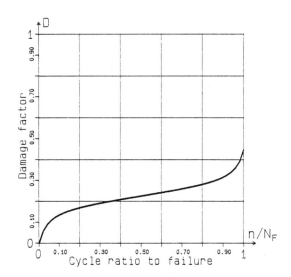

FIGURE 2 : Damage factor in terms of cycle ratio to failure (1).

For plain concrete subjected to repeated stress reversals, Tepfers (3) assumes that the tensile part and the compressive part of the loading cycle do not interfere with each other. Therefore, the number of cycles to failure can be derived from Eq.4 :

$$S_{max} = Max \left| \frac{\sigma_{max}}{f_c} ; \frac{\sigma_{min}}{f_t} \right| = 1 - 0.0685 \ Log \ N_f \qquad (5)$$

Though this approximation can be controversed, it does not affect the principles of the theory that leads to our conclusions in this work.

Under variable loading levels (the level and the range of the cyclic loading change from one cycle to the next), the number of cycles to failure can be estimated using the Miner's rule :

$$\Sigma \ \frac{n_i}{N_{fi}} = 1 \qquad (6)$$

Although its applicability to repeated sequences of identical loading cycles is not ensured, recent advances assert its validity for random cyclic loading (4). For concrete subjected to stress reversals, we assume that the cycle ratio to failure accumulates separately for the compressive part of the cycle on the one hand, and for the tensile part on the other hand. In the latter case, we introduce a tensile damage factor increasing as the cumulative cycle ratio to failure. Thus, tensile failure occurs as soon as the tensile damage factor reaches unity.

APPLICATION TO PLAIN CONCRETE PRISMS

According to the theory of damage, the "actual area" $d\tilde{S}$ of any area element, part supposed undamaged of the initial area dS, is related to the local value of the damage factor attaching to this element :

$$d\tilde{S} = (1 - D)\ dS \tag{7}$$

The actual area is supposed to support the "actual stress" $\tilde{\sigma}$ related to the nominal stress σ in such a way that $\tilde{\sigma}\ d\tilde{S} = \sigma\ dS$:

$$\tilde{\sigma} = \frac{\sigma}{1-D} \tag{8}$$

The modulus of elasticity E_0 remains unchanged over the undamaged part of the area element. Restricted to the actual area, Eq.1 turns to :

$$\tilde{\sigma} = E_0\ (\varepsilon - \varepsilon_n) \tag{9}$$

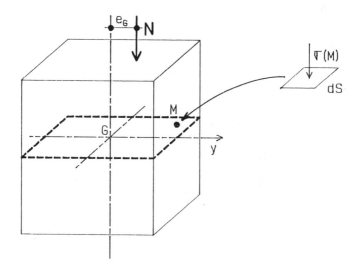

FIGURE 3 : Plain concrete prism subjected to eccentric compressive load.

We have applied our model to plain concrete prisms subjected to uni-axial eccentric compressive load (Fig.3). At any time, the damage factor gains the value $D(M)$ due to the local stress history at each location M belonging to a damaged cross-section. Taking Eq.9 into account, the following expressions are easily derived from the equations of equilibrium expressed in terms of actual stress (see (1) for more details) :

$$N + N_n = \int_{\tilde{\Omega}} \tilde{\sigma}_e\ d\tilde{S} \qquad \text{with} \quad N_n = \int_{\tilde{\Omega}} E_0\ \varepsilon_n\ d\tilde{S} \tag{10}$$

$$N \, e_G + M_{nG} = \int_{\tilde{\Omega}} y_G \, \tilde{\sigma}_e \, d\tilde{S} \qquad \text{with} \qquad M_{nG} = \int_{\tilde{\Omega}} y_G \, E_o \, \varepsilon_n \, d\tilde{S} \qquad (10)$$

$$\text{with} \qquad \tilde{\sigma}_e = E_o \, \varepsilon \qquad \text{at each location M.}$$

These equations express the equilibrium of the "actual part" of the damaged cross section subjected to a fictitious loading depending on the residual strain distribution. As assumed above, the modulus of elasticity is constant throughout the section. Finally, the obvious solution of Eq.10 leads to the stress distribution, having regard to Eq.1 and Eq.2 :

$$\sigma(M) = (1 - D(M)) \, (\tilde{\sigma}_e - E_o \, D(M) \, \varepsilon_{nf}) \qquad (11)$$

$$\text{with} \qquad \tilde{\sigma}_e(M) = \frac{N + N_n}{\tilde{\Omega}} + \frac{N \, e_{\tilde{G}} + M_{n\tilde{G}}}{\tilde{I}} \, y_{\tilde{G}}(M)$$

In Eq.10 and Eq.11, $\tilde{G}$, $\tilde{I}$ and $\tilde{\Omega}$ denote respectively the center of gravity, the moment of inertia and the area of the actual part (union of the $d\tilde{S}$ areas) of the damaged cross section.

The computation is performed after discretization of the cross section of the prism as also the test duration. For each computation step, $\tilde{G}$, $\tilde{\Omega}$ and $\tilde{I}$ are evaluated taking into account the local values previously gained by the damage factor. The extreme stress levels reached at any location M during a common loading cycle are obtained when the cyclic load N gains its extreme values in Eq.11 :

at cycle n : $\qquad \sigma_{min} \leq \sigma(M) \leq \sigma_{max} \qquad$ when $\qquad N_{min} \leq N \leq N_{max}$

As assumed before, the expended cycle ratio to failure is supposed to accumulate by simple addition. Local failure occurs when the local Miner's summ reaches unity (Eq.6). When all of the area elements have failed, the failure of the prism is completed. The validity of our model has been established in a previous paper (1).

The main characteristics of the computed prisms reported in the present paper are listed below :

cross section : $b \times h = 102mm \times 152mm$

load eccentricity : $e_G \leq h/6$
    (no tensile stress throughout the cross section)

tangent modulus of elasticity : $E_o = 1000 \, f_c$

tensile strength : $f_t = 0.085 \, f_c$

initial stresses at extreme fiber : $\sigma_{min} = 0.097 \, f_c$
    $\sigma_{max} = 0.65$ to $0.85 \, f_c$

We present the result of our computations with $\sigma_{max} = 0.75 \, f_c$ for two eccentricities (Fig.4). Due to the softening of the concrete located in the most stressed zone of the prism, the neutral axis enters the cross section and progresses in the direction to the opposite face of the prism. This causes the initially compressive stress at the less stressed edge of the section to turn progressively to tensile-compressive stress. The local tensile damage factor attaching to each concerned area element increases at each subsequent loading cycle. If we assume that any area element cracks as soon as the re-

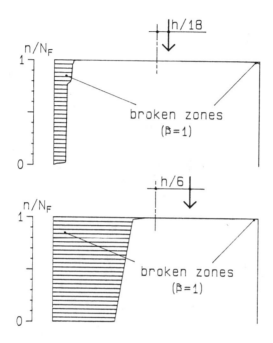

FIGURE 4 : Crack growth in terms of cycle ratio to failure.

levant tensile damage factor reaches unity, then the depth of the shaded area on our diagramms represents the tensile crack length in terms of the cycle ratio to failure. n denotes the number of applied cycles whereas N_f is the fatigue life of the prism. According to our figures, the tensile crack grows sharply during the very first stage of the loading, probably due to the S-shape of the damage curve (Fig.2). Finally, the computation predicts the breaking of the prism, both by crushing of the most stressed concrete elements and by tensile fracture on the opposite side of the prism. This unexpected mode of failure (the cross section of the prism being initially under compression) has been observed experimentally by other authors (5).

CRACK GROWTH IN R-C FLEXURAL MEMBERS

For the prediction of crack growth in reinforced concrete beams subjected to fatigue, our model must take the behaviour of the tension reinforcement into account. In a previous work (6) we derived, from a model based on fracture mechanics, S-N relationship for embedded ribbed bars located in the tension zone of flexural members :

$$N_f = K(R) \ (\sigma_{max} - \sigma_{min})^{-m} \qquad \text{with} \quad R = \frac{\sigma_{min}}{\sigma_{max}} \qquad (12)$$

the number of cycles to failure being the larger of the two, considering successively :

$$(1) : \qquad m = 5.136 \qquad K = 10^{\ 20.58 \ - \ \frac{3.89}{1.952-R}}$$

$$(2) : \qquad m = 37.59 \qquad K = 10^{\ 111.4 \ - \ \frac{24.16}{1.762-R}}$$

Moreover, we assume that the stress-strain relationship remains unchanged until the fatigue failure occurs.

Eq.11 remains usable under the following conditions :

- the bars cross-sectioned area is multiplied by $n_0 = E_s/E_0$, E_s being the modulus of elasticity of the steel;

- D(M) is equal to unity at any location M within the tensile cracked zone of the cross section;

- when attaching to reinforcing steel area, D(M) remains equal to zero (undamaging material), but Eq.11 becomes :

$$\sigma(M) = n_0 \ \tilde{\sigma}_e(M)$$

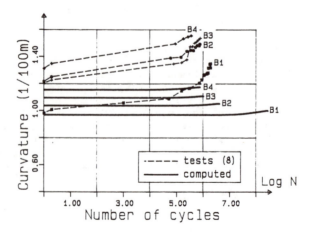

FIGURE 5 : RC-beams curvature versus number of applied cycles.

In a previous paper (7), the so built up model is used to predict the fatigue failure of RC-beams subjected to repeated loading. The calculated numbers of cycles to failure prove in fair agreement with available tests figures.

We have applied our model to the prediction of the behaviour of four identical RC-beams tested in pure bending fatigue by other authors (8). The circumstances of the tests are to be found in their paper. The computation is performed after discretization of the cross section of the beam, the number of applied cycles being increased step by step up to the failure. The maximum curvature, computed at a cracked cross-section of the beam, shows a sligth increase during the last stage before failure (Fig.5). As a matter of fact, the stiffness remains almost constant, due to the low compressive stress on the extreme concrete fiber (σ_{max} ranging from 0.493 f_C for B1 to 0.588 f_C for B4). Though they prove similar, the tests figures and the computed curves do not fit very well, partly due to lack of informations on tests circumstances (as can be seen for N = 1), partly due to creep strain and bond strength degradation that are omitted in our model.

To complete this work, we apply our model to the prediction of tensile crack progression in RC flexural members subjected to fatigue. The main data of the calculation are listed below :

- percentage of reinforcement : $A_S/bh = 0.008$ for B5
                                              0.02 for B6

- yield stress of reinforcement : $f_e = 11.45$ f_C

- modulus of elasticity : $E_S = 460$ f_e

- compressive stress at extreme fiber : $\sigma_{max} = 0.75$ f_C
      (for both B5 and B6) $\sigma_{min} = 0.10$ f_C

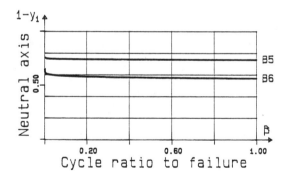

FIGURE 6 : Tensile crack depth in RC-beams subjected to bending fatigue.

The depth y_1 of the compression zone enlarges with the cycle ratio to failure (Fig.6 - h is assumed equal to unity). The neutral axis overlaps the initial crack tip, causing the uncracked tension zone to vanish early. Thereby, it may be conclude that initial tension cracks cannot lengthen in RC flexural members subjected to pure bending fatigue.

CONCLUSIONS

The behaviour of concrete subjected to fatigue is accounted for using a model based on the theory of damage. When subjected to compressive loading the progressive degradation of concrete is related to a unique damage factor. A separate tensile damage factor is introduced to account for the degradation of the concrete subjected to tensile fatigue in the vicinity of a tension crack tip. Aas-Jakobsen formula along with Miner's rule are used to predict the failure.

According to this model, the crack propagation results from a redistribution of the internal forces due to progressive softening of the concrete behaviour, along with a strength decrease of the concrete due to the tensile damage accumulated in the vicinity of the crack tip. Applied to plain concrete prisms, the so built up model predicts developing tensile damage leading to crack propagation into the cross section of the prism initially entirely subjected to compression. Under various loading circumstances, the computed lives are in good agreement with available tests figures. The observed modes of breaking square with the computed damage distribution at failure throughout the cross section.

The model is completed to account for RC members subjected to bending fatigue. For embedded tension reinforcement, a S-N relationship is used, that was previously derived from a model based on fracture mechanics. The behaviour of the bars is assumed to remain unmodified up to the failure. The so built up model accounts fairly well for the fatigue strength of normally reinforced beams. Computed curvature evolutions prove similar to available tests results. Noticed differences could be attributed to creep strain and bond degradation which are omitted in our model. Concerning tension crack propagation, the computation results suggest that initial cracks would not lengthen in RC flexural members subjected to pure bending fatigue.

REFERENCES

1. DESTREBECQ, J.F., Modèle de comportement du matériau béton sous chargement répété de compression. 4th int. Conf. on Durability of Building materials and Components, Singapore, 1987, pp.455-62.

2. TEPFERS, R., Tensile fatigue strength of plain concrete. J. Amer. Concr. Inst., 1979, 8, 919-33.

3. TEPFERS, R., Fatigue of plain concrete subjected to stress reversals. In Fatigue of concrete structures, Amer. Concr. Inst., 1982, SP75, pp. 195-215.

4. SIEMES, A.J.M., Miner's rule with respect to plain concrete variable amplitude tests. In Fatigue of concrete structures, Amer. Concr. Inst., 1982, SP75, pp. 343-72.

5. OPLE, F.S., HULBOS, C.L., Probable fatigue life of plain concrete with stress gradient. J. Amer. Inst., 1966, 1, 59-81.

6. DESTREBECQ, J.F., BRESSOLETTE, Ph., Comportement des barres crénelées pour béton armé sollicitées en fatigue. Coll. Association Universitaire de Génie Civil, Clermont-Ferrand, 1988, 4p.

215

7. DESTREBECQ, J.F., Application du concept d'endommagement à la fatigue du béton armé. CNRS/GRECO Rhéologie des Géomatériaux, Rapport Scientifique, 1987, pp. 176-82.

8. LOVEGROVE, J.M., SALAH EL DIN, Deflection and cracking of reinforced concrete under repeated loading and fatigue. In Fatigue of concrete structures, Amer. Concr. Inst., 1982, SP75, pp. 133-52.

INFLUENCE OF THE MATERIAL STRUCTURE ON THE LONG TERM PROPERTIES OF THE CONCRETE COMPOSITES IN BENDING

ANDRZEJ BURAKIEWICZ´, LESLAW HEBDA˜, JERZY PIASTA˜
´Institute of Fundamental Technological Research
Polish Academy of Sciences, 00 049 Warsaw, POLAND
˜Institute of Technology and Organization of Construction,
Technical University of Kielce, 25 314 Kielce, POLAND

ABSTRACT

The aim of the paper is to present the long term properties of Steel Fibre Reinforced Concretes (SFRC), as a function of their structure and load level. The test results reported here are based on data obtained during 280 days of loading.
The tests were performed using SFRC beams of dimensions 50 x 50 x 1000 mm, loaded in four point bending. The structure of the tested SFRC beams was changed by varying the type of aggregate and the type of steel fibre reinforcement, also the fibre content was changed. As a coarse aggregate crushed basalt and limestone of a maximum size of 8 mm have been used. Two types of fibres were applied: plain round fibres and HAREX fibers which cross section is very irregular. The fibre content by volume was for both fibre types 0.7, and 1.3%. The tested beams were loaded at the age of 90 days, applying two levels of loads: 0.7 and 1.0 times the force at failure of plain concrete beams. The deformations were recorded using the LVTD transducer for measuring the mid-span deflections of the beams. In the paper the results of tests up to 280 days of load applying are discussed.

INTRODUCTION

SFRC elements show very good properties in short time tests, but to obtain information about theirs durability it is necessary to perform different types of long time tests. Since 1984 on the Technical University of Kielce with the cooperation of Institute of Fundamental Technological Research in Warsaw a research program is in progress in which the bending creep in beams and the crack propagation in elements with initial notch are tested [1,2]. The aim of the tests, presented in this paper, is to determine the influence of the material structure - type of aggregate and fibre reinforcement - on the long time properties of Steel Fibre Reinforced Concrete (SFRC) in bending. A special interest is given to the limestone aggregate, because in the region of Kielce the production of concrete is based on this type of aggregate.

SPECIMENS AND THE EXPERIMENTAL PROCEDURE

In the tests beams of size 50 x 50 x 1000 mm were used. To obtain a good fibre distribution and to avoid the wall effect during casting of separate specimens, they have been cut out from bigger plates of 1000 x 600 x 70 mm. For the preparation of the test specimens ordinary portland cement type "35" and crushed basalt and limestone aggregate were used, both of 8 mm maximum size. The sand was taken from the sand pit. As a fibre reinforcement two types of fibres were applied: plain round fibres cut from mild steel wire (0.4 x 40 mm) of own production, and HAREX fibres of German origin. The HAREX fibres are produced by milling process and as result of the fabrication procedure their cross section is very irregular, the length being 25 mm.

The mix proportion for cement:sand:aggregate were: 1:1.07:3.56 for concrete composite with basalt aggregate (B) and 1:0.9:3.24 for concrete composite with limestone aggregate (L). Both proportions were designed for equal mix density with 400 kgs of cement for 1 m^3 and water/cement ratio was equal to 0.6. The volumetric contents of fibres were 0.0%, 0.7% and 1.3%.

The details of the specimens preparation and also the mechanical and chemical characteristic of the aggregates could be find in the paper presented on the Euromech Colloquium 204 [3].

On basis of the short time tests the level of the long time load was established. The tested beams were loaded at the age of 90 days, applying two levels of loads: 0.7 and 1.0 times the force at failure of plain concrete beams. The long time test were performed in constant environmental conditions, 60% RH and at the temperature $18^0 + 2^0$ C. The mid-span displacements of the beams were measured using a linear variable transducer (LVTD).

THE TEST RESULTS

The four points bending test were performed at the age of specimens 28 to 30 days. For each mix composition three specimens were tested and load mid-span displacement curve was recorded, fig 1. At this age also the compressive strength was tested using cubes 150 x 150 x 150 mm. For each type of mixture on the basis of average load-displacement curves the ACI toughness index was calculated. The result of short time bending and compressive tests and the calculated toughness indices are presented in table 1. They shown an evident influence of the internal structure of tested composite materials on their short time mechanical properties. The bond between matrix and basalt grains is lower then between limestone and the cement matrix [3,4]. This is represented in lower mechanical properties of similar concrete and SFRC composites. The influence of fibres in short time tests on the flexural and compressive strength is similar for all types of tested composites, but looking on the post cracking behaviour of the tested materials , it is visible that the fibres improve more the concrete composites mixtures with basalt aggregates. For the mixtures without fibres, the ACI toughnes indices are similar 1.94 and 1.98, but for the tested specimens with 1.3 % of fibres, the difference is significant 6.05 and 3.75 for basalt and limestone aggregate mixtures respectively, table 1.

TABLE 1.
Short time properties of tested concrete composites.

Type of mix	Max. bending force {N}	Flex.strgh. {MPa}	ACI tough.	Compr.strgh. {MPa}
B −0.0%	275	1.98	1.94	26.8
B −0.7%	405	2.91	5.85	33.1
B −1.3%	492	3.54	6.05	28.3
L −0.0%	410	2.95	1.98	40.1
L −0.7%	535	3.85	3.59	44.0
L −1.3%	791	5.70	3.75	42.8
LH−0.7%	571	4.11	5.32	45.0
LH−1.3%	548	3.95	5.14	29.7

B − basalt aggregate, L −limestone aggregate, H −HAREX fibres,
0.0%, 0.7%, 1.3% − fibres volumetric fraction.

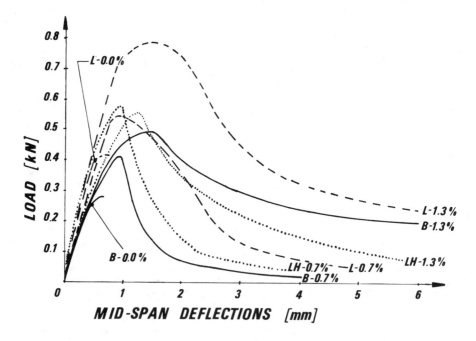

Figure 1. Load mid-span deflections curves for bending tests at the age of 28 days for all types of mixes.

The long time tests started at the age of specimens 90 days. For each type of a mixture and the load level, three specimens were loaded. The average mid-span deflection measured during the long time tests for different type of aggregate fibre reinforcement and load level are presented in the tables below. The mid-span deflections presented in this paper were recorded up to 280 days of load duration. The tests are still in progress and the results of longer load application will be reported in the future.

TABLE 2.

The average mid-span deflections for specimens with $v_f = 0.0\%$ of steel fibre reinforcement

Type of mix/load level x Pmax	Mid-span deflections after days of load duration								
	0	1	7	14	28 [mm]	60	90	150	280
B /0.7	0.121	0.131	0.189	0.230	0.264	0.408	0.463	0.540	0.640
L /0.7	0.117	0.120	0.127	0.140	0.157	0.267	0.334	0.400	0.510

B- basalt aggregate, L- limestone aggregate.

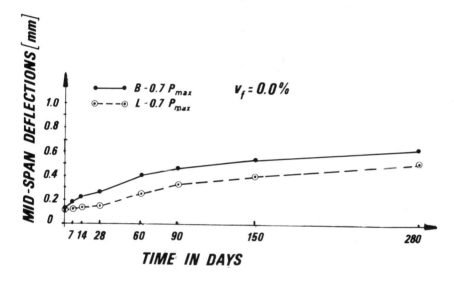

Figure 2. Mid-span deflections of plain concrete specimens under long-term bending loading.

TABLE 3.

The average mid-span deflections for specimens with $v_f = 0.7\%$ of steel fibre reinforcement

Type of mix/load level x Pmax	Mid-span deflections after days of load duration								
	0	1	7	14	28 [mm]	60	90	150	280
B/0.7	0.562	0.579	0.555	0.641	0.754	0.840	1.113	1.512	1.690
B/1.0	0.581	0.840	1.183	1.334	1.442	1.778	2.058	2,411	2.991
L/0.7	0.409	0.496	0.654	0.736	0.820	0.980	1.076	1.207	2.028
L/1.0	0.490	0.610	0.817	0.900	1.000	1.227	1.362	1.580	2.263
LH/0.7	0.397	0.507	0.697	0.793	0.910	1.110	1.225	1.380	1.677

B- basalt aggregate, L- limestone aggregate, H- HAREX fibres.

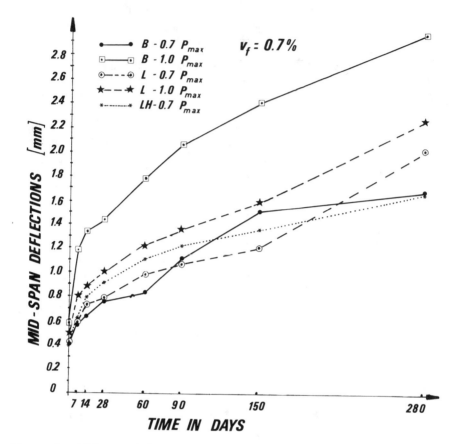

Figure 3. Mid-span deflections of SFRC specimens with 0.7% volume fraction of fibres under long-term bending loading.

TABLE 4.
The average mid-span deflections for specimens with v_f = 1.3% of steel fibre reinforcement

Type of mix/load level x Pmax	Mid-span deflections after days of load duration								
	0	1	7	14 [mm]	28	60	90	150	280
B /0.	0.463	0.530	0.740	0.883	0.993	1.323	1.635	2.027	2.549
B /1.0	0.478	0.685	0.843	0.964	1.096	1.446	1.850	2.329	3.184
L /0.7	0.207	0.250	0.320	0.357	0.387	0.443	0.488	0.538	0.693
L /1.0	0.460	0.486	0.513	0.566	0.618	0.724	0.830	0.942	1.168
LH/0.7	0.457	0.513	0.580	0.657	0.760	1.060	1.220	1.420	1.785

B- basalt aggregate, L- limestone aggregate, H- HAREX fibres.

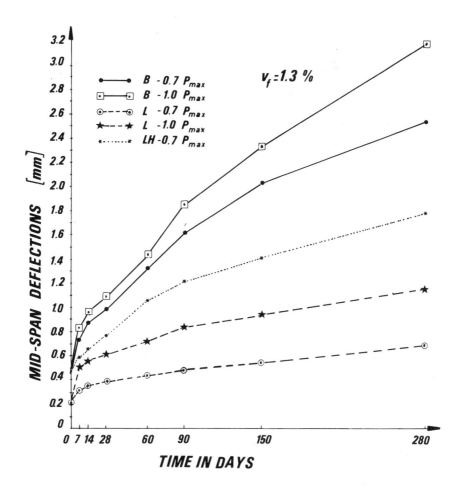

Figure 4. Mid-span deflections of SFRC specimens with 1.3% volume fraction
of fibres under long-term bending loading.

To compare in relative terms the tests results, apparent mid-span
deflection - U_a, have been used. Their have been calculated as a percentage
increase of mid-span deflections - ΔU, compared with the deflections - U_0,
at the beginning of the load duration, formula 1.

$$U_a = \frac{\Delta U}{U_0} \times 100 \%$$

(1)

The results of this calculated apparent mid-span deflections are presented
in the tables 5, 6, 7 and fig.5.

TABLE 5.

The average apparent mid-span deflections for specimens with
v_f = 0.0% of steel fibre reinforcement

Type of mix/load level x Pmax	Mid-span apparent deflections after days of load duration								
	0	1	7	14	28	60	90	150	280
					[%]				
B /0.7	0	8	56	90	118	237	283	346	428
L /0.7	0	3	9	20	34	128	186	242	336

B- basalt aggregate, L- limestone aggregate.

TABLE 6.

The average apparent mid-span deflections for specimens with
v_f = 0.7% of steel fibre reinforcement

Type of mix/load level x Pmax	Mid-span apparent deflections after days of load duration								
	0	1	7	14	28	60	90	150	280
					[%]				
B /0.7	0	3	12	14	34	49	98	151	201
B /1.0	0	45	104	130	148	206	254	315	414
L /0.7	0	21	60	80	101	140	163	195	396
L /1.0	0	25	67	84	104	151	178	223	362
LH/0.7	0	28	76	100	129	180	209	247	323

B- basalt aggregate, L- limestone aggregate, H- HAREX fibres.

TABLE 7.

The average apparent mid-span deflections for specimens with
v_f = 1.3% of steel fibre reinforcement

Type of mix/load level x Pmax	Mid-span apparent deflections after days of load duration								
	0	1	7	14	28	60	90	150	280
					[%]				
B /0.7	0	14	60	91	115	185	253	338	451
B /1.0	0	43	76	102	129	203	287	387	566
L /0.7	0	21	55	73	87	114	136	160	235
L /1.0	0	6	12	23	34	57	80	105	154
LH/0.7	0	12	27	44	66	132	167	211	291

B- basalt aggregate, L- limestone aggregate, H- HAREX fibres.

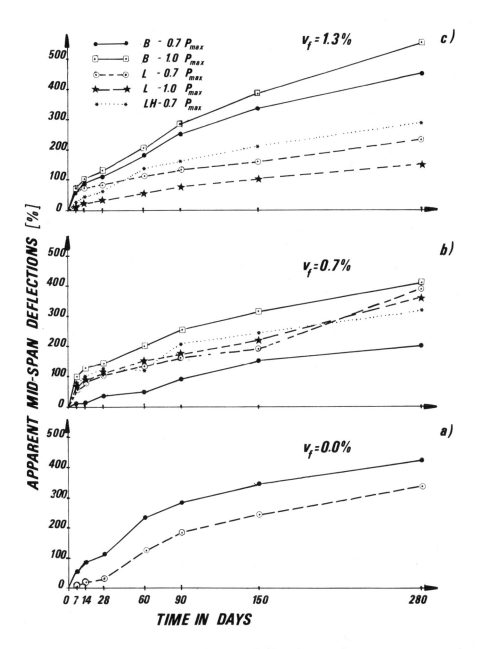

Figure 5. The apparent mid-span deflections of concrete composites specimens, tested in long time bending; B – specimens with basalt aggregate, L – specimens with limestone aggregate, LH – specimens with limestone aggregate and HAREX fibre; a) apparent mid-span displacements for plain concrete beams, b) apparent mid-span displacements for SFRC beams with 0.7 fibre volume fraction, c) apparent mid-span displacement for SFRC beams with 1.3 fibre volume fraction.

DISCUSSION OF THE EXPERIMENTAL RESULTS

After 280 days of bending load duration in all tested type of specimens there was not observed stabilization of mid-span deflections. For all types of mixture and all load levels the specimens with the basalt aggregate have shown bigger mid-span deflections compared to specimens with limestone aggregate. The lowest increase in deflection in absolute terms was for the plain concrete specimens, fig.2. For the specimens with limestone aggregate and plain steel fibre reinforcement the higher increase in deflections was for the lower fibre content, fig.3. For the specimens with basalt aggregate the higher fibre content gives higher deflections, fig.4. Specimens with HAREX fibres and limestone aggregate have shown higher deflection compared with the specimens containing plain steel fibres.

The difference in the obtained results are due to difference in the structure of the tested material, particular in the difference in the contact zones between the aggregate and cement paste and the cement paste and the fibres. The surface of the crushed basalt aggregate grains are smooth and that reduced the mechanical bond. The limestone has a different microstructure ie. the surface is porous and rough. This gives a better mechanical bond, but the most important factor is the chemical bond between cement paste and the aggregates [3,4]. It is interesting that the addition of the fibres results in softening of the concrete composites. The SFRC specimens are showing bigger displacement then the plain specimens under this same loads. But for the apparent mid-span deflections it is observed that the specimens with the limestone aggregate reinforced with steel fibres present lower increase in the apparent displacement compared with plain specimens.

CONCLUSIONS

The following conclusions can be formulated on the basis of presented test results and discussion:

- The type of the structure of the tested specimens strongly influence their long term properties.
- Specimens with basalt aggregate have bigger increases in mid-span deflection then the specimens with limestone aggregate.
- SFRC specimens show bigger increase in mid-span deflections as compared with plain concrete specimens.
- Specimens with HAREX fibres show bigger increases in mid-span deflection as compared with specimens reinforced with straight plain steel fibres.
- All tested specimens up to reported 280 days of load duration did not show stabilization of mid-span deflections.

REFERENCES

1. Burakiewicz, A., Hebda, L., and Piasta, J., The time-dependent properties of SFRC as a function of their structure. Proceedings of the 3rd Int. Symp. on Development in FRC Composites, Sheffield 13 - 17 July 1986.

2. Brandt, A.M., and Hebda, L., Example of the experimental design method

in the long term testing of SFRC. Proceedings of the 3rd Int. Symp. on Development in FRC Composites, Sheffield 13-17 July 1986.

3. Burakiewicz,A., Energy absorption of Steel Fibre Reinforced Concrete as a function of its structure, In Brittle Matrix Composites 1, ed., A.M.Brandt and I.H.Marshal, Elsevier Applied Science Publishers, London, 1987, pp. 331-40.

4. Odler, I., and Zurz, A., Structure and bond strength of cement-aggregate interfaces. Mat.Res.Soc.Proc. 114, pp.21-27, 1988 Materials Research Society.

MODELS FOR COMPRESSIVE BEHAVIOR LAWS
OF FIBER REINFORCED CONCRETE

C. DOAN, G. DEBICKI, P. HAMELIN

CERMAC - Centre d'Etudes et de Recherches sur les
Matériaux Composites
20, Avenue Albert Einstein, 69621 VILLEURBANNE CEDEX
FRANCE

ABSTRACT

We study two cases of models for satisfying description of the compressive behavior law. The first approach takes into consideration the expressions of Mazars corresponding to $\sigma_1 = E_0 (1 - D_C) \epsilon_1$ where $D_C = F(\epsilon_M)$ and $(\epsilon_M) = \epsilon_2 \sqrt{2}$. In this case we define a damage law D. The second model is a viscoélastoplastic model which can be described by the microslip model. For each model, we study the variation of the model's constants as a function of fibers direction, fibers nature, fibers lenght... We can evaluate the capacity of each model for describing anisotropic characteristics of fiber reinforced concrete and micro-mechanism of tensile cracking of matrix and slip of fibers.

INTRODUCTION

An experimental approach [1] shows us that fiber reinforced concretes present non linear behaviour law (σ-ϵ), anisotropic properties which vary as a function of fiber nature and moulding process. Micro-mechanisms of failure are detected by strain gage, ultrasonic pulse velocity and acoustic measurements. We distinguish several stages of cracking as shrinkage cracking, crack initiation, stable crack propagation and particularly slip of fibers and mechanisms which pull out fiber. The main interest of fiber reinforced concrete depends on the increase in the energy of deformation which is very interesting in impact behavior.

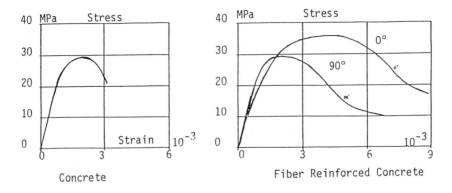

Figure 1. Behavior law of concretes

The behavior law of the concrete can be considered as composing of four phases (fig. 1). In the first stage ($0 < \sigma < 0.5\, \sigma_R$, σ_R - failure stress) the microfissures don't develop, the material remain elastic. The microfissures begin to increase in the second stage ($0.5\, \sigma_R < \sigma < 0.8\, \sigma_R$) they meet each other and create macrofissures. In the third stage ($0.8\, \sigma_R < \sigma < \sigma_R$) the macrofissures develop in the load direction. The material becomes progressively orthotropic. The last stage is denoted by a process of failure.

Consider the behavior law of fiber reinforced concretes. The first elastic stage is little shorter ($0 < \sigma < 0.4\, \sigma_R$). The second stage is modified because of presence of fiber. It depends on fiber direction. In the third stage the fiber direction is very important on creating macrofissures. For exemple, fiber at 0° prevent the appearance of macrofissures. The failure process continuos as long as phenomen pull-out of fiber. It depends on effects of friction and adhesion between matrix and fiber.

The experimental method for caracterizing brittle materials is presented by Debicki and al [6]. Our perpose is modelisation of the non-linear behavior of fiber reinforced concretes.

VISCO-ELASTIC-PLASTIC MODEL

It is common to represent the friction at contact surface by Coulomb dampers. Assuming that the bodies in contact are rigid and the friction force is proportional to the corresponding normal force .

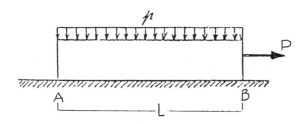

Figure 2. Coulomb damper.

According to the Coulomb law gross slip of the bar will begin if P exceeds a critical value $\mu p L$. μ is the coefficient of friction . Contrary, there will be no motion if P is less than this critical value. But experiments have shown that the Coulomb damper is acceptable as a consequence of the small normal load. In the case of high normal load, howerever, only partial slip may be expected. Meng and al [2] have proposed a new microslip model for present partial slipping on the friction interface.

Microslip model
The model is composed of an elastic bar, an attached spring and a shear layer with thickness next to zero. The bar has a constant cross-

sectional area A and Young's modulus E.

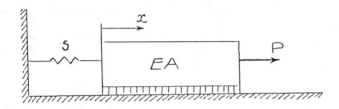

Figure 3. Microslip model.

The friction force in the shear layer, τ, per unit length is given by :

$$\tau = \begin{cases} ku \ , & |u| \leq \tau_m/k \\ \tau_m \ , & \text{otherwise} \end{cases} \qquad (1)$$

where k is the stiffness per unit length, τ_m is the limit level (usually $\tau_m = \mu p$) and u(x) is the displacement at a point of distance x from the left end of the bar.

The equilibrium equation will be :

$$EAu" - \tau = 0, \ 0 \leq x \leq L \qquad (2)$$

with the boundary conditions :

$$EAu'(0) - SU(0) = 0 \quad \text{and} \quad EAu'(L) = P \qquad (3)$$

Solution of the equilibrium equation
There are three separate cases to be considered depending on the deformed state of the shear layer : purely elastic, partial slip and total (or gross) slip :

Elastic : The system will deform elastically as long as the displacement at A remaing below le value τ_m/k. That means $\tau = ku$. The equation (2) and boundary conditions (3) under the dimensionless form become :

$$EAu"(z) - kL^2u(z) = 0 \ , \quad 0 \leq z = x/L \leq 1 \qquad (4)$$

$$\frac{EAu'(0)}{L} - Su(0) = 0 \quad \text{and} \quad \frac{EAU'(1)}{L} = P \qquad (5)$$

The solution of this equations is :

$$u(z) = \frac{P}{kL} \ [\frac{1}{f(\alpha,\beta)}] \ [\frac{\sinh \alpha z + \cosh \alpha z}{\beta \sinh \alpha + \alpha \cosh \alpha}] \qquad (6)$$

where α and β are two normalized parameters :

$$\alpha = (kL^2/EA)^{1/2} \quad , \quad \beta = SL/EA$$

$f(\alpha,\beta)$ is a dimensionless function of α and β :

$$f(\alpha,\beta) = \frac{1}{\alpha} \left[\frac{\alpha \tanh \alpha + \beta}{\beta \tanh \alpha + \alpha} \right] \tag{8}$$

Partial slip :

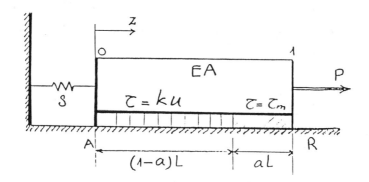

Figure 4. Model under partial slip condition

For the partial slip situation illustrated in Fig. 3 we have the equilibrium equations following :

$$EAu'' - ku = 0 \quad \text{for } 0 < x < (1-a) L$$
$$EAu'' - \tau_m = 0 \qquad \text{for } (1-a)L < x < L \tag{10}$$

Together with a boundary conditons (3) one obtains the continuity conditions across the transition point $(1-a)L$:

$$u^l = u^r \quad \text{and} \quad (u')^l = (u')^r \tag{11}$$

The solution of the system of equations (10) with (2) and (11) is

$$u(z) = \begin{cases} \dfrac{\tau_m}{k.} \dfrac{\beta \sinh \alpha z + \alpha \cosh \alpha z}{\beta \sinh \alpha(1-a) + \alpha \cosh \alpha(1-a)} \quad , \quad 0 < z < 1-a \\[3mm] \dfrac{\tau_m}{K} + \dfrac{\tau_m L^2}{EA} [z (\dfrac{z}{2} - 1) + \dfrac{1}{2} (1-a^2)] + \dfrac{PL}{EA}(z-1+a)] \quad , \end{cases} \tag{12}$$

$$1-a < z < 1$$

Gross slip :. While the entire contact surface becomes plastic we have the following equilibrium equation :

$$EAu'' - \tau_m = 0 \quad , \quad 0 < z < 1 \tag{13}$$

The solution for this equation with the boundary conditions (2) will be :

$$u(x) = \frac{L^2 \tau_m}{2EA} z^2 + (P - \tau_m L) z + \frac{P - \tau_m L}{SL} AE \qquad (14)$$

Remarks

For our purpose we express the force P as a function of the displacement U at the right end R.

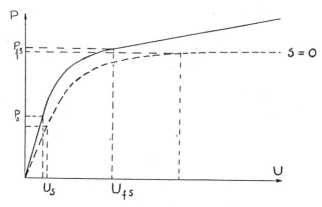

Figure 5.   Force/Displacement Relationship

We distinguish three stages of the system :

Elastic behavior :

$$P = k_1 U \quad , \quad 0 \le P \le P_S \qquad (15)$$

where

$$k_1 = \frac{F(0,\alpha,\beta) \; F_2(1,\alpha,\beta)}{F_1(1,\alpha,\beta)} \cdot \frac{P_{fs}}{U_{fs}}$$

Partial slip :

$$P = f_1(a) \; P_{fs} \qquad \qquad P_S \le P \le P_{fs}$$
$$\qquad \qquad \text{for} \qquad \qquad \qquad \qquad (16)$$
$$U = f_2(0) \; U_{fs} \qquad \qquad U_S \le U \le U_{fs}$$

where $f_1(a) = F_1(a,\alpha,\beta)/F_1(1,\alpha,\beta)$ and $f_2(a) = F_2(a,\alpha,\beta)/F_2(1,\alpha,\beta)$

Gross slip :

$$P = P_{fs} \; [1 + k_2 \; (\frac{U}{U_{fs}} - 1)] \qquad (17)$$

where $k_2 = F_2(1,\alpha,\beta)/[(1+\beta)\alpha^2 \; F_1(1,\alpha,\beta)]$

The functions F, F_1 and F_2 are defined as following :

$$F(a,\alpha,\beta) = \frac{1}{\alpha} \; [\beta + \alpha \tanh \alpha(1-a)] \; / \; [\alpha + \beta \tanh \alpha(1-a)] \qquad (18)$$

$$F_1(a,\alpha,\beta) = a + F(a,\alpha,\beta) \tag{19}$$

$$F_2(a,\alpha,\beta) = 1 + \alpha^2\ a\ [\frac{a}{2} + F(a,\alpha,\beta)] \tag{20}$$

If we know two limits : beginning of the partial slip and begin of the gross slip. We can determine two normalised parameters α and β of the system.

Application

Experimences shows that fiber reinforced concretes present non linear behavior law ($\sigma-\epsilon$). We try to describe that materials by the model of microslip. The failure of the specimen denotes the beginning of gross slip (σ_{fs}, ϵ_{fs}), while the end of the linear part denotes the beginning of partiel slip (σ_s, ϵ_s). It is reasonable to set $S = 0$. That reduces $\beta = 0$.

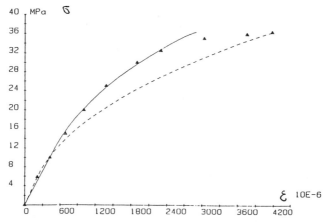

Figure 6. Fiber direction is 0°.

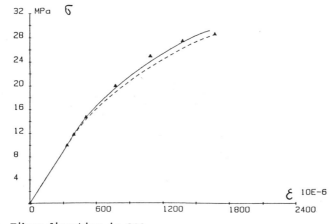

Figure 7. Fiber direction is 90°.

From the quations (16) we obtain :

$$P_s/P_{fs} = f_1(0) = \tanh\alpha \: / \: \alpha \tag{21}$$

$$U_s/U_{fs} = f_2(0) = 1/(1 + \frac{\alpha^2}{2}) \tag{22}$$

The figures 6 and 7 show the behavior laws of the fibre reinforced concretes. The orientations of fiber are 0° and 90° respectively. The points ▲ ▲ are experimental values. The lines are theorical curves determined from the model with α calculated from the equation (21) (curve ——) and α calculated from the equation (22) (dash curve ‑ ‑). The parameter α determined from the equation (21) gives the best approach.

The figure 8 shows the variation of the ratio σ_1/σ_{fs} and of the coefficient α versus the fiber direction.

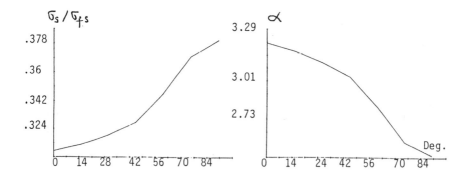

Figure 8. σ_s/σ_{fs} and α versus the fiber direction.

For count the damping of the concretes we adapt a dashpot to our model (fig. 9). An dynamic experience is necessary to identify the material (fig. 10).

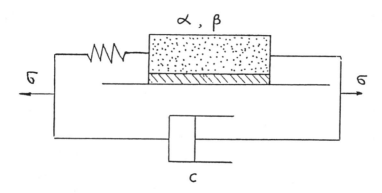

Figure 9. Viscoelasto-plastic model.

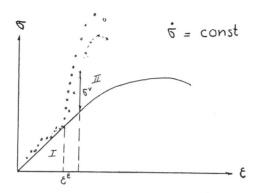

Figure 10. Dynamic experience.

DAMAGE MODEL

We have used the concept of the effectif stress proposed by Kachanov [3], [4].

Assumed, there exists an extention $\epsilon > 0$. There is two groupes of variables characterizing an irreversible thermodynamic process : observable variables (ϵ_{ij} and associated variables σ_{ij}) and internal variables).

Bazant [7] introduced the notation :

$$\dot{\sigma}_{ij} = \dot{\sigma}_{ij}{}^e - \dot{\sigma}_{ij}{}^d \tag{23}$$

with
$$\dot{\sigma}_{ij}{}^e = (\dot{\sigma}_{ij})_0 \, (1 - D)$$
$$\dot{\sigma}_{ij}{}^d = (\sigma_{ij})_0 \, \dot{D} \tag{24}$$

where $(\sigma_{ij})_0$ are the initial stress.

By assuming the law of damage development $D = f(\hat{\epsilon}).\dot{\hat{\epsilon}}$ we obtain :

$$D\,(\hat{\epsilon}_M) = \int_0^{\hat{\epsilon}_M} f(\hat{\epsilon}) \, d\,\hat{\epsilon} = F(\hat{\epsilon}_M) \tag{25}$$

where $f(\epsilon)$ is a positive continu function of ϵ defined as :

$$\hat{\epsilon}^2 = \sum_i < \epsilon_i >^2 \quad \text{for} \quad \begin{array}{l} <\epsilon_i> = \epsilon_i \text{ si } \epsilon_i > 0 \\[4pt] <\epsilon_i> = 0 \text{ si } \epsilon_i < 0 \end{array} \tag{26}$$

Three particular cases of sollicitation have been studied.

Damage by tension : the extension and the stress are in on

direction

$$\sigma_1 = E_0 \ (1 - D_T) \ \epsilon_1$$

$$D_T = F_T \ (\hat{\epsilon}_M) \tag{27}$$

$$\hat{\epsilon}_M = \epsilon_1$$

Damage by compression : The extension and the stress are perpendicular

$$\sigma_1 = E_0 \ (1 - D_C) \ \epsilon_1$$

$$D_C = F_C \ (\hat{\epsilon}_M) \tag{28}$$

$$\hat{\epsilon}_M = \epsilon_2 \ \sqrt{2}$$

Multiaxial sollicitations :

$$D = \alpha_T \ D_T + \alpha_C \ D_C \quad \text{with condition } \alpha_T + \alpha_C = 1 \tag{29}$$

($\alpha_C = 0$: single tension ; $\alpha_T = 0$: single compression) . Because of $\epsilon_i > 0$ we can write :

$$\epsilon_i = \epsilon_{Ti} + \epsilon_{ci} \tag{30}$$

where

$$\epsilon_{Ti} \text{ due to } \sigma_i > 0$$
$$\epsilon_{ci} \text{ due to } \sigma_i < 0 \tag{31}$$

Hence

$$\hat{\epsilon}^2 = \sum_i \ (< \epsilon_{Ti} + \epsilon_{ci}>)^2 \tag{33}$$

$$\alpha_T = \sum_i \ (\epsilon_{Ti}^2 + \epsilon_{Ti} \ \epsilon_{ci})/\hat{\epsilon}^2$$

$$\alpha_C = \sum_i \ (\epsilon_{ci}^2 + \epsilon_{Ti} \ \epsilon_{ci})/\hat{\epsilon}^2 \tag{34}$$

The equations (34) satisfy the condition (29).

Expression for D
One assumes a commun formula for D_T et D_C :

$$D = 1 - \epsilon_{Do} \ \frac{1-A}{\hat{\epsilon}_M} - \frac{A}{\exp \ [B(\hat{\epsilon}_M - \epsilon_{Do})]} \tag{35}$$

where ϵ_{Do} is the limit of initial damage. A and B are material caracters. For simplify the exploitation we have used the expressions of D proposed by Masars [5].

$$\sigma = Eo \ [(\epsilon_{Do}(A - 1)/\nu\sqrt{2} + A\epsilon e^{B(\nu\sqrt{2}\epsilon + \epsilon_{Do})}] \tag{36}$$

Aplication
The exploitation of the experimental curves showned in the figure 6 and 7 gives a good approach by the damage model (fig. 11) :

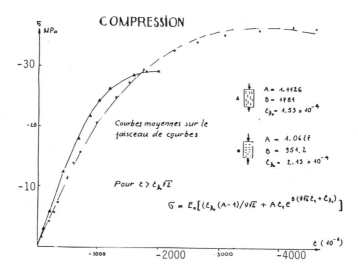

Figure 11. Damage modelling

The constants of the model are determined as :

$A = 1.1126$
$B = 1781$ for 0°
$\epsilon_{Do} = 1.53 \times 10^{-4}$

$A = 1.0677$
$B = 951.2$ for 90°
$\epsilon_{Do} = 2.13 \times 10^{-4}$

We have found out that the rapports ϵ_{Do}^{C} (90°)/ϵ_{Do}^{T} (0°) ϵ_{Do}^{C} (0°)/ϵ_{Do}^{T} (90°) are the same (1.52 and 1.59 respectively). That means damage mode under compression is an extension mode wich confirmes the hypothesis of the theory.

CONCLUSION

The visco-elasto-plastic model can describe differents phases of surface damage of materials. It is possible to establish relationships between the constants and phase limits of the model and the effects of friction, of fiber/matrix of fissures. A such presentation allows to describe the fiber role and to optimise lately the fabrication of materials.

The damage model is too adequate as the visco-elasto-plastic model for describing the non-linearity of the behavior law.

The damage model is easier to introduce in structural calculation but it does not take into consideration whith actual formulation the irreversibility and the residual strain after loading neither the influence of the deformation rate.

REFERENCE

1. Debicki G., "Contribution à l'étude du role de fibres dispersées anisotropiquement dans le mortier de ciment sur les lois de comportement, les critères de résistance et la fissuration du matériau". <u>Thèse d'état</u>, INSA de LYON - Université Claude Bernard LYON I, Octobre 1988.

2. Meng C-H., Griffin J.H, Bielak J., The influence of microslip on vibratory response. Part 1 : A new microslip model". <u>J. of Sound and Vibration</u>, 107(2), 1986, pp. 279-293.

3. Kachanov L.M., Time of the rupture process under creep conditions, <u>Izv Akad Nauk SSR Otd Tekh Nauk</u>, n°8, 1958, pp. 26-31.

4. Lemaitre J., Chaboche J.L., Aspect phénoménologique de la rupture par l'endommagement, <u>Journal de Mécanique Appliquée</u>, Vol. 2, n° 3, 1978, pp. 317-365.

5. Mazars J., Description du comportement multiaxial du béton par un modèle de matérfiau élastique endommagemen", <u>Internation Conference on Concrete under Multiaxial Conditions</u>, Toulouse - France, 22-24 mai, 1984.

6. Debicki G., Raclin J. Hamelin P., Influence of fiber orientation on the fracture mechanical properties of a fiber reinforced mortar. The Second International Symposium on Brittle Matrix Composites, 12-22 Sept., Cedzyna-Pologne, 1988.

7. Bazant Z.P., Kih S.S., Plastic Fracturing theory for concrete, <u>Journal of the Engineering Mechanics Division</u>, ASCE, Vol. 105, NEM 3, 1970, pp. 407-442.

A PROBABILISTIC MODEL FOR DAMAGE OF CONCRETE STRUCTURES

DENIS BREYSSE

Lecturer
Laboratoire de Mécanique et de Technologie (L.M.T.)
61 Avenue du Président Wilson, 94230 CACHAN , France.

Abstract

A probabilistic approach has been developed to model the   behaviour of the material. Assuming a randomly elastic brittle behaviour at the micro-level, continuous damage mechanics is then used to infer the macroscale behaviour. At each time of the process, and for any point in the material, the constitutive law may be computed and the probability of damage is known. Assuming homogeneity of the material at the macro-level, this law has been used in finite element computations and several tests have been performed. The numerical results show that an experimental characteristic such as scattering becomes a new parameter of the behaviour. The influence of heterogeneities is then discussed for two classical examples (pure compression and buckling).

INTRODUCTION

Following Kachanov (5), the development of continuous damage theory has been very useful to model the behaviour of ductile or semi-brittle materials. Lemaitre and Chaboche (7,8), Krajcinovic (6) and Mazars (9) developed it and they showed that the mean behaviour of concrete structures is correctly inferred using damage theory for the material.

The behaviour can be summarized as follows :

A controlling parameter ε is chosen (related to strains for instance), and its variation leads to evolution of damage D (the damage is seen as an internal variable, related to the density of voids or microcracks). Then damage affects the elastic characteristic E_0 of the material and the constitutive law is written :

$$\sigma = E_0 . (1-D(\varepsilon)).\varepsilon$$

where σ is the stress tensor and E_0 the initial tensor of elastic moduli.

If the constitutive law and the evolution of damage D (ε) is given, the process is deterministic and a unique response is possible for a set of data. But when one proceeds to make experiments on structures (compression or tension tests, bending of beams), then always appears a certain scatter in the experimental results. Care taken in the experimental procedure may reduce it, but this scatter always exists and its magnitude depends on many parameters : material characteristics (porosity, density and size of the heterogeneities...), geometry and size of the structure, experimental equipment, etc...

It is well known that a great part of this macro-scale variance (scatter in the load-displacements curves for instance) is related to micro-scale heterogeneities. Various approaches are valid to understand this phenomenon and to infer the macro-behaviour from the micro-level information.

Krajcinovic (6) develops a micromechanical approach and, with a deeper knowledge of how and when microcracks can first initiate, then propagate and finally coalesce, he is able to infer macro-scale conclusions.

Others (Roelfstra (11)) used a numerical approach, designing a finite element mesh which represents a certain volume of material, with its constitutive elements (cement paste and aggregates randomly distributed) and have studied the behaviour of such a structure.

But the more commonly developed approach uses statistical tools. One can say that the scatter results from a great number of parameters which are not known exactly, but only by their mean values, and which are randomly distributed within the structure. It is then sufficient to assign to any point of the material, before the computations, randomly distributed properties (Young's modulus or strength, for instance). Only a distribution function of these parameters and a random process (the Monte-Carlo method, for instance) are needed, and the following process is entirely deterministic (12).

The approach developed here is quite different. We think that the behaviour of an elementary volume of material cannot be given a priori. Of course it depends on some randomly distributed parameters (such as geometrical characteristics of the aggregates, or of the local cement paste strength...) but it also depends on the history of the loading : A neighbouring crack which is propagating or healing will seriously modify the local path. In summary, we can say that the local instantaneous behaviour may be forecast, but that its evolution cannot, it depends on too many unknown factors.

So, extending a classical approach which was first introduced by Weibull (15), we will build a kind of "probabilistic constitutive law" which will be implemented according to the following scheme :

At each step of the loading process, and for any point of the structure :

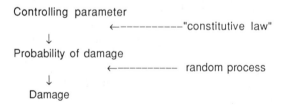

The model will be used in finite element calculations and some results will be discussed.

1. ESTABLISHMENT OF THE PROBABILISTIC MODEL

1.1 What does the probabilistic process involve ?

Let us consider an elementary volume dV located around a point M (x,y,z) within the structure.
The state at initial time t_0 of the volume dV may be described with :
- strains ε_0 (M) (or stresses σ_0 (M))
- damage parameter d_0 (M), value taken at the time $t = t_0$ by the random variable D(M) , where the scalar d_0 (M) belongs to the interval [0,1]. For practical reasons, we will assume that d_0 may be written as the ratio n/p of two integers.
Let us suppose that a change occurs in the external conditions such that it will follow a local change of strains $\Delta\varepsilon_0$ (M) (or of stresses $\Delta\sigma_0$(M)). We have to write that the damage may be modified and that its further value will depend on a given distribution function f_D (d, ε_0, $\Delta\varepsilon_0$,d_0) (probability density function) where D is the random variable, ε_0 and d_0 represent the present state (of strain and damage), and $\Delta\varepsilon_0$ is the increase in the strains :

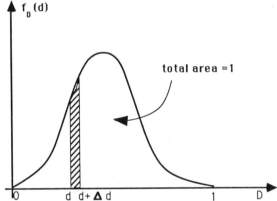

Figure 1. Probability density function f_D (d).

It is also possible to use the cumulative function F_D :

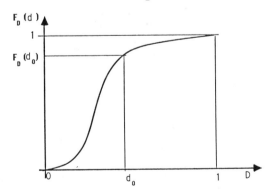

Figure 2. Cumulative density function F_D (d).

240

We then build the finite probabilistic space Ω, an ensemble of elementary events E_i and of their probabilities p (E_i) as follows :

E_0 : $\qquad\qquad 0 \le D \le d_0 \qquad\qquad$; $\quad p(E_0) = F_D(d_0)$

....

....

E_1 : $\qquad\qquad d_0 \le D \le d_0 + \Delta d_0 \qquad$; $\quad p(E_1) = F_D(d_1) - F_D(d_0)$

....

....

E_i : $\quad (i\text{-}1) \times \Delta d_0 + d_0 \le D \le d_0 + i \times \Delta d_0$; $\quad p(E_i) = F_D(d_i) - F_D(d_{i-1})$

....

....

E_n : $\quad (n\text{-}1) \times \Delta d_0 + d_0 \le D \le d_0 + n \times \Delta d_0$; $\quad p(E_n) = F_D(d_n) - F_D(d_{n-1})$

or $\quad (n\text{-}1) \times \Delta d_0 + d_0 \le D \le 1 \qquad$; $\quad p(E_n) = 1 - F_D(d_{n-1})$

with $\Delta d_0 = (1 - d_0 / n)$ where n is an integer.

In this space Ω, an "event" is defined by the fact that the random variable D takes a value between two limits.

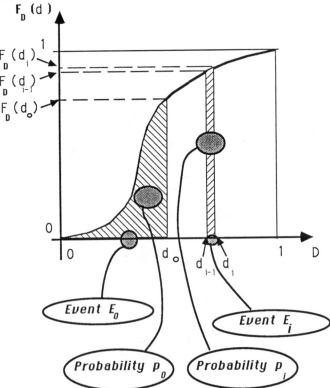

Figure 3. Definition of the random events and of their probability.

Then the process will be the following :

At each step of the process, and for any point within the structure :
- State of the elementary volume dV $\quad\quad$ d_0, ε_0 $\quad$ **KNOWN**
- Change in the external conditions $\quad\quad$ $\Delta\varepsilon_0$ $\quad$ **KNOWN**
- Probability of all the events belonging to Ω $\quad$ $E_i, p(E_i)$ **SPECIFIED**
- Realization of ONE event by a random simulation $\quad$ E_i $\quad\quad$ **TO BE DETERMINED**
- New value of damage $\quad\quad$ $d_0{}'$ $\quad$ **TO BE CALCULATED**

We can see that the probabilistic process acts "in real time", the function f_D being related to the present and local state of the material.

Of course, we have now to build this function in a correct way before proceeding to the computations.

1.2 Description of micro- and macro- levels

Following Weibull's works, Jayatilaka (4), Freudenthal (2) and Mihashi(10)have developed the weakest link concept or the parallel bar concept (bundle), assessing that the strength of the bars follows a given distribution function. This allows us to predict the scatter of the global strength and the size effect, but no information is available except the value of the failure load for the whole structure. Using the same "loose bundle" concept (cf fig. 4.), we will build our distribution law, f_D.

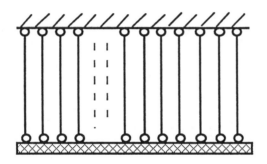

Figure 4. Loose bundle (parallel) concept.

Two different levels have to be considered :
- The microscale : It is the level of the defects (voids, micro-cracks...). Here, it is very difficult to know what is the real behaviour (complex microstructure analysis would be required) and phenomenological considerations are more practical. We will assume a very simple behaviour for the defects : elastic-brittle, with a randomly distributed probability of fracture $p_F(\varepsilon)$ under a given strain . This law $p_F(\varepsilon)$ is supposedly known (we could take the same form as Weibull :

$(p_F(\varepsilon) = 1 - \exp(-(\varepsilon/\varepsilon_0)^m)$ with two parameters, or a more complicated one).
The micro-level is described by
- the strain ε
- the stiffness k_0 of the defects (supposed to be the same for every defect)
- the stress results, equal to $k_0 \varepsilon$ if the bar is unbroken, and equal to zero if the bar is broken.

- The macroscale : It is the level of the representative volume element (at this scale, the material can be considered as homogeneous and the continuous damage theory may be used). It contains a great number of defects, which are supposed to be independent, and the macro-level behaviour appears as an integration of these local behaviours.
The macro-scale level is described by :

- the macro strain ε, identical to the micro one (an uniform state of strain in the whole bundle is assumed)
- the total number of defects N_t
- the number of fractured defects N_f
We will define the damage D as the ratio $D=N_f/N_t$ (where D belongs to [0,1]).

1.3 Constitutive law.

Let us consider the structure in an initial state (I) $[\varepsilon, D_i]$ of strains and damage (this couple of values is given and known).

When ε is changed to $\varepsilon + \Delta\varepsilon$, each bar in the structure may (or may not) break between the initial and final times. Then the number of fractured defects in the final state (F) is only probabilistic, as the value of resulting damage D_j (where D_j belongs to $(D_i,1)$).

We will write $p(D_j: D_i)$ the probability that exactly (j-i) bars, among those which are unbroken in (I), will be broken in (F).

Knowing $p_F(\varepsilon)$, we have the conditional probability for an unbroken bar in (I) to be broken in (F) :

$$P(F\backslash I) = \frac{P_F(\varepsilon + \Delta\varepsilon) - P_F(\varepsilon)}{1 - P_F(\varepsilon)}$$

Then it follows :

$$p(D_j\backslash D_i) = C_{N_t-i}^{j-i}. \, [P(F\backslash I)]^{j-i} . \, [1-P(F\backslash I)]^{N_t-j}$$

So we can write with a binomial law **B** : $p\,(D_j\backslash D_i) = B(\,N_t-i, P(F\backslash I)\,)$

If the initial state is $\varepsilon = 0$ and $D_i = 0$ (no strain, no damage), we have :
$p(1\backslash 0) \;=\; [P(F\backslash I)]^{N_t}$

which is the classical form of Weibull, which takes into account the volume of the structure (here the parameter N_t), and which allows some size effect.

We can add that, under the hypothesis $P(F\backslash I)$ greater than 0.15 and $N_t \times P(F\backslash I)$ greater than 15, the binomial law can be approximated by a Gaussian law **G** with :

$$B(n,p) \approx G(np, \sqrt{np(1-p)})$$

This Gaussian distribution may furnish an approximate formulation if N_t takes larger values. It would bemore practical for numerical simulations.

The constitutive law described above has been implemented in a finite element code (CESAR), and an iterative process has been worked up during which, at each step, the conditional probabilities of damage are computed and random events occur. In such a process, two identical sets of data given at the initial time will lead to two different results, the local paths followed by the elementary volumes being different.

Some examples will follow.

2. NUMERICAL SIMULATIONS. DAMAGE OF CONCRETE STRUCTURES

The numerical results presented in this paper were obtained using a probabilistic formulation (1) slightly different from that described above. We will not give details about it, but the principles are nearly the same and all following remarks will remain accurate.

The model is developed in these examples for the purpose of describing the behaviour of concrete (plain or reinforced concrete). The values of the local parameters were choosen to fit with the "mean macro-behaviour" expected for concrete. The overall tendencies are, in the first stage, more important than the exact values...

2.1. Pure compression of plain concrete

The test specimen for the numerical simulation is a cube of 1000 cubic centimeters. The displacements on two opposite sides are controlled, giving a "macro-state" of compression .

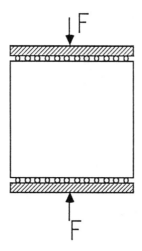

Figure 5. Simulation of plain compression.

At the initial time , the state of strains is homogeneous in the whole specimen and each elementary volume possesses the same probability of damage, but some random events occur. They create weaknesses in the specimen, and a nonhomogeneous state then results.

When we follow the evolution of the damaged zone (see fig. 6 and 7), we can see that, even if the appearance of the first defects is completely random, the growth of the damaged zone is related to the fact that defects find quickly a certain "order" and contribute to the creation of few macro-defects. This phenomenon becomes pronounced near the peak on the global load-displacement curves.

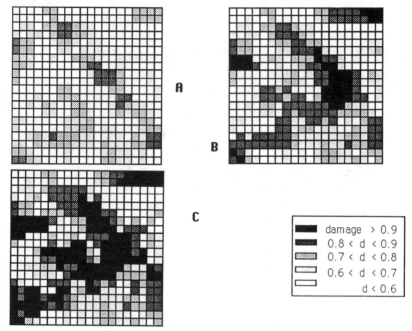

■	damage > 0.9
▨	$0.8 < d < 0.9$
▤	$0.7 < d < 0.8$
▭	$0.6 < d < 0.7$
▭	$d < 0.6$

Figure 6 . Spread of the damage zones at three steps (A,B,C) of the loading.

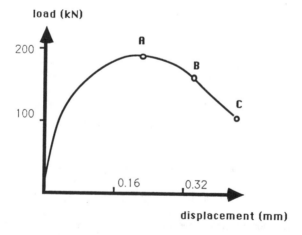

Figure 7. Load-displacement curve in compression.

The approach developed here leads to a very heterogeneous internal state in the test specimen, even under an external homogeneous set of conditions, and this description may somewhat fit the real behaviour. We find also the same kind of results as several others authors (Van Mier (14), Torrenti (13)), who wrote that the post-peak behaviour, even in pure compression, expresses a structure behaviour (few bits of the initial piece being separated by few macro-cracks), and not a material behaviour.

Finally, we can add that this model allows redistribution of stresses and "closing" of the cracks : In the classical probabilistic approach, the values (of Young's modulus for instance) are given at the beginning , with a Monte-Carlo process, and weak and strong regions are known before the loading. They will not change during the loading. But, in our approach, the probability for damage to take any value at a given stage of the loading process is computed, and a highly damaged region (which one may compare to a crack) may always keep its value of damage unchanged, while the damage increases in the neighbouring region.

2.2 Compression on a thin specimen

A thin bar of 200x10x10 cm is meshed with four nodes elements and subjected to uniform displacements at one end, the second end being fixed. The probabilistic damage model is used and the load increases, while all the elements are in the same state at the initial time.

Table 1 summarizes some results and the displacements of the bar are shown in fig 8..

Table 1 : Axial and Lateral displacements of a thin bar under compression

axial displacement (cm)	compressive load (kN)	lateral displacement (cm)
-0.05	41.4	0.0015
-0.15	92.5	0.0107
-0.25	111.6	0.0327
-0.35	124.5	0.0790
-0.45	102.1	0.1294

Two facts may be noted.

- Some defects appear during the process and locally weaken the structure. They provoke a change in the global behaviour. The maximum compressive load is about 102 kN instead 247 kN, which is expected for a cubic specimen with the same cross section and the same model. This means that the influence of the defects (weaknesses or potential weaknesses) in the structure will depend on many factors, such as the geometry of the structure, the limit conditions, the location of the defects in the structure, etc...

- In this simulated experiment, we can note that a geometrically non linear process is initiated (between the fourth and the fifth step, the lateral displacement increases by 63 % while the axial one increases by only 29 %). Although the finite element computations are performed here with a linear formulation, this fact may explain the beginning of geometrically nonlinear processes such as buckling.

It is well known that, in real structures, buckling is due to defects (in the external conditions such as the eccentricity of the compressive load, or in the structure, such as the residual stresses, etc...). Our model seems capable of describing the part due to the material defects.

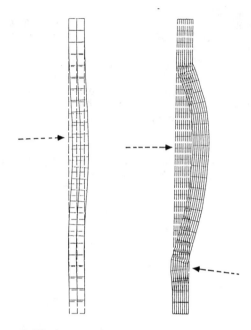

Figure 8. Displacements in a thin piece under compression.

CONCLUSION

A model based on a statistical approach has been developed. A constitutive law gives at any time of the process the probability for any elementary volume to reach a given value of damage as a function od the present state of strains and of damage. This law has been implemented in finite element calculations.

The model allows a good description of phenomena such as the breaking of a cube under boundary conditions of pure compression, or the nonlinear behaviour of a slender bar subjected to compression.

The influence of the defects is taken into account, in a manner one may expect for real materials. The next objective should be to better determine how various parameters (the intensity of the defects, their location in the structure, the geometry of the specimen or the external loads, etc...) can modify this influence.

Furthers researchs are intended to study the use of different constitutive laws (for instance, the hypothesis of independence between the local defects is quite good in tension, but micromechanical studies show that other parameters may be taken into account under different states of stress), and their use in computations.

Finally, the influences of the mesh has to be studied, since a defect is necessarily located at a Gauss point. But this is a topic of broader interest, of the primary importance for every finite element computations.

REFERENCES

(1)   Breysse D., Introduction à un modèle d'endommagement probabiliste pour les structures composites à matrice fragile (in french), 1988, Rapport Interne n. 83. L.M.T., Cachan, France.

(2)   Freudenthal A., Statistical approach to brittle fracture. In Fracture, vol 2. ed. Leibowitz, Academic Press, New York, 1968.

(3)   Horii H., Nemat-Nasser S., Brittle failure in compression : splitting, faulting and brittle-ductile transition, Phil. Trans. of the Royal Soc. of London, 1986, 319, pp. 337-374.

(4)   Jayatilaka A., Fracture of engineering brittle materials, Applied Science Publ. London , 1979.

(5)   Kachanov L.M. , Time of the rupture process under creep conditions. Izv. Akad. Nauk. S.S.R. Otd. Tekh. Nauk., 1958, 8, pp 26-31.

(6)   Krajcinovic D., Da Fonseca G.U., The continuous damage theory of brittle material, J. of Appl. Mech.1981, 48, pp. 809-815.

(7)   Lemaitre J., Chaboche J.L, Aspect phénoménologique de la rupture par endommagement (in french) , J. de Méc. Appliquée, 1978, 2, pp 317-365.

(8)   Lemaitre J., Damage modelling for prediction of plastic or creep fatigue failure in structures. 5 th conference on structural mechanics in reactor technology, S.M.I.R.T. 5, 1979.

(9)   Mazars J. , Application de la mécanique de l'endomagement au comportement non linéaire et à la rupture du béton de structure (in french), Thèse de Doctorat d'Etat, L.M.T. Paris 6, 1984.

(10) Mihashi H., A stochastic theory for fracture of concrete. In Fracture Mechanics of Concrete, Wittmann, Elsevier Science Publ., Amsterdam, 1983, pp. 301-339.

(11) Roelfstra P.E., Sadouki H., Fracture process in numerical concrete. In Fracture toughness and fracture energy of concrete , Wittmann, 1986, pp. 105-116.

(12) Rossi P., Richer S., Numerical approach of concrete cracking based on a stochastic approach, Materials and Structures, 1987, 20, pp. 334-337.

(13) Torrenti J.M., Some remarks upon concrete softening, Matériaux et Structures, 1986, 113.

(14) Van Mier, Multiaxial strain-softening of concrete, Matériaux et Structures.1986, 111.

(15) Weibull W., A statistical distribution of wide applicability, J. of Appl. Mec., 1951, 18, pp 293-297.

TENSILE STRENGTH OF BRITTLE MATERIALS
IN THE DEFORMATION FIELD WITH A GRADIENT

KRZYSZTOF M. MIANOWSKI
Professor
Institute of Building Research, Warsaw
ul. Goraszewska 1 m.1, 02-910 Warszawa Polska

ABSTRACT

The problem defined in headline is subjected to an analysis with reference to the model proposed by the author in his earlier papers (1,2). This model has been modified in order to suit the new purpose of analysis. The deformations of the brittle material are presented as a sum of deformations produced by the stress of long interaction (eccentric tension, flexion) and deformations due to the stress of short interaction type (connected to the damage). The phenomenon of the variability of the tensile strength, observed in the deformation fields with a gradient, is due to the influence of the stress of short interaction. Applying the above hipothesis, a comparative analysis of the strength of rectangular elements of varying depths in bending has been carried out.

MODEL OF THE BRITTLE MATERIAL

The structure of the model is presented in the Figure 1a. It consists of two grains with a contact layer containing bonds, between them.

The damage of the model can arise in the area of the contact layer only and consists in the breakage of bonds, while the grains representing stronger parts of material, are liable to deformation but not damage.

The system of bonds in the contact layer coincides with the ortogonal $a \times a$ mesh, which is situated in the plane of action of bending moment.

Up to the instant of breakage, the bonds follow the rules of elasticity. The modulus of elasticity for all bonds is the same but their individual ultimate elongations ε differ.

The distributions of the strength of bonds within each strip of the area (of the thickness "a") are identical and determined by a function of density of the ultimate elongations $\Psi(\varepsilon)$ defined within certain range of ultimate strains $\langle \varepsilon_{min}, \varepsilon_{max} \rangle$.

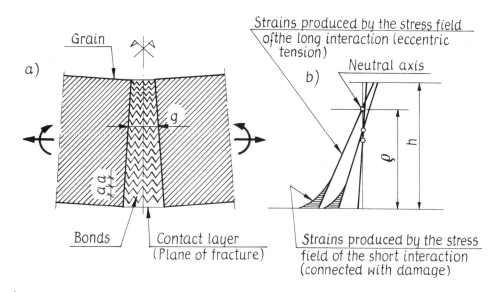

Figure 1. Model of brittle material, a) Structure, b) Components of contact layer strains

DEFORMATIONS OF CONTACT LAYER

Deformations of a contact layer are the sum of deformations produced by the stress field of the long interaction (eccentric tension or compression, bending) and by the stress field of the short interaction, due to the damage and local transmission of forces from the broken bonds to the adjacent bonds (Figure 1b).

The value of the former is calculated on the assumption of the plane sections. On the other hand, at the estimation of the local deformations due to the damage it is necessary to take into account the warping of the contact layer at the points of damage.

Our considerations concerning the local loss of flatness of the deformation area refer to the edge zone of that area.

The fact of a break at a single strip of a group of bonds of the ultimate strains is identical with the effect of the acting at that strip (after the breaking of this bonds) a force P_j (Figure 2), the value of which can be defined by a relation[j]

$$P_j = [\Psi(\varepsilon)]_{\varepsilon = \varepsilon_j} \ \Delta\varepsilon \ \varepsilon_j \ E \quad , \tag{1}$$

where the product $[\Psi(\varepsilon)]_{\varepsilon = \varepsilon_j} \Delta\varepsilon$ expresses the number of bonds of the ultimate elongations ε_j , and $\Delta\varepsilon =$constant represents the difference between the ultimate elongations of bonds belonging to two successive classes of strength.

Nominal stresses σ_d acting along the axis of the one-but-last strip of the area, produced by the force P_j can be determined approximately by a known relation

$$\sigma_d \simeq \frac{P_j}{\pi} \ [\ \frac{1}{a} + \frac{a^3}{(a^2 + g^2)^2} \ ] \quad , \tag{2}$$

where g is the thickness of the contact layer assumed to equal a .

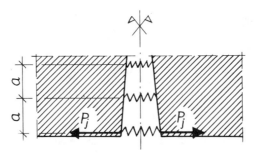

Figure 2. Explanatory sketch to the relations (1) and (2)

In the above circumstances, $\Delta\Delta\varepsilon_j$ – the component of the deformation of the contact layer, evolved on the axis of the one-but-last strip due to the breakage on the edge of the bonds class ε_j

$$\Delta\Delta\varepsilon_j = \frac{1}{E} \ \frac{[\sigma_d(\varepsilon)]_{\varepsilon = \varepsilon_j}}{\int\limits_{\varepsilon_j}^{\varepsilon_{max}} \Psi(\varepsilon)d\varepsilon} \tag{3}$$

The second term appearing on the right-hand side of the formula (3) represents the actual stresses in the one-but-last strip. The dominator of the fraction expresses the area of the strip surface with the damage duly allowed for.

COMPENSATING PROPERTIES OF THE MODEL IN THE DEFORMATION FIELDS WITH A GRADIENT CRITERION OF THE LOCAL COMPENSATION

The compensating properties of the model in the deformation fields with a gradient manifest themselves by the fact that owing to the bearing capacities of the less strained and less damaged portions of the contact layer and to the stiffness of grains, certain parts of the contact layer remain in the state of equilibrium in spite of their bearing capacity being locally exceeded.

In the analysis of this problem we assume that, at the critical stage of the process, the damage arising in two adjacent strips of the area differs by one class of bonds only.

The loss of compensating properties consists in the fact that at a certain stage of the process, immediately after the destruction of bonds class ε_r in the strip under the highest strain (the edge one), there follows without any increment in loading, the destruction of the same class of bonds in the one-but-last strip.

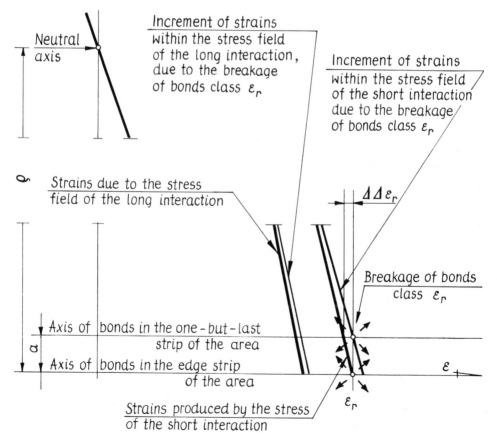

Figure 3. Explanatory sketch to the relation (4)

In Figure 3 there is presented the deformation of the contact layer at the stage of destruction of bonds class ε_r in the last strip of the area.

In connection with the above damage, the increments of deformation develop in the last and one-but-last strips of the area. Each one is equal to the sum of two components. The first one quantity is evolved by the stress field of the long interaction and it arises from the elongation and rotation of the section. It is very small, as the reaction due to the damage is distributed over the whole section. The second component evolved by the local loading produces local impression on the grain surface and therefore this component factor has to be considerably larger than the former one.

Consequently it has been assumed that the factor responsible for the loss of the local compensating properties of the area in question is the component of strains due to stress field of the short interaction (ignoring the effects of the first component).

In the conditions presented above, the relation describing the state of the local loss of the compensating properties of the area (criterion of the local compensation), takes the form

$$\Delta\Delta\varepsilon_r = \frac{a}{\rho}\,\varepsilon_r\;,\tag{4}$$

where: $\Delta\Delta\varepsilon_r$ - is a component of the strain of one-but-last area strip caused by the breakage of bonds class ε_r in the last strip of the area,

$\dfrac{a}{\rho}\,\varepsilon_r$ - is the difference between strains of the one-but -last and the last (the edge) strips at the instant when they reached the value of ε_r, established on the assumption of principle of plane sections.

After substituting, the relation (4) can be written as follows

$$\frac{\frac{1.25}{\pi}\,[\,\Psi(\varepsilon)\,]_{\varepsilon=\varepsilon_r}\,\Delta\varepsilon}{\displaystyle\int_{\varepsilon_r}^{\varepsilon_{max}}\Psi(\varepsilon)\,d\varepsilon} = \frac{a}{\rho}\;.\tag{5}$$

The value to be obtained from the formula (5) is the elongation of the most strained strip of the area. After reaching it the area loses locally the compensating properties.

In view of complicated character of the relation (5) any discussion can refer to its detailed form only. We assume therefore that the function of ultimate elongations density can be presented in the form

$$\Psi(\varepsilon) = \frac{1}{\varepsilon_{max}}\tag{6}$$

On substituting formula (6) to the equation (5) we obtain the relation

$$\frac{\varepsilon_r}{\varepsilon_{max}} = 1 - (\frac{1,25}{\pi a} \frac{\Delta\varepsilon}{\varepsilon_{max}})\rho , \tag{7}$$

hence, the shorter the distance between the edge of the area and neutral axis is, the larger is the deformation ε_r.
The expression in parantheses constitutes the constant of the model, which shall be denoted by A. The deformation field with a very small gradient does not possess compensating properties and its ultimate strains are determined by means of the dynamic strength criterion. For the assumed density function the ultimate elongation $\varepsilon = 0,367\varepsilon_{max}$ (2). Substituting $\varepsilon_r = 0,855\varepsilon_{max}$ -value of the ultimate strain for the state of the maximum compensation (see the next chater) in relation (7), we find that the limiting distance between the edge of the area and the neutral axis $\rho_* = 0,145/A$. Consequently the relation (7) can be written in the form

$$\frac{\varepsilon_r}{\varepsilon_{max}} = 1 - 0,145 \frac{\rho}{\rho_*} . \tag{8}$$

The attainment of the state of strains ε_r, when the area in its edge loses its compensating properties proves that the process has attained the semi-spontaneous phase, characterized by the fact that the damage is due in part to the action of spontaneous effects and in part to the increasing of loading (2). However, such a state is very close to the ultimate state of the bearing capacity of the model. On account of that the criterion of the local compensation can be used at comparative analysis of the strength of the model at various strain gradients.

ANALYSIS OF ELEMENTS OF A RECTANGULAR CROSS SECTION IN BENDING

The general relation, from which the tensile stresses can be obtained, has the following form

$$\sigma = E\varepsilon(1 - \frac{\varepsilon}{\varepsilon_{max}}) . \tag{9}$$

Considering that $\varepsilon = \varepsilon_k \frac{y}{\rho}$ (Figure 4) we obtain

$$\sigma(y) = E\varepsilon_k \frac{y}{\rho} (1 - \frac{y}{\rho} \frac{\varepsilon_k}{\varepsilon_{max}}) . \tag{10}$$

We can calculate the sum of stresses in the tensile zone

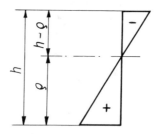

Figure 4. Deformations of the element in bending

$$\int_0^\rho \sigma(y)dy = E\varepsilon_k\rho(\frac{1}{2} - \frac{1}{3}\frac{\varepsilon_k}{\varepsilon_{max}}) \,. \qquad (11)$$

Further on, from the conditions of equilibrium of the axial forces, we can find the position of neutral axis, on assumption that the diagram of stresses in the compressive zone has the triangular shape

$$\frac{1}{2}E\varepsilon_k\frac{(h-\rho)^2}{\rho} = E\varepsilon_k\rho(\frac{1}{2} - \frac{1}{3}\frac{\varepsilon_k}{\varepsilon_{max}}) \qquad (12)$$

hence

$$\frac{2}{3}\frac{\varepsilon_k}{\varepsilon_{max}}\rho^2 - 2h\rho + h^2 = 0 \,. \qquad (13)$$

The above relation enables to define ρ when the depth of the section "h" and the deformation at the edge ε_k are given.
We can evaluate the moment as a function of the strains at the edge ε_k.

$$M(\varepsilon_k) = bE\{\int_0^\rho [\frac{y}{\rho}\varepsilon_k - (\frac{y}{\rho}\varepsilon_k)^2 \frac{1}{\varepsilon_{max}}]ydy + \frac{(h-\rho)^3}{3\rho}\varepsilon_k\} =$$

$$= bE[\frac{\rho^2}{3}\varepsilon_k - \frac{\rho^2}{4}\frac{\varepsilon_k^2}{\varepsilon_{max}} + \frac{(h-\rho)^3}{3\rho}\varepsilon_k] \,. \qquad (14)$$

We can find now the edge strains at which the moment attains its maximum

$$\frac{\partial M(\varepsilon_k)}{\partial\varepsilon_k} = \frac{\rho^2}{3} - \frac{\rho^2}{4}\frac{2\varepsilon_k}{\varepsilon_{max}} + \frac{(h-\rho)^3}{3\rho} = 0 \,, \qquad (15)$$

hence

$$\frac{\varepsilon_k}{\varepsilon_{max}} = \frac{2}{3} \left[1 + \frac{(h-\rho)^3}{\rho^3} \right] . \tag{16}$$

Solving the equations (13) and (16) we find that $\rho=0,604h$ and $\varepsilon_k= 0,855 \; \varepsilon_{max}$. These parameters corresponding to the maximum possible bearing capacity of a rectangular section have been evaluated without taking into account the criterion of the local compensation. The solution therefore holds good only, if the said criterion is fulfilled.

Substituting now the least ultimate strains $(\varepsilon_k=0,367\varepsilon_{max})$ into the relation (8) we find that ρ_k, the maximum depth of the tensile zone that enables to obtain such edge strains, equals $\rho_k=4,36\rho_*$. In what follows we shall carry out the calculations of the bearing capacity for rectangular cross-sections whose tensile zone varies within limits $\rho_* \ll \rho \ll 4,36\rho_*$.

Having assumed ρ/ρ_* we can locate the position of the neutral axis from the equation (13), which after substituting in the formula (8) will take the following form

$$\left(\frac{h}{\rho}\right)^2 - 2 \; \frac{h}{\rho} + \frac{2}{3} \left(1 - 0,145 \; \frac{\rho}{\rho_*} \right) = 0 . \tag{17}$$

Then we evaluate the ratios ρ/h and h/ρ_*. Having these values as well as ε_r - the ultimate edge strain (obtained from criterion of compensation) we can calculate the value of M_u - the fracture moment, from the relation (14).

The results of the above analysis are shown in the table. At the same time, Figure 6 shows the graphs representing the ultimate states of stresses and strength of material as a function of the depth of section (to be understood as a coefficient of filling up of the block of stresses at the ultimate bearing capacity $R = M_u/bh^2 E\varepsilon_{max}$).

The obtained results of analysis, regarding the rectangular cross-section, can be interpreted as follows.

With the depth of section h increasing from $1,66\rho_*$ to $4,92\rho_*$ the value of the ultimate edge strain decreases from $0,855\varepsilon_{max}$ to $0,367\varepsilon_{max}$, while the strength (the ratio of coefficients of the filling of the block of stresses at the ultimate state) decreases from 1.48 to 1. Also, position of neutral axis moves down (from 0,604 h to 0,535 h).

TABLE 1
Results of the analysis of strength for rectangular cross-section

ρ/ρ_*	1	2	3	4	4,36
ρ/h	0,604	0,579	0,559	0,541	0,535
$\varepsilon_r/\varepsilon_{max}$	0,855	0,709	0,564	0,418	0,357
$R = M_u/bh^2 E\varepsilon_{max}$	0,0716	0,0675	0,0628	0,0535	0,048
h/h_{min}	1	2,08	3,24	4,46	4,92
R/R_{min}	1,48	1,40	1,30	1,11	1

By reducing the depth of a section below its lower ultimate depth (h = $1,66\rho_*$) we shall not achieve the growth in strength of material above its upper ultimate value (1,48). Similarly, by increasing the depth of a section above its upper ultimate depth ($4,92\rho_*$) we shall not obtain a drop in strength below its lower ultimate value (1).

FINAL CONCLUSIONS

The results presented above are in accordance with the observations made so far in the course of tests carried out on brittle materials. It can be therefore concluded that the suggested model of the brittle material and the hypothesis concerning the mechanism of destruction, put forward rightly explain the essence of the phenomenon of the growth in strength of brittle materials in the strain field with a gradient.

The suggested method can be also applied to the analysis of the effects of the shape of the cross-section on the strength of the elements in bending and to the analysis of strength in other structures made from brittle materials.

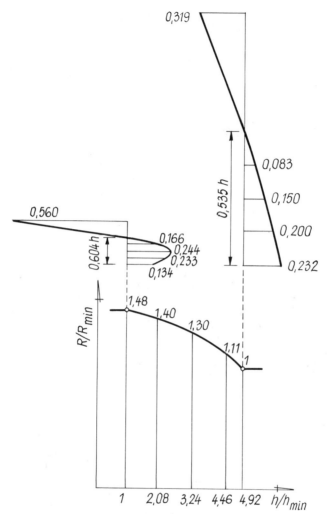

Figure 5. Results of analysis on elements of rectangular cross-section in bending: a) Ultimate states of stresses as a function of the sectional depth, b) Strength as a function of the sectional depth

REFERENCES

1. Mianowski, K.M., Dynamic Aspects in Fracture Mechanisms. Proc. of the Euromech 204 Colloquium "Structure and Crack Propagation in Brittle Matrix Composite Materials", Brandt A.M. Marshall I.H. eds., Elsevier Applied Science Publishers, London-New York, 1986, 81-91
2. Mianowski, K.M., Effect of Dynamic Factor in Mechanics of Fracture in Brittle Materials, Archive of Civil Engineering, Warszawa, 1987, 331-357 (in Polish)

AUTOGENOUS HEALING OF PLAIN CONCRETE

ANDRZEJ MOCZKO
Institute of Building Science
Technical University of Wrocław
Plac Grunwaldzki 11, 50-370 Wrocław, Poland

ABSTRACT

This paper reports a study of the nature of the healing action and presents the results of research aimed at determining the possibility of applying the acoustic emission to the evaluation of the ability of plain concrete to self healing. In particular, it was found that the measurement of acoustic emission during initial and repeated loading permits to determine the extend of this phenomenon. Among others it is also shown that the ability of plain concrete to autogenous healing is closely related to the magnitude of initial loading and the time between loading and reloading.

INTRODUCTION

The fact that plain concrete has the inherent ability to heal cracks has been known since a long time (1,2,3). However, this phenomenon is so far unexplored. The predominating opinion is that this effect is caused by formation of crystals of calcium carbonate and calcium hydroxide in the cracks (3,4). Magnesium carbonate, magnesium hydroxide and alkali carbonates are present only in trace amounts or not at all. Healing process occurs far more rapidly under water than in air (4,5).

The hypothesis of the mechanism of healing is that the free calcium oxide in the cement and the calcium hydroxide liberated by the hydration of the tricalcium silicate of the cement, is carbonated by the carbon dioxide in the surrounding air and water. The carbon dioxide reacts with a solution of calcium hydroxide on the surfaces of the cracks. As the concentration of calcium hydroxide is reduced at the surfaces, more of it migrates from the interior of the material. Likewise, as the concentration of the soluble carbonates is reduced in the cracks, more diffuses in from the water phase. Calcium carbonate crystals precipitate and grow out from the surfaces of the cracks. The rate of diffusion of the carbonates is much greater into the cracks than into the solid, relatively nonporous cement paste. Therefore, the calcium carbonate crystals accumulate in the cracks

and on the exposed surfaces of the specimens. The strength developed may be caused by the mechanical interpenetration of the
crystals and the surface of the paste, by the interlacing of
cristals, by their twinning, and by the establishment of polar
and Van der Waal forces between the crystals and the surfaces
of the paste and aggregate. There seems to be a direct correlation between the percentage of the surface area of the crack
faces covered with crystals and the ensuing healing strength.
    It is an evidence that getting to know the nature of the
healing action, most of all depends on the development of the
available methods of investigation, out of which the acoustic
emission method seems very promising. This paper presents the
research oriented at an evaluation of the extend of the autogenous healing in plain concrete.

EXPERIMENTAL DETAILS

The basic research has been carried out on 90-days old concrete
of 20 MPa strength, under quasi-axial compression. The following
procedure was adopted.
    Concrete specimens were preliminary subjected to selected
load   levels corresponding to three different stages of concrete destruction. Then after unloading the specimens were reloaded to the previously obtained levels of stress. In order to
determine autogenous healing effect the periods between loading
and reloading were different. The following values for the pauses in loading were assumed:
- T_1 = 0 hours (immediate reloading)
- T_2 = 2 hours
- T_3 = 24 hours
- T_4 = 48 hours
- T_5 = 168 hours.

The following values for primary load were assumed:
- P_1 = 50 kN - level of initiation of cracks
- P_2 = 150 kN - level of stable crack propagation
- P_3 = 200 kN - level of unstable crack propagation.

All examinations were carried out on cube specimens of the dimensions 10 x 10 x 10 cm. The specimens were compressed in an
INSTRON universal testing machine. To measure the AE signals
was used the acoustic emission equipment set, made by the Acoustic Emission Technology Corporation. The ring down counting
method in which a signal is counted each time it exceeds a pre-
-set threshold was used. During the experiments the following
parameters of acoustic emission were recorded:
- total AE counts
- root mean square (RMS) as a measure of the energy of emitted
  signals.
These acoustic emission parameters were recorded using two-chan-
nel graphic recorder made by Riken Denshi from Japan. The
friction between the specimen and the surface of the loading
beam was eliminated by polishing the specimens surface and

smearing it with smear STP.

The load was applied perpendicularly to the seting direction. All tests were carried out at 20°C temperature and relative humidity of 55%. During the break in loading the specimens were stored in climatic chamber with constant temperature of 18°C and relative humidity about 95%.

EXPERIMENTAL RESULTS

The results obtained have shown that it is possible to determine the autogeneous healing effect by means of the acoustic emission method. For instance, figures 1 and 2 show the results obtained for the primary load 150 kN and 200 kN, respectively. In both of these figures the total AE counts are plotted against the load for different pauses in loading.

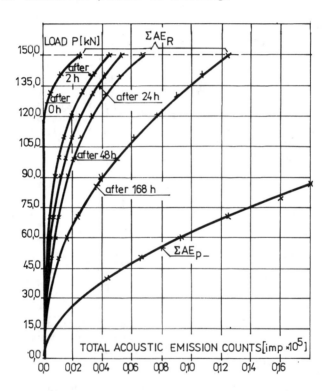

Figure 1. Influence of the duration of pause between loading and reloading for the primary load 150 kN.

It is an evidence of the important influence of healing effect. It is visible, that as the duration of pause between primary and secondary loadings increases the acoustic emission effects appear relatively sooner and the total number of counts increases.

In order to determine the range of this phenomenon the

coefficient "K" was postulated as an evaluating criterion which was defined as the degree of realization of the Kaiser´s effect (6).

$$K = (1 - \frac{\sum AE_R}{\sum AE_P}) \times 100\% \qquad (1)$$

where:

$\sum AE_R$ - total AE counts recorded during re-loading

$\sum AE_P$ - total AE counts recorded during primary loading

K - degree of the realization of the Kaiser´s effect

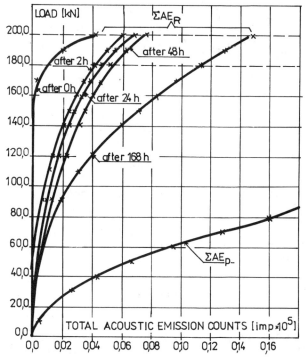

Figure 2. Influence of the duration of pause between loading and reloading for the primary load 200 kN.

Basing on the research and assuming the above evaluating criterion the evaluation of the self-regeneration capability of concrete was carried out. The obtained values of the coefficient "K" for the various values of the primary load and differentiated periods between loading and reloading are shown in Fig.3.

It may be observed that when the primary load is rather small and when the period between loading and reloading increases then there is a distinct tendency to a considerable autogenous healing of the concrete structure. For higher values of primary loading the capability of this phenomenon significantly decreases.

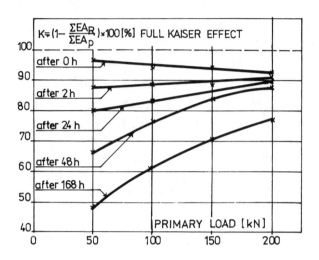

Figure 3. Degree of realization of the Kaiser´s effect.

CONCLUSIONS

- Acoustic emission permits to determine the extend of autogenous healing of plain concrete.
- The capability of the healing action of concrete structure
increases both with the decrease of the primary load level
and with the increase of the period between primary loading
and reloading.

ACKNOWLEDGEMENTS

The research reported here was carried out with the co-operation
between the Institute of Fundamental Technological Research,
Polish Academy of Sciences and the Institute of Building Science
Technical University of Wrocław.

REFERENCES

1. Glikey, H.J., The autogenous healing of concretes and mortars. Proceedings, ASTM, vol. 26, Part II, 1926, p. 470 and
vol. 29, Part II, 1929, p. 593.

2. Sorokev, V.J. and Desov, A.J., Autogenous healing of concrete. Zement, vol. 25, 30, 1936, p. 505.

3. Lauer, Kenneth R. and Slate, Floyd O., Autogenous healing
of cement paste. ACJ Journal, Proceedings, vol. 27, 10, 1956,
pp. 1083-1098.

4. Report of ACJ Committee 224, Causes, evaluation and repair of cracks in concrete structures. ACJ Journal, 5, 1984, pp. 211-230.

5. Zamorowski, W., The phenomenon of self-regeneration of concrete. The International Journal of Cement Compositier and Lightweight Concrete, vol. 7, 2, 1985, pp. 199-201.

6. Hoła, J. and Moczko, A., The analysis of destruction process of selected structures of concrete using ultrasonic pulse velocity and acoustic emission (in Polish). Ph.D. Thesis, Institute of Building Science, Technical University of Wrocław, Wrocław, December 1984.

THE INFLUENCE OF PRESENCE OF STRUCTURE DEFECTS ON THE DIAMOND-CONTAINING COMPOSITE FRACTURE TOUGHNESS

Maistrenko A.
Institute for Superhard Materials,
the Academy of Sciences of the UkrSSR,
Kiev, USSR

Kromp K., Müller M.
Max-Planck-Institut für Metallforschung,
FRG

Unlike structural steels and alloys wherein the energy dissipation occurs at the tip of a running crack when a metal is
being plastically deformed, the mechanism of the plastic deformation in its classic understanding is absent in brittle
materials (e.g. ceramics, polycrystalline SHM and rocks), but
the energy dissipation takes place here as a result of microcracking in the vicinity to the crack tip (Fig.I). This
region is named "dissipative" (I,2).

The diamond-containing composite material developed at
the Institute for Superhard Materials is used to fit drilling
bits. This material has a dual-phase structure with cracked
hard particles in which the energy dissipation in the process
of microcracking occurs prior to its deformation. Therefore,
we have a composite of dissipative structure wherein microcracks influence its fracture mechanism deterioratively.

Consider the diamond-containing composite material (DCM)
as a dissipative medium using statements of Pompe and
Krecher (2).

It was shown in (3) that the DCM also contains microcracks which are distributed within the composite volume and

that they have been generated in diamond grains in the process of hot pressing. Consequently, we have on hand an initial damage in the material occurring in its dissipative region. Thus, the effective elastic energy release rate G_c^* can be given by the following superposition

$$G_{Ic}^* = 2\gamma_p - \frac{dU_d}{da} + \frac{dU_s}{da}, \qquad (I)$$

where $\gamma_p = \gamma_m \left\{ I - [I - (\frac{h}{2d})^2] \frac{\rho}{(E^*/E_0)^{I/2}} \right\} \left\{ I + (\gamma_d - \gamma_m) C_d / \gamma_m \right\}$

γ_d, γ_m are the effective specific works of the diamond and matrix fracture, respectively; γ_p is the effective fracture energy in the "process zone"; h is the width of the "process zone" (Fig.I); d is the given grain size; ρ is the density of microcracks in the dissipative region; E^*, E_o are the effective elastic moduli for the damaged and the initial structures of the composite respectively; C_d is the diamond concentration by volume; U_d is the energy of dissipation in the dissipative region; U_s is the elastic energy introduced into the dissipative region by the residual stresses.

The values of the energy dissipated in the composite structure in the process of diamond grains cracking can be defined as:

$$U_d = \frac{G_{Ic}^* (I-C_d) \varepsilon^*}{3 (I - \varepsilon^*)} \qquad (2)$$

$(\varepsilon^* = \frac{\pi^3}{240} (5 - 4\nu^*)(I + \nu^*) kNd^3$ is the parameter indicating the level of the structure damage), and the value of the elastic energy introduced into the dissipative region by the residual stresses can be found by

$$U_s = \frac{5}{6} \frac{G_{Ic}^* F_o^2}{\sigma_o^2} \frac{C_d(I - C_d^2)(\Delta \alpha \Delta T)^2}{2 - \varepsilon^* + C_d \varepsilon^*}, \qquad (3)$$

where σ_o is the fracture stress of the matrix, $\Delta\alpha = \alpha_m - \alpha_d$; ΔT is the temperature gradient in the process of the composite sintering.

By inserting the Eqns.(2) and (3) into the Eqn.(I) and carrying out the operations necessary we obtain the equation

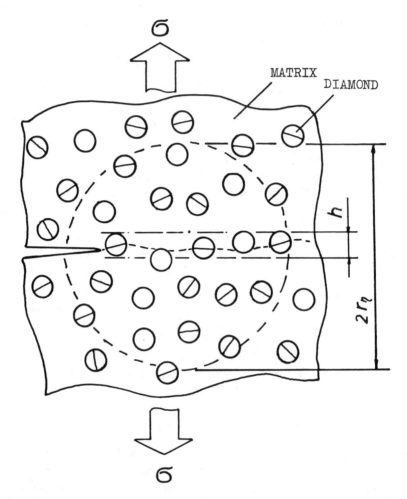

Figure I. The schematic view of the model for the crack development mechanism in a dual-phase composite allowing for the initial damage level of its structure.

in its final form:

$$\frac{G_{Ic}^{\ast}}{2\gamma_m} = \frac{\gamma_p}{\gamma_m}/(I + \frac{(I - C_d)\varepsilon^{\ast}}{3(I - \varepsilon^{\ast})} + \ldots$$

$$\ldots + \frac{5}{6}\frac{E_o^2}{G_o^2}\frac{C_d(I - C_d^2)(\Delta\alpha\Delta T)^2\varepsilon^{\ast}}{2 - \varepsilon^{\ast} + C_d\varepsilon^{\ast}}). \qquad (4)$$

By using the known Irwin-Orowan ratio between the elastic energy release and the stress intensity factors the Eqn.(4) can be solved with respect to the values of the effective fracture toughness for the composite:

$$\frac{K_{Ic}^{\ast}}{K_{Ic}^m} = \left\{ \frac{G_{Ic}^{\ast}}{2\gamma_m}\frac{E^{\ast}(I - \nu_m^2)}{E_m(I - \nu^{\ast 2})} \right\}^{I/2} \qquad (5)$$

(K_{Ic}^m is the fracture toughness for the matrix).

The computer-assisted numeric simulation has shown that the increase in the diamond concentration in the DCM entails an increase in the tensile residual stresses in the matrix (3) and the enhancement of the structure damage caused by the increased number of diamond grains being cracked. The comparison between the theoretical curves and the experimental $K_{Ic}^{\ast}$ values shows a good conformity what bears witness to the validity of the factors chosen influencing the DCM damage mechanism.

When equalizing ΔT to 0 in the Eqn.(4) we assume the residual stresses to be absent in the structure. The corresponding theoretical diagram is given in Fig.2 whence it follows that by lowering the residual stresses we positively influence the resistance of the composite to the fracture, the effect growing higher with the content of diamond in the composite, and it can amount to 26%. However, if in the Eqn. (4) $\varepsilon^{\ast}=0$, i.e. all the diamond grains would be undamaged after the DCM sintering and, subsequently, the whole structure would be free of the damage, then in this case the

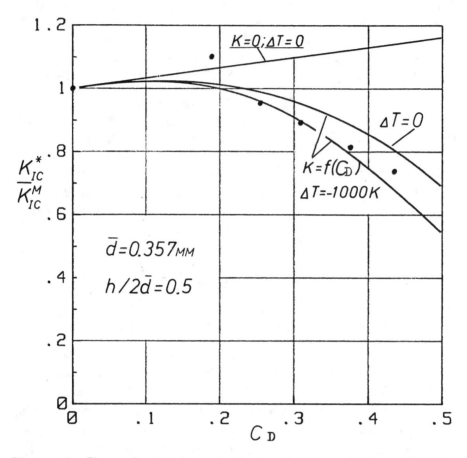

Figure 2. The calculation of the variation of the effective
fracture toughness values in a real composite and also in a
composite which is free of the residual stresses and the
initial damage in its structure.

theoretical results shown in Fig.2 indicate at the increase
of the effective fracture toughness in the DCM, this increase
amounting to 86% as compared to the corresponding real
conditions.

ACKNOWLEDGEMENT

Authors express their gratitude to professor V.Gerold for his
attention and support when carrying out this study.

REFERENCES

I. Claussen, N. and Rühle, M., Design of transformation toughened ceramics. In Advances in Ceramics, ed., N.Claussen, Amer. Ceram. Soc. Inc., Columbus, 1986, I2, pp. I37-63.

2. Pompe, W. and Kreher, W., Theoretical approach to energy-dissipative mechanisms in zirconia and other ceramics. In Advances in Ceramics, ed., N.Claussen, Amer. Ceram. Soc. Inc., Columbus, 1984, I2, pp.283-92.

3. Maistrenko, A.L., The strength and fracture of composite diamond-bearing tool materials. In Advances in Fracture Research, ed., S.Valluri, Perg. Press, India, 1984, 4, pp.302I-28.

SINTERING OF A SILICON NITRIDE MATRIX COMPOSITE REINFORCED WITH SILICON CARBIDE WHISKERS

TOSHIYA KINOSHITA, MASANORI UEKI AND HIROSHI KUBO
Materials Research Lab.-I/R&D Laboratories-I,
Nippon Steel Corporation
1618 Ida, Nakahara-ku, Kawasaki 211, JAPAN

ABSTRACT

Slurry processing methods for whisker-reinforced ceramics are investigated which are applicable to pressureless sintering. The major procedures involve adjustment of slurry pH followed by spray drying. The relative density, sintering shrinkage, fracture toughness and flexural strength of the resultant composites were compared to those of materials prepared by a conventional ball-milling process.

INTRODUCTION

Since the beginning of the 1980's, research and development of whisker reinforced ceramics[1-10] has been carried out with the aim of increasing the inherent toughness. In most cases, the methods used to produce these ceramics, were exclusively pressurised sintering such as hot pressing[1-7] and hot isostatic pressing(HIP)[8], due to their poor sinterability. Such pressurised sintering processes do not allow formation of complex shapes and result in high costs for machining parts from the pressed bodies. Although a few trials[9] for pressureless sintering have been carried out, densification was only possible for composites in which a maximum of 10 vol% whiskers was added. Much higher volume fraction loading of the whiskers is expected to be required for significant toughening in the composites, however, sufficient densification of such composites has not yet been achieved.

In the present study, a pressureless sintering method is tried for a 20 wt% SiC whisker-reinforced Si_3N_4 ceramic with special attention to the powder processing such as pH adjustments to the slurry and use of spray drying. Sintering shrinkage, mechanical properties and microstructure of the sintered composite are investigated with special emphasis on their improvements by the present processing.

EXPERIMENTAL PROCEDURES

Raw Materials

Table 1 indicates purity and grain size of the starting Si_3N_4 powder as well as those for sintering aids, and the purity, diameter and length of SiC whisker are also shown in the table. The compositions of the matrix and the composite materials are shown in Table 2. The procedures to prepare separate slurries of both matrix powders and SiC whiskers are shown in Fig. 1. The matrix powder was mixed by ball-milling in acetone as a medium for 24 h. The SiC whiskers were also crushed to adjust the aspect ratio by ball-milling in acetone for 48 h. A SEM micrograph of the crushed whiskers is shown in Fig. 2. The average aspect ratio of the whiskers was in range of 6~8. After ball-milling followed by drying, the matrix powder and whiskers were separately mixed with distilled water to form slurries.

Adjustment of Slurry pH and Spray Drying

Figure 3 illustrates the procedure to blend SiC whiskers with the matrix powders. To form a homogeneous dispersion of whiskers in the green body, careful treatment of the powder-whisker mixture is necessary to avoid agglomeration of whiskers in the slurry. Immediate drying of the slurry is also required to retain the homogeneous whisker dispersion. After ball-milling, the surface state of both the matrix powders and the whiskers were modified to that favorable for good dispersion by adjustment of the slurry pH. Nitric

TABLE 1
Starting material specifications

	Material	Purity(%)	Diameter(μm)
Matrix powder	Si_3N_4(Ube SN-E-10)	>97	0.1~0.3
	AlN (Anzon)	>98	3.2
	Al_2O_3(Sumitomo Chem. AKP-50)	>99.995	0.3
	Y_2O_3 (Japan Yttrium)	>99.99	0.4
Whisker	SiC (Tokai Carbon)	>99	diameter:0.1~1.0 length:5~50μm

TABLE 2
Composition of the matrix and composite(wt%)

Material	Si_3N_4	AlN	Al_2O_3	Y_2O_3	SiC whisker
Matrix	77.5	6.25	6.25	10	——
Composite	62	5	5	8	20

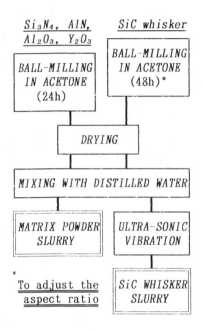

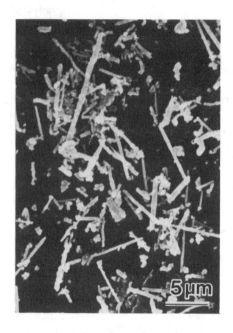

Figure 1. Preparation of matrix powder and SiC whisker slurries.

Figure 2. SEM micrograph of SiC
whiskers after crushing by ball-
milling.

acid and diethylamine were added to the slurry to increase acidity and alka-
linity respectively. Then, the slurry was ultrasonically vibrated, followed
by spray drying. The resultant spray-dried blended raw materials were granular in shape with a diameter of $5 \sim 30 \mu m$, as shown in Fig. 4, in which the
SiC whiskers were surrounded by the matrix powders.

Sintering Conditions and Evaluation of Sintered Body
The spray dried blended material shown in Fig. 4, was uni-axially pressed in
metal dies with a pressure of 100 MPa. The green body obtained was then
isostatically pressed at room temperature(CIP) with a pressure of 700 MPa
followed by sintering at 1770°C for 5 h in a N_2 gas atmosphere.
    The dimensions of the sintered plate (40×40×6mm) were measured to
detect the shrinkage by sintering. Namely the dimensions were compared with
those measured in the green state after CIP, with special emphasis on the
difference in the shrinkage for directions perpendicular and parallel to the
pressing axis prior to CIP. Test pieces were then cut from the plate for
mechanical testing. Using these specimens, density was measured by the
Archimedes method, then fracture toughness(K_{IC}) and three-point flexural
strength were measured at room temperature by means of the single edge pre-
cracked beam(SEPB) method[11] and by procedures specified in JIS-R1601, respectively. The microstructures of the sintered bodies were observed to
correlate with mechanical properties, using optical and scanning electron
microscopy.

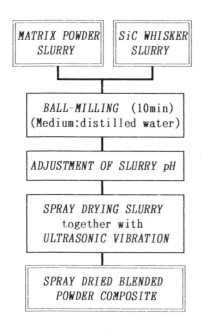

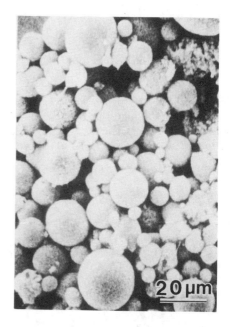

Figure 3. Procedure to blend the
matrix powder and SiC whisker.

Figure 4. SEM micrograph of spray
dried blended powder composite.

RESULTS AND DISCUSSION

Sintering Shrinkage

Table 3 compares the ratios of sintering shrinkage normal and parallel to
the uni-axial pressing direction for sintered bodies without whiskers (i.e.
the same composition as the composite matrix) and composites prepared using
pH adjustment and ball-milling. Although the ratio for the composite prepared by pH adjustment was almost equivalent to that of the sintered body
with matrix composition, the ratio for the composite was slightly smaller
than that for the matrix composition. Namely, anisotropy of sintering

TABLE 3
Comparison of the anisotropy in the sintering shrinkage

Material & Processing	Matrix	Composite(20 wt% SiC whisker)	
		pH adjustment	Ball-milling
Ratio of sintering shrinkage normal and parallel to the uni-axial pressing axis	0.92	0.88	0.64

shrinkage was observed in the composites.

Figure 5 shows optical micrographs of polished surfaces perpendicular to the uni-axial pressing direction of sintered bodies prepared by (a) pH adjustment and (b) ball-milling. The whiskers in (a) oriented randomly, while those in (b) are oriented with the length of the whisker perpendicular to the pressing-direction. Hence, sintering shrinkage in the direction perpendicular to the pressing-axis was small in the sintered composite prepared by ball-milling and the composite exhibited significant shrinkage anisotropy. On the other hand, the spray dried blend with less-oriented whiskers exhibited a smaller degree of shrinkage anisotropy during sintering.

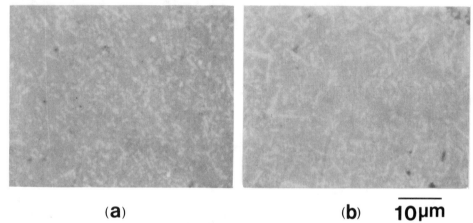

(a) (b) 10μm

Figure 5. Optical micrograph of a surface perpendicular to the pressing axis for a 20 wt% SiC whisker-reinforced Si_3N_4 composite prepared by (a) adjustment of the slurry pH and (b) conventional ball-milling.

Densification through Adjustment of Slurry pH

Figure 6 shows the effect of pH adjustment for the slurry on the relative density of sintered composites. The calculated, theoretical density (TD) of the composite was 3.31 g/cm^3 for the present composition. For comparison, the relative densities of the composite prepared by ball-milling treatment are also shown in the figure as the dashed zone with a density around 85% of theoretical. In these cases, distilled water or acetone were used as media and the slurry was dried by rotary evaporator.

By adjusting the slurry pH, the sintered body with a relative density higher than 91%TD was obtained in the alkali region pH = 7～11.5. In this range the higher the pH value, the higher the sintered density. Sufficient densification (≧95%TD) was achieved with a slurry higher than 9. In the acid region (pH = 2～7), densification was also achieved by lowering the pH value, with densities higher than 95%TD being attained at pH≦ 3. Consequently, the relation between relative density and the pH exhibited a "V" shape curve having a minimum at pH = 7.

Densification by pH adjustment in the alkali region must be attained by improved dispersibility of the whiskers in the matrix in the green state.

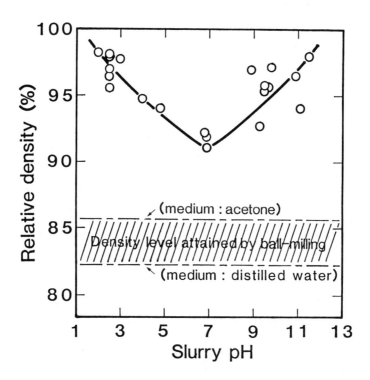

Figure 6. Effect of adjustments to slurry pH on the relative density (%TD) of sinterd 20 wt% SiC whisker-reinforced Si_3N_4 composites.

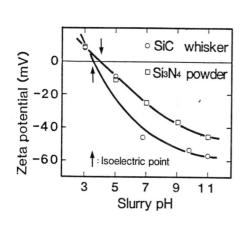

Figure 7. Zeta potential vs. slurry pH for the SiC whiskers and ·Si_3N_4 powders.

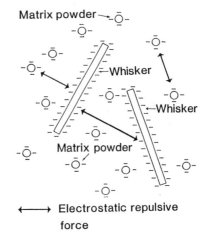

Figure 8. A schematic of a model for dispersion by electrostatic repulsive forces which may operate under alkaline conditions.

Figure 7 shows the relationship between zeta potential and pH value for slurries of SiC whisker and Si_3N_4 powder measured by a micro-electrophoresis method. Both curves exhibited an isoelectric point where the zeta potential equals zero near pH = 4. For pH values in the alkali region, the zeta potentials for whiskers and powders have high absolute values with the same sign. It can be expected that electrostatic repulsive forces are exerted between whiskers and powders as schematically shown in Fig. 8. Through this mechanism, homogeneous dispersion may be achieved.

As seen in Fig. 7, in the region of pH < 3, both zeta potentials for whisker and powder have very low absolute values. Therefore, strong repulsive forces should not be expected. Figure 9 shows the relation between pH value and viscosity of SiC whisker and Si_3N_4 powder slurries measured by a B-type viscometer. The viscosity of such a slurry must be one of the measures of dispersibility of the whisker or powder in the slurry. When whiskers become agglomerated in the slurry, high viscosities must be recorded. The pH value-viscosity relations for both slurries exhibited an "inverted V" shape having a maximum value at pH = 7 for SiC whisker and at pH = 5 for the Si_3N_4 powder, respectively. The high dispersibility observed in the acid region must be due to some surface modification occurring on the SiC whiskers and/or Si_3N_4 powders. The detailed mechanism for densification by pH adjustments in this region is however still unknown.

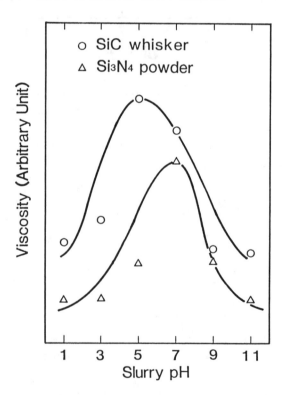

Figure 9. Effect of the slurry pH adjustments on the viscosity of water-based slurries of SiC whisker and Si_3N_4 powder.

Mechanical Properties and Microstructure

In Fig. 10, the fracture toughness, K_{IC} and the flexural strength are correlated with the relative density of the sintered bodies prepared by various
slurry treatments such as ball-milling, and alkalinity/acidity adjustments.
Both fracture toughness and flexural strength were apparently improved by
densification attained by pH adjustments. Either higher or lower pH values
for alkali and acid slurries respectively, have resulted in a significant
increase in densification. A 1.5 to 2.0 times improvement in fracture toughness and flexural strength, compared to those for materials treated by
ball-milling, was also observed. The higher flexural strength in the sintered bodies through pH adjustment is due mainly to the increased density
and partly due to the decrease in size of whisker cluster from which fractures originate. As mentioned previously, the maximum size of spray dried

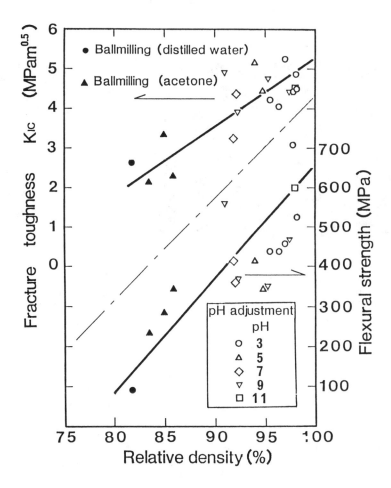

Figure 10. Correlation of the fracture toughness, K_{IC} and the flexural
strength with the relative density(%TD) of the sintered 20 wt%SiC
whisker-reinforced Si_3N_4 composites prepared by various processings.

composite powder is about $30\,\mu\mathrm{m}$ in diameter. Even if there are whisker clusters in the spray dried material their size is still smaller than those observed in sintered bodies prepared by a different method[10]. These are usually between $30{\sim}200\,\mu\mathrm{m}$ in diameter.

The fracture toughness was measured by introducing a pre-crack into the specimen(SEPB method)[11], followed by three-point bend testing at room temperature. Figure 11 shows a typical SEM image of a fracture surface after such a test for a specimen which contains 20 wt% SiC whisker (relative density: 98.0%TD). In the SEM observations, it was hard to find evidence for whisker pull-out. On the other hand, Fig. 12 shows a SEM micrograph for the same material, in which a crack was introduced by Vickers indentation on the polished surface. Deflection of the crack is clearly observed in the micrograph. From these results, the operative toughening mechanism in the composite at room temperature is considered to be crack deflection.

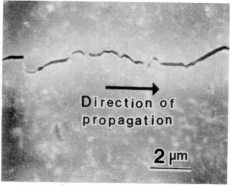

Figure 11. SEM fractograph after K_{IC} measurement by the SEPB method for a SiC whisker-Si_3N_4 composite (density: 98%TD).

Figure 12. SEM micrograph of a crack originating from a Vickers indentation on the polished surface of the composite.

CONCLUSIONS

Pressureless sintering has been successfully applied to produce 20 wt% SiC whiskers-reinforced Si_3N_4 ceramics. Adjustments to the slurry pH in both alkali and acid environments were effective for densification of the sintered composite. In the alkali region, sufficient densification($\geqq$95%TD) was achieved at a slurry pH higher than 9 due to the improved dispersibility of the whiskers in the matrix. Such densification was also attained in the acid region pH$\leqq$ 3.

Although anisotropy of sintering shrinkage was observed in the composites, the degree was smaller for composites prepared by pH adjustments compared to those prepared by ball-milling.

Fracture toughness and the flexural strength of the composites were 1.5 to 2.0 times higher than those of materials prepared by conventional ball-milling processes due to containing less-oriented whiskers. The operative toughening mechanism in the composite at room temperature was considered to be crack deflection.

REFERENCES

1. Ueno, K. and Toibana, Y., Mechanical properties of silicon nitride cera-mic composite reinforced with silicon carbide whisker, J. Ceram. Soc. Japan, 1983, 91, pp. 491-97.

2. Becher, P.F. and Wei, G.C., Toughening behavior in SiC-whisker-reinforced Alumina, J. Am. Ceram. Soc., 1984, 67, pp. C267-69.

3. Shalek, P.D., Petrovic, J.J., Hurley, G.F. and Gac, F.D., Hot-pressed SiC whisker/Si$_3$N$_4$ matrix composites, Am. Ceram. Soc. Bull., 1986, 65, pp. 351-56.

4. Buljan, S.T., Baldoni, I.G. and Huckabee, M.L., Si$_3$N$_4$-SiC composites, Am. Ceram. Soc. Bull., 1987, 66, pp. 347-52.

5. Homeny, J., Vaughn, W.L. and Ferber, M.K., Processing and mechanical properties of SiC-whisker-Al$_2$O$_3$-matrix composites, Am. Ceram. Soc. Bull., 1987, 66, pp. 333-38.

6. Samanta, S.C. and Musikant, S., SiC whisker-reinforced ceramic matrix composites, Ceram. Eng. Sci. Proc., 1985, 6, pp. 663-72.

7. Claussen, N. and Petzow, G., Whisker-reinforced oxide ceramics, J. de Physique, 1986, 47, pp. C1-693-702.

8. Lundberg, R., Kahlman, L., Pompe, R., Carlsson, R. and Warren, R., SiC-whisker-reinforced Si$_3$N$_4$ composites, Am. Ceram. Soc. Bull., 1987, 66, pp. 330-33.

9. Tiegs, T.N. and Becher, P.F., Sintered Al$_2$O$_3$-SiC-whisker composites, Am. Ceram. Soc. Bull., 1987, 66, pp. 339-42.

10. Hayami, R., Toibana, Y. and Ueno, K., Fractography of inorganic fiber-reinforced ceramics, J. Mater. Sci. Soc., Japan, 1983, 19, pp. 335-40.

11. Nose, T. and Fujii, T., Evaluation of fracture toughness for ceramic materials by a single-edge-precracked beam method, J. Am. Ceram. Soc., 1988, 71, pp. 328-33.

YOUNG'S AND SHEAR MODULI OF LAMINATED CARBON/CARBON COMPOSITES BY A RESONANT BEAM METHOD

Alexander Wanner   and   Karl Kromp
Max-Planck-Institut für Metallforschung
Institut für Werkstoffwissenschaften
Seestraße 92, D-7000 Stuttgart 1, Federal Republic of Germany

ABSTRACT

The room temperature in-plane shear, interlaminar shear, and axial Young's moduli of beams cut from plates of two different C/C laminates have been determined by a resonant beam method. The technique and evaluation method are described in detail. The results are discussed and compared to data from static tests.

INTRODUCTION

Carbon/carbon composites are artificial carbons reinforced with carbon fibres. The properties of these composites are very different from those of monolithic carbons and graphites. Due to the interaction of matrix and fibres the typical brittleness of these materials is reduced and the strength is raised, which makes them suitable for a variety of new applications [1,2].

Sophisticated carbon and graphite fibres are extremely anisotropic [3], which generally results in a complex mechanical behaviour of the composite on both microscopic and macroscopic scale. The geometrical arrangement of the fibres can have an even greater influence on the properties of the composite than other key points, such as the choice of the component materials or the control of the pyrolyzation process.

In this work the macroscopic elastic behaviour of two commercial C/C laminates with different stacking sequences is investigated with an elaborate resonant beam method with which both the axial Young's modulus and the axial transverse shear moduli of the specimen can be determined. The results are discussed and compared to static data from earlier work on the same materials.

MATERIALS AND METHOD

The commercial C/C-laminates[*] investigated in this work are composed of orthogonally woven fabrics of yarns of PAN based high modulus fibres. Each yarn consists of approximately 3000 fibres with an average diameter of 7μm. The preimpregnated fabrics are stuck together in a bidirectional (0°/90°, Material A) or multidirectional (0°/±45°/90°, Material B) stacking order. After several pyrolyzation/reimpregnation cycles, during which the binder is transformed to a roughly isotropic carbon matrix, the composites contain microcracks and residual stresses caused by anisotropic fibre contraction during the cooling process. The volume fraction of the fibres is about 45%, the densities are 1.37 g/cm^3 (Material A) and 1.33 g/cm^3 (Material B). For our investigations, plates of 36 fabric layers were available. This relatively high number of layers enables us to regard the composites quasi-homogeneous. Due to the symmetry of their stacking orders the effective elastic properties of both the bidirectional and the multidirectional composite should be transversal isotropic, which means the principal material axes are orthogonal and

$$E_{11} = E_{22} \neq E_{33}$$

$$G_{13} = G_{23} \neq G_{12}$$

E_{11}, E_{22}: in-plane Young's moduli
E_{33} : Young's modulus perpendicular to laminate plane
G_{13}, G_{23}: interlaminar shear moduli
G_{12} : in-plane shear modulus
(see also Fig.1)

Of course, the values of these elastic constants are expected to be different for the bidirectional and the multidirectional material.

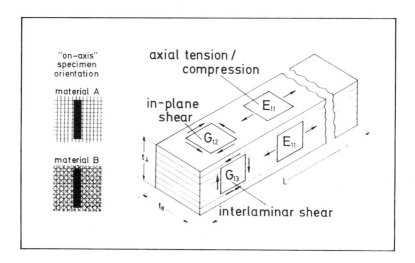

Figure 1. The principal material axes coincide with the geometrical axes of the specimens. The moduli E_{11} and G_{12} or G_{13} can be determined from flexural vibrations parallel (for G_{12}) or perpendicular (for G_{13}) to the laminate plane.

[*] The materials were kindly supplied by SIGRI GmbH, Meitingen, FRG.

The axial Young's modulus can easily be obtained from a quasi-static tensile test, whereas the determination of the interlaminar shear modulus G_{13} and the in-plane shear modulus G_{12} is more difficult. In principle both E_{11} and the shear muduli can be obtained from quasi-static bending experiments, but this method is quite time-consuming, since the span to depth ratio has to be varied over a wide range in order to separate the bending and shearing parts of the deflections. Moreover, the method is complicated by frictional effects and stress concentrations at the loading points, so that very often only rough estimations of the shear constants are obtained. This method and a variety of other quasi-static methods of investigating interlaminar and in-plane shear, e.g. square plate twist and torsion of prismatic bars, are reviewed in [4]; it is demonstrated that each method has specific experimental difficulties and that the results obtained from different tests can scatter considerably. Obviously there is no method that has been proved to be remarkably superior.

In this work values of E_{11}, G_{12}, and G_{13} are obtained from resonant beam experiments. It will be demonstrated that most useful results can be obtained with a single, simply shaped specimen and with comparatively little effort.

Theory of the Resonant Beam Method

Flexural vibrations: As a first approximation, the only elastic constant affecting the flexural vibrations of a quasi-homogeneous beam is the axial Young's modulus. The Bernoulli-Euler (BE) theory of flexural motions gives the relationship between the Young's modulus and the resonant flexural frequencies of the beam. Provided compressive and tensile moduli are equal and independent of the strain rate, the relationship for prismatic bars of rectangular cross-section and "free-free" end conditions is as follows [7]:

$$E_{BE} = \frac{48 \, \rho \, \pi^2 \, L^4}{t^2 \, m_n^4} \, (f_n^{flex})^2 \qquad (1)$$

ρ	: material density
L	: specimen length
t	: specimen thickness parallel to flexure plane
f_n^{flex}	: flex. frequency of mode n
m_n	: constant (see Tab.1)

The BE theory is based on the assumption of pure bending. Consequently, it does not account for transverse shear deformation effects. These are known to increase with the ratio of the axial Young's modulus (E_{11}) to the axial-transverse shear modulus (G_{12} or G_{13}), the thickness to length ratio of the beam ($t_\parallel/L$ or $t_\perp/L$), and the vibration mode number.

In laminated, i.e. highly anisotropic beams, in which the ratio of the axial Young's modulus to the interlaminar shear modulus can be high (>20), transverse shear deformation can become significant at low modes of vibration even for beams of t/L<0.01 [5].

The more exact Timoshenko differential equations of flexural beam motion correct for these shear effects and moreover for the effects of rotatory inertia. Huang [6] has solved these equations for beams of various shapes and end conditions. The solution relevant for our problem is:

$$E_{Tim} = E_{BE} \cdot (1 + (t/L)^2 (a_n + b_n (E/G) - c_n (t/L)^2 (E/G))) \qquad (2)$$

a_n, b_n, c_n are constants listed in table 1 (n = 1..6). Equation (2) can be written as:

$$E_{Tim} = A_n + B_n \cdot E/G \qquad (2a)$$

substituting $A_n = E_{BE} \cdot (1 + a_n (t/L)^2)$ $\qquad (2b)$
$B_n = E_{BE} \cdot (b_n (t/L)^2 - c_n (t/L)^4)$ $\qquad (2c)$

TABLE 1
Constants in Equations 1 and 2

Mode No. n	m_n	a_n	b_n	c_n
1	4.7300	4.12	1.23	4.20
2	7.8532	9.08	4.60	32.0
3	10.996	15.6	9.89	122
4	14.137	23.7	17.2	333
5	17.279	33.5	26.4	746
6	20.420	45.0	37.6	1449

At least two independent linear equations with the form of Equation 2a are required to obtain E and E/G. These can be set up by exciting different modes n of vibration and/or by variation of t/L. If Equation 2 is applicable even in the case of high E to G ratios it should be possible to obtain not only the axial Young's modulus, but also the shear modulus in the respective flexure plane.

In the ranges of 0.0081 $\leq$ t/L $\leq$ 0.0121 and 1 $\leq$ n $\leq$ 5 the corresponding equations of vibrating cantilever beams ("clamped-free" end conditions) have been applied successfully to a unidirectional graphite fibre reinforced epoxy composite revealing an E_{11} to G_{12} ratio of 40 [5]. In the case of unidirectional on-axis specimens, however, G_{12} is the "longitudinal transverse shear modulus" and the terms "in-plane shear" and "interlaminar shear" are not defined.

Longitudinal Vibrations: The frequencies of longitudinal vibration of a uniform beam are a function of the Young's modulus E_{11} in the axial direction; in case of "on-axis" specimens (see Fig.1) they are almost independent of the transverse shear constants G_{12} and G_{13}. The longitudinal frequency equation is given e.g. in [7] ("free-free" end conditions):

$$E^{long} = (4 L^2 \rho (f_n^{long})^2 K) / n^2 \qquad (3)$$

K is a correction factor which depends on the shape of the specimen, the mode number, and the Poisson's ratios ν_{12} and ν_{13}. In this work the Poisson's ratios were assumed to be 0.3, leading to corrections of less than 1% for the first three modes.

Specimen Preparation and Experimental Procedure

Specimens: Of each material three prismatic specimens of rectangular cross-section were cut from plates of 9 mm nominal thickness. In order to avoid difficulties arising from torsional coupling effects connected with off-axis flexure, only on-axis specimens were prepared, i.e. the geometrical axes of each specimen coincided with the principal axes of the respective material (Fig.1). The nominal cross-section of the six specimens was 9 x 18.3 mm^2, the length varied between 200 mm and 265 mm.

Apparatus and principle of measurement: For generation and detection of sonic and ultrasonic signals in the range of 0.5-135 kHz, an ELASTO-MAT 1.024 (made by Institut Dr. Förster, Reutlingen, FRG) and self constructed piezo-ceramic transmitter/receiver systems were used. A detailed description of the measurment technique is given in e.g. [7,8].

The mechanical arrangement is shown schematically in Fig.2. The specimen is supported by two thin and light yarns of carbon fibre. One end of the specimen is connected to the sonic transmitter. At the other end, its actual motion is detected by the receiver. For good coupling, both transmitter and receiver are connected to the specimen with a thin wire. The vibrational spectrum is registered by plotting the amplitude of specimen vibration against the applied frequency. Peaks in this plot can be identified as resonant frequencies.

By varying the geometrical arrangement of transmitter, receiver and specimen the three partial spectra (i.e. longitudinal as well as flexural parallel and perpendicular to laminate plane) can be registered separately.

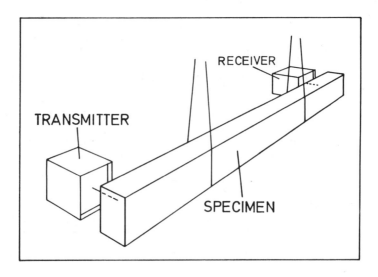

Figure 2. Arrangement for the determination of flexural resonant frequencies. The longitudinal spectrum can be registered by connecting the transmitter and receiver faces to the end surfaces of the beam.

Evaluation of the vibrational spectra

From the flexural resonant frequencies f_n coefficients A_n and B_n (Eqations 2b/c) were calculated. Thus from each set of frequencies a set of straight lines in a E vs. E/G plot was obtained according to Equation 2a (Fig.3/4). The coordinates of the intersections in these plots are graphical solutions of E and E/G.

For practical reasons only the results of one specimen per material are presented. The results obtained from the other specimens turned out to be identical within the limits given in Table 4. Exact specimen dimensions, densities and vibrational frequencies as well as the graphically determined E/G-results and the E-values calculated therefrom are given in Tables 2 and 3.

In order to check the influence of transverse shear effects on the flexural vibrations, all flexural data were also evaluated with the simple Bernoulli-Euler frequency equations (1), which, as already described above, apply only if these effects are negligible. The resulting Young's moduli, referred to as "E_{11}^{BE}", are also given in Tables 2 and 3.

Finally, the axial Young's moduli obtained from longitudinal frequencies according to equation (3) are listed in the same tables.

RESULTS AND DISCUSSION

General remarks concerning the method: Well defined cross-over points in the E-E/G-planes are obtained in evaluating the flexural spectra (Figures 3 and 4). This shows that the flexural frequency equations based on the Timoshenko theory can be applied even to extremely anisotropic and comparatively compact specimens.

The axial Young's moduli measured from flexural frequencies match quite well with those obtained from longitudinal frequencies. Consequently, it can be concluded that the E/G ratios (and thus the shear moduli) determined simultanously should also be reliable.

A summary of the final results obtained in this investigation is given in Table 4.

Comparison with static data: Very little static data are available on the same materials. Wieninger [9] obtained E_{11} =48 GPA from on-axis tensile test of material B, which is identical with our result.

Haug [10] performed a series of bending tests on material A. By variation of the span to depth ratio, he obtained 61.5 GPa < E_{11} < 67 GPa and 2.6 GPa < G_{13} < 3.7 GPa, depending on the way of evaluating the force-deflection curves measured, which turned out to be non-linear. If there is a non-linear elastic behaviour, which is very often the case with composite materials, the elastic constants measured depend on the applied strain range. The tensile and compressive strains during beam vibration are small compared to those applied e.g. in static bending tests. Consequently, the precondition that tensile and compressive moduli be equal is not satisfied to the same degree in bending tests as it is in flexural vibration.

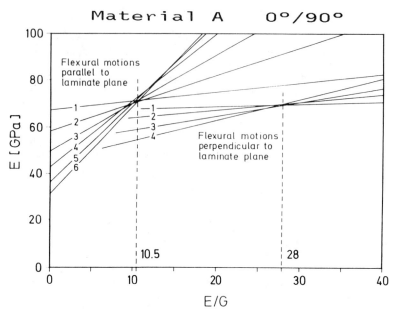

Figure 3. The straight lines in this E vs. E/G plot are calculated from the flexural resonant frequencies of a material A specimen according to Equation 2a. The numbers indicate the vibration modes. A graphical solution for E_{11} and E_{11}/G_{12} or E_{11}/G_{13} is given by the intersections of the lines.

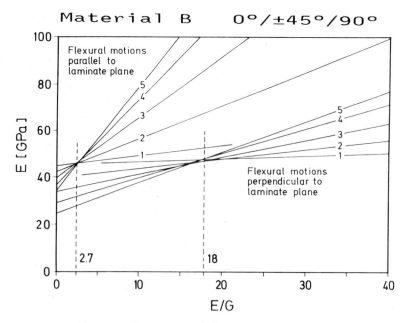

Figure 4. Same as figure 3 for a material B specimen.

TABLES 2 AND 3

Material A (0°/90°)

L: 265.1 mm $t_\parallel$: 18.35 mm $t_\perp$: 9.11 mm ρ: 1.371 g/cm^3

	MODE	FREQUENCY [kHz]	YOUNG'S MODULUS E_{11} BE* [GPa]	Tim** [GPa]	E/G value used for calculation of E_{11} (Tim)
Flexural vibr. parallel to laminate plane	1	1.860	65.9	71.2	
	2	4.720	55.8	70.7	
	3	8.415	46.2	71.2	E_{11}/G_{12} = 10.5
	4	12.68	38.4	72.8	
	5	17.09	31.2	72.1	
$t_\parallel$/L = 0.0692	6	21.64	25.7	70.7	
Flexural vibr. perpendic. to laminate plane	1	0.927	66.4	69.4	
	2	2.431	60.1	69.8	E_{11}/G_{13} = 28
	3	4.424	51.7	69.4	
$t_\perp$/L = 0.0344	4	6.748	44.1	69.8	
Longitudinal vibrations	1	13.49	70.1		
	2	26.90	69.9		
	3	40.29	69.8		

Material B (0°/±45°/90°)

L: 201.1 mm $t_\parallel$: 18.41 mm $t_\perp$: 9.76 mm ρ: 1.325 g/cm^3

	MODE	FREQUENCY [kHz]	YOUNG'S MODULUS E_{11} BE* [GPa]	Tim** [GPa]	E/G value used for calculation of E_{11} (Tim)
Flexural vibr. parallel to laminate plane	1	2.675	43.3	46.0	
	2	7.027	39.3	46.2	
	3	13.01	35.1	46.7	E_{11}/G_{12} = 2.7
	4	20.08	30.6	46.6	
$t_\parallel$/L = 0.0915	5	27.99	26.6	46.2	
Flexural vibr. perpendic. to laminate plane	1	1.441	44.7	47.5	
	2	3.687	38.5	46.7	
	3	6.642	32.5	48.0	E_{11}/G_{13} = 18
	4	10.09	27.5	48.1	
$t_\perp$/L = 0.0485	5	13.69	22.7	48.2	
Longitudinal vibrations	1	15.12	49.1		
	2	29.97	48.3		
	3	44.97	48.5		

* Axial Young's modulus E_{11} obtained from Bernoulli-Euler equations.
** Axial Young's modulus E_{11} obtained from Timoshenko equations.

This might be the main reason why the static data measured by Haug show deviations from our results. Nevertheless, the agreemend is quite good.

TABLE 4
Summary of final results

		material A 0°/90°	material B 0°/±45°/90°
axial Young's modulus	E_{11}^{long} [GPa]	70 ± 0.8	48.5 ± 0.8
	E_{11}^{flex} [GPa]	71 ± 1.7	47 ± 1.3
in-plane shear modulus G_{12} [GPa]		7.0 ± 0.4	19.3 ± 1.2
E_{11} / G_{12}		10.5 ± 0.5	2.7 ± 0.1
interlaminar shear modulus	G_{13} [GPa]	2.5 ± 0.15	2.7 ± 0.25
E_{11} / G_{13}		28 ± 1.5	18 ± 1.5

Comparison of the two materials: The distribution of fibre orientation within the laminate plane has a high influence on the axial Young's modulus E_{11} and the in-plane shear modulus G_{12}.

The Young's modulus of material A exceeds that of material B by 40%. This is not surprising, if we consider that PAN-based carbon fibres are extremely anisotropic. They have a high Young's modulus (up to 1000 GPa) in the fibre direction and a relatively low effective modulus (< 20-40 GPa) in off-axis directions deviating more than 10° from the fibre axis [3]. In material A every second fibre is parallel to the 11-direction, in material B only every fourth one.

On the other hand, the more uniform distribution of fibre orientation in the laminate plane leads to an approximately isotropic in-plane behaviour of material B. The E_{11} to G_{12} ratio of 2.7 is in the range found for truly isotropic materials, for which generally E to G ratios between 2 and 3 are obtained according to:

$$\nu = E/2G - 1 \quad <=> \quad E/G = 2 (1 + \nu) \qquad ; \qquad \nu = 0 \ldots 0.5$$

Since "diagonally" bracing ±45°-fibres do not exist in material A, the in-plane shear modulus of this composite turns out to be only 37% of that of material B.

By comparison, the influence of the fibre orientation distribution on the interlaminar shear modulus G_{13} is small; almost identical values are obtained for both materials.

CONCLUSION

The effective elastic properties of two different C/C laminates (0°/90° and 0°/±45°/90°-types) have been investigated at room temperature. The investigations were performed with a resonant beam method, which has rarely been applied on extremely anisotropic materials.

As expected, the moduli of in-plane tension/compression and in-plane shear turned out to be very different, while the interlaminar shear moduli are almost identical.

It could be shown that the resonant beam method is a good method to get a quick, reliable survey of the elastic properties of a material, even if it is as anisotropic as the C/C composites investigated here.

REFERENCES

1. Fitzer, E., The future of carbon-carbon composites. Carbon, 1987, 25, 163-190.

2. Diefendorf, R.J., Continuous carbon fiber reinforced carbon matrix composites. In Engineered Materials Handbook, Vol.1 - Composites, ASM International, Metals Park, Ohio, 1985.

3. Johnsson, W., The structure of PAN based carbon fibres and relationship to physical properties. In Handbook of Composites, Vol.1 - Strong Fibres, ed., W. Watt and B.V. Perov, North-Holland Publ. Comp., 1985.

4. Tarnopol'skii, Yu.M. and Kincis, T., Methods of static testing for composites. In Handbook of Composites, Vol.3 - Failure Mechanisms of Composites, ed., C.G. Sih and A.M. Skudra, Elsevier Publ., 1985.

5. Dudek, T.J., Young's and shear moduli of unidirectional composites by a resonant beam method. J. Compos. Mater., 1970, 4, 232.

6. Huang, T.C., The effect of roratory inertia and shear deformation on the frequency and normal mode equations of uniform bars with simple end conditions. J. Appl. Mech., 1961, 28, 579-584.

7. Spinner, S. and Tefft, W.E., A method for determining mechanical resonance frequencies and for calculating elastic moduli from these frequencies, Proc. ASTM, 1961, 61, 1221-1238.

8. Standard test method for Young's modulus, shear modulus, and Poisson's ratio of glass and glass-ceramics by resonance, Book of ASTM Standards, 1971 Annual Report, C623.

9. Wieninger, H., Institut für Festkörperphysik, Universität Wien, Austria, unpublished data, 1988.

10. Haug, T., Max-Planck-Institut für Metallforschung, Stuttgart, FRG, unpublished data, 1985.

THE INFLUENCE OF THE DISPERSED PHASE PARTICLES CRACKING ON A CHANGE OF THE ELASTIC MODULUS IN A BRITTLE COMPOSITE

Novikov, N.V., Maistrenko,A.L.,
Bogatyryova, G.P. and Simkin, E.S.
Institute for Superhard Materials,
the Academy of Sciences of the UkrSSR,
Kiev, USSR

Intensive works aimed at establishing regularities in an influence which various kinds of imperfections and structural defects exercise in new composites resulting in a change of their physico-mechanical properties are now underway in practice of developing and investigating such materials. In scientific literature well-known are the works of Vanin G.A. (I), Budiansky B. and O'Connell R.(2), Gross D.(3), Fujii T. and Zako M.(4) and others which are dedicated to the analysis of an influence which regularly or stochastically distributed discontinuities or crack-like defects exercise on a variation of physico-mechanical properties of the media in question. In these publications the most general cases and approaches have been analysed. They can be used for describing imperfections of the composite structures reinforced with fibers, bars or with dispersed particles.

Among composites produced by modern technology there exist diamond-bearing composite materials (DCM) consisting of a homogeneous cemented carbide matrix and grains of natural or synthetic diamond stochastically distributed in it (5). The deficient character of composites of this type is revealed in the form of diamond grains cracking caused by the action of thermal stresses which are due to the thermoelastic mismatch between the metal catalyst inclusions found in dia-

mond grains and the diamond lattice and also in the form of
grains fracturing under the action of contact stresses when
the composite is hot pressed.

As a composite of this type in question represents a
two-component medium containing stochastically distributed
solid particles of the second phase, the computation of the
effective elastic modulus values was carried out at the first
stage of these studies by the methods based on the elasticity
theory applied to stochastically nonhomogeneous media (6,7).
Simultaneously, it was assumed that the composition can be
considered as a perfect structure free of any defects. The
results of computing the effective elastic modulus E^{*} of a
composite related to the elastic modulus E_m of the matrix
(BK6, E_m=640 GPa) and as a function of the volumetric diamond
content are shown in Fig.I (k=0). However, the experimental
elastic modulus values for this type of DCM measured on specimens of 5x5x35 mm by the resonance method using the instrument "Zvuk-I07" and by the ultrasonic method using the instrument USM-3 are considerably lower by their absolute magnitude compared with those computed (Fig.I) and this discrepancy grows with the volumetric diamond content. It is obvious that this discrepancy can appear only because the initial composite structure imperfection mentioned above was
left out of consideration.

The number of diamond grains fractured in a composite in
the process of hot pressing was defined by the method of re-
covery followed by the fractioning of fragments on sieves. As
a result, the fraction of fractured grains in the reference
to their total number within a volume unit at a given volumetric diamond concentration in a composite was defined. Furthermore, the fraction of fractured synthetic diamond grains
of different grades (from AC20 to ACI32T) and grit sizes in
the range of 200/I60 to I000 μm was defined when heat treatment of starting powders was performed in a hydrogen furnace
without pressure wherein the parameters of temperature and
time were identical to those used in an actual process of the
composite hot pressing. As a result, the dependence of the

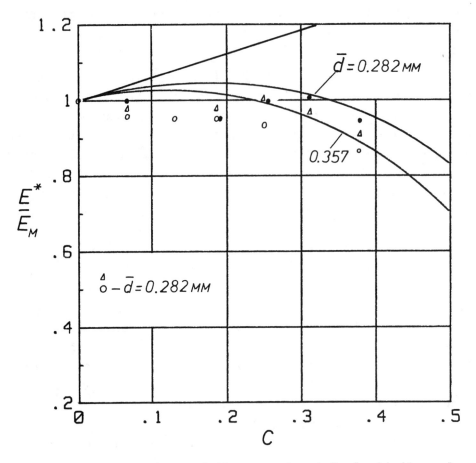

Figure I. The comparison of the experimental elasticity modulus values with the theoretical values E⁕ at different fixed
diamond integrity coefficients with various diamond
concentrations.

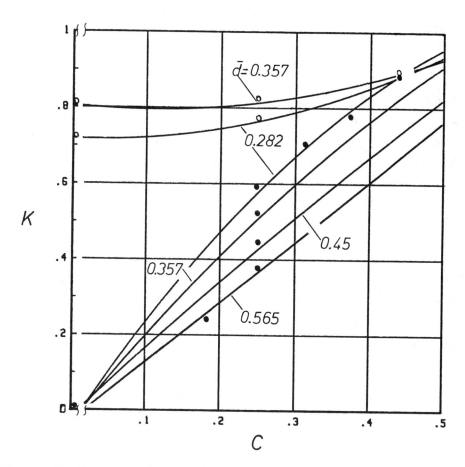

Figure 2. The dependence of the second phase particle damage
parameter on the grade of diamonds, the reduced diameter value
and the diamond concentration in the composite.

"damage degree for particles of the second phase, K" a new concept characterizing a ratio between the fraction of fractured grains and the total number of grains in the composite volume, on the volumetric diamond concentration, C_D was established (Fig.2). From the Figure it follows that at the volumetric diamond concentration of 0.25 which is mostly used in commercial DCMs a larger part of grains contain cracks propagating mainly in the ⟨III⟩ plain. They can be conditionally presented as cracks occupying the full diametrical cross-section of a grain with a reduced diameter d (8). Thus, having in hand an information about the number of grains in a volume unit as a function of a reduced diameter and a volumetric concentration and also the value K for the respective diamond grade and grit size (see Fig.2) one can readily define the number of cracks (with the reduced area of a free surface $2\pi \bar{d}^2$) in a volume unit of the composite:

$$N^* = K \cdot N \qquad (I)$$

where $N = m^{3/2}$; $m = (5 + 0.5 \cdot \bar{d}) \cdot C_D / \bar{d}$.

To take the existence of cracked grains distributed within a volume into account, let us make use of the approach proposed by Budiansky and O'Connell (2) where it is written that:

$$\frac{E^*}{E_0} = I - \pi^2 (I + \nu^*)(5 - 4\nu^*) \, \epsilon/30, \qquad (2)$$

where $\epsilon = 2 N^* A/P$, A,P are the area and the perimeter of a crack, E_0 is the reduced elasticity modulus for a composite of perfect structure, and ν^* is the reduced Poisson coefficient for DCM.

When dealing with composites whose components have elasticity moduli slightly differing from another it is reasonable to apply the known "mixture rule"

$$E_0 = E_m + (E_D - E_m)C_D. \qquad (3)$$

Otherwise, when computing the value E_0 for a corresponding reinforcement type one should resort to strict solutions described in the above-mentioned works (I,4,6,7).

In the case of interest if we consider cracked grains as spheres cut by diametrical cracks in accord with the above assumptions the value will take the following form: $\epsilon = KNd^3/8$. Thus, the eqn.(2) in its general form can be written as follows, when computing the effective elasticity modulus for a composite comprising dispersed solid grains containing cracks:

$$E^* = (I - \mathcal{E}^*)E_0, \qquad (4)$$

where $\mathcal{E}^* = \pi^3(I + \nu^*)(5 - 4\nu^*) \, K \, N \, d^3/240$ is the structure damage parameter.

The numerical solution of the approach described was realized in the form of a programme for the computer EC I035. Fig.I shows theoretical curves characterizing a change of the effective elasticity modulus of DCM containing synthetic diamonds of different grades (AC50 and AC82T of 3I5/250 grit size). The comparison of the experimental values E^* with the computed ones convincingly proves the validity of the model representation chosen here. Naturally, one may question the validity of the structure damage parameter E^* used in the computations as only the free surface area of a crack in a grain has been taken into consideration and the phenomenon of a grain sliding at the absence of adhesion is left unaccounted. An attempt to consider this component of area in a computation scheme led us to the attainment of values by far lower than those obtained in the experimental way even in those cases when the observation of the fracture surfaces of broken specimens of some DCM types unambiguously showed the absence of adhesive bonds between the diamond and the matrix. This apparent contradiction can be eliminated by analysing the distribution of residual thermal stresses due to the thermo-elastic mismatch of components in the composite. As the thermal expansion coefficient of the matrix which is $6.2 \cdot 10^{-6}$ K^{-I} is considerably higher than that of diamond which equals to $\alpha = 3.7 \cdot 10^{-6} \cdot K^{-I}$ it becomes obvious that a diamond grain is in the state of volumetric compression. The residual stresses of high level squeeze the particle so tightly that

a free area at the interface is practically absent. It is
this singularity that permitted to establish the independency
of the effective elasticity modulus of bonding state at the
phase interface in DCM though this contradicts to the known
results of the basic works dedicated to composites. Nonethe-
less, the property established in this work is not a contradiction but an exception from the general rule formulated for
composites without taking the initial stresses of a structure
into consideration.

The described computational method for estimating the
effective elasticity modulus E^{*} of DCM allowed to discover
the main reason of the elasticity modulus lowering in diamond-
bearing composites in comparison with that theoretically predicted. This allowed not only to make the computational
method more exact but also permitted to substantiate the re-
commendations directed at the elimination of the established
deficiency consisting in the necessity of applying those
technological processes of the DCM sintering which leave the
integrity of diamond in composites intact.

ACKNOWLEDGEMENTS

Authors express their gratitude to prof. N.Claussen (TUHH,
F.R.G.) for his support when carrying out this study.

REFERENCES

I. Vanin, G.A., Micromechanics of composite materials. Kiev,
   "Naukova Dumka", I985, 302 p. (in Russian).

2. Budiansky, B. and O'Connell, R., Elastic moduli of a
   cracked solid. Int. J. Solid Struct., I976, I2, 8I-97.

3. Cross, D., Spannungsintensitätsfactoren von Rissystem.
   Inge. Arch., I982, 5I, 30I-40 (in German).

4. Fujii, T. and Zako, M., The fracture mechanics for composite materials. (Translated from Japanese, ed. Burlayev,
   V.I.), Moscow, "Mir", I982, 232 p. (in Russian).

5. Novikov, N.V., Tsypin, N.V. and Maistrenko A.L., Composite
   diamond-bearing materials based on cemented carbides.
   "Sverkhtverdye materialy", I983, 2, 3-5 (in Russian).

6. Khoroshun, L.P. and Maslov, B.P., Methods of automated computation of physico-mechanical constants in composite materials. Kiev, "Naukova Dumka", 1980, 156 p. (in Russian).

7. Maslov, B.P., Maistrenko, A.L. and Drobyazko, V.V., On the estimation of the strength in composite diamond-bearing materials. "Sverkhtverdye materialy", 1982, 5, 3-9 (in Russian).

8. Maistrenko, A.L., Kulakovsky, V.N. and Simkin E.S., The distribution of diamonds in composite materials designed for the rock crushing application. "Sverkhtverdye materialy", 1985, 5, 24-9 (in Russian).

STRUCTURE AND PROPERTIES OF SILICON AND TITATIUM NITRIDES-BASE MATERIALS

Ivzhenko V.V., Kuzenkova M.A.,
Svirid A.A., Dub S.N.
Institute for Superhard Materials,
the Academy of Sciences of the UkrSSR,
Kiev, USSR

Silicon nitride-base materials offer promise as a high-temperature structural ceramics, they exhibit, however, relatively high brittleness and low strength. To improve fracture toughness and strength of ceramics a number of methods aiming at rise of energy to failure by second-phase particles is used (I). Physico-mechanical properties of ceramics can be improved by incorporating ceramic inclusions into a matrix which have a higher thermal expansion coefficient and a higher Young's modulus (2).

The present investigation deals with the stress state of the Si_3N_4-TiN-MgO system and with the influence of structure on physico-mechanical properties of the material.

Depending on properties of starting powder compositions materials with weak interaction at the phase interface (composition I), those with sphere-shaped TiN inclusions (composition 2) and those with aggregated TiN inclusions (composition 3) were fabricated. It was determined by X-ray diffraction that β-Si_3N_4 and TiN are the main phases in the materials fabricated. Titanium nitride lattice spacing is slightly influenced by the properties of the starting compositions.

At the silicon nitride-titanium nitride interface a diffusion zone of interaction is observed.

In the heterophase materials of the Si_3N_4-TiN system
distinctions in thermal expansion coefficients for the matrix
and the inclusion ($\alpha_{Si_3N_4-MgO} = 3.5 \cdot 10^{-6}$, $\alpha_{TiN} = 9.35 \cdot 10^{-6}$)
as well as in their elastic properties ($E_{Si_3N_4-MgO} = 304$ GPa,
$E_{TiN} = 350$ GPa) induce thermal stresses during cooling.

To evaluate the effects of TiN on the stress state of
the materials under study X-ray diffraction analysis (3) was
used which allows stress state evaluation on the ~ 30 µm
surface layer in silicon nitride specimens.

The values achieved show that the increase in TiN content in the starting mixture up to 5-10 wt.% induces a considerably higher resultant stresses in the materials surface
layers depending on the concentration. At higher TiN concentrations $\Delta\sigma$ value drops to 0. For materials fabricated from
the composite 3 $\Delta\sigma$ value is weakly TiN concentration-depen-
dent. The TEM and SEM observations of the materials under
study revealed no microcracks in the structure. As the stress
is applied, however (in preparing a microsection), the phase
interface in the materials I often becomes an initiator of
peripheral microcracking around large TiN inclusions (~ 10
µm), peripheral microcracking is not detected in the materials 2, in the materials 3 intensive microcracking of inclusions is observed.

The test results suggest that the materials I are composites with a fairly high strength of inclusions and rather
low strength of the phase interface , the materials 2 are
composites with rather high strength of inclusions as well as
of phase interface and that the materials 3 are composites
with low strength of aggregated inclusions. Variations in
ratio between inclusion and phase interface strength values
are due to growth of diffusion region at the phase interface (4).

Experimental study of a crack extension near the inclusions in the hot-pressed materials of the Si_3N_4-TiN-MgO system shows that after Vickers pyramid indentation the crack
detours around the inclusion and deflects from it at low

stresses. Fractographic studies have shown that the amount of secondary cracks increases with TiN particles concentration. The main crack branching appears to be particularly intensive in the materials with aggregated TiN inclusions.

We can see from an analysis of the equation giving the trajectory to which the fracture surface in the fields of external tensile and thermal stresses at inclusions approaches 4 that the main crack tends to deflect from the inclusion. The degree of the deflection should increase with stress state of the matrix phase and with decrease in external stresses. When the crack extends at an inclusion under the action of crack-tip stress field in the region of maximum tensile stresses (at the phase interface or in the inclusion) microcracking may occur. The stress state of the matrix layers which defines the crack deflection disappears. The main crack connects with a sphere-shaped microcrack and tends to grow in the direction which is defined by external stress. In addition to this, due to the local blunting of the crack tip when interacting with a sphere-shaped microcrack secondary cracks may form. The tensile stresses also disappear in the matrix layers in the case of an inclusion failure, the crack connects with microcracks in the inclusion and tends to extend in the main direction. With this type of the interaction the crack branching causing considerable increase in the energy to failure is most likely to occur.

The enlargement of the zone of the interaction at the phase interface and the formation of the structure with sphere-shaped inclusions allowed to raise the energy to the material failure by 20 to 30%. The formation of the structure with aggregated TiN inclusions provided a 30 to 40% increase in the energy to failure.

Thus, the results obtained have suggested the presence of the stress state in the hot-pressed materials of the Si_3N_4-TiN-MgO system induced by the distinction in thermal expansion coefficients for the matrix and the inclusion. It is also shown that the most efficient method to improve physico-mechanical properties of the materials under study is a combination of the two known methods for improving toughness of brittle materials: the incorporation of inclusions with high thermal expansion coefficients and the incorporation of inclusions with a high ability to microcracking.

REFERENCES

I. Rice, R.W. Mechanisms of toughening in ceramic matrix composites. Ceram. Eng. Sci. Proc. 1981, N2, 661-701.

2. Mah-Fai-H. Mendiratta Madan, G. and Zipsitt, Harey A. Fracture toughness and strength of Si_3N_4-TiC composites. Amer. Ceram. Soc. Bull., 1981, 60, N11, 1229-31, 1240.

3. Komyak, N.I. and Myasnikov, Yu.G. X-Ray Methods and Devices for Stress Measurement. Mashinostroyeniye, Leningrad, 1972.

4. Ivzhenko, V.V. and Shcherbakov, A.I. Study of physico-mechanical properties of hot-pressed materials silicon nitride- and titanium nitride-base hot-pressed materials. In: Superhard and High-Melting-Point Materials. ISM AN Ukr.SSR, Kiev, 1985, pp. 106-109.

Solid state investigation of ceramic-metal interface bonding

U. Jauch, G. Ondracek
Rheinisch-Westfälische Technische Hochschule
Institut für Gesteinshüttenkunde
- Glas, Keramik, Binde- und Verbundwerkstoffe -
Mauerstr. 5, D-5100 Aachen

ABSTRACT

Cermets are a potential class of composites, in which, however, the question of bonding at the ceramic-metal interface plays a key role: non bonding reduces the composite with respect to properties as Young's modulus, electrical conductivity or thermal expansion to those of the porous matrix phase material. This is why a new dilatometric technique is used to study the bonding between ceramic and metal phases selected to produce cermets.

The phases used as an example are aluminiumtitanate eutectic and nickel, the method is based on the effect,

- that before debonding the measured effective thermal expansion coefficient of the composite lies between the thermal expansion coefficients of its phases.

- that after debonding the effective thermal expansion coefficient drops to that of the matrix phase.

INTRODUCTION: BOUNDS AND PROPERTIES

As known by theory the properties of a multiphase material as a function of phase concentration may be determined by bounds of different order [6].

Supposing nothing more than that the material is two-phased, ultimate *I. order bounds* include all properties being realized by the two-phase system, as shown in fig. 1 for cermets. There is no possibility to find properties outside first order bounds by theory. Knowing, however, that the material is two-phased and isotropic, closer *II. order bounds* follow according to *two* assumptions about the system: being two-phased and isotropic (fig. 1).

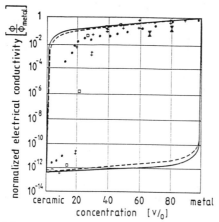

Figure 1. Electrical conductivity of cermets
(—— I. order bounds; --- II. order bounds; experimental values) [6]

Continuing that way by increasing the number of presuppositions - or, in other words, becoming more and more definite about the materials microstructure - leads finally to total convergence of the bounds resulting in one single curve, which describes exactly the dependence of the properties on the microstructure of a two-phase material (fig. 2).

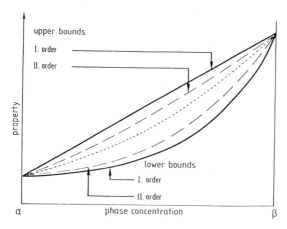

Figure 2. Bounds of different order

Again as theory points out, *5 parameters* control dominantly
the microstructural effect on the properties of a multi-phase
material (fig. 3) [6].

implicite parameters	1. number of phases	e.g. two-, three-, multiphased premise: thermochemical equilibrium
	2. type of microstructure matrix phase microstructure interconnecting phase microstructure	premise: continuum principle
explicite parameters	3. volume fraction of phase (concentration factor) volume × number of the phase particles 4. shape of phase particles (shape factor) 5. orientation of phase particles (orientation factor)	premise: spheroidal model mean values

Figure 3. Microstructural parameters and model concept
premises

Others than these parameters may be neglected in the frame of
an engineering approach to the problem.
For a two-phase material this statement means, that the five
factors control definitely, where a *certain* property value
inbetween the bounds has to be expected (fig. 4).

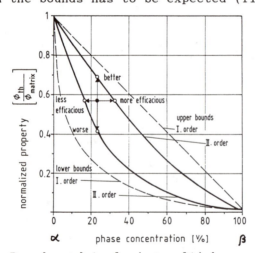

Figure 4. Bounds and tayloring multiphase materials

Based on the bound concept it may now be defined, how to
change the in-situ microstructure of a two-phase material in
order to construct taylor-made materials with desired
properties.

If, for example, *the property of the in-situ material*
lies at the spot inbetween the bounds as shown in fig. 4

- one may either save ß-phase, assumed to be rare and
  expensive, without changing the property

- or one may improve the property - as Youngs modulus or the
  thermal conductivity - at constant phase concentration

by precalculated changes in the microstructure.

Those precalculated changes - however - will only lead
to the desired effective properties of the composite, if
bonding exists between its phases. This is, why phase bonding
plays a key role in tayloring multiphase materials. And this,
exactly, is the major problem with composites as cermets, in
which - preferentially - *incoherent phase boundaries* occur
between the phases (fig. 5).

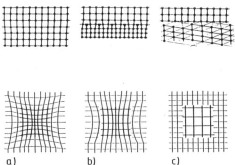

Figure 5. Coherence at phase boundaries schematically
          a) coherent   b) partly coherent   c) incoherent

Additionally, these phases may have large differences in
their thermal expansion coefficients, which complicates the
bonding problem.

BOUNDS AND BONDING

Sintering, for instance, a cermet material consisting of a
nonbonding ceramic matrix phase and a nonbonding included
metal phase one obtains a porous ceramic material the pores
of which are filled by loose metal particles. The same is
true for concentrations, where the metal phase forms the
matrix whilst ceramic particles fill the pores. That
situation, for example, exists in the cermet system uranium
dioxide-molybdenum (fig. 6).

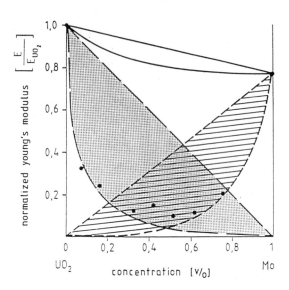

Figure 6. Phase bonding and bounds for Young's modulus of elasticity

As a consequence of the debonded phase boundary the Young's moduli of these cermets measured by tensile tests do not fit into the bounds for the UO_2-Mo-system but do fit inbetween the bounds for porous UO_2 or porous Molybdenum respectively.

Constructing taylor-made materials therefore needs to observe two conditions:

- the properties of the selected phases have to result in the desired effective properties of the composite

and

- phase bonding has to take place between the selected phases.

How to find out criteria about bonding is the content of the following chapter.

DILATOMETRIC BONDING TEST
As pointed out before the comparison between measured property values and calculated bounds provides a first, qualitative approach to the problem of bonding. In order to determine phase bonding quantitatively one once more starts from microstructure-property-correlations:

Considering the theoretical equation about the dependence of the thermal expansion coefficient of two-phase materials [1, 3]

$$\alpha_C = \alpha_M + (\alpha_D - \alpha_M) \frac{E_D [E_C (1 - 2v_M) - 3E_M (1 - 2v_C)]}{3E_C [E_D (1 - 2v_M) - E_M (1-2v_D)]} \qquad (1)$$

(α = thermal expansion coefficient; E = Young's modulus of elasticity; v = Poissons ratio of transversal contraction; C, D, M = Subscripts for Composite, Dispersed and Matrix phase)

and assuming the second phase to be pores one notices, that thermal expansion does not depend on porosity. In the equation subscript "D" denotes the included phase. If "E_D", the Youngs modulus for the included phase, becomes zero for pores, the second term of the equation disappears and the thermal expansion coefficient of the porous material is equal to the thermal expansion coefficient of the nonporous, compact material. This is an important result in so far,

- as any property as rupture strength, Young's modulus or conducitvity decreases with porosity except the thermal expansion coefficient (fig. 7) and
- as this result provides the basis to measure phase bonding by dilatometry.

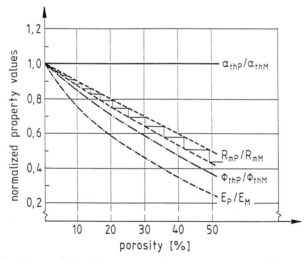

Figure 7. Normalized properties as a function of spherical porosity

α_{th} = thermal expansion coefficient
R_m = rupture strength
Φ_{th} = thermal conductivity
E = Young's modulus of elasticity
P,M = subscripts for porous (P) and compact material (M)

To demonstrate the method an aluminium titanate-
titaniumoxide eutectic as ceramic matrix phase and nickel as
the included metal phase of a cermet has been chosen. The
thermal expansion coefficients of nickel and the ceramic
phase have been measured separately in a Netzsch dilatometer
(fig. 8).

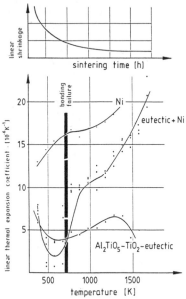

Figure 8. Shrinkage by sintering (above) and thermal
expansion and phase bonding (below)

Afterwards a pressed pellet of Al_2TiO_5-TiO_2 eutectic and
Nickel powders was sintered isothermally in the same
dilatometer. Where the slope of the shrinkage curve pointed
out, that the sintering process had been taken place, thermal
contraction of the cermet - which means the thermal expansion
coefficient during cooling - was measured dilatometrically
(fig. 8). This thermal expansion coefficient of the cermet at
elevated temperatures was found, as expected, between the
thermal expansion coefficients of its two phases. By
continued cooling however thermal stresses are induced at the
phase boundaries, since the ceramic matrix phase contracts
less than the metal inclusions do. If these stresses exceed
the phase bonding strength debonding at the phase boundary
takes place and the system changes from a cermet to a porous
ceramic, where the pores contain loose metal particles. And
since, as pointed out above, the thermal expansion
coefficient of a porous ceramic will be identical with the
one of the compact ceramic material, the thermal expansion
coefficient of the chosen cermet system drops to the thermal
expansion coefficient of the ceramic phase, when phase
debonding occurs. In the present system the phase boundaries
crack at about 700 degrees.

Knowing the temperature interval (ΔT) to cause debonding as well as the thermal expansion coefficients and bulk moduli of the phases we may now calculate the bonding strength by existing equations [3, 5]

$$\delta = 3\Delta T \ (\alpha_D - \alpha_M) \ \frac{K_m K_D}{K_D - K_M} \underset{K_D > K_M}{\approx} 3\Delta T \ (\alpha_D - \alpha_M) \ K_M \tag{2}$$

which, of course, gives only a first engineering approach for two reasons:

1) In technical systems the phase boundaries are not smooth according to the rough surface and irregular shape of the inclusions. This may lead to additional mechanical bonding. For exact measurements the inclusions should therefore have smooth surfaces (or the surface roughness would have to be known quantitatively).

2) Whilst the brittle, ceramic phase will not suffer plastic deformation such plastic deformation will certainly take place in the metal inclusions above their yield strength and therefore partly compensate the thermal induced stresses. This is why for more accurate calculations of the bonding strength one needs to know the temperature function of the yield strength of the included metal phase.

Both sources of inaccuracy, however, can be eliminated by respective experimental measures.

Phase bonding strengthes achieved on this way are obviously correlated with the "WORK OF ADHESION" (σ)

$$\Delta G_{D/M}^{Gr} - \Delta G_D^{Gr} - \Delta G_M^{Gr} \approx \sigma \tag{3}$$

Since this "WORK OF ADHESION" is a function of the surface energies of the phases (ΔG_D^{Gr}, ΔG_M^{Gr}) and the interface energy ($\Delta G_{D/M}^{Gr}$) and since this energy may be determined by wetting experiments and by interference microscopy (fig. 9) to measure the contact angles between the phases via the Young-Dupré-equation [4, 5, 7] an alternative way to investigate phase bonding in solid state, the region denoted still by question marks in fig. 10, is available, being started now supplementing the already existing results about wetting in ceramic-metal-composites (fig. 10).

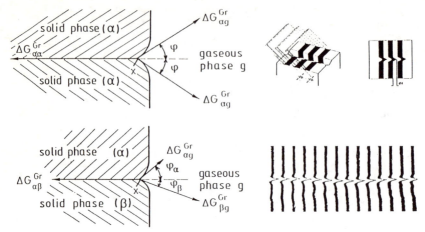

Figure 9. Phase boundary equilibrium and interference microscopy

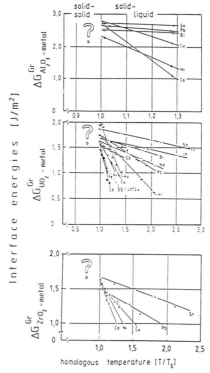

Figure 10. Interface energies between solid ceramics and liquid metals as function of homologous temperature (T = measuring temperature; Ts = metal melting point)

REFERENCES

1. Gribb, J.L., <u>Nature</u> <u>220</u>, 1968, 576.

2. Ledbetter, H.M. and Austin, M.W., Proc. Conf. on Residual
    Stresses, Garmisch-Partenkirchen, ed. by E. Macherauch,
    V. Hauk, DGM-Oberursel, 1987.

3. Nazare, S. and Ondracek, G., <u>Z.</u> <u>Werkstofftechnik</u>, 1978,
    9, 140.

4. Nikolopoulos, P. and Ondracek, G., <u>Z.</u> <u>Werkstofftechnik</u>,
    1982, **13**, 60 and <u>J.</u> <u>Nucl.</u> <u>Mat.</u>, 1981, **98**, 306.

5. Nikolopoulos, P., Sotiropoulou, D., Jauch, U. and
    Ondracek, G., Ceramics and Metals, Bilateral Seminars of
    the Int. Bureau, Kernforschungsanlage Jülich, 1987, 281.

6. Ondracek, G., <u>Review</u> <u>on</u> <u>Powder</u> <u>Metallurgy</u> <u>and</u> <u>Physical</u>
    <u>Ceramics</u>, **3-3/4**, 1988, 1.

7. Ondracek, G., <u>Werkstoffkunde</u>, expert verlag Sindelfingen,
    1986, 58 and 254.

ACKNOWLEDGEMENT

The present work was substantially supported by the Deutsche
Forschungsgemeinschaft Bonn and partly performed at
Kernforschungszentrum and Universität Karlsruhe. This support
- as well as the assistance of Mrs. Peters, Mrs. Schmitz and
Mr. Trechas - is gratefully acknowledged by the authors.

DETERMINATION OF ELASTIC PROPERTIES OF FIBRE REINFORCED COMPOSITES AND OF THEIR CONSTANTS

JAN PIEKARCZYK, ROMAN PAMPUCH

Institute of Materials Science AGH, Cracow (Poland), al.Mickiewicza 30

ABSTRACT

Theoretical analysis shows that measurements of the velocity of propagation of ultrasonic waves enable to determine with a relatively high precision all elasticity constants and material constants of anisotropic materials such as fibre re-inforced composites and ceramic polycrystals containing anisotropically aligned pores. Experiments have been made with widely differing materials of this type. They confirm that such a determination is feasible for materials the elasticity and material constants of which have values ranging from a few to several hundreds of GPa and gives results which agree well with equations proposed for a design of elastic properties of composites on the basis of properties of the composite constituents. The results of the present work indicate that by using this method even small, undetectable under the microscope, deviations from a given anisotropic packing of the constituent phases (fibres, pores) can be identified.

INTRODUCTION

Real materials often show an anisotropy of elastic constants due to a pre-ferential alignment of the constituent phases, for instance of fibres in fibre re-inforced composites or of pores in porous ceramics. In such cases a determination of at least five elastic constants with materials of a hexa-gonal symmetry is required [1,2,3]. Basing on concepts presented in [4,5] it is the purpose of the present paper to show that the all the elastic

constants of anisotropic materials may be conveniently determined by measuring the velocity of propagation of longitudinal and transverse ultrasonic
waves in different directions of the tested material samples. In comparison
with static and resonance methods the ultrasonic method requires the least
number of samples, because all the elastic constants may be determined here
already on a single sample while the former methods require several samples
cut out at different angles from the tested material.

The relations between the elastic constants and the velocities of longitudinal and transverse ultrasonic waves in composites with a uni-directional
alignment of the fibres and in materials having a hexagonal symmetry are
given by [4,6]:

$$\varrho \cdot v_L^2 \, (\theta) = 1/2 \left[A(C_{ij}, \theta) + B(C_{ij}, \theta) \right] \tag{1}$$

$$\varrho \cdot v_{\parallel}^2 \, (\theta) = 1/2 \left[A(C_{ij}, \theta) - B(C_{ij}, \theta) \right] \tag{2}$$

$$\varrho \cdot v_{\perp}^2 \, (\theta) = 1/2 (C_{11} - C_{12}) \sin^2(\theta) + C_{44} \cos^2(\theta) \tag{3}$$

where:

$$A(C_{ij}, \theta) = (C_{11} + C_{44}) \sin^2(\theta) + (C_{33} + C_{44}) \cos^2(\theta)$$

$$B(C_{ij}, \theta) = \left\{ (C_{11} - C_{44})^2 \sin^4(\theta) + (C_{33} - C_{44})^2 \cos^4(\theta) + \right.$$

$$\left. + 2 \sin^2(\theta) \cos^2(\theta) \left[(C_{11} - C_{44})(C_{44} - C_{33}) + 2(C_{13} + C_{44})^2 \right] \right\}^{1/2}$$

and v_L is the velocity of longitudinal ultrasonic waves, $v_{\parallel}$ is the velocity
of transverse waves polarized parallel to the plane determined by the highest axis of symmetry and the direction of propagation of the wave, $v_{\perp}$ is
the velocity of transverse waves polarized normal to this plane, θ - is the
angle between the symmetry axis of the material and the direction of wave
propagation (Fig.1) and ϱ - is the density of the material.

The relations hold for wave-lenghts which are much greater than the dimensions of the inhomogeneities of the material. Moreover, the sample should
realise the conditions of a three-dimensional medium. The latter requirement is met when all dimensions of the sample are approximately equal or
when the smallest dimension of the sample (e.g. a cross-section with bars)is
several times greater than the wave-length used [7].

Table 1 shows relations between the elastic constants of materials of a hexagonal symmetry and the velocity,direction of propagation and plane of pola-
risation of the ultrasonic waves. The relations between the material constants (Young´s modulus, stiffness modulus and Poisson number) and the elastic constants, derived from the equations quoted in Table 1 are as follows:

TABLE 1

Relations between the velocity of ultrasounds, v_i, and the elastic constants, C_{ij}, of materials

Mode	Direction of propagation	Direction of polarization	Velocity	Equation	
L	1 or 2 ($\theta=90°$)	–	v_1	$C_{11} = \varrho\, v_1^2$	(4)
L	3($\theta=0°$)	–	v_2	$C_{33} = \varrho\, v_2^2$	(5)
L	$\theta = 45°$	–	v_3	$C_{13} = \frac{\varrho}{2}\left\{\left[(4v_3^2 - v_1^2 - v_2^2 - 2v_4^2)^2 - (v_1^2 - v_2^2)^2\right]^{1/2} - 2v_4^2\right\}$	
T	3($\theta=0°$) 1 or 2 ($\theta=90°$)	1 or 2 3	v_4	$C_{44} = \varrho\cdot v_4^2$	(7)
T	1($\theta=90°$) 2($\theta=90°$)	2 1	v_5	$C_{12} = \varrho\,(v_1^2 - 2v_5^2)$	(8)
T	$\theta = 45°$	x	v_6	$C_{13} = \frac{\varrho}{2}\left\{\left[(v_1^2 + v_2^2 + 2v_4^2 - 4v_6^2)^2 - (v_1^2 - v_2^2)^2\right]^{1/2} - 2v_4^2\right\}$	
T	$\theta = 45°$	y	v_7	$C_{12} = \varrho\,(v_1^2 + 2v_4^2 - 4v_7^2)$	(10)

Remarks: L-longitudinal, T-transverse; x-polarization in the plane determined by the symmetry axis and the direction of wave propagation; y-polarisation normal to the former plane
ϱ -density of the material.

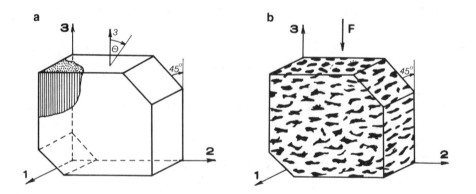

Figure 1. The directions (1,2,3, 3+θ) of measurements of velocity of ultrasounds in composite (a) and graphite (b) samples. The axis 3 is the symmetry axis (C^3) of the samples.
  Remarks: F is the direction of load application during the forming of graphite material before firing.

$$E_{33} = C_{33} - \frac{2C_{13}^2}{C_{11}+C_{12}} \tag{11}$$

$$E_{11} = E_{22} = \frac{(C_{11}-C_{12})(C_{11}C_{33} + C_{12}C_{33} - 2C_{13}^2)}{C_{11}C_{33} - C_{13}^2} \tag{12}$$

$$G_{31} = C_{44} \tag{13}$$

$$G_{12} = C_{66} = 1/2 \, (C_{11} - C_{12}) \tag{14}$$

$$\mu_{31} = \frac{C_{13}}{C_{11}+C_{12}} \tag{15}$$

$$\mu_{13} = \frac{C_{13}(C_{11}-C_{12})}{C_{11}C_{33}-C_{13}^2} \tag{16}$$

$$\mu_{12} = \frac{C_{12}C_{33}-C_{13}^2}{C_{11}C_{13}-C_{13}^2} \tag{17}$$

Among the above seven constants two are interdependent, namely:

$$E_{33}\mu_{13} = E_{11}\mu_{31} \tag{18}$$

$$G_{12} = \frac{E_{11}}{2(1+\mu_{12})} \tag{19}$$

The indices i and j at the Poisson numbers μ, denote, respectively, the stress direction and the direction of deformation.

As well known, for isotropic materials the number of independent elastic constants reduces to two, C_{11} and C_{44} while for thin fibres which conform to conditions of a uni-dimensional medium [8] the Young's modulus may be determined directly from equation:

$$E = \varrho \cdot v_L^2 \, . \tag{20}$$

where E is the Young's modulus, ϱ is the density, and v_L is the velocity of the longitudinal ultrasonic wave.

EXPERIMENTAL PART

Samples of the following materials were tested:

1. A non-porous composite (GFRE) uni-directionally re-inforced by E glass fibres with an epoxide matrix (Epidian 5) and its components; the volume

fraction of fibres was 70.0 ± 1.1% (at a 95% confidence interval);

2. A composite (SWRA1) uni-directionally re-inforced by steel wires
(H25N20S2) with an aluminium alloy (AK 11) matrix [*] and its components [9].
The volume fraction of the aluminium matrix was 62.5%, the one of the wires
was 32.6%, 2.6% of the composite was constituted by the diffusion layer and
2.3% by pores.

3. A SiC-Si composite of the SILCOMP type obtained by reaction of unidirectionally aligned carbon fibres with liquid silicon [10] of a phase composition given in Table 2 [11].

4. Polycrystalline small-grained graphite (PG) of different densities
between 1.786 and 1.880 g/cm^3 [12].

TABLE 2
Phase composition of the SiC-Si composite

Sample code	Density (g/cm^3)	Volume fraction (%)			
		SiC	Si	C	Pores
1	2.62+0.03	60.5+2.5	16.0+2.5	22.3+0.2	1.16+0.05
2	2.71	51.7	39.6	7.9	0.42
3	2.71	41.7	56.0	0.3	1.97

The elastic constants were determined by using a materials tester MT-541
equipped with ultrasonic 1 MHz transducers for longitudinal waves and an
ultrasonic defectoscope DI-4T (INCO,Poland) equipped with 2 MHz transducers
for transverse and 4-10 MHz transducers for longitudinal ultrasonic waves.
The Young's modulus of glass fibres and steel wires was determined by using
the MT-541 materials tester with specially adapted transducers. The velocity
of propagation of ultrasonic waves in the composites was measured normal and
parallel to the fibre axis and additionally under and angle of 45 deg to the
axis. These measurements provided seven different velocities of propagation.
This enabled to calculate each of the elastic constants C_{13} and C_{12} from two
different equations and each of the material constants (E_{33}, E_{11},μ_{31}, μ_{13},
μ_{12}) from four different measurements. However, the velocity v_7 in case of
the composite GFRE was not determined due to an excessively high damping of
the ultrasounds and the velocity v_3 measured in the SiC-Si composite was not
taken into account since negative values of C_{13} were obtained in this case
which cannot be explained at present.

[*] The composite and its matrix material were kindly supplied by
J.Myalski from the Silesian Polytechnics, Katowice.

The elastic constants and material constants calculated from the measurements for the composite and the graphite samples are given, respectively, in Tables 3-4 and Figure 2.

TABLE 3
Experimentally determined elasticity and materials constants in composites

Constant	GFRE		SWRA1		SiC-Si		
					Sample 1	Sample 2	Sample 3
C_{33} (GPa)	58.2	$\pm$ 0.1	152.7	$\pm$ 1.6	309.7	314.1	285.4
C_{11} (GPa)	24.0	0.3	169.9	2.8	165.5	245.1	265.6
C_{44} (GPa)	9.2	0.1	42.3	0.7	89.0	100.6	103.4
C_{12} (GPa)	8.5	0.6	102.7	6.1	41.8	54.0	68.4
C_{13} (GPa)	9.4	0.9	83.1	4.3	63.0	59.5	69.1
E_{33} (GPa)	52.8	0.9	102.0	7.5	271.0	290.4	256.8
E_{11} (GPa)	20.3	0.4	98.2	2.7	147.0	226.1	238.1
G_{31} (GPa)	9.2	0.1	42.3	0.7	89.0	100.6	103.4
G_{12} (GPa)	7.8	0.2	33.6	0.7	61.9	95.6	98.6
μ_{31}	0.289	0.020	0.305	0.022	0.30	0.20	0.21
μ_{13}	0.111	0.010	0.293	0.038	0.16	0.15	0.19
μ_{12}	0.311	0.020	0.461	0.041	0.16	0.15	0.21
Density (kg/m^3)	2069	5	4366	10	2620	2710	2710

Remarks: Data based on ten measurements in each direction in each sample; the absolute error was calculated at a 95% confidence interval.

TABLE 4
Material constants of the matrix and reinforcing fibres

Constant		Component	
	Matrix	Epoxy resin	AK11-Alloy
E_o (GPa)		4.56 $\pm$ 0.19	78.5 $\pm$ 0.9
G_o (GPa)		1.66 $\pm$ 0.09	29.4 $\pm$ 0.4
μ_o		0.372 $\pm$ 0.01	0.334 $\pm$ 0.004
density (kg/m^3)		1206 $\pm$ 10	2685 $\pm$ 10
	Fibres	Glass E fibre	Steel wire
E_w (GPa)		72.10 $\pm$ 0.17	206.7 $\pm$ 2.8
density (kg/m^3)		2480 $\pm$ 30	7850

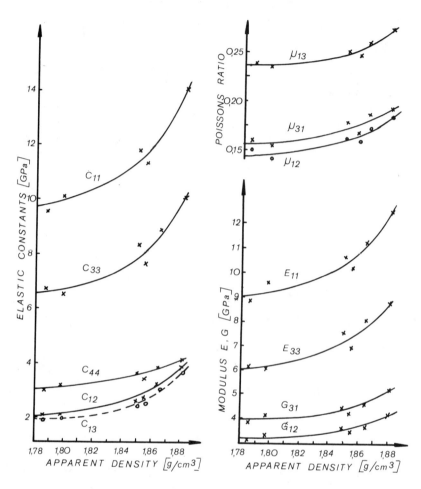

Figure 2. Elastic and material constants vs. apparent density of the polycrystalline graphite samples.

The highest values of the maximum relative error (up to 7.5% for E_{33}) appear
with the elastic and material constants of the SWRAl composite where there
occur elongated pores between the wires and a diffusional layer at the ma-
trix-wires interface which is only slightly adhering to the remaining components of the composite. Both components give rise to a strong damping of the
ultrasounds, especially of the transverse ones, which require for their
transmission a high resistance to shearing at the interfaces. Additionally,
the non-uniform alignement of the wires contributes to a large maximum error
in this case. With the remaining composite materials the maximum relative
errors of the elastic and materials constants do not exceed 1%, except

for μ, C_{13}, C_{12} of the SiC-Si composite. With the polycrystalline graphite this error is generally lower than 1% (except for C_{12} and C_{33}).

DISCUSSION

The dependence of the elastic and materials constants of polycrystalline graphite vs. its density is given in Fig.2. Besides the obvious increase of all constants with density an important anisotropy of elastic and material constants is observed. E.g. the Young's moduli E_{11} and E_{33} differ here by 30% while the stiffness moduli G_{31} and G_{12} - differ by 19%. This elastic anisotropy is most probably due to a preferential alignment of disc-shaped pores which occur in the graphite studied. As shown in Fig.1 one part of these pores is aligned with their equatorial axis normal to the direction of the load applied during forming of this material before firing. When stresses are applied to such a material parallel to the (longer) equatorial axis of these pores lesser stress concentrations and lesser deformations occur than is the case with stresses applied in the direction of the polar axis of the pores. This gives rise to lower Young's moduli in the latter direction and higher moduli in the former direction.

The SiC-Si matrix composite contains besides SiC fibrous particles some residual non-reacted carbon fibre material. Due to the highly anisotropic microstructure and anisotropy of elastic properties of the carbon fibres their presence gives rise to a high anisotropy of the Young's moduli. At a 22.3% content of carbon fibres the ratio $1-E_{11}/E_{33}$ is as high as 45.8% while at a 0.3% content of carbon fibres this ratio is lower than 7.5%.

Due to the defective microstructure of the SWRA1 composite no attempts have been made in this case to compare the experimentally determined elastic and materials constants with theoretical calculations of these constants which may be e.g. based on the rule of mixtures. This has been, however, possible with the non-porous and uniform GFRE composite. The calculations have been based on data obtained for the whole composite and its separate components. According to [5,13] the following relations hold for the relations between the constants of a composite consisting of an isotropic matrix reinforced with continuous isotropic fibres and the constants of the separate constituents:

$$E_{33} = E_w V_w + E_o (1 - V_w) \tag{21}$$

$$E_{11} = \frac{E_o (1 + A \cdot N \cdot V_w)}{1 - N \cdot V_w} \qquad (22)$$

where:

$$N = \frac{E_w/E_o - 1}{E_w/E_o + A}$$

and A- parameter of the packing of the fibres which takes the value A=1 for a hexagonal packing and A=2 for a regular packing. Furthermore,

$$\mu_{31} = \mu_w \cdot V_w + \mu_o \cdot (1 - V_w) \qquad (23)$$

$$\mu_{12} = 1 - \frac{E_{11}}{2K} - 2 \cdot (\mu_{31})^2 \cdot \frac{E_{11}}{E_{33}} \qquad (24)$$

where:

$$K = \frac{(K_w + G_o)K_o + (K_w - K_o)G_o \cdot V_w}{K_w + G_o - (K_w - K_o) \cdot V_w}$$

$$G_{31} = \frac{G_o \left[G_w \cdot (1 + V_w) + G_o \cdot (1 - V_w) \right]}{G_w(1 - V_w) + G_o(1 + V_w)} \qquad (25)$$

$$\mu_{13} = \mu_{31} \frac{E_{11}}{E_{33}} \qquad (26)$$

$$G_{12} = \frac{E_{11}}{2(1 + \mu_{12})} \qquad (27)$$

E_{33} and G_{31} are the Young's and stiffness modulus, respectively, calculated in the direction parallel to the fibre axis; E_{11} and G_{12} are, respectively, the appropriate moduli normal to this axis; K is the bulk modulus of the composite in the stress plane; V_w is the volume fraction of fibres. The indices w and o denote, respectively, the fibres and the matrix while the indices i and j at the Poisson number denote, respectively, the direction of applied load and the direction of the deformation brought about by this load. Insering into equations (21-25) the experimentally determined material constants of the epoxy matrix and glass fibres (see Table 4) and assuming that A=1.3, i.e. assuming that the packing of the fibres is close to the hexagonal packing, the values of the constants shown in left column of Table 5 have been obtained at a fibre volume fraction of 70%. The material constants determined experimentally for the composite GFRE containing 70% of volume of glass fibres are quoted in the right column of Table 5.

Comparing these data it may be seen that they agree within the limits of the error of measurements, except for the stiffness modulus G_{31} the measured value of which is higher than the one calculated from equation (25) by as much as 22%. We suppose that this is due to the fact that equation (25) is not adequate for this material, where regions of a weak adherence of fibres to the matrix probably occur. The effect of such inhomogeneities upon the stress concentration and deformation has been discussed before for polycrystalline graphite.

TABLE 5

Comparison of material constants: (1) calculated on the basis of constants determined for separate components and (2) determined for the composite

Constant	(1)	(2)
E_{33} (GPa)	51.85	52.8
E_{11} (GPa)	20.85	20.3
G_{31} (GPa)	7.19	9.2
G_{12} (GPa)	7.94	7.8
μ_{31}	0.272	0.289
μ_{12}	0.312	0.311
μ_{13}	0.109	0.111
density (kg/m^3)	2098	2069

REFERENCES

1. Ney J.F., Physical Properties of Crystals, Clarendon Press, Oxford 1957.
2. Biagi E., Frosini S., Masotti L., Borchi E., The Propagation of acoustic waves in polycrystals with axial texture. Phys.Stat.Sol, (a), 1968, 93, 151-162.
3. Schreiber E., Anderson O.L., Soga N., Elastic constants and their measurement, McGraw-Hill, New York, 1973.
4. Musgrave M.J.P., On the propagation of elastic waves in aelotropic media. II Media of hexagonal symmetry. Proc.R.Soc., London, 1954, A226,356-366.
5. Piekarczyk J., Własności sprężyste kompozytu jednokierunkowego włókno szklane-żywica epoksydowa. Inżynieria Materiałowa, 1987, 8, 85-90.
6. Zimmer J.E., Cost J.E., Determination of the elastic constants of a unidirectional fiber composite using ultrasonic velocity measurements. J.Acoustic.Soc.Amer., 1969, 47, 795-803.

7. Piekarczyk J., Hennicke H.W., Pampuch R., On determining the elastic constants of porous zinc ferrite materials. CFI/Ber.DKG., 1982, 59, 227-232.

8. Piekarczyk J., Zastosowanie metody ultradźwiękowej do pomiaru modułu Younga włókien ceramicznych. Inżynieria Materiałowa, 1984, 5, 9-13.

9. Hyla I., Śleziona J., Technologia otrzymywania i własności kompozytu Al-drut stalowy. Inżynieria Materiałowa, 1984, 5, 153-158.

10. Pampuch R., Walasek E., Białoskórski J., Reaction mechanism in carbon - liquid silicon systems at elevated temperatures. Ceram.Intern., 1986, 12, 99-106.

11. Piekarczyk J., Białoskórski J., Walasek E., Phase composition and elastic properties of composites in SiC-Si-C system. Proc.IX-th Graphite Conference. Zakopane (Poland), September, 1988.

12. Piekarczyk J., Pampuch R., Tekstura i własności sprężyste tworzyw grafitowych. Ceramika 24, PAN O/Kraków, 1976.

13. Whitney J.M., Daniel I.M., Pipes R.B., Experimental mechanics of fiber reinforced composite materials. Society for Experimental Stress Analysis. New Jersey, 1982.

FORMATION OF GLASS FIBER-CEMENT PASTE INTERFACIAL ZONE AND ITS EFFECT ON THE MECHANICAL PROPERTIES OF GLASS FIBER REINFORCED MORTAR

MITSUNORI KAWAMURA and SHIN-ICHI IGARASHI
Department of Civil Engineering
Kanazawa University
2-40-20, Kodatsuno, Kanazawa, Ishikawa, 920,JAPAN

ABSTRACT

Microhardness measurements and EDXA anlyses were made in the regions around glass fiber strands embedded in the cement paste to elucidate the effects of curing temperature and mineral admixture on the microstructure of the inter-facial zone between a glass fiber strand and cement paste matrix phase. The characteristics of flexural strength and toughness of glass fiber reinforced mortars(GRC mortars) were also discussed relating them to the results obtained by the microhardness measurements and EDXA analyses. Curing temperature significantly influenced the microhardness distribution patterns in the interfacial zone. Both the softest and the hardest region exsist within the interfacial zone. The discontinuity in microhardness in the regions in samples cured at a high temperature(38°C) appears to greatly affect the toughness of the corresponding GRC mortars. Overall micro-hardness distribution patterns were not changed by the addition of the mineral admixtures.

INTRODUCTION

Glass fiber reinforced mortars(GRC mortars) lose their strength and toughness with time when exposed to wet environments or natural weather. Two explanations have been proposed to this age-embrittlement of GRC mortars. The one is based on the deterioration of glass fibers due to a chemical attack of alkaline pore solution [1]. The other suggests that the gradual reduction in flexural strength and toughness of glass fiber reinforced cement results from the growth of dense hydration products of $Ca(OH)_2$ among filaments in glass fiber strands [2,3]. Namely, according to the latter,the formation of the dense hydration products around glass fibers increases the fiber-matrix bond strength, leading to brittle behavior of GRC mortars. The reduction in flexural strength and toughness at relatively early ages in GRC mortars is generally explained by the latter mechanism, while in later ages by the degradation of glass fibers due to a chemical attack of alkaline pore solution in mortar.

Several measures have been proposed to improve the durability of GRC mortars. Blended cements have been used for modifying the microstructure of cement paste matrix in the vicinity of glass fibers. The effects of the incorporation of blastfurnace slag on fracture toughness of GRC mortar were investigated by Mills [4]. He concluded that the substitution of blast-furnace slag for portland cement failed to yield any improvement in strength and toughness of GRC mortars, attributing the failure to the release of overwhelming quantities of $Ca(OH)_2$ from the hydration of portland cement component or to the stronger affinity of $Ca(OH)_2$ for AR glass than for blastfurnace slag. The replacement of cement by flyash for improvement of age-embrittlement in GRC mortar has been investigated by some workers, but they did not obtain any satisfactory results for the improvement of the durability of glass fiber reinforced cement [5,6]. However, it should be noted that the effectiveness of flyash in improving the durability of glass fiber reinforced cement varies widely with the type of flyash despite of minor differences in their pozzolanic activity and chemical compositions [5]. An investigation on possible beneficial effects of silica fume in glass fiber cement composites was carried out by Bentur and Diamond [7]. They reported that pre-treatment of glass fiber rovings with a dispersed silica fume slurry was found to be a most effective method to improve the durability of GRC paste made with AR glass.

Generally, the mechanical properties of fiber reinforced composites with brittle matrices are considered very sensitive to the microstructure of the interfacial zone between matrix and inclusions. It has been revealed that the microstructure of the interfacial zone between cement paste matrix and inclusion phase such as sand grains and fibers is fairly different from that of bulk cement paste phase away from the interface [8]. Bentur,Diamond and Mindess [9] showed that cracks appeared to change its course at the porous region away from the actual interface in steel fiber-cement paste system. As for the interfacial zone formed around glass fibers, the micro-structure of "actual interface" or "spaces among individual filaments in a strand" were examined in detail by SEM observation for fractured section of glass fiber reinforced cement specimens [3]. However, generally speaking, there were found only a few informations available concerning the whole interfacial zone containing the region from the interface to bulk cement paste phase, although the microstructure of the whole zone formed around glass fiber strands is considered to be related to the mechanical behavior of GRC mortars. The microstructural features of glass fibers-cement paste interfacial zone must depend on curing condition and the type of mineral admixture incorporated into the glass fiber-cement paste system.

In this study, microhardness measurements and EDXA analyses were made in the regions around glass fiber strands embedded in the cement paste to elucidate the effects of curing temperature and mineral admixture on the characteristics of microstructure in the interfacial zone. The reduction in flexural strength and toughness in GRC mortars was also discussed being related to the informations obtained from the microhardness measurements and EDXA analyses.

EXPERIMENTAL DETAILS

Materials and Mix Proportion of GRC Mortars

The cement used was ordinary Portland cement. A river sand, the specific gravity of which is 2.64, was used as a fine aggregate. Chopped strands of commercial AR glass fibers were incorporated into mortars. The length of AR glass fibers was 13mm. The physical properties of the flyash and the silica fume used are given in TABLE 1. 10% and 20% of cement by weight were replaced by silica fume and flyash in admixture-bearing mortars. Superplasticizer of 0.8% by weight of the cement was added to silica fume-bearing mortars. The mix proportions of GRC mortars are given in TABLE 2.

TABLE 1
Phsical Properties of Admixture

	Specific Gravity	Surface Area(m^2/g)
Flyash	2.23	2.1
Silica Fume	2.33	24.2

TABLE 2
Mix Proportion of GRC Mortars

	W/C	Cement : Sand : Flyash or Silica Fume
Admixture-free	0.55	1 : 2 : 0
Flyash-bearing	0.55	0.8 : 2 : 0.2
Silica Fume-bearing	0.55	0.9 : 2 : 0.1

Procedure

Flexural strength test of GRC mortars : The GRC mortar specimens for
the flexural strength test were 40 x 40 x 160 mm prisms which were produced
by the premixing method. Specimens which had been placed in a moist room
for 24 hours at 20°C, were demolded and stored under three different
conditions:(a) in water at 5°C, (b) in water at 20°C, (c) in a moist
atmosphere at 38°C, 100% R.H. Flexural strength tests were conducted under
the three-point bending. In order to determine the toughness of GRC mortars,
deflections of a prism at loading points were measured. The toughness was
evaluated by calculating the area enclosed by load-deflection curve and
deflection axis up to 1 mm of deflection.

Microhardness measurement and EDXA analysis : As shown in Fig.1, nine
glass fiber strands were embedded in the cement paste with the water:cement
ratio of 0.55 corresponding to the cement paste phase of the mortars. The
cement paste prisms containing glass fiber strands were cured under the same
condition as mortar specimens for the mechanical tests. After curing for
prescribed times, slices were cut from the paste prisms by using a diamond
wheel saw. The slices were polished with a diamond slurry. The microhardness tester with Vickers indenter (Minimum load:0.1g) was used to measure
the microhardness around a glass fiber strand. After microhardness
measurements were made, the samples were dried in vacuum at room temperature,
and then their polished surfaces were coated with gold for EDXA analysis.
EDXA analyses were made at about 8 spots within the regions from the interface toward the bulk cement paste. The spectrum for the Kα peaks of Ca and
Si were accumulated for a period of 100 seconds of counting. A polished
surface of a sample was shown in Fig.2. It is found from Fig.2 that the
spaces among individual glass filaments within the glass strand are densely
filled.

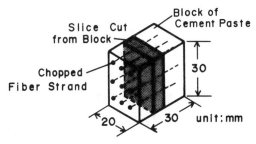

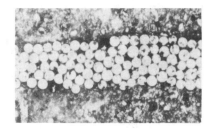

Figure 1. Sample for microhardness
measurement and EDXA analysis

Figure 2. Micrograph of polished
surface of a sample for microhardness
measurement and EDXA analysis

RESULTS

GRC Mortars without Admixture
    Effects of curing temperature on flexural strength and toughness :
Fig.3 shows the flexural strengths of GRC mortars with various fiber volume
fractions cured at three different temperatures. The flexural strength of
GRC mortars stored in water at 5°C increased with age up to 180 days.
However, considerable reductions in flexural strength occurred in GRC mortars
with the fiber volume fraction of 2.0 and 2.5% between 180 and 360 days.
The flexural strength of GRC mortars stored in water at 20°C also
considerably decreased between 180 and 360 days. The flexural strength of
specimens cured at 38°C little changed with age. Differences in flexural
strength between specimens with different fiber contents cured at 38°C were
extremely small compared to the differences in flexural strength between
different specimens cured at 5°C.
    The changes with age in toughness of GRC mortar specimens are shown in
Fig.4. It is found from Fig.4 that the toughness of most GRC mortar
specimens stored at 5°C increased with age up to 180 days. While GRC mortar
specimens with the fiber volume fraction of 2.0 and 2.5% cured at 20°C
started losing their toughness after 60 days, the ones with the fiber volume
fraction less than 1.5% showed a slight reduction in toughness after 90 days.
Extremely rapid age-embrittlement occurred in all of the specimens cured in
a fog box at 38°C by the age of 28 days. There were little differences in
toughness between mortars with different fiber contents in the curing
temperature of 38°C. Attention should be paid to the fact that the increase
in toughness by the addition of glass fibers was completely lost in all of
the mortar specimens cured at 38°C by the age of 360 days.

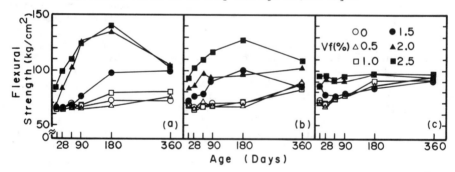

Figure 3. Flexural strength of GRC mortars without admixture:
(a) in water at 5°C,(b) in water at 20°C,(C) in a fog box at 38°C

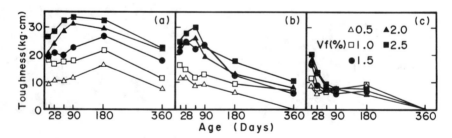

Figure 4. Toughness of GRC mortars without admixture:
(a) in water at 5°C,(b) in water at 20°C,(c) in a fog box at 38°C

Effects of curing temperature on microhardness in glass fiber-cement
paste interfacial zone : The values of microhardness measured in the interfacial zone around a glass fiber strand in samples cured at 5°C and 38°C are
presented in Fig.5(a) and (b), respectively. The microhardness in the interfacial zone in samples cured in water at 5°C decreased with distance from
the interface up to about 50µm. In the regions greater than about 50µm away
from the interface, the microhardness increased with distance up to about
100µm, beyond which microhardness decreased again toward bulk cement paste
phase.

The microhardness in the interfacial zone as well as in the bulk cement
paste phase in GRC mortar specimens cured at 5°C changed a little with curing
time by at least 60 days. On the other hand, comparison between microhardness distribution patterns around a glass fiber strand given in Fig.5(a)
and (b) apparently shows that the values of microhardness in the immediate
vicinity of a glass fiber strand are greater in samples cured at 38°C than
in the ones cured at 5°C. Another conspicuous difference in the features of
the microhardness distribution patterns between the two different curing
temperatures is that the values of microhardness around 100µm away from the
interface are considerably greater in samples cured at 38°C than in the ones
cured at 5°C. Thus, relatively rapid hardening of the regions around 100µm
from the interface in samples cured at 38°C results in discontinuous rises
in microhardness as shown in Fig.5(b).

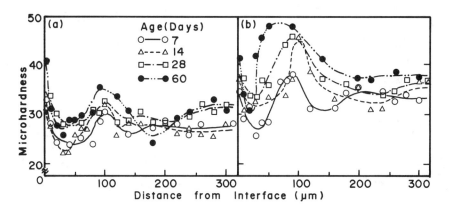

Figure 5. Microhardness distribution patterns around a glass fiber strand
in admixture-free cement paste :(a) in water at 5°C,(b) in a fog box at 38°C

Ca/Si ratio in the interfacial zone : Fig.6 shows the Ca/Si ratio at
8 spots in the region up to about 100µm from the interface. Little change
in the Ca/Si ratio with distance from the interface was found in samples
stored in water at 5°C. However, as shown in Fig.6(c), great variations of
the Ca/Si ratio in the interfacial zone in samples cured at 38°C contrast
with little change of the Ca/Si ratio found in other samples stored in water
at 5°C. High Ca/Si ratios found in several spots in samples cured at 38°C
may mean the existance of abundance of Ca-rich spots in the interfacial zone
formed at a high curing temperature of 38°C. Relation between microhardness
and the Ca/Si ratio in the interfacial zone was not unambiguous.

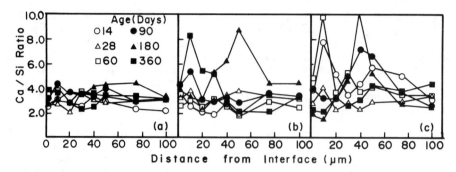

Figure 6. Ca/Si ratio in the interfacial region within admixture-free cement paste: (a) in water at 5°C, (b) in water at 20°C, (c) in a fog box at 38°C

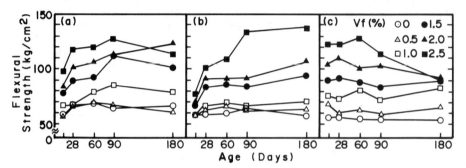

Figure 7. Flexural strength of GRC mortars with flyash:
(a) in water at 5°C, (b) in water at 20°C, (c) in a fog box at 38°C

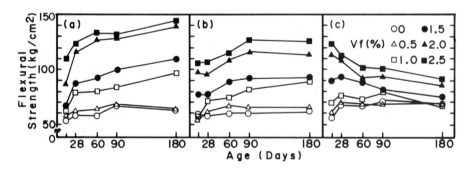

Figure 8. Flexural strength of GRC mortars with silica fume:
(a) in water at 5°C, (b) in water at 20°C, (c) in a fog box at 38°C

GRC Mortars with Admixture

Effects of curing temperature   on   flexural strength and toughness :

Fig.7 and 8 show the changes of flexural strength with age in GRC mortars with flyash and silica fume, respectively. As a whole, the time-dependent tendency in the flexural strength of admixture-bearing GRC mortars is not so greatly different from that of admixture-free ones. Silica fume-bearing GRC mortars made at relatively low fiber contents show a little greater reinforcing effect than other GRC mortars.

As shown in Fig.9 and 10, the toughness of GRC mortars with flyash and silica fume changes with age. Little reduction in toughness was found in most flyash-bearing GRC mortars stored at 5°C and 20°C until 90 days, and thereafter, the toughness of flyash-bearing specimens cured at 5°C slightly decreased with time. Comparison between toughness vs. age curves given in Fig.4 and 9 apparently shows that the addition of the flyash considerably delayed age-embrittlement in GRC mortars cured at 38°C. It is found from Fig.10 that silica fume-bearing mortars cured at 5°C completely maintained their initial toughness until at least 180 days. However, after 90 days, significant loss of toughness was found in silica fume-bearing mortars with the fiber content of 2.0 and 2.5% cured at 20°C.

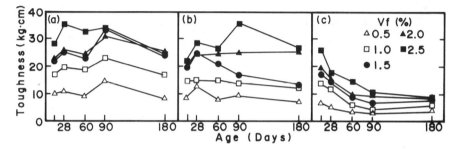

Figure 9. Toughness of GRC mortars with flyash:
(a) in water at 5°C, (b) in water at 20°C, (c) in a fog box at 38°C

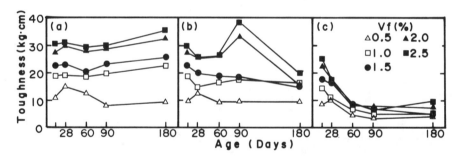

Figure 10. Toughness of GRC mortars with silica fume
(a) in water at 5°C, (b) in water at 20°C, (c) in a fog box at 38°C

Effects of curing temperature on microhardness in glass fibers-cement paste interfacial zone : Fig.11 shows the microhardness distribution patterns around glass fiber strands embedded in the flyash-bearing paste stored in water at 5°C and in a fog box at 38°C. The microhardness distribution patterns indicate that the softest region exists around a few tens μm away from the interface, followed by the regions where microhardness increases with distance from the interface toward bulk paste phase. However, as shown in Fig.11, great variations in microhardness in the interfacial zone embedded in the flyash-bearing paste cured at 38°C contrast with gradual changes in microhardness with distance within about 100μm from the interface in the corresponding samples cured at 5°C. Furthermore, Fig.11 shows that the differences in microhardness between the softest and the

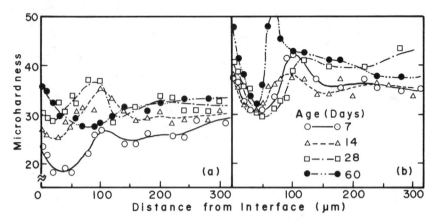

Figure 11. Microhardness distribution patterns around a glass fiber strand in flyash-bearing cement paste : (a) in water at 5°C, (b) in a fog box at 38°C

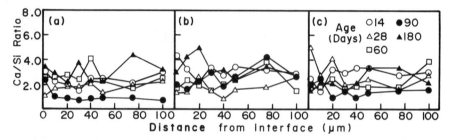

Figure 12. Ca/Si ratio in the interfacial region within flyash-bearing cement paste: (a) in water at 5°C, (b) in water at 20°C, (c) in a fog box at 38°C

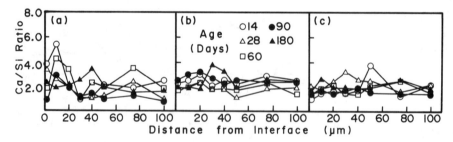

Figure 13. Ca/Si ratio in the interfacial region within silica fume-bearing cement paste: (a) in water at 5°C, (b) in water at 20°C, (c) in a fog box at 38°C

hardest regions increased with increasing curing time at the curing temperature of 38°C. Abrupt changes in microhardness vs. distance curves in samples cured at 38°C (Fig.11(b)) were also found in samples without admixture (Fig.5(b)).

Ca/Si ratio in the interfacial zone: The Ca/Si ratios in the inter-facial zone in samples with flyash and silica fume are shown in Figs.12 and 13. Ca-rich spots were not found in the samples with the flyash and silica fume except that the Ca/Si ratios in the vicinity of the interface in several samples with the silica fume cured at 5°C are relatively high.

DISCUSSION

As mentioned above, GRC mortars with and without admixture cured at 38°C showed embrittlement even at early ages, but the ones cured in water at low temperature of 5°C maintained their toughness even in later ages. Embrittlement of glass fiber reinforced cement in early ages has been attributed to changes in the microstructure of the interfacial regions between glass fibers and cement paste phase by several workers [10]. The characteristic features of microhardness distribution patterns in the interfacial zone given in Figs.5 and 11 suggest that age-embrittlement in GRC mortars is related to the characteristic microstructures formed in the limited regions between glass fibers and bulk cement paste phase. Microhardness decreased with increasing distance from the interface toward the softest regions. Beyond this softest region, the value of microhardness increased with distance, and then decreased again toward the bulk cement paste phase. The overall characteristic trend in microhardness vs. distance curves within the interfacial zone is not greatly different from one another. However, two differences in the features of microhardness distribution patterns in the interfacial region between the sample cured at 5°C and 38°C are noticeable. The one is the occurrence of discontinuous changes in microhardness in samples cured at 38°C. Microhardness in the interfacial region discontinuously changes at a position within about 100μm from the interface in samples cured at 38°C, while such abrupt changes in microhardness are not found in microhardness vs. distance curves in samples cured at 5°C. The degree of the discontinuous difference in microhardness increased with age in samples cured at 38°C. This fact shows that, in samples cured at 38°C, the microhardness of bulk cement paste phase increased with time with the rapid progress of the hydration of cementitious components despite of only a little increase in the microhardness of the softest region in the vicinity of a glass fiber strand. The other difference in microhardness distribution patterns in the interfacial zone between the samples cured at 5°C and 38°C is that the values of microhardness at the immediate vicinity of the interface are considerably greater in samples cured at 38°C than in the ones cured at 5°C. The increase of microhardness in the region adjacent to the interface may indicate the increase of stiffness at that region. This increase of stiffness at the immediate vicinity of the interface appears to mean the increase of the shear resistance against fibers pull-out. The increase in bond strength between glass fiber strands and cement paste matrix is supposed to bring about breaks of fibers themselves without debonding.

On the other hand, discontinuous changes of microhardness in the interfacial regions indicate the formation of heterogeneous regions. This heterogeneity in stiffness in the interfacial regions might raise stresses in the matrix around glass fiber strands. The discontinuity of stiffness within the interfacial zone presumably influenced the process of cracking leading to the alteration of the mechanical behavior of GRC mortars before and after the initiation of cracks in the matrix. As a whole, the characteristic features in microhardness distribution patterns within the interfacial zone in the samples with an admixture were not so greatly different from those in the admixture-free samples. Remarkable Ca-rich spots were not found in admixture-bearing samples cured at 38°C. These facts suggest that the age-embrittlement in GRC mortars should be related to the formation of the discontinuous regions in the vicinity of glass fiber strands rather than to the growth of Ca(OH)2 around glass fiber strands. In relation to the results obtained in this study, the finding that a rock-cement paste interfacial zone showing a rise in microhardness has lower resistance to the propagation of crack deserves attention [11].

332

CONCLUSIONS

The results obtained in this study are summarized as follows:

(1) Flexural strength of GRC mortars stored in water gradually increased with age up to 180 days. The flexural strength of GRC mortars cured at 38°C slowly changes with age up to 180 days. The time-dependent tendency in the flexural strength of admixture-bearing GRC mortars was almost same as that of admixture-free ones.

(2) The reduction in toughness occurred at later ages in GRC mortars cured at 5°C. Age-embrittlement occurred in mortars with and without an admixture cured at 38°C even at early ages. However, the addition of the flyash delayed the age-embrittlement in GRC mortars cured at 38°C.

(3) The characteristic trend in microhardness vs. distance curves within the interfacial zone was not changed by the addition of an admixture. However, curing temperature significantly influenced the microhardness distribution patterns within the interfacial zone.

(4) The value of microhardness discontinuously changed within about 100μm from the interface. The discontinuity in stiffness in the interfacial zone may influence stress fields or cracking processes within the interfacial zone. Thus, age-embrittlement in GRC mortars appears to relate to the discontinuous changes in microhardness.

(5) Ca-rich spots were not found in the vicinity of glass fiber strands in admixture-bearing samples cured at 38°C. However, the age-embrittlement occurred in these GRC mortars.

REFERENCES

1. Litherland, K.L. and Proctor, B.A., Proc. Symp. Durability of Glass Fiber Reinforced Concrete. Chicago, 1985, pp.124-135.
2. Cohen, E.B. and Majumdar, A.J., Proc. RILEM. Symp. 1975, pp.315-325.
3. Stucke, M.S. and Majumdar, A.J., Journal of Materials Science, 11(1976), pp.1019-1030.
4. Mills, R.H., Cement and Concrete Research, vol.11.,1981, pp.421-428.
5. Leonard, S. and Bentur, A., Cement and Concrete Research, vol.14.,1984, pp.717-728.
6. Singh, B. and Majumdar, A.J., The International Journal of Cement Composites and Lightweight Concrete, vol.7, No.1, 1985, pp.3-10.
7. Bentur, A. and Diamond, S., The International Journal of Cement Composites and Lightweight Concrete, vol.9, No.3.,1987,pp.127-135.
8. Diamond, S., Proc. 8th International Congress on the Chemistry of Cement, vol.1,1986,pp.122-147.
9. Bentur, A., Diamond, S. and Mindess, S., Journal of Materials Science, 20(1985), pp.3610-3620.
10. Diamond, S., Proc. Symp. Durability of Glass Fiber Reinforced Concrete, Chicago, 1985, pp.199-209.
11. Saito, M. and Kawamura, M., Cement and Concrete Research, vol.16, No.5, 1986, pp.653-661.

A STUDY ON FRACTURE PROCESS ZONES IN CONCRETE BY MEANS OF LASER SPECKLE PHOTOGRAPHY

XU SHILANG and ZHAO GUOFAN
Department of Civil Engineering
Dalian University of Technology

ABSTRACT

In this paper, the whole processes from the appearances of micro-cracks to the slow growth of main macro-cracks and the deformation characters of the notched concrete beams of two different dimensions before the occuring of unstable fracture were researched by means of laser speckle photography.

INTRODUCTION

The early researches were mainly concentrated on the calculation of the fracture toughness of concrete in Mode I by testing notched beam specimens. It has been observed that when K_{Ic} is calculated from the measured values of the maximum load, the initial notch length and the formulas developed from linear elastic fracture mechanics, the value of K_{Ic} is dependent on the dimensions of the beam. This size and geometry dependency can be attributed to slow crack growth and nonlinearity due to random and geometrical interlock effects.

Fracture process zones in concrete materials are the important factors influencing the fracture behaviour. This emphasizes the necessity to visualize the process of crack initiation and development. For fitting this need, this paper attemptes to find an available observation method to identify cracks on concrete and monitor the displcement fields across the cracks and regions surrounding them. The crack tip opening displcement $CTOD_c$ of concrete material was measured by using this observation method also.

----

*   Project supported by the National Science Fundation of China

EXPERIMENTAL TECHNIQUE OF SPECIMENS

The fracture process zone in concrete may involve two parts of a slow subcritical crack growth and a microcracks nucleation, The part of a slow subcritical crack growth is assumed to have a capability of aggregate bridging. It is assumed that the traction-free crack initiation will occur when the crack-front displcement at the tip of the traction-free crack equals the critical crack-displcement value. That is the critical crack tip opening displacement value $CTOD_c$. This value is often obtained from the descending part of the uniaxial stress-displcement curve when a fictitious crack model presented by Hillerbory was used, but, it was seldom gained from the direct measurement. In order to scan the fracture process zone and to measure the critical crack tip opening displacement CTODc, the laser speckle photography technique is used in this study.[1] This testing technique is accurate enough to detect and measure the displacement associated with microcracks and has full-field measurement capability. Moreover, it has to be non-contacting so that the cracking process is not affected.

In our study, three-point bending notched beam specimen was used, as shown in Fig.1. The specimen dimensions are given

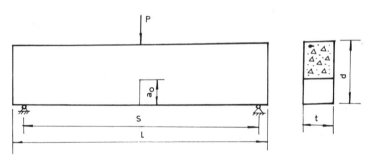

Fig.1 Three-point bending notched beam specimen

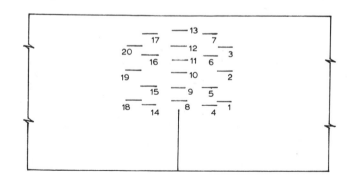

Fig.2 The measured point positions of strain gauges in front of notch in the specimen S-1, with gague length of 10 mm

in Table 1. The intial crack/depth ratios a_0/d are 0.4 and the span/depth ratios s/d are 4.

These concrete specimens were made with mix proportions of 1:1.87:3.36 and w/c=0.57 . The maximum grain size is 20 mm. All the specimens were tested between 28-32 days after they were casted. The cubic compressive strength f_{cu} is 35.2 MPa, and the splitting tensile strength f_t is 2.45 MPa. The modulus of elasticity E_c is 31000 MPa.

For potentiating the reflectivity some aluminite powder were applied and a net of 10X10 mm was drawed on the laser exposed side surface of the specimen. The loading point, the right point and left point of notch tip were marked. For specimen S-1, twenty strain gauges were attatched on the another surface in front of notch in the specimen, as shown in Fig.2.

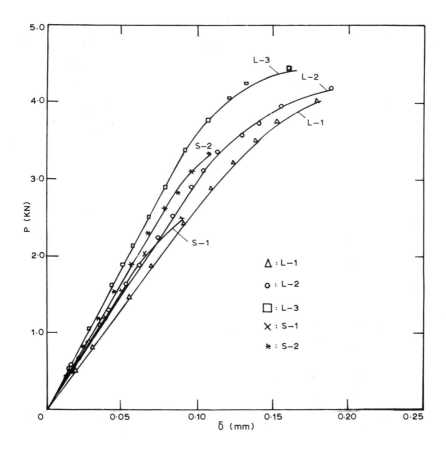

Fig.3 The load vs. loading-point displcement P-δ curves of notched beams of concrete

CRITICAL CRACK TIP OPENING DISPLACEMENT OF CONCRETE

The point-by-point analysis was used to gain the loading-point displacements, i.e. deflections of the notched beams, of the photographic plates of all load intervals of the specimens. The elastic deformations of the loading frame and the supports and the contact deformations between the specimens and supports were deducted from the total deformations measured by using the speckle photography and the point-by-point analysis. As the results, the load vs. loading-point displacement P-δ curves of all tested specimens were obtained. The P-δ curves were drawn in Fig.3.

It can be found that the loads Pnone corresponding to the occurence of the nonlinearity in the P-δ curves are 73.3 percent and 70.9 percent of their maximum loads Pmax for specimens with depths 150 mm and 100 mm separately. So, it can be known that the nonlinearity in the P-δ curves should be

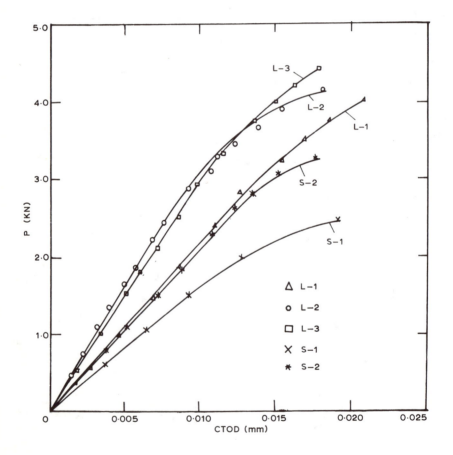

Fig.4 The load vs. crack tip opening displacement P-CTOD curves
of notched beams of concrete

TABLE 1
The measured values of $CTOD_c$, K_{Ic} and Gc
of all specimens in this study

No. of Specimen	l×d×t (mm)	P_{max} (N)	$CTOD_c$ (mm)	δ_0 (mm)	G_c (N/m)	K_{Ic} (MPa√m)
L-1	700X150X100	4000	0.0207	0.169	138.6	0.998
L-2	700X150X100	4158	0.0181	0.178	154.8	1.001
L-3	700X150X100	4413	0.0177	0.162	104.1	1.076
Average		4190	0.0188	0.170	132.5	1.025
S-1	515X100X100	2435	0.0191	0.090	50.3	0.770
S-2	515X100X100	3302	0.0176	0.109	65.1	0.989
Average		2869	0.0184	0.100	60.5	0.880

independent on the depth of specimen within the range of this test.

The load vs. crack tip opening displcement P-CTOD curves of all specimens were drawn in Fig.4. It can occurence the nonlinearity in the P-CTOD curves are 71.0 percent and 70.9 percent of their maximum loads Pmax for specimens with depth 150 mm and 100 mm separately. the nonlinearity in the P-CTOD curves should be independent on the depth of specimen within the range of this test also.

Recent investigations showed that the critical crack tip opening displacement $CTOD_c$ could be taken as a material parameter to judge the unstable fracture in concrete. It is very convenient that the laser spekle photography is used to measure the critical crack tip opening displacement $CTOD_c$. The measyred values of critical crack tip opening displacement $CTOD_c$, fracture toughness K_{Ic} and critical strain energy release rate G_c of all specimens in this study were given in Table 1. The results show that fracture toughness K_{Ic} and critical strain energy release rate G_c are dependent on the dimensions specimens, but, critical crack tip opening displcement CTOD might be independent on the dimensions of specimens. Hence, the critival crack tip opening displacement $CTOD_c$ could be taken as a material parameter to judge the fracture of concrete.

CRACK PROPAGATION AND FRACTURE PROCESS ZONES IN CONCRETE

It is well known that crack proparation and fracture process zones in concrete are the important factors influencing on the fracture behaviour of concrete. In this study, it is mainly done to visualize the process of crack initiation and development. Using the point-by-point analysis, the cracks on the concrete specimen surface at different loads were positioned by detecting the regions which were manifested by a jump in the orientation and the period of young's fringes see Fig.5

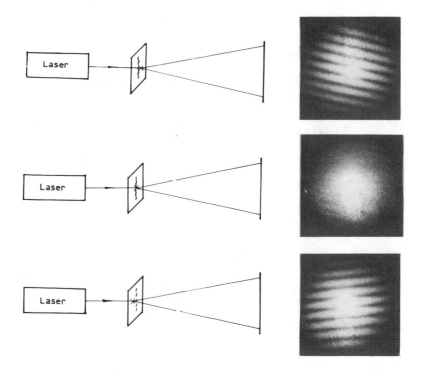

Fig.5 Visulization of the crack path and subsequent measurement
using speckle photography

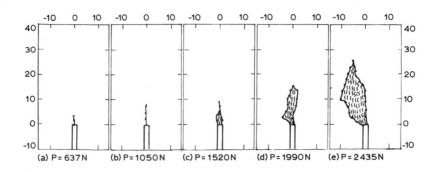

(a) P= 637N   (b) P= 1050N   (c) P= 1520N   (d) P= 1990N   (e) P= 2435N

Fig.6 The process of crack propagation and development of
fracture process zone in a surface
of concrete specimen S-1

339

By this way , the process of crack propagation and development of fracture process zone on the concrete surfaces of a typical specimen was carefully drawn in Fig.6. From the obsevations of all specimens, it is found that there is an evident subcritical growth process for main crack before unstable fracture occurs. The shapes of the fracture process zones are irregular, long and narrow bands. And the average maximum width of the fracture process zones of these specimens is 9.6 mm which is equal to about 50 percent of the maximum aggregate particle size.

DISTRIBUTION OF STRATION FIELD IN FRONT OF NOTCH

The depth of the specimen S-1 is 100 mm, with a precast notch

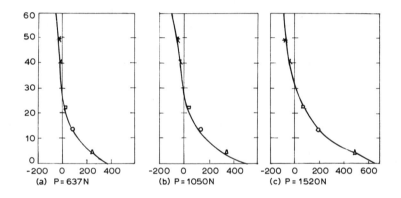

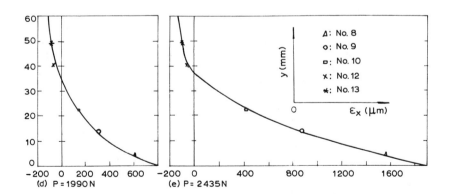

Fig.7 The distribution of strain field in front of notch in specimen S-1

length of 40 mm. The maximum load Pmax measured before occuring of unstable fracture is 2435 N. The positions of twenty strain gauges attaching on a surface in front of notch in the beam can be seen in Fig.2. The lengths of all gauges are 10 mm. The measured results showed that there is an evident strain concentration within a band region in front of crack along crack axis. The distance from notch tip is smaller, the concentration of strain is intenser. But, the change degrees of the strain values which were measured from measurement points No.1 to No.7 and No.14 to No.20 with the distances of 10 mm and 15 mm from the span middle of the beam are very small, and these values of strain are further lower than the limit tensile strain value of the concrete. So, the subcritical crack propagation and formation of micro-cracked zone only occured within the band region. These observations are in agreement with the speckle photographic observation.

The measured results of horizontal direction strain ε_x of measurement points No.8 to No.13 were shown in Fig.7. Contrasting Fig.6 and Fig.7, it was known that the limit tensile strain value of the concrete of this beam is 250 μ. If the maximum tensile strain is taken as a cracking criterion, then, when ε_x of a point is larger than 250 μ, the material of this point will crack. Analysing the experimental results of strain values from Fig.7, the lengths of crack propagation in the specimen surface for various loading stages were listed in Table 2 using this criterion. as a contrast, the observed results of crack propagation lengths in the another surface of the same specimen for various loading stages obtained by speckle photography were listed in table 2 also. It can be found that the observed results of lengths of crack propagation using above mentioned two experimental methods are very similar.

TABLE 2

Observed results of crack propagation length
in a surface of specimen S-1

Ratios of loads Pi/Pmax	Observed results by speckle photography from Fig.6 (mm)	Observed results by measurements of strain gauges from Fig.7 (mm)
0.262	4.0	4.5
0.431	8.0	7.0
0.624	10.0	10.5
0.817	15.0	16.5
1.000	26.0	27.5

From Fig.7, it was found that when load was small, the neutral axis of net section of the beam coincided with the horizontal axis of centre of net cross section of the beam. Before fracture the position of neutral axis of beam had an upward moving from 70 mm to 77mm at a distance from bottom of the beam of which the depth was 100 mm and length of precast notch was 40 mm. It means that there was a subcritical main crack propagation in this specimen.

CONCLUSIONS

1. It was observed that there is an evident subcritical growth process for main crack before unstable fracture. The shapes of the fracture process zones are irregular; long and narrow bands. And the maximum widths of the fracture process zones are equal to about 50 percent of the maximum aggregate particle size.

2. The properties of the fracture process zones are in agreement with the Dugdale-Barenblatt crack model.It is confirmed again that a fictious crack model presented by Hillerborg and a blunt crack band theory presented by Bazant are available for analyzing the cracking and fracture in concrte .

3. Critical crack tip opening displcement $CTOD_c$ measured within the region of this test is independent of the size of specimens. Therefore the parameter $CTOD_c$ could be taken as a material parameter to judge the unstable fracture in concrete.

4. There is an evident strain concentration within a band region in front of crack along the crack axis. The position of the beam neutral axis had an upward movement before unstable fracture occurred.

REFERENCE

1   Jacquot, P. and P.K. Rastogi, Speckle Metrology and Holographic Interferometry Applied to the Study of cracks in concrete, in Fracture Mechanics of Concrete (edited by F.H. Wittmann), Elsevier Science Publishers B.V., 1983, The Netherlands, pp. 113-156.

STRENGTHENING THE INTERFACIAL ZONE BETWEEN STEEL CEMENT PASTE

CHEN ZHI YUAN and WANG NIAN ZHI
Department of Materials Science and Engineering
Tongji University, Shanghai, China

ABSTRACT

The interfacial zone between cement paste and steel bar is the weakest zone in the reinforced concrete. In this paper the method for strengthening the interfacial zone was proposed. After strengthening the interfacial zone was examined by XRD, SEM, and Microhardness tester. The results show that in the zone after strengthening the portlandite is finer in size, less in quantity and denser in structure consisting of a large quantity of C-S-H gel than that in origeneral one. So, strengthening the interfacial zone is an effective way to improve the durability and mechanical properties under loading.

INTRODUCTION

Specialists said for sure that reinforced concrete (R.C.) would still be the one of the main building materials in 21st century. The strength of R.C. depends not only on the strength of the concrete and steel, but also on the bonding strength between them because the bonding enables stresses to be transferred through the interface between concrete and steel so as to increase the tensile and flexural strength of R.C. The bonding becomes more important in the cementitious composite including steel with smooth surface such as steel fibre reinforced concrete (SFRC) and floor and slab systems.

Evidences show that (1) under loading the cracks in R.C. or SFRC tend to propagate along the weakest region in the interfacial zone between steel and concrete. Strengthening the interficial zone, under common loading rate, will increase the strength of the components and not considerably influence the toughness.

As to the durability of R.C. or SFRC, the anticorrosion properties of components will depend to a great extent on the microstructure of interficial zone, expecially for concrete which contains aggressive ions in themselves. In the porous interfacial zone the rust gets easily to diffuse and preferentially deposit, making the corrosion rate of steel quicker. Densifying the interfacial zone will greatly restrain the corrosing process.

The interfacial zone is weak because there is a higher W/C ratio within the zone than that in the bulk cement paste. And within the zone the hydration of cement ends fundamentally in the early age, then the interfacial structure will not be denser and strengthened further and forms a porous region. Moreover, the surface of steel is rich in portlandite with preferred orientation (3-5). In this paper the methods to strengthen the interfacial zone and the strengthened structure were studied.

MATERIALS AND METHODS

Materials

Portland cement and high-quality mild steel were used for preparing the samples. The steel rod used was 5 mm and 10 mm in dimeters. The former was in rolled surface; the latter was in polished surface. Three kinds of samples used are shown in Fig.1, in which sample (a) was used for XRD, pulling-out tast and electrochemical measurements, sample (b) and (c) were used for SEM and microhardness test. Each kind of samples had two types. Type A was in ordinary moulding process (W/C = 0.35); in type b the interfacial zone was strengthened, others were entirely the same as type A. All samples after demoulding were cured in water up to 28 days.

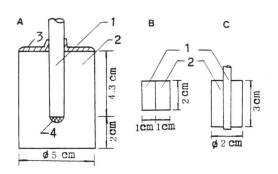

(A) 1. 10 mm steel bar
2. cement paste cylinder
3. paraffin
4. epoxy
(B) 1. steel block 20 x 20 x 10 mm
2. cement paste 20 x 20 x 10 mm
(C) 1. 5 mm steel bar
2. 20 mm cement paste cylinder

Figure 1. Three kinds of samples used.

The method to strengthen interfacial zone

Now, there are three common methods used for improving the interfacial zone between aggregate and cement paste: using silica fume as additive, decreasing W/C ratio of bulk paste and the Sand Enveloped with cement (S.E.C) technology. The addition of silica fume into concrete will decrease the pH value of the solution in pores of hydrated cement paste (hcp), so as to increase the

corrosion rate of steel (6). And the decrease of W/C ratio of bulk
paste will make the concrete more expensive and the porous region
in interfacial zone cannot be entirely eliminated. The idea of
S.E.C. technology is useful here for strengthening the interfacial
zone between steel and cement paste, but we must take into account
that bonding of steel with smooth smooth surface to cement paste
is much weaker than aggregates, and steel bar can not be mixed
with paste. So, in our experiment a molding method was used as
follows:
1. Coating a thim layer of fresh cement paste with the same W/C
ration as that in the bulk paste on the smooth steel bar;
2. Applying dry cement power on it so carefully that the whole
surface of steel can be coated by a layer of power.
3. Knocking it gently, dropping the superfluous power on it and
 making the layer of power very thin and even;
4. Putting it in mold and casting very carefully.

EXPERIMENTS AND RESULTS

1. Microhardness test

Measure the microhardness both on the polished cross section of
cylinder sample (c) and the surface of cement paste side of the
interface in cubic sample (b), which had been cut into two halves
just on the interface between steel and cement paste. On the
former, a distribution curve of microhardness along a
perpendicular line to the interface was plotted (Fig.2); on the
latter, and average microhardness value from the values measured
on 10 spots on the surface was obtained which was 32.52 kg/mm2 for
type a and 46.67 kg/mm2 for type B.

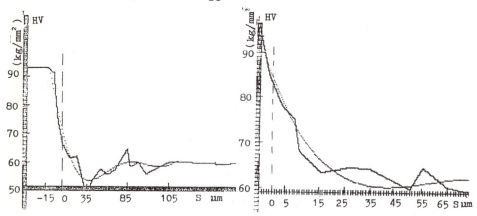

with type A inferface   with type B interface

-- -- -- interface ------ line joining the spots measured .....
line imitated by computer

Figure 2. the microhardness values across the cross section of
composite sample

According to the microhardness - porosity plot of hcp established by C.J. Pinchin (4), the porosity on the interface here, corresponding to the microhardness on the surface of interface in cement paste side, is 29% for type A and 21% for type B.

The distribution curve of microhardness shows that the microhardness values for type B is higher than that for type A at every spots in interfacial zone in cement paste side up to the spots 80 um away from the interface, where the values for both samples reach to the values of bulk about 59.3 kg/mm2. What deserves mention is that there is a weakest region in interfacial zone for type A, but not for type B. All spots in the interfacial zone of type B are stonger than that in the bulk paste.

2.Counting the number of pores

Count the number of pores in a 250 μm x 5 mm area on the surface of cement paste side of the interface with the optical microscope with magnifying of 4 x 20. the results were showed on Tab.1. In addition, choose 20 lines on the surface, each 500 um long and 100 um apart from each other. Count the total length of each line which goes throught the pore (L), the porosity P = L/500.

TABLE 1
Pore size distribution within 250 x 5000 μm in cement paste side

Pore size, μm		5	5-10	10-25	25-50
count of pores	Type A	128	568	333	199
	Type B	410	380	117	16

The average porosity for 20 lines is 30.15% for type A and 19.7% for type B. These values are essentially in accordance with that derived from the microhardness values metioned above, showing again that the porosity of interface for type A is much lower than that for type B. Tab. 1 also shows that there are very few big pores and relatively more small pores for type B.

It must be pointed out that the optical microscope cannot distinguish the pores with less than 1 μm diameter, but this method is still useful for comparing the porosity of interface approximately, especially for comparing the count of big pores.

3. SEM observation

Take a series of photograghs for the cross section and both sides of the interface of composite samples with SEM, observe the density at low magnifications and the hydration products at high mignifications.

The microphotographs show obviously that the interface of type b of composite sample is much denser than that of type A and has much fewer big pores. There is a long crack along the interface in type A sample, probably due to the dry shrinkage of sample during the experimental process. While the type B sample has no crack along the interface, indicating the stronger interfacial bonding. Moreover, there are many microcracks in the region 15-50 um away from the interface in type A sample, but few in type B sample as shown in Fig.(3). A lot of photos shows the similar results. The location of the weakest region is in accordance with that derived from microhardness test.

Fig.(4) shows that the interfacial zone of type B is mainly composed of flocculent CSH gel and is very compact, but the one of type A is composed of a little fibriform CSH gel with pores in it and many CH crystals. On the surface of steel side of interface, there are many oriented big CH sticking on steel surface for type A and dense gel for type B as shown in Fig.(5).

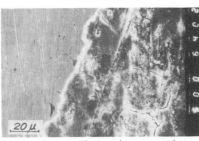

Sample A: many microcracks   Sample B: few microcracks
         near the interface   near the interface

Fig.3: SEM photograghes of cross section.

Sample A: big CH and fibriform   Sample B: flocculent CSH gel
         CSH gel

Fig.4: SEM photograghs on the hcp side of interface

Sample A: big CH   Sample B: dense CSH

Fig.5: SEM photograghs on the steel side of interface

 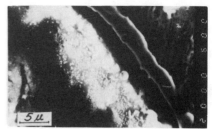

Fig 6: Big CH crystals on
Sample A interface,

Fig 7: A CH ribbon along the
interface in sample A
cross section,

We can see a ribbon consiting of big CH crystals along the surface
of steel on type A on the Fig.(6) and Fig.(7).

4. XRD test

Take the hcp of the 10 um thickness adjacent to the surface of
steel after the steel bar was pulled out with an adjustable
reamer, (see ref.(3)) then ground and mix the hcp with pure Al2O3
powder as a internal standard in the ratio of 5:2 by weight. Put
the mixed powder to the XRD test.
a. The amount of CH on the interface can be compared semi –
   quantitatively with the parameter Q(CH) (3):

$$Q(CH) = H(CH) + H(CaCO3) / H(Al2O3)$$

H(CH): height of diffraction peak for the (001) faces of CH
crystals (diffration angle α = 9.05) H(CaCO3): height of the
strongest peak for CaCO3 and Al2O3 respectively.
b. The average diameter of CH crystals in the direction
   perpendicular to (001) faces can be compared with the parameter
   D(CH):

$$D(CH) = K\lambda / \beta \cos\alpha$$

K,λ: constant; α = 9.05 ; β : the radian corresponding to the
width at half height of the diffraction peak for (001) faces.

c. The degree of hydration for hcp can be compared with Q(C3S):

$$Q(C3S) = H(C3S) / H(Al2O3)$$

H(C3S): height of the strongest peak for C3S.

d. Measuring and calculating the orientation index I for CH (7) on
   the interface. The results are listed in Tab.2. In the
   interfacial zone of type B, compared with that of type A, both
   quantity and size ofCH decrease substantialy; the orientation
   of CH at the surface of steel is essentially eliminated; the
   degree of hydration of hcp is lower. All parameters mentioned

above for type B are similar as that for the bulk cement paste

TABLE 2
Contents average diameter and preferred orientation of CH
and contents of C3S

	QCH	D001(A)	QC3S	I
in bulk paste	1.75	241	0.58	---
type A interface*	2.71	301	0.47	2.81
type B interface*	2.00	215	0.55	1.33

* on cement paste side.

5. Measuring the electrochemical properties

The main test circuit for the measurement of electrochemical
properties is shown in Fig.8.

The samples tested at a particular age were immersed in a
saturated Ca(OH)2 solution for 2-4 hours, to lower the resistance
of the cement paste. then the electrochemical measurements were
carried out in that solution.

Before testing, the content of dissolved Cl- ions within 10
um thickness of hcp adjacent to the interface was determined. Both
type A and B samples (in 10 C water) were measured; the results
were the same. When 1% or 2% were added, the contents of Cl- were
0.43% amd 0.24% by weight, respectively. The aim is to eliminate
the possibility of electrochemical change caused by the different
concentrations of dissolved Cl-.

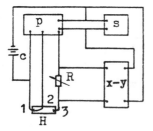

P: Potentiostate
S: ultra low frequency generator
C: condenser
R: resistance
X-Y: X-Y function recorder
H: constant temperature liquid
   container
1: steel embedded in cement
   paste
2: platinum electrode
3: saturated calomel electrode

Fig.8. Testing circuit for electrochemical properties.

In general, the service life of steel bar in rainforced
concrete is divided into two stages: the initiation stage of
corrosion and the propagation stage. The initiation stage depends
upon the rate of Cl- diffusion towards the anode and of
penetrating the passive film. The acceleration rate of the
corrosion phopagation stage depends upon additional Cl- ions in

the interfacial zone migrating towards the steel, and the rate of
diffusion on the corrosion products. Althought there isa corrosion
during the propagation, the concrete can still be used in service
until the rusts get so much as to make the concrete crack.

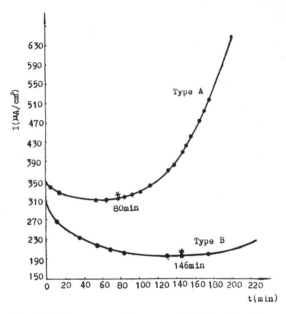

Fig.9: The composite sample (steel and paste,
W/C = 0.35 NaCl 1%, 28 days) tested at
1100 mV contant potential.

Under a constant potential, the samples underwent an
accelerated corrosion test. The results of composite samples
tested at 1100 mV constant potential are shown in Fig.9.   In
figures, the steady value of current indicates the corrosion rate
of steel for the passive state. The time at which the current
starts to increase marks the rupture of the passive film and the
end of the initiation state. From then on, the rate of current
increase can measure the corrosion acceleration. From the figures,
the steady current density response I of the B-type sample is
lower than that of type A. After the rupture of the passive oxide
film, the acceleration of current increase is far slower in type B
than that in type A. The benefits of the type B interface are
obvious.
     The samples both type A and B containing 3% NaCl were
clivaged after constant potential test at 750 mV for 50 minutes
and the steel surface of comosite sample were examined. Covering
transparent film on the steel surface, the rust area was traced.
the typical rust patterns for type A and B are shown in Fig.10

 Type A 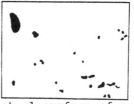 Type B

Fig.10. The rust pattern on the steel surface of composite sample after constant potential test at 750 mV for 50 minuts.

From the figure the rust area of type A sample is much larger than that of type B, and the former connectes together and forms a large extent, while in type B interface the rust forms in spots and little extent.

6. Pulling out test

The embedded steel bars in sample (a) were pulled out at the water-saturated condition and 12 mm/min pulling rate. The bonding stress which is obtained by dividing the maxium load by the interfacial area is 90.4 N/cm2 for type B, 29% greater than that for type A (70.2 N/cm2), indicating again that the interface of type B is denser.

DISCUSSIONS

A good coincidence in all results mentioned above is a firm base to analyze for the characteristics of the interfacial zone. The interfacial zone of type B is more compact than that of type A, even than the bulk, and unlike type A, has no weakest region in interfacial zone.

As to the hydration products, in the interfacial zone of type A, there are many big CH crystals. CSH gel structure is porous and fibriform or reticulate in form consisting of I and II types gel. There are few unhydrated cement. This is because of a layer of water film which is convered on interface during the early age of hydration. The W/C ratio in interfacial zone is then very high. CH crystals get very big and CSH gel also grows into fibriform products outside the anhydrous cement through solution-precipitation reaction because there is enough space for them to grow, thus leading to the weak and porous interfacial zone. However, in interfacial zone of type B, CH crystals are very small and CSH gel becomes flocculent and very dense due to absorption of water by an hydrous cement and hydration in site because there is lower W/C ratio in the interfacial zone and not enough space for them to grow up, lending to a compact microstructure. It can be expected that the microstructure of interfacial zone will become much denser at the later age due to the hydration of unhydrated cement remaining.

A hypothesis called " electrical double layer overlapping" is maybe able to explain the mechanism, of formation of interfacial zone (8).

CONCLUSION

Strengthening the interfacial zone between steel and cement paste of reinforced concrete with our technology is a effective way to improve the durability and mechanical properties under loading.

REFERENCES

1. Bentur, A., Mindess, S., Diamond, S., "Cracking processes in steel fibre reinforced cement paste" Cement and Concrete Research (C.C.R.) 1985, 15(2), PP331-342.

2. Chan Zhi Yuan, Wang Nian Zhi "The Effect of Cement Paste-Steel Interfacial Zone on the Electrochemical Properties of Reinforcing Rod" In MRS Symposium Proceeding Vol. 114 "Bonding in Cementitions Composites". Editors S. Mindess and S.P. Shah (Boston, 1987) p.277.

3. Pinchin, D.J., Tabor, D., "Interfacial phenomena in steel fibre reinforced cement I: Structure and Strength of Interfacial region" C.C.R. 1978, 8(1) pp15-24.

4. Rinchin D.J., and Tabor, D. "Mechanical Properties of the Steel / Cement Interface: Some experimental Results" In "Fibre Rinforced Cement and Concrete" Vol.2, P.521 RILEM Synposium 1975, General Editor, Adam Neville in association with D.J. Hannant, A.J. Majumdar, C.D. pomeroy, R.N. Swamy.

5. Page C.L., Khalaf, M.N.Al., ritchie, A.G.B., "Steel / Mortar InterfacesL mechanical Characteristics and Electrocapillarity" C.C.R. 1978, 8(4), PP481-490.

6. Montiro, P.J.M., Gjørv, D.E., Mehta, P.K. "Microstructure of the Steel - Cement paste interface in the presence of chloride" C.C.R. 1985 15(5), PP781-784

7. Grandet, J., Ollivier, J.P., "New Method for the Study of Cement - Aggregate InterfaceS" 7th International Congress On the Chemistry of Cement, Vol.III, VII-85 Paris, 1980.

8. Wang Nian Zhi, "Relationship between Microtructure of transition Zone between steel and hcp and Electeochemical Performance of steel Reinforcement" "Papers Abstracts of the Third National Colloid and Interface Chemistry Symposium" (shanghai, China, Nov. 1987) B30, (in Chinese).

DEBONDING PROCESSES BETWEEN STEEL FIBRE
AND CEMENT MATRIX

JANUSZ POTRZEBOWSKI
Institute of Fundamental Technological Research
Polish Academy of Sciences
00-049 Warsaw, Swietokrzyska 21, Poland

ABSTRACT

The aim of the presented research is to investigate the interaction between steel fibre and cement matrix during the loading up to the level when the ultimate pull-out load is reached. To assure a very high accuracy the speckle photography method was applied. The process of the fibre debonding and the propagation of the crack at the interfacial zone along the fibre have been observed. The position of the crack front at consecutive levels of loading has been exactly determined. Using the least squares method the distribution along the fibre of the fibre-matrix slip has been described by analytical formulae and the best fit continuous curves have been found out. Then, the distribution of the fibre strain and stress has been derived. Finally, the effective fibre-matrix bond distribution at each level of loading has been determined along the fibre.

   The general model of the bond phenomenon in SFRC in which friction is the main factor being responsible for the fibre-matrix interaction was proposed.

INTRODUCTION

There are several mechanisms of fracture which may be distinguished in fibre reinforced concrete (FRC) elements under loading such as: crack appearance and propagation in the brittle matrix, fracture of the chemical, physical and mechanical fibre-matrix bond and the pulling-out of the fibres across the cracks. This last phenomenon is connected to the yielding of steel fibres and to the local crushing of the matrix, (1). The most important among these mechanisms seems to be the debonding and the pulling out of fibres. These two mechanisms are mainly responsible for the strength and toughness of the FRC. Their separation and quantitative description are indispensable for any analytical treatment of interaction between fibre and matrix in SFRC.

   In several studies ((2),(3),(4)) the bond phenomena in the fibre-matrix interface were estimated after pull-out test results but in that approach the distributions of bond and stress in the interface along the fibre could be only assumed as a hypothesis.

To obtain reliable information on real distribution of bond only these test methods seem to be appropriate which do not interfere in the above mentioned phenomena but which give enough precise results. Such methods were applied by the author to specimens with fibre-matrix interface exposed to external observation (5): i) by optical microscope the interface was observed during pull-out loading and ii) relative fibre-matrix displacements were measured by the speckle photography method.

In the paper the application of the method ii) to a series of specimens is described and an approach to analyse the obtained results is proposed.

TEST METHOD

The aim of the performed tests was to investigate the interaction between steel fibre and cementitious matrix during loading from an initial unloaded stage to a level when the ultimate pull-out load has been reached.

The applied test method is described in detail in (5). It is enough here to mention that the paste specimens were prepared with a single fibre exposed on an external face. The loading has been carried out stepwise and the increments of fibre and matrix displacements have been measured at the successive loading steps. The distribution of the measurement points along the fibre is given in Fig 1.

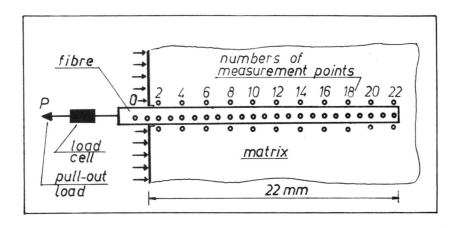

Figure 1. Distribution of measurement points.

More information of specimens preparation, the description of the speckle photography method and details of the procedure may be found out in (5),(6) and (7).

TEST RESULTS

The relative displacements caused by the increments of the pull-out load were measured at the previously marked points on fibre and matrix. The

applied distribution of measurement points assured the precise
determination of displacement increments of the fibre and of the matrix.
The density of the measurement points along the fibre was double in
comparison to these on the matrix because the essential changes of
displacements between two points of the fibre were expected.

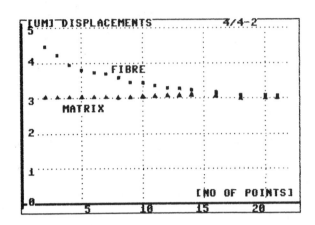

Figure 2. Relative displacements of the fibre and the matrix caused by an
increment of the load.

An example of results of the measurements is shown in Fig.2. The
increment of displacements (relative displacements) of the fibre and the
matrix caused by the one step of the loading is shown. The very small rigid
shift of matrix was introduced intentionally to get a basis for estimation
of a slip between fibre and matrix.

ANALYSIS OF TEST RESULTS

As the first step the relative fibre—matrix slips at the marked measurement
points for each stage of loading were calculated. The discrete relative
values of the slip at these points obtained for successive step of loading
are shown in Fig.3.

To obtain the global values of fibre—matrix slip a superposition of
values received at successive steps of loading should be made. The simple
superposition involving the addition of values measured at marked points
cannot give positive results because:
- the influence of both measurement errors and scatters might completely
  spoil the results,
- the small global load relaxation between the successive loading steps
  could be omitted.

To avoid the summing up errors and scatters the discrete slip values were
approximated at each level of specimen deformation by continuous curves
using the least squares method. As the result of several attempt to obtain
the best fit curves two types of analytical function finally have been
employed:
- exponential function type Y=EXP(a*X+b) applied to the most of the loading
  levels and

- linear function type Y=c*X+d applied in a few cases together with the exponential functions.

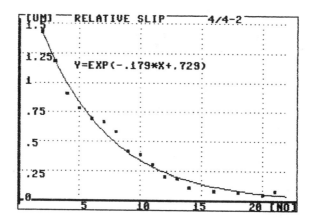

Figure 3. Approximation of discrete slip values by an exponential curve.

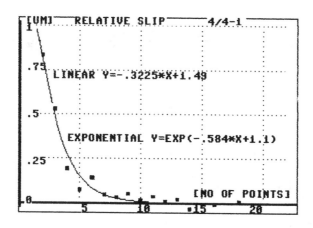

Figure 4. Approximation of discrete slip values by linear and exponential function.

An example of the approximation of measured values by a curve representing an exponential formula is shown in Fig.3. The connection of two mentioned above curves may be observed in Fig.4 where the first step of loading from P=0 N to P=15 N is shown.

The small relaxations of the pull-out force during interruptions between successive steps of loading were taken into account and considered as the decrement of the slip proportional to the load relaxation. The interruptions were caused by the requirements of the speckle photography method itself and lasted several minutes each. The degree of load relaxation was relatively low (as average 4% of the global pull-out force

at each loading level). The pull-out load relaxations were produced
probably by small deformation of the grip in which the specimens were
maintained and possible by other plays in the loading system. In Fig.5 the
procedure of the slip superposition in connection with the slip reduction
caused by the load relaxation is shown. In similar way the slip
distributions along the fibre at each level of loadind in the range from
zero until ultimate pull-out load were separately obtained for each tested
specimen . The example of the slip distributions for the whole range of
loading is shown in Fig.6.

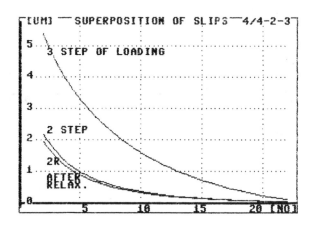

Figure 5. An example of slip superposition.

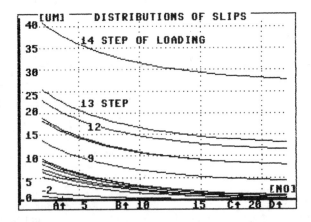

Figure 6. Distributions of slips along the fibre for one specimen at whole
range of the pull-out load.

357

The analysis of obtained curves of the global slips allows to estimate precisely the crack length and propagation of the crack tip during loading. It was supposed that the crack tip was situated at a point where no more fibre-matrix slip was observed. In Fig.6 the crack propagation along the fibre in interfacial layer may be observed.The position of the crack tips at consecutive steps of loading are marked by letters "A", "B", "C", "D". Then the full debonding of the fibre occured, although the pull-out load still increased. Such a behaviour indicates the substantial role of friction in transferring the forces from fibre to matrix during fracture of the SFRC.

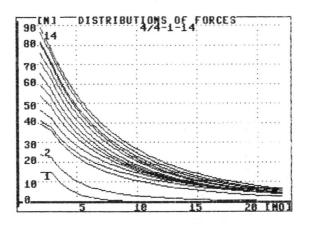

Figure 7. Distribution of force in the fibre along its length.

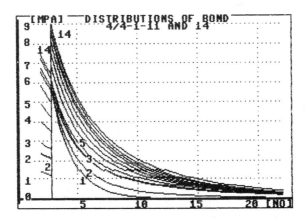

Figure 8. Distributions of bond along the fibre for one specimen.

The analytical descriptions of the fibre-matrix slips were used for further considerations. Applying the procedure of differentiation to the function describing the slip the distribution of the strain along the fibre has been derived. Knowing the Young's modulus and the dimensions of the fibre the distributions of stresses and forces at the fibre have been calculated (Fig.7). Then the fibre—matrix bond distributions were determined by double differentiation of the slip distribution with respect to the fibre, (Fig.8). The lower bond in small region at the front of the fibre was caused by the lack of fibre—matrix interaction in this region at the first step of loading. The interaction in that region probably was damaged during demoulding of the specimen. At the next step of loading the interaction in this region of the fibre has the form of friction. One can easily distinguish regions of different bond conditions which appeared in spite of the assumption of smooth and continuous curves of slip distributions.

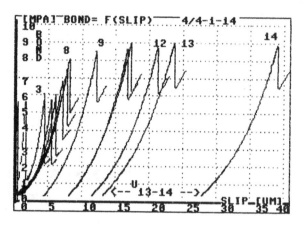

Figure 9. Relation bond versus slip.

The analytical description of the real bond distribution is a step forwards to the construction of the fibre—matrix interaction models and allows to obtain the reliable bond—slip relations. In the existing

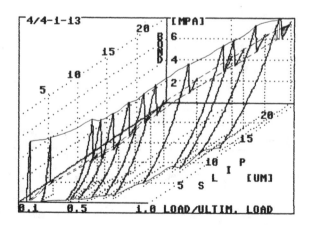

Figure 10. Bond as a function of fibre—matrix slip and ratio n

analytical models these relations were frequently assumed as simply linear, (4),(8). The real bond-slip relation is shown in Fig.9 as a set of curves representing successive steps of loading of one specimen. There is no simple functional relation between bond and slip because for one value of the slip several values of the bond do exist. It suggests that besides the slip a second factor exists which influences the bond value, namely the ratio n=(pull-out load)/(ultimate pull-out load), expressing how advanced is the pull-out load. The bond presented as an function of the slip and of the ratio n is shown in Fig.10. Points of plotted surface represent the real bond conditions during the process of loading up to the ultimate load.

In Fig-s 9 and 10 records of the rigid slip increments caused by the fibre motion may be observed (for instance U_{13-14}). As it is seen these slip increments do not influence the fibre-matrix bond. The existing interfacial friction allows to distinguish following components of the pull-out process: deformation of the fibre, fibre debonding and motion, repeated fibre anchorages. The global fibre-matrix slip is a sum of the rigid fibre slip and the slip caused by the strain of the fibre.

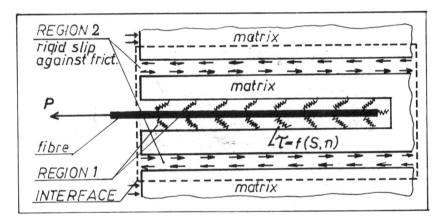

Figure 11. Scheme of the fibre-matrix interface model.

Fig. 11 shows a scheme of the fibre-matrix interface model based on above considerations. Two regions of interactions are distinguished: region 1 where the interactions are modelled by springs and region 2 - characterized by the frictional interaction. The constitutive relation describing the behaviour of the region 1 is presented in the form of the graph in Fig.10 as the function

bond=F(slip, (pull-out load)/(ultimate pull-out load))

The pull-out force is transferred from the fibre over the interfacial region 1 (deformation of the springs) to the region 2. When a temporary force equilibrium assured by friction is exceeded because of the pull-out load increment, then the motion of the deformed fibre starts in region 2. The rigid slip is observed till the moment when a new equilibrium in region 2 is reached and then further deformations of springs starts.

CONCLUSIONS

The carried out investigation is an attempt to combine the speckle photography with the pull-out method applied to the SFRC. From the obtained results the distribution of the fibre strain and stress as well as the effective fibre-matrix bond distribution at each level of loading have been determined along the fibre. To confirm the high accuracy of the results the compliance between values of pull-out load measured by means of a load cell and from force distribution in the fibre along its length (Fig.7) is presented in Fig.12.

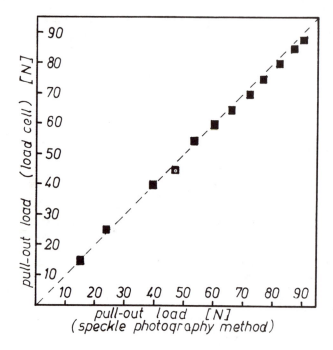

Figure 12. The compliance between pull-out load measured and deduced from the force distributions

The results of the tests have confirmed the main role of friction in load transfer between steel fibres and cement matrices in SFRC. The chemical and physical bond is essential only at the initial phase of loading. Then the debonding of the whole fibre occurs far before the ultimate pull-out load is reached. t is believed that the proposed general model of the bond phenomenon will allow to take into account the bond processes in the designing of SFRC.

REFERENCES

1. Brandt, A.M., Influence of the fibre orientation on the energy absorption at fracture of SFRC specimens. In Brittle Matrix Composites 1, eds; A.M. Brandt, I.H. Marshall, Elsevier Applied Science Publishers, London, 1986, pp. 403-420

2. Pinchin, D.J. and Tabor, D., Interfacial contact pressure and frictional stress transfer in steel fibre cement. In: Testing and Test Methods of Fibre Cement Composites, ed., R.N. Swamy, The Construction Press Ltd, Lancaster, 1978, pp. 337–341.

3. Stroeven, P., de Haan, Y.M., Bouter,C. and Shah,S.P., Pull-out tests of steel fibres. In: Testing and Test Methods of Fibre Cement Composites, ed., R.N. Swamy, The Construction Press Ltd, Lancaster, 1978, pp. 335–365.

4. Burakiewicz, A., Testing of fibre bond strenght in cement matrix. In: Testing and Test Methods of Fibre Cement Composites, ed., R.N. Swamy, The Construction Press Ltd, Lancaster, 1978, pp. 335–365.

5. Potrzebowski, J., Behaviour of the fibre/matrix interface in SFRC during loading. In: Brittle Matrix Composites, eds: A.M. Brandt, I.H. Marshall, Elsevier Applied Science Publishers, London, 1986, pp. 455–469.

6. Potrzebowski, J., Investigation of steel fibre debonding processes in cement paste. In: Bond in Concrete, ed., P. Bartos, Applied Science Publishers, London, 1982, pp. 51–59.

7. Potrzebowski, J., Speckle photography for measurement of fibre/matrix displacements in SFRC, In: Mechanics and Technology of Composite Materials, (ed.), Y. Simeonov, Publishing House of the Bulgarian Academy of Sciences, Sofia, 1985, pp. 598–601.

8. Sokolowski, M., On one-dimensional model of the fracture processes. Engineering Transaction, 1977, 25, pp. 369–393.

THE STATE OF CRYSTALLIZATION OF THE SECOND PHASE IN YTTRIA AND ALUMINA-DOPED HPSN AND ITS INFLUENCE ON THE STRENGTH AT HIGH TEMPERATURES

C. T. Bodur and K. Kromp
Max-Planck-Institut für Metallforschung
Institut für Werkstoffwissenschaften,
Seestraße 92, D-7000 Stuttgart 1, FRG

ABSTRACT

An Y_2O_3-Al_2O_3 (8 wt% - 2wt%)-doped hot pressed silicon nitride material was heat treated in argon at high temperatures (1300 - 1400°C) for various heat treatment times (4 - 40 hours) in order to crystallize the second phase which is assumed to be partly amorphous and partly crystalline. These heat treated specimens were then tested to analyze the effects of the various heat treatments on the high temperature mechanical properties (slow crack growth behaviour, fast fracture behaviour and elastic moduli). Electron microscopy (SEM, TEM) and X-ray diffraction were performed to observe the fracture appearance, microstructure and to analyze the phases before and after the heat treatments. No improvements in the mechanical properties of the heat treated specimens could be observed although the heat treatments have succeeded at least partially in crystallizing the amorphous second phase.

1. INTRODUCTION

Excellent high temperature properties [1]: Strength, creep, fracture toughness, hardness, corrosion/erosion resistance, good thermal schock resistance make silicon nitride ceramics ideal materials for highly stressed structural applications at high temperatures such as internal combustion engines and gas turbines [2,3].

HPSN ceramics are fabricated by mixing the powders of α-Si_3N_4 and additives (commercially the most used additives are: MgO; Y_2O_3; or Y_2O_3-Al_2O_3) and hot pressing $\approx$ 30 MPa at high temperatures $\geq 1700°C$ in nitrogen atmosphere for times $\geq 0.5h$ [1]. Sintering of Si_3N_4 powders occurs by a liquid phase which is formed by the reaction additives (i.e. MgO; Y_2O_3; or Y_2O_3-Al_2O_3) and silica or oxinitride phases which are always present in the starting Si_3N_4 powders [4,5]. All the other impurities (e.g. Ca, Al, Fe, Na), in the starting mixture and also developed during the fabrication, are contained in the sintering liquid [6]. α-Si_3N_4 is thermodynamically unstable at high temperatures (>1400°C), so it will transform to β-Si$_3$N$_4$ which is the primary phase of HPSN [7,8]. The final product consists of hexagonal β-Si$_3$N$_4$ grains and partly crystalline partly amorphous second phase, depending on the kinds of additives, impurities in the starting powder and the processing conditions.

High temperature (≥1000°C) properties of the HPSN's is controlled by the second phase [9-11]. In order to improve the high temperature properties (e.g. creep, subcritical crack growth, oxidation) of the HPSN materials, the effects of the weak amorphous second phase must be minimized. Research, done by others in order to improve the second phase effects, involved:

i) Using sintering aids which can form more refractory second phases; this will improve the high temperature properties of the HPSN's. Y$_2$O$_3$ additions to HPSN's have improved the high temperature strength [12-14], critical stress intensity factor [13] and the oxidation and thermal schock behaviour [14] as opposed to MgO-doped HPSN's.

ii) Heat treatments in various atmospheres (air, oxygen, nitrogen and inert) after the fabrication of HPSN's have succeeded in crystallizing the amorphous second phase:
-oxidation and creep behaviour of an Y$_2$O$_3$-doped Si$_3$N$_4$ material have been improved by a pre-oxidation heat treatment [15],

-secondary crystalline phases have formed during heat treatments in air of an Y$_2$O$_3$-Al$_2$O$_3$-doped Si$_3$N$_4$ [16],

-in nitrogen, heat treating Y$_2$O$_3$-doped HPSN has improved the oxidation resistance [17],

-in an Y$_2$O$_3$-Al$_2$O$_3$-doped HPSN material, heat treated in nitrogen, the amorphous second phase has changed to a crystalline phase which in turn increases the strength of the material [18],

-although a heat treatment of an Y$_2$O$_3$-doped HPSN has resulted in a secondary crystalline phase, no change in the stength of the material was noticed [19],

-heat treatment in argon has caused the disappearance of the glassy phase of an MgO-doped Si$_3$N$_4$ [20].

In this work various heat treatments were applied to an Y$_2$O$_3$-Al$_2$O$_3$-doped HPSN material in argon atmosphere. The goal was to partly crystallize the amorphous phase, which should increase the high temperature properties of the material. The effects of the heat treatments were investigated by various mechanical tests (controlled crack growth experiments, fast fracture tests and elastic moduli measurements) at high temperatures in an inert atmosphere. Electron microscopic (SEM, TEM) and XRD investigations of the materials were done before and after the heat treatments in order to analyze the effects of the heat treatments on the microstructures.

2. EXPERIMENTAL PROCEDURE

2.1 Material

The material, for this investigation, was an Y_2O_3-Al_2O_3-doped HPSN. The data of the starting powders were made available by the manufacturer, ESK Kempten, FRG and are listed in Table 1. The hot pressed silicon nitride specimens were rectangular bars of 3.5 by 7 by 35 mm.

TABLE 1
Manufacturer's data of the starting powder

Chemical analysis of the Si_3N_4 powder:							
Element	N_2	O_2	C	Ca	Al	Fe	Mg	Si
Wt.%	37.9	1.9	0.3	0.25	0.22	0.2	0.08	rest

Grain size:
97% < 10.6 μm,   50% < 1.6 μm,   6% < 0.6 μm

Specific surface area: 6.1 m^2g^{-1}

Modification: 96% α-Si_3N_4,   4% β-Si_3N_4

Additives: 8 wt.% Y_2O_3,   2 wt.% Al_2O_3

2.2 Heat Treatments

In this work different heat treatments in argon atmosphere were performed. They are listed with labels in Table 2. Details of the heat treatments and the equipment are given in [21].

TABLE 2
Heat treatment conditions in argon (°C, hours)

A	B	C	D	E	F
1400, 4	1400, 8	1400, 10	1400, 10 + 1300, 10	1300, 40	1360, 30

2.3 Density Measurements

The densities of the specimens (as received and the heat treated) were measured by the liquid immersion method in CCl_4. Two specimens from each heat treatment condition and five specimens of the as recieved material were used for density measurements. The density of each specimen is an average of three measurements.

2.4 Mechanical Testing

The mechanical properties of the heat treated and the as recieved specimens were tested in various experiments (i.e. controlled crack growth, fast fracture and elastic moduli) in order to analyze the effects of the heat treatments. Details on these experiments are given in [21]. In this paper only the results will be discussed.

2.5 Electron Microscopic and XRD Analyses
SEM and TEM investigations combined with EDX and X-ray diffraction (by a Guinier camera) were performed for all the heat treated and the as received specimens.

3. RESULTS AND DISCUSSION

Almost all the heat treatments changed the morphology of the surface of the specimens, especially the ones at higher temperatures $\geq 1360°C$. The surfaces of these specimens are quite porous and contain whiskers, which was also found in MgO-doped HPSN's [20]. The color of the specimens has changed from dark gray or black to grayish-green. The surface became very soft, similar to blackboard chalk. A thin surface layer has formed on some of the specimens and in some cases, they had detached from the bulk of the specimens. Such a surface layer is shown in Figure 1 of a 1400°C, 8h heat treated specimen. Its thickness is ≈0.27mm and it is quite porous.

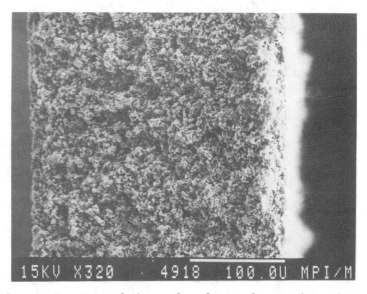

Figure 1. Porous structure of the surface layer of a specimen, heat treated in argon (1400°C, 8h)

Detail pictures of Figure 1 are given in Figures 2 and 3 which show the porous structure and whiskers in the surface layer. This morphology of the surface was not observed in lower temperature heat treated specimens, i.e. $\leq 1300°C$

The complete list of the measured densities of the specimens is given in Table 3.

Figure 2. A detail picture of Figure 1 -Porous structure

Figure 3. A detail picture of Figure 1 -Whiskers

Only a very small change can be observed in the densities after the heat treatments. Densities of the C, D and E heat treated specimens have decreased ≈ 0.15% from that of the as received material. This may be explained by the porous structure of the surface layers of the heat treated specimens.

TABLE 3
Densities of the as received and heat treated HPSN's [g/cm^3]

As received	A	B	C	D	E	F
3.277±.002	3.276±.001	3.277±.0	3.274±.002	3.273±.001	3.272±.001	3.277±.0

A list of the elastic constants of all the heat treated and as received specimens is given in Table 4. The measurements were done in the temperature range 20 - 1250°C by the resonance method [21]. This listing only contains the room temperature and 1200°C values of the elastic constants, shear modulus, Young's modulus and the Poisson's ratio. Almost no change can be seen from these results, if one considers the complexity of the measurements.

TABLE 4
20°C and 1200°C elastic constants of the as received and heat treated specimens

T[°C]	As received	A	B	C	D	E	F
20	117[1] 309[2] .32[3]	117 307 .31	117 308 .31	116 307 .32	117 306 .31	118 308 .31	117 307 .32
1200	103 273 .32	103 271 .32	103 274 .33	103 273 .33	105 276 .32	104 277 .33	104 276 .33

(1) G: shear modulus [GPa]   (2) E: Young's modulus [GPa]
(3) ν: Poisson's ratio.

TABLE 5
Stress intensity factors (K_{Ic}'s) of the as received and "C" and "E" heat treated specimens [MPa√m].

T[°C]	As received	C	E
1200	5.02±0.28[1]	5.59±0.10	5.39±0.14
1200	5.85±0.08[2]	5.66±0.11	5.65±0.06
20	8.39±0.01[2]	-	-

(1) From the controlled crack growth experiments in argon
(2) From the fast fracture tests (i.e. critical stress intensity factor) in vacuum.

The listing of the stress intensity factors is given in Table 5. The stress intensity factor experiments are explained in detail in [21]. Here only the results are presented. If one compares the K_{IC}'s of the heat treated and the as received spesimens from the controlled crack growth experiments, an increase of $\approx$ 11% for "C" and $\approx$ 7% for "E" after the heat treatments is seen. Although controlled crack growth experiments were successfully performed on this material at 1200°C, experimental uncertainties affect the computer evaluation of the K_{IC}'s. If one assumes no errors occur during the experiments and the calculation, the increase of the K_{IC}'s of the heat treated specimens can be attributed to the crystallization of the amorphous intergranular phase. In contrast, there is a small decrease in the K_{IC}'s of the heat treated specimens in comparison to those of the as received material from the fast fracture tests. This may be explained as the effect of crystallization of the amorphous phase is negligible in fast fracture. i.e. the fast fracture tests were performed at stress rates of 10^4 MPa/s in vacuum to achieve " inert-K_{IC} values ". At these high loading rates the weak second phases behave " quasi-brittle " and do not show much influence on the strength.

Elongated β-Si$_3$N$_4$ can be observed on the fracture surfaces of the as received material, tested by the controlled crack growth experiments in argon at 1200°C (displacement rate of 5μm/min and a span length of 30 mm), Figure 4.

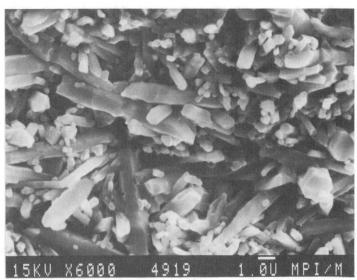

Figure 4. SEM of a typical fracture surface of an as received specimen, fractured in a displacement controlled experiment ($\dot{\delta}$ = 5μm/min), T= 1200°C in argon.

The fracture is predominantly intergranular. The elongated β-grains have a hexagonal shape with a length $\approx$ 5 to 10μm and aspect ratios of $\approx$ 5 to 10. In Figure 5 a heat treated (1400°C, 10h + 1300°C, 10h) specimen's fracture surface is presented. Although the experimental conditions of the specimen

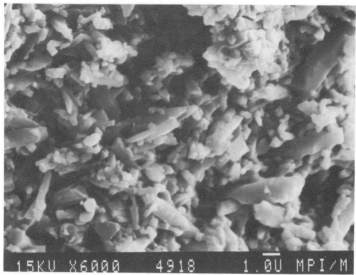

Figure 5. SEM of the fracture surface of a heat treated specimen (1400°C, 10h + 1300°C, 10h), fractured in a displacement controlled experiment ($\dot{\delta}$ = 5μm/min), T= 1200°C in argon.

of Figure 5 are exactly the same as that of the specimen of Figure 4, the morphology of the fracture surface is different. The 1400°C, 10h + 1300°C,

10h heat treated specimen's fracture surface shows not only single β-Si$_3$N$_4$ grains but also larger grains, probably formed by the small single grains of β-Si$_3$N$_4$. Cavities can also be observed in Figure 5; this may be a consequence of the disappearance of the intergranular amorphous phase [20].

A result of the EDX analysis of the surface layer (see Figure 1) of a heat treated (1400°C, 8h) specimen is shown in Figure 6. This data shows the elemental composition of the big grain (white color), in Figure 2. Altough Si is dominating, the other elements (Y and Al) could also be observed over the thickness of the surface layer. This behaviour of Y and Al concentrations was not observed in the bulk of the specimen by the EDX analysis. This may be a result of moving the amorphous phase, rich in additives and impurities, to the surface and creating cavities in the bulk of the material.

All the heat treated and as received materials were investigated by XRD (Guinier method) in powdered form to identify the crystalline phases. The results were the same for all the specimens in terms of the phases contained. All the specimens have hexagonal β-Si$_3$N$_4$ grains as the primary phase ($\approx$ 90-95 vol%) with lattice parameters a= 7.608 Å and c= 2.9109 Å. A second crystalline phase ($\approx$ 5-10 vol%) were also existed in all the specimens, the heat treated and the as received. The phases are listed below and also references are given of the same phases, previously observed.

- Y$_5$N(SiO$_4$)$_3$ Yttrium Nitride Silicate, hexagonal with a= 9.41 Å, c= 6.76 Å (ASTM 33-1459) [22],

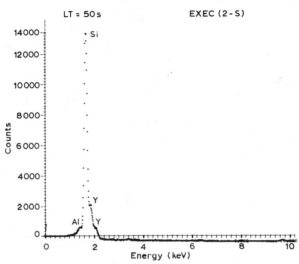

Figure 6. EDX spectrum of the large grain in Figure 2., specimen heat treated 1400°C, 8h in argon.

- $Si_3N_4.10Y_2O_3.9SiO_2$ Yttrium Silicon Oxide Nitride, hexagonal with a= 9.40 Å, c= 6.80 Å (ASTM 30-1462) [5],

- $Y_{4.67}(SiO_4)_3O$ Yttrium Oxide Silicate, hexagonal with a= 9.347 Å, c= 6.727 Å (ASTM 30-1457) [23].

These secondary crystalline phases can easily be differentiated from the primary phase by their XRD lines.An example of the diffraction lines observed is given in Table 6. The lines of this second crystalline phase fit to the three phases listed above. The lattice parameters of these phases are close together, therefore a unique phase could not be identified in the materials. Another problem in this investigation was that the intensity of the diffraction lines could not be measured.As a result, the three secondary crystalline phases found in all the heat treated specimens were also in the as received material. Summarizing, it can be concluded that by the XRD analyses, the state of the crystallization of the amorphous second phase, as a result of the heat treatments, could not be determined.

As already mentioned above, the diffraction lines of the three secondary phases were observed both in all the heat treated specimens and in the as received specimen. Therefore, it was evident that TEM investigations combined with EDX and EELS had to be performed for further identification of the state of crystallization. Up to now, only the specimens in the as received state and specimens in a state after a load cycling experiment at 1200°C have been investigated by TEM. The load cycled specimens thus underwent a heat treatment of ≈ 5 hours at 1200°C in argon. A transmission electron micrograph of an as received specimen is shown in Figure 7. The black spots indicate the positions investigated by the EDX (the picture was taken after the EDX investigation). The EDX analysis of a grain, black spot above left, is shown in Figure 8 -the Si peak indicates that it is a Si_3N_4 grain, which was expected. The EDX analysis for a position between the grains in Figure 7 is presented in Figure 9 -besides Si, the peaks of Y, Ca and Al

appear. These peaks arise from the impurity content (Ca and Al) and additives (Y and Al), see Table 1. The Y and the Al peaks indicate that at this
intermediate position the second phases exist. A selected area diffrac-
tography at this same position is shown in the left corner of Figure 7,
pointing out that the phases are in the amorphous state.

TABLE 6

XRD data by Guinier method (Guinier-Film CuKα_1 4h) of a heat treated
specimen (1400°C, 10h + 1300°C, 10h) in argon.

$d_o(Å)$	I_o	β-Si$_3$N$_4$ - hkl (33-1160)	Y$_5$N(SiO$_4$)$_3$ - hkl (33-1459)
6,6	40	100	-
4,05	5	-	200
3,80	40	110	-
3,38	5	-	002
3,29	100	200	-
3,12	10	-	102
3,08	10	-	210
2,80	20	-	211
2,74	10	-	112
2,71	10	-	300
2,66	100	101	-
2,49	100	210	-
2,31	10	111	-
2,19	10	300	-
2,18	30	201	-
1,93	10 diff.	-	222 (30-1462)
1,90	10	220	-
1,89	5	211	-
1,827	10	310	-
1,760	40	301	-
1,590	20	221	-
1,546	10	311	-
1,510	20	320	-
1,453	20	002	-
1,437	10	410	-
1,432	5	401,102	-
1,358	5	112	-
1,340	40	321	-
1,330	10	202	-
1,317	10	500	-
1,288	20	411	-
1,267	10	330	-
1,255	20	212	-

For a specimen cycled 5 hours at 1200°C in argon, the transmission electron
micrography is shown in Figure 10. The dark area between the Si$_3$N$_4$ grains
obviously is in a crystallized state -in the as received specimen such
areas could not be seen.

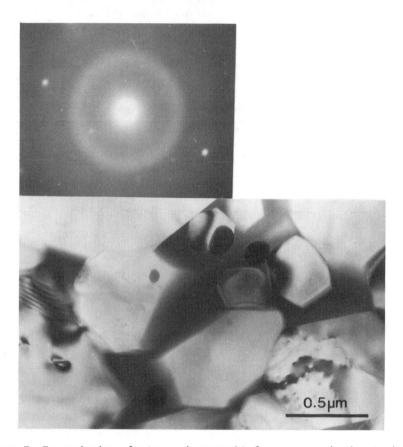

Figure 7. Transmission electron micrograph of an as received material.

From this first result of TEM investigation, it can be concluded that a heat treatment of 5 hours at 1200°C in argon partially crystallizes the amorphous second phase.

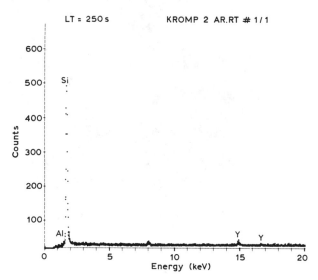

Figure 8. EDX spectrum for a Si_3N_4 grain in Figure 7.

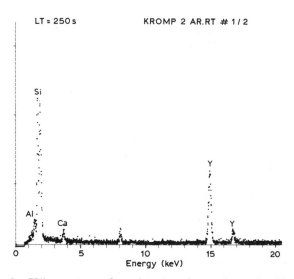

Figure 9. EDX spectrum for the amorphous phase in Figure 7.

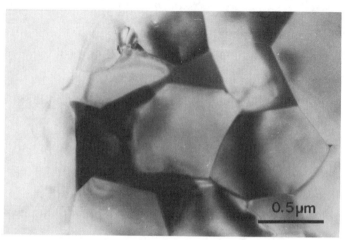

Figure 10. Transmission electron micrograph of a specimen, cycled in argon at 1200°C for 5 hours.

4. CONCLUSIONS

The first microstructural investigation by the TEM pointed out that the crystallization of the second phase in the investigated Yttria and Alumina-doped HPSN took place, even after 5 hours of heating at a temperature of 1200°C.

The XRD investigations exhibit three phases with very similar lattice parameters that could be expected to crystallize in larger amounts after the different heat treatments. From the results, it could not be distinguished, which of these phases prevails in the material and to what degree the crystallization has proceeded after the heat treatments. All these points must be cleared by further TEM investigations combined with EDX and EELS.

Concerning the mechanical qualities before and after the heat treatments, it should be mentioned that an expected improvement in the strength by partly crystallizing the second phase was not found. This expected improvement could be degraded by the changes in the grain structure and by the growth of a surface layer, which both were pointed out by the SEM investigations of the fracture surfaces and by the density measurements.

The fast fracture test (i.e. inert K_{Ic}-test) obviously is not suitable to point out any improvement in the strength behaviour in this material. At high temperatures, the fast fracture test is not sensitive to changes in the second phase due to crystallization, because at these high crack velocities the second phase behaves brittle anyway. The controlled slow crack growth test seems to be more sensitive to these changes, as the K_{Ic} values calculated from these experiments point out. A more striking result is to be expected in creep experiments, which will be performed.

Concluding, it can be remarked that the heat treatments performed with this Yttria and Alumina-doped HPSN did not succeed to improve the strength qualities at 1200°C.

ACKNOWLEDGEMENT

The authors gratefully acknowledge the support of the Deutsche Forschungs-gemeinschaft under contract No: Kr 970/2.

REFERENCES

1. Ziegler, G., Heinrich, J. and Wötting, G., Review: Relationships between processing, microstructure and properties of dense and reaction-bonded silicon nitride. J. Mater. Sci., 1987, **22**, pp. 3041-86.
2. Godfrey, D. J., The use of ceramics in diesel engines. In Nitrogen Ceramics, ed., F. L. Riley, Noordhoff, Leyden, 1977, pp. 647-52.
3. Godfrey, D. J., The use of ceramics for engines. In Science of Ceramics, ed., P. Vincenzini, Faenza, Italy, 1984, **12**, pp. 27-38.
4. Lange, F. F., High temperature deformation and fracture phenomena of polyphase Si_3N_4 materials. In Progress in Nitrogen Ceramics, ed., F. L. Riley, Martinus Nijhoff Publishers, Boston/The Hague/Dordrecht/ Lancaster, 1983, pp. 467-90.
5. Wills, R. R., Holmquist, S., Wimmer, J. M. and Cunningham, J. A., Phase relationships in the system Si_3N_4-Y_2O_3-SiO_2. J. Mater. Sci., 1976, **11**, pp. 1305-9.
6. Clarke, D. R., The microstructure of nitrogen ceramics. In Progress in Nitrogen Ceramics, ed. F. L. Riley, Martinus Nijhoff Publishers, Boston/The Hague/Dordrecht/Lancaster, 1983, pp. 341-58.
7. Bowen, L. J., Weston, R. J., Carruthers, T. G. and Brook, R. J., Hot-pressing and the α-β phase transformation in silicon nitride. J. Mater. Sci., 1978, **13**, pp. 341-50.
8. Knoch, H. and Gazza, G. E., On the α to β phase transformation and grain growth during hot-pressing of Si_3N_4 containing MgO. Ceramurgia Int., 1980, **6**, pp. 51-6.
9. Dutta, S., Fabrication, microstructure, and strength of sintered β'-Si_3N_4 solid solution. Am. Ceram. Soc. Bul., 1980, **59**, pp. 623-25,634.
10. Wiederhorn, S. M. and Tighe, N. J., Structural reliability of yttria-doped hot-pressed silicon nitride at elevated temperatures. J. Am. Ceram. Soc., 1983, **66**, pp. 884-9.
11. Clarke, D. R. and Thomas, G., Microstructure of Y_2O_3 fluxed hot-pressed silicon nitride. J. Am. Ceram. Soc., 1978, **61**, pp. 114-8.
12. Gazza, G. E., Effect of yttria additions on hot-pressed Si_3N_4. Am. Ceram. Soc. Bull., 1975, **54**, pp. 778-81.
13. Knickerbocker, S. H., Zangvil, A. and Brown, S. D., High-temperature mechanical properties and microstructures for hot-pressed silicon nitrides with amorphous and crystalline intergranular phases. J. Am. Ceram. Soc., 1985, **68**, pp. C-99-101.
14. Weaver, G. Q. and Lucek, J. W., Optimization of hot-pressed Si_3N_4-Y_2O_3 materials. Am. Ceram. Soc. Bull., 1978, **57**, pp. 1131-4,1136.
15. Lange, F. F., Davis, B. I. and Graham, H. C., Compressive creep and oxidation resistance of an Si_3N_4 material fabricated in the system Si_3N_4-Si_2N_2O-$Y_2Si_2O_7$. J. Am. Ceram. Soc., 1983, **66**, pp. C-98-9.

16. Falk, L. K. L. and Dunlop, G. L., Crystallization of the glassy phase in an Si_3N_4 material by post-sintering heat treatments. J. Mater. Sci., 1987, **22**, pp. 4369-76.
17. Gazza, G. E., Knoch, H. and Quinn, G. D., Hot-pressed Si_3N_4 with improved thermal stability. Am. Ceram. Soc. Bull., 1978, **57**, pp. 1059-60.
18. Tsuge, A. and Nishida, K., High strength hot-pressed Si_3N_4 with concurrent Y_2O_3 and Al_2O_3 additions. Am. Ceram. Soc. Bull., 1978, **57**, pp. 424-26,431.
19. Pierce, L. A., Mieskowski, D. M. and Sanders W. A., Effect of grain-boundary crystallization on the high-temperature strength of silicon nitride. J. Mater. Sci., 1986, **21**, pp. 1345-48.
20. Clarke, D. R., A comparison of reducing and oxidizing heat treatments of hot-pressed silicon nitride. J. Am. Ceram. Soc., 1983, **66**, pp. 92-5.
21. Bodur, C. T. and Kromp, K., Investigation of the strength behaviour of HPSN with yttria at high temperatures, (in German).In Fortschritts-berichte der Deutschen Keramischen Gesellschaft (Cfi: ceramic forum international), to be published.
22. Jack, K. H., Review: Sialons and related nitrogen ceramics. J. Mater. Sci., 1976, **11**, pp. 1135-58.
23. Wills, R. R., Cunningham, J. A., Wimmer, J. M. and Stewart, R. W., Stability of the silicon yttrium oxynitrides. J. Am. Ceram. Soc., 1976, **59**, pp. 269-70.

INFLUENCE OF THE TECHNOLOGY AND MICROSTRUCTURE ON ELASTIC PROPERTIES OF SINTERED COPPER DETERMINED BY MEANS OF RESONANCE METHOD

LESZEK RADZISZEWSKI, JERZY RANACHOWSKI
Physical Acoustic Department,
Institute of Fundamental Technological Research
of the Polish Academy of Sciences, Warsaw, Poland

ABSTRACT

The subject of this paper is acoustic investigation of porous media, using sintered copper as an example. The method of investigation is described and results of measurements of moduli of elasticity versus porosity and temperature are given. The comparison was than carried out of the data obtained with theoretical relationships according to Mackenzie and Rossi models of porous media.

INTRODUCTION

This work was aimed at formulation of the relationship between moduli of elasticity of porous media, such as Poisson ratio ν and Young modulus E , and porosity and temperature of measurement. Sintered copper was chosen as investigated medium. In order to obtain a change of porosity of specimen powdered copper was pressed and sintered one, two or three times. For determination of moduli of elasticity was used the method of measurement of resonance frequency of circular plate transverse vibrations. Using this method ν was determined with an accuracy of 0.1% and E with an accuracy of 0.25%. The relationships found between Young modulus and porosity E = f(P) were compared with analogical relationships anticipated in the Mackenzie and Rossi theories. Measurements were carried out within the porosity range of 2 - 11% and temperature range of 83 - 473 K.

PREPARATION OF SAMPLES

As a raw material was used 99.9% pure electrolytic copper powder. Samples were supposed to have isotropic structure and 2 - 11% porosity. In order to obtain such a medium following technology was applied:

Preliminary pressing in the die block

|

Hydrostatic densification

|

Sintering

|

Specified quantities of copper powder have been pressed from both sides in the die block with feeding chamber of 30.5 mm diameter at 50 MPa load, without any lubricant. In this way samples were obtained of dimensions ϕ30.5x3.5 mm and porosity 30%. For futher compaction already obtained compacts have been placed in the elastic matrices made of natural rubber and placed in the pressure chamber filled with the mixture of rape seed oil with methyl alcohol. Hydrostatic densification was carried out at 315 MPa through 15 minutes. Sintering was carried out at the atmosphere of dissociated ammonia NH_3 at temperature 1223 K through 1 hour. In this way a set of 40 specimens of dimensions ϕ30x3 was prepared. Specimens from this first series featured porosity within 9 - 15% range.

For further compaction specimens have been placed in the pressure chamber and subjected to hydrostatic densification under pressure amounting to 315 MPa through 15 minutes. Then compacts were sintered at 1223 K through 1 hour under vacuum. Porosity of this second series of specimens was comprised within 4 - 9% limits. For final compaction 14 samples were selected, that were placed in the pressure chamber and pressed under pressure equal to 600 MPa through 15 minutes. Then specimens were sintered under vacuum at 1223 K through 1 hour. Porosity of specimens after last cycle of pressing and sintering was comprised within 2 -4% limits.

The technological process carried out in this way enable to get specimens of isotropic structure and porosity from 2 to 9%. Pores had almost spherical shape of diameter ranging from 2 μm up to 6 μm and in 1mm^3 of material there were from 15x10^5 up to 17x10^5 pores, depending on porosity value.

For analyzing influence of shape of pores on elastic properties of material special specimens were prepared using following procedure: single densification was carried out as described above, but sintering in the atmosphere of dissociated ammonia NH$_3$ according to program given in Table 1.

TABLE 1

Program of sintering of samples under different conditions

Temperature of sintering		Duration of sintering	Number of specimens
°C	K	min	
900	1173	60	3
		15	3
		30	3
950	1223	45	3
		60	3
		120	3
		15	3
1000	1273	30	3
		45	3
		60	3

METHOD OF MEASUREMENTS

In this work Young modulus and Poisson ratio were found for sintered copper using circular plate torsial vibrations method [1] . Diagram of measuring equipment is displayed in Fig.1.

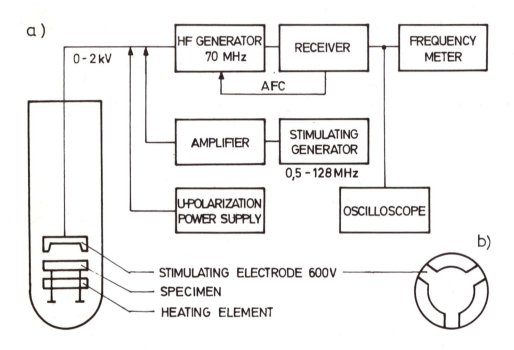

Figure 1. Diagram of measuring apparatus for resonant investigation of moduli of elasticity and shape of electrode

Investigated specimens of $\phi30\times3$ dimensions were electrostatically stimulated to resonant torsial vibrations at second and third resonance frequencies. In order to enable free vibrating those specimens were fixed in three points, located at nodal lines of torsional vibrations [2,3].

After placing the specimen on supports and closing the vacum chamber with quartz tube the vacuum was produced equal to 10^{-4}Tr. This vacuum was maintained through all time of measurements. Specimen has been electrostatically stimulated to torsional vibrations. At specimen vibrations frequency equal to resonance frequency this frequency value was read out with 1Hz accuracy. At the same time specimen temperature was determined. After analysis of the obtained results, a part of specimens have been selected for investigation of changes of moduli of elasticity versus measurement temperature. Temperature investigations were carried out on 16 specimens obtained after II and III operation of pressing and sintering. Measurement method was analogous with this difference, that the specimen was chilled with liquid nitrogen down to 80 K. Total time of chilling was equal to 8 hours. After carrying out the measurement of specimen resonant frequencies at this temperature specimen was heated in a very slow manner at temperature gradient equal to 0.7 K/min, up to 473 K. Starting from 80 K measurement of resonant frequencies have been carried out every 10 K up to 473 K. From the frequencies found during those measurements for each specimen Young modulus and Poisson ratio were calcualted and $E = f(T)$ and $\nu = f(T)$ curves were plotted. Error in determination of Poisson ratio amounts to 0.1% and for Young modulus to 0.25%. Meassuring error is influenced mainly by following factors: method of fixing, shape of specimen, accuracy of measurements of specimen temperature and resonant frequencies.

RESULTS

The investigation performed have proved, that for all tested specimen groups the Young modulus is linear function of porosity (P). For the group after second densification and sintering this relationship is given by the formula

$$E = E_T - 0.0315P \qquad (1)$$

while for the group after third densification and sintering

$$E = E_T - 0.0437P \qquad (2)$$

where E_T is a coefficient dependent on the temperature of measurements and methods of preparation of specimens.

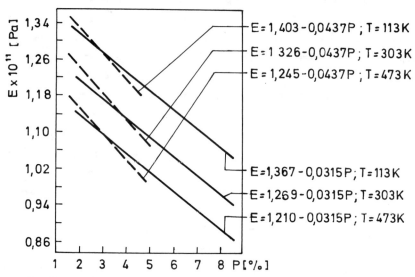

Figure 2. Young modulus versus porosity for specimens after II and III pressing and sintering, measured at temperatures T = 113K, 303K and 473K.

After carrying out statistical analysis of $\nu = f(P)$ at fixed measurement temperature, it follows that

$$\nu = \nu_T - 0.0024P \tag{3}$$

where ν_T is a coefficient dependent on temperature of measurements

$$\nu_T = 0.34682 \qquad \text{for } T = 113K$$
$$\nu_T = 0.35873 \qquad \text{for } T = 473K$$

Investigations did not prove the influence of the method of densification on the values of Poisson ratio. It was found out, that the shape of pores affects the elastic properties of the specimen compacted or sintered under various conditions. Below are given relationships $E = f(P)$ for three groups of samples sintered at different temperatures using various times:

- specimens sintered through 60 minutes

$$E = E_T - 0.0269P \tag{4}$$

$$E_T = 143.6734 - 0.0537T \tag{5}$$

- specimens sintered through 30 minutes

$$E = E_T - 0.0265P \tag{6}$$

$$E_T = 139.8579 - 0.0505T \tag{7}$$

- specimens sintered through 15 minutes

$$E = E_T - 0.0457P \tag{8}$$

$$E_T = 160.4760 - 0.0781T \tag{9}$$

The investigations performed for $E = f(T)$ relationship proved, that for all investigated groups of specimens the linear relation $E = E_P + \alpha T$ may be accepted, where E_P is a coefficient dependent on porosity and conditions of preparation of specimens and α is a coefficient of temperature changes. For instance, for specimens subjected to many cycles of pressing and sintering

$$E_P = 146.4402 - 3.67P \quad (GPa) \tag{10}$$

For specimens sintered at different temperatures using various times this relationship is given by following formulae:
- for specimens sintered through 60 minutes

$$E_P = 142.9803 - 3.105P \quad (GPa) \tag{11}$$

- for specimens sintered through 30 minutes

$$E_P = 140.1641 - 3.105P \quad (GPa) \tag{12}$$

- for specimens sintered through 15 minutes

$$E_P = 143.1515 - 3.105P \quad (GPa) \tag{13}$$

Investigations of influence of temperature of measurements on Poisson ratio proved that increase of temperature causes proportional increase of Poisson ratio. Analysis of equation $\nu = f(T) = \nu_p + \beta T$ proved that coefficient ν_p is function of porosity

$$\nu_p = 0.00243P + 0.341796 \qquad (14)$$

and β is a coefficient of temperature changes.

CONCLUSIONS

The investigations performed have proved, that elastic properties of sintered copper depend mainly on porosity, however not only on volume fraction of pores, but as well on their microstructure i.e. shape of pores.

Variation of Young modulus versus porosity is by an order of magnitude greater than changes of Poisson ratio. Change of porosity within the range from 2% up to 10% results in change of Young modulus by 20% and change of Poisson ratio by 6%. Increase of temperature of measurement from 113 K up to 473 K causes decrease of Young modulus by 15% and increase of Poisson ratio by 0.4%.

Theoretical relationships $E = f(P)$ for sintered copper, consistent with the Mackenzie model [4] are presented in Table 2.

TABLE 2

Theoretical relationships E = f(P) according
to Mackenzie model

No	Tempe- rature	E_o [GPa]	ν_o	a_E	Mackenzie equation
1	113K	133.0177	0.41163	1.9748	E=133 (1-1.97P)
2	303K	123.0794	0.42944	1.9711	E=123 (1-1.97P)
3	473K	119.1477	0.42354	1.9690	E=119 (1-1.97P)
4	113K	138.0000	0.34682	1.9970	E=138 (1-2.00P)
5	303K	128.5000	0.35548	1.9950	E=128 (1-2.00P)
6	473K	121.8000	0.35873	1.9940	E=122 (1-2.00P)

E_o - Young modulus of sintered copper matrix

ν_o - Poisson ratio of sintered copper matrix

a_E - slope of a straight line in Mackenzie equation

In Table 2 E_o the values for No.1, 2 and 3 were taken from resonance measurements of solid 99.99% pure polycrystalline copper, carried out within the framework of this research. The values of E_o for No.4, 5 and 6 were found from approximation of equations (1), (2) and (3) for porosity P = 0%. Both those procedures are to some extent in error and the problem of determination of E_o values for the matrix of sintered material requires further investigations. When comparing experimental relationships E = f(P) with equations from Table 2 it is evident, that there exists a good consistency of theoretical and experimental results and that the least discrepancies occur at low temperatures and at low porosities.

Influence of the shape of pores on elastic properties was determined on the basis of the theory elaborated by Rossi for concentration of stresses on pores, [5]. According to investigations published in [6], for sphericalpores slope of the curve E = f(P) is the least for a given medium and axial ratio a/c=1. Following deviation from spherical shape to prolate spheroid increases also the curve slope, what was confirmed experimentally. On their basis one can determine axial ratio a/c for pores and influence of technological parameters, i.e. temperature of sintering and duration of sintering of specimens on obtained shape of pores (Figs. 4 and 5).

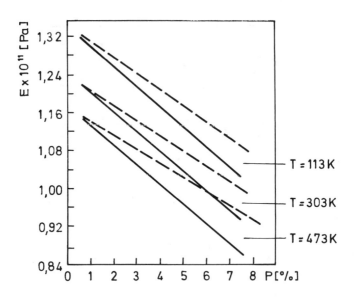

Figure 3. Comparison of relationship E = f(P) according Mackenzie model and experimental curves found at various measurement temperatures T = 113K, 303K, 473K.

In Fig. 4 straight line for which axial ratio is equal to 0.99 was found for specimens sintered at 1273 K (according to Table 1). When temperature of sintering and duration of sintering are variable (according to Table 1), then for an avarage specimen the axial ratio a/c = 1.23, i.e. the shape of pores is slightly different from spherical. On the other hand, when as a parameter is used the duration of sintering, then from Fig. 5 it is evident that following increase of duration of sintering shape of pores in samples approximates spherical shape.

From comparison of Fig. 4 and Fig. 5 and Table 1 it follows also that:

1. samples sintered at 1273 K have spherical pores,
2. samples sintered at 1173 K through 60 minutes and at 1223 K

through 15 or 30 minutes have non-spherical pores,
3.   temperature   of   sintering   has   greater   influence   on
     properties of sintered materials than duration of sintering.

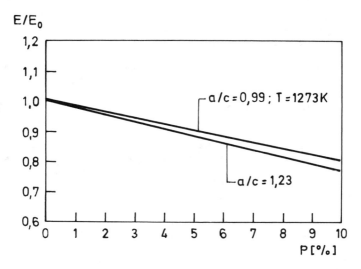

Figure 4. Normalized Young modulus versus porosity at various
axial ratios of spheroidal pores at temperature of T = 303K,
     when as a parameter is used temperature of sintering.

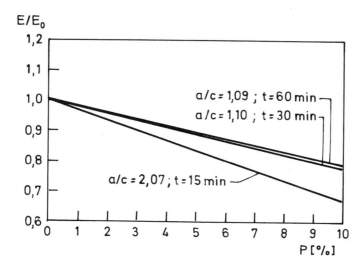

Figure 5. Normalized Young modulus versus porosity at various
axial ratios of spheroidal pores at temperature of T = 303K,
     when as a parameter is used duration of sintering.

REFERENCES

1. Ryll-Nardzewski J., Application of circular plate resonant properties for determination of moduli of elasticity, IFTR Reports 44/1973 (in Polish).

2. Jemielniak R. and Pilecki Sz., Measurement of moduli of elasticity and of internal friction using the acoustic resonance method, Transactions of the Conference "Electric and Acoustic Methods of Material Testing", Jabłonna 1982 (in Polish).

3. Pilecki Sz. and Jemielniak R., Investigation of internal friction of plastically strained metals, within temperature range from liquid nitrogen temperature up to 230°C, Transaction of the Conference "Electric and Acoustic Method of Materials Testing" Jabłonna 1983 (in Polish).

4. Mackenzie J.K., The elastic constants of a solid containing spherical holes, Proc. Phys. Soc. 863, 2, 1950.

5. Rossi R.C., Prediction of the Elastic Moduli of Composites, J. Amer. Soc., 51, 234-237, 1968.

6. Kreher W., Ranachowski J., Rejmund F., Ultrasonic waves in porous ceramics with non-spherical holes, Ultrasonics, 1977, 3.

PROPERTIES OF γ'-α EUTECTIC COMPOSITE

KATARINA KOVACOVA

Institute of Materials and Machine Mechanics
of Slovak Academy of Sciences, ul. Februárového
víťazstva 75, Bratislava 836 06, Czechoslovakia

ABSTRACT

A γ'-α eutectic reaction is produced from ternary Ni-Al-Mo eutectics by a suitable quarternary addition. Thequaternary Ni-21Mo-9Al-6Ta (wt.%) alloy forms γ'-α eutectics with a sufficiently narrow freezing range, $\Delta T=(11\pm1,3)K$, to allow aligned structures to be produced by unidirectional solidification. For a chosen constitution , the eutectic composits of the γ'- matrix and of 15 vol.% α-Mo reinforcing fibres. The coupled growth of γ'-α composite was evident as far as the growth rate interval from 0.5 cm/hr to 2 cm/hr was concerned. Tantalum has a strong effect on the strengthening the γ'-matrix at elevated temperature. Due to Ta addition, tensile strength of γ'-α eutectics is higher as in ternary γ/γ'-α composite.

INTRODUCTION

Aligned composites of Ni-Al-Mo system received considerable attention for futher development as advanced turbine blade materials.This alloy system is extensively studied with additional alloying to improve high temperature strength, creep resistance and corrosion resistance. Tantalum is one of the refractory metal causing a pronounced of high temperature specific strength of material, having also a relatively high solubility in intermetallic γ' phase,[1]. The planarity of the solid-liquid interface is anecessary condition of the coupled

growth of eutectic phase mixture.   This   means   that   for   the
freezing range (ΔT) the following condicion must be imposed:
$\Delta T \leq G.D.R^{-1}$, [2].

The purpose of the present paper is to define the coupled
growth of the γ'-α eutectic composite in the quaternary
Ni-13Mo-20Al-3Ta (at.%) alloy.

EXPERIMENT

The quarternary alloys of the Ni-Al-Mo-Ta system needed for
directional solidification were prepared by melting under argon
in ceramic crucible and   then   cast   into   copper   mold.   The
freezing range ΔT of the   quarternary   alloy   Ni-13Mo-20-Al-3Ta
was determined by DTA measurements on simultaneus thermal
apparatus STA 401 (Netzsch). High purity Al$_2$O$_3$ powder was used
as the standard substance. Heating and cooling within the
temperature interval from 1493 K to 1653 K was at the rate of 5
K/min. The rods were directionally solidified in a vertical
resistance furnace with high-purity protective argon   atmosfere
[3]. All experiments were performed at the constant temperature
gradient in the liquid at the solid-liquid interface, G= 150
K/cm and the solidification rate varied from 0.5 cm/hr to 3
cm/hr. The samples for the study of the morphological
properties were cut by a diamond wheel. Distribution of the
solutes between the co-existing phases was determined by the
electron probe microanalysis using a JEOL JEM-100C ASID-4D
instrument with ZAF4 computer program software supplied by
LINK. The relative error of the composition analysis is $\leq$ $\pm$1%
of individual element.

To   specify   mutual   crystalographic   orientation   of   the
phases, thin foils were   prepared,being   cut   parallel   to   the
solid-liquid interface. The thin foils were examined in the
electron microscope JEM-100C at acceleration voltage 100 KV.
Selected area electron difraction (SAD) patterns were taken
from   various   areas   of   the   matrix-fibre   interphase.
Crystallographic orientation relationship of the γ'-α was
determined by the complex evaluation of the electron difraction
patterns. Scanning electron microscopy (SEM) was applied   to
quantitative analysis of reinforcing fibres growth, the growth
of lamelae at the eutectic grain boundaries and cell growth.
Vickers hardness measurements were performed on apparatus HPO
VEB, Werkstoffprüf- maschinen Leipzig.

RESULTS

As the morphological stability of the solid- liquid interface is controling by solidification parameters then in the first the size and accurary of the freezing range ΔT must be determined. For studying alloy this interval $\Delta T = (11\pm1.3)$ K was evaluated. For a chosen constitution, eutectic composite consists of the matrix (intermetallic γ' phase) and of (14.8 ±0.3) vol.% α-molybdenum reinforcing fibres. The coupled growth of γ'-α eutectic composite was evident as far as the growth rate interval from 0.5 cm/hr to 2 cm/hr was concerned. Due to crystallization rate increase above the upper limit of the interval R>2 cm/hr cellular morphology begins to grow at curved solid- liquid interface, accompanied simultaneously by transformation of reinforcing molybdenum fibres to lamelae.

In bivariant γ'-α eutectic, the matrix has a FCC structure and α-Mo fibre has a BCC structure. This relationship of crystal structures in co-existing phases assumes a relatively simple preferred orientation relation between them. Using the TEM method, crystallographic orientation relationship of the co-existing phases was obtained :

$$\begin{array}{lll}\text{growth direction} & \| \langle 001\rangle\gamma' & \| \langle 001\rangle\alpha \\ \text{interphase} & \| \{110\}\gamma' & \| \{100\}\alpha\end{array}$$

The distribution of solutes in co-existing phases was specified. It was found that there is a relatively low solubility of all elements (Ni, Al and Ta) in Mo fibres, even though their solubility is high in correspoding binary equilibrium diagrams with molybdenum. On the other hand, Mo solubility in intermetallic γ' phase in ternary γ/γ'-α eutectic is 4 at.% [2]. In bivariant γ'-α eutectic this value is reduced to a half. The presence of Ta decreases Mo solubility in intermetallic γ' phase.

Under chosen load the mean hardness value calculated from minimum 10 measurements on each sample HV 10= (342$\pm$2) and it is not a function of the growth rate.

Results of ultimate tensile strength measurements as a function of temperature were compared with the ternary γ/γ'-α eutectiç. The observed increase in ultimate strength of 600 MPa at 800 C ilustrates the strengthening due to Ta. Longitudinal ductility is very good (~ 18%).

CONCLUSIONS

1. Quarternary monovariant Ni-13Mo-20Al-3Ta (at.%) alloy   has
   the freezing interval ΔT = (11$\pm$1.3) K. The alloy consist of
   the γ' matrix and of 15   vol.% α-Mo   fibres.   The   coupled
   growth of the eutectic composite was evident as far as   the
   growth interval from 0.5 cm/hr to up 2 cm/hr is   concerned.
   For R > 2 cm/hr cellular morphology is   growing   at   curved
   solid liquid   interface.   Using   the   TEM   method,
   crystallographic   orientation   relationship   of   the
   co-existing
   phases was obtained :
   growth direction $\parallel$ $\langle 001 \rangle_{\gamma'}$, $\parallel$ $\langle 001 \rangle_{\alpha}$
   interphase $\parallel$ $\{110\}_{\gamma'}$, $\parallel$ $\{100\}_{\alpha}$
   The solutes show relatively low solubility in Mo fibres. Ta
   decreases Mo solubility in the matrix.
2. The hardness value is HV 10 = (342 $\pm$2)
3. Ta increases the ultimate tensile strength of   600   MPa   at
   800 C. Longitudinal ductility is very good ($\sim$ 18%).

REFERENCES

1. Ochiai, S., Oya, Y. and Suzuki,   T.,   Alloying   of Ni_3Al,
   Ni_3Ga, Ni_3Si. Acta metall., 1984, 32, 289-99.

2. Pearson,   D.D.,and   Lemkey,   F.D.,   Solidification   and
   properties   of $\gamma'/\gamma-\alpha$ Mo   ductile/ductile   eutectic
   superalloy.   In Solidification and Casting of Metals,
   London, 1979, 526-32.

3. Kováčová, K.,   Unidirectional   solidification   of   Ni-Al-Mo
   alloy. J.Crystal Growth, 1984,66, 426-30.

APPLICATION OF HIGH STRENGTH CEMENT & CARBON FIBER COMPOSITE

KUNIYUKI TOMATSURI
KOUICHI NAGASE
TAISEI Corporation/Technology Research Center
344-1, Nase-machi, Totsuka-ku, Yokohama, Japan 245

ABSTRACT

A precast concrete curtain wall with carbon fiber matrix
(CFM) was studied and produced, to obtain light, rather
complicated and free shaped and easy-to-fix one. Mix
proportion, mixing sequence, mixing methods, consolidation
methods, curing methods and others that we should know to
produce it were examined and a suitable method was decided.
Through these investigation we found that we could produce
CFM curtain wall, using a general mixer which was most
available in precast concrete factories and that the order of
putting materials into a mixer was important.

INTRODUCTION

Wood, brick, concrete and metal have been used as a perimeter
wall material of buildings. Wood was the most widely used
material before a modern age, and it is still widely used in
the field of the personal housing. But glass became a main
material after the curtain wall was introduced into the
building construction. Metal was mainly used for the curtain
wall because glass could easily be fixed in the metal. And a
little later, improvement of the joint between concrete and
glass enabled concrete to be used as a curtain wall. Metal
and concrete with glass are the major materials now as
building facade.
The metal curtain wall with glass is light, easily to be
fabricated, and simple to be fixed to the columns and the
beams of the buildings. But its disadvantage is large
deflection and great heat transmission of the wall due to
high thermal conductivity. So generally it is used with heat
insulating materials and it induces high cost in many cases.
On the other hand, precast concrete (PC) curtain wall has
advantages in thermal insulation, in cost and in ability of
producing of complicated or irregular shape of facade.

Although it has heavy weight and it needs complicated
connection system to beams or floors, which sometimes brings
difficulties to fix it. Besides it, recently cracks on
concrete surface, due to alkali silicate reaction phenomena,
became closed up, and it has been a major problem that the
durability of concrete should be taken more careful attention
and be revalued.
So we noticed carbon fiber matrix (CFM). Using it, we can
make a complicated and free shaped curtain wall like the PC
curtain wall with less cost than the metal curtain wall. And
it seems to be durable enough to offer good satisfaction.
First of all, the main motivation to make a curtain wall with
the CFM is its lightness. Generally, the average thickness
of PC curtain wall is 150 ∿ 160 mm. If we use the CFM
instead, it will become around 50 mm and the curtain walls
weight not less than half times as much. It contributes to
make a building light and also make fasteners simple and easy
to use.

MATERIALS AND MIX PROPORTIONS

Materials and mix proportions were examined. Several papers
about the CFM were referred. In this study, the Omni mixer
was employed, because it was said that the CFM can be mixed
well only with the Omni mixer.
Main results are shown in Fig. 1 and 2, and following results
were obtained.

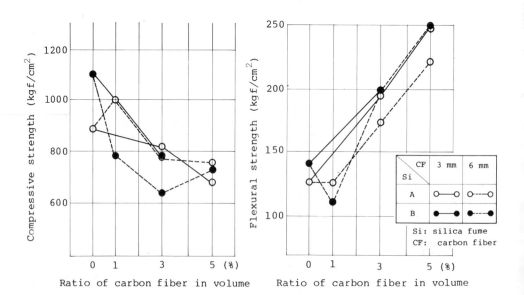

Figure 1. Test result 1 (28 days strength)

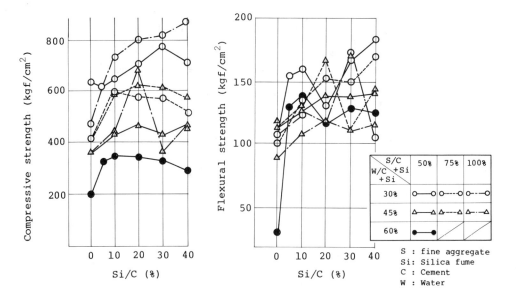

Figure 2. Test result 2 (28 days strength)

1. CFM with higher carbon contents showed higher flexural strength, but lower compressive strength. It can be considered that higher fiber contents will make mixing difficult and will bring more air into the mortar.
2. Thinking of dispersion of the fiber and strength, the most suitable length of the carbon fiber is 3 mm, and this conclusion can be brought from relation between bond and strength of the fiber.
3. Influence of silica fume content on strength was not clear, but mortar with higher silica fume content showed relatively higher flexural strength.
4. When sand content was less than 75% of the cement content, mortar with lower sand content showed lower compressive strength and constant flexural strength. When sand content was greater than 75%, mortar with higher sand content showed higher compressive strength and lower flexural strength.

TABLE 1
Mix proportion ratio

$\dfrac{W}{C + Si + Sl}$	$\dfrac{Si + Sl}{C}$	$\dfrac{W}{C}$	$\dfrac{S}{C + Si + Sl}$	CF
30%	37.5%	41%	61%	3%
weight				volume

In addition to these tests, mixing time, mixing sequence, and mixing method were examined, and the final mix proportion ratio shown in Table 1 was selected. Properties of the materials selected are shown in Table 3. Very fine slug, finer than cement but coarser than silica fume, was used. In Table 2, Properties of the CFM whose mix proportion was selected through these tests were shown.

TABLE 2
Properties of the carbon fiber matrix

values			expressed by	Notes (test piece)
fresh mortar	flow	130 ∿ 140	mm	
	slump	10 ∿ 13	cm	
	air content	6 ∿ 7	%	
compressive strength		676	kgf.cm^2	size: φ50 x 100 h
flexural strength		167	kgf/cm^2	size: 40 x 40 x 160
tensile strength		34	kgf/cm^2	section: 45 x 45
elastic modulous	tension	2.10×10^5	kgf/cm^2	section: 45 x 45
	compression	2.34×10^5	kgf/cm^2	size: φ50 x 100 h
coefficient of air permeability		2.05×10^{-10}	cm/sec	size: φ50 x 25 t
coefficient of permeability		2.28×10^{-4}	cm^2/sec	size: φ50 x 150 h
drying shrinkage		13.5×10^{-4}	at 28 days	size: 100 x 100 x 400
freezing and thawing		good	300 ycle	size: 100 x 100 x 400
conduction of heat	dry	0.569	kcal/m·h·°C	size: 200 x 200 x 25
	wet	0.424		
linear expansion coefficient	dry	8.01	x 10 /°C (0 ∿ 25 °C)	
	wet	5.62		
specific heat	dry	0.25	cal/g·°C	
	wet	0.27		

TABLE 3

Characteristics of carbon fiber, silica fume, slug selected

material	specific gravity	fineness (cm^2/g)	unit weight (kg/ℓ)	diameter (μm)	tensile strength (kg/mm^2)	elastic modulous (kg/mm^2)	heat resist- ance (°C)	alkali resist- ance
Carbon fiber	1.65			18	75	3200	800	good
silica fume	2.02	21,600		1.37 (mean)				
slug	2.91	8000	0.6	4.00 (mean)				

MIXING TESTS

Types of the mixer were examined. Generally, it is said that the CFM can be mixed only with the Omni mixer, but during the tests, it was found that the carbon fiber would be dispersed well when silica fume existed. Available pan type mixer can be used. Since the Omni mixer is not used in usual factory, employment of the special mixer will bring higher production cost. Production by generally available mixer was desired. Mixing sequence was also examined. Mixing methods tested to decide the mixing sequence and time, were shown below. The forced pan type mixer employed in this study was shown in photo 1.

Case 1: All materials are put into the mixer at one time. Easy and good productivity.

Case 2: Good dispersion of the carbon fibers was expected when fine materials were with them.

Case 3: Fibers were mixed in the Omni mixer, and others were mixed in the forced mixer.

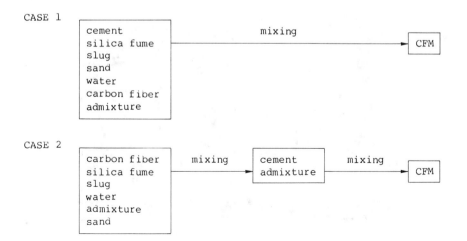

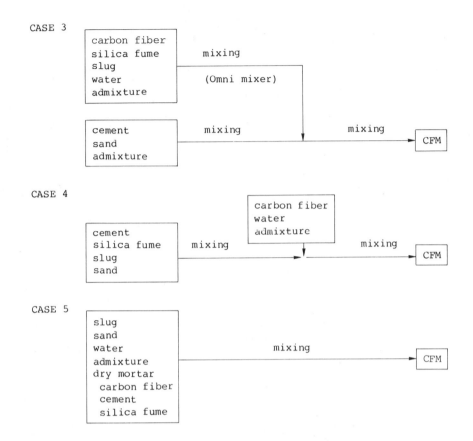

CASE 3

| carbon fiber
silica fume
slug
water
admixture | mixing

(Omni mixer) |

| cement
sand
admixture | mixing | mixing | CFM |

CASE 4

| carbon fiber
water
admixture |

| cement
silica fume
slug
sand | mixing | mixing | CFM |

CASE 5

| slug
sand
water
admixture
dry mortar
  carbon fiber
  cement
  silica fume | mixing | CFM |

Case 4: All materials but water and fibers were mixed, and
         fibers were dispersed by water.
Case 5: Silica fume was added instead of fine aggregate in a
         factory to dry mortar which was created by us and
         discussed later. Sand and slug were mixed with it in
         a mixer.

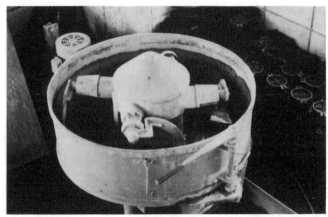

Photo 1.   100 ℓ forced mixer

Following results were obtained.

Case 1: Fiber balls were observed, and fibers were not dispersed well. Since fiber balls were produced in the mixing with both cement and water, cement should be mixed with water last.

Case 2: Mortar was mixed without problem.

Case 3: The first stage of the mixing was done in an Omni mixer, and the slurry from the Omni mixer was mixed with cement, sand and admixture in the general mortar mixer. Although good dispersion was obtained, special equipment, the Omni mixer, was required.

Case 4: Fiber balls were observed, and fibers were not dispersed well. Longer mixing time showed better, but still not good results.

Case 5: Fiber balls of about 20 mm were observed. After cement meets water and hydration starts, good dispersion can hardly be obtained. It was considered that water should be poured last.

Test results of strength were shown in Table 4.

TABLE 4
Result of mixing tests

case	mix proportion data			mixing time (sec)	fresh mortar properties				strength (kgf/cm²) upper: cure in 20 °C water lower: steam cure			
	w/c (w/c + Si + Sl)	Si/c (si + sl/c)	s/c + si + sl		air content (%)	flow (mm)	slump (cm)	unit weight (kg/ℓ)	compressive		flexural	
									1 week	4 weeks	1 week	4 weeks
1	0.41	0.25	0.60	270	6.3	158	12.2	1.96	493	666	131	138
	0.30	0.37							394	466	120	121
2	0.41	0.25	0.60	210	8.9	154	11.3	1.90	437	601	127	139
	0.30	0.37							444	520	139	151
3	0.41	0.25	0.60	210	9.4	148	9.7	1.94	459	686	136	159
	0.30	0.37							343	414	127	144
4	0.41	0.25	0.60	240	4.9	166	12.0	2.00	501	679	162	174
	0.30	0.37							364	403	164	155
5	0.41	0.25	0.60	210	8.8	180	16.6	1.92	410	557	144	138
	0.30	0.37							398	487	135	131

CFM by Case 1, in which all materials were mixed at one time and fibers were not dispersed well, showed lowest flexural strength. CFM by Case 4, in which some fiber balls were observed, showed highest flexural strength. CFM by Case 2 showed almost same properties as CFM by Case 3, in which the Omni mixer was used.

These results indicated that production of actual curtain wall panel could be conducted with the Case 2.

This indication was also confirmed by the followed tests for consolidation method, finishing method and curing method examination.

TRIAL PRODUCTION OF THE CURTAIN WALL

The CFM was cast in the form shown in Fig. 3. Workability
and properties of the CFM were examined totally. A 1.5 m^3
pan type mixer (Figure 4) was employed for this study.
Mixing proportion and mixing method were same as shown
before. The CFM was cured in the 40 °C steam for 8 hours.
The results of the test by the 100 litter mixer were
confirmed through this test. Carbon fibers as well as other
materials were dispersed well and same workability was got.
Both the internal and the table vibrator were employed.
The precast concrete panel produced in this study has been
laid in the open air to examine the change along time, and
drying shrinkage of the CFM are shown in Fig. 5.

PRODUCTION OF THE CURTAIN WALL

The CFM curtain walls were produced and applied on the actual
building on trial bases. These curtain walls were applied on
the wall of the connecting part of two buildings. Good
appearance was obtained. Photo 2 shows them.

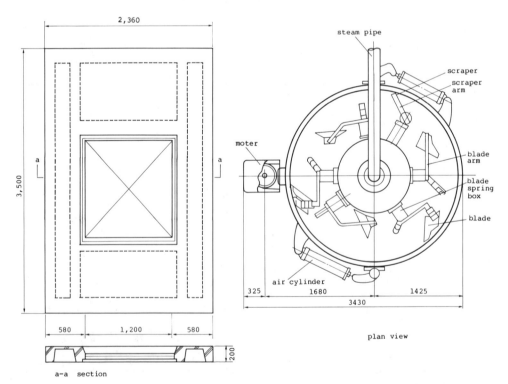

Figure 3. A curtain wall slab
for test

Figure 4. 1.5 m^3 pan type
mixer

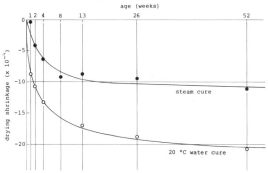

strength at 4 weeks upper: steam cure lower: cure in 20 °C water		
compressive	flexural	tensile
773 kgf/cm^2 611 kgf/cm^2	155 kgf/cm^2 125 kgf/cm^2	39.7 kg/cm^2 37.0 kg/cm^2

Figure 5. Drying shrinkage (test piece) of CFM for a test curtain wall

Photo 2. CFM curtain wall applied

DRY MORTAR WITH THE CARBON FIBER

Through this study, dry mortar system, in which mortar with carbon fiber was produced at site by adding only water, was created. All materials but water, such as cement, carbon fibers, fine aggregate and admixture, were mixed at a factory. Mortar with carbon fiber can get at site with a usual mortar mixer. Mix proportion and properties are shown in Table 5. This dry mix system are being used as a finishing material of block, wooden surface and concrete.

TABLE 5
Mix proportion and properties of dry mortar

mix proportion (kg)				
cement	fine aggregate	carbon fiber	acrylic emulsion	water
100	150	1	13.3	46
properties				
unit weight	compressive strength	flexural strength	adhesion to concrete	water absorption for 24 hours	permeability
1.88 kg/ℓ	329 kgf/cm^2	80 kgf/cm^2	14 kgf/cm^2	2.5% vol.	0.3% vol.

CONCLUSION

The CFM curtain wall was carefully and sufficiently investigated and applied. We are going to observe the CFM curtain walls fixed to a real building. This is a trial application for a facility of our own company. We may change mix proportion and production method next. It depends on result of the observation. But it was confirmed that CFM curtain walls can be produced with a general forced type mixer.

HIGH STRENGTH CEMENT PASTE

HANS-ERIK GRAM

KERSTIN OLSSON

SWEDISH CEMENT AND CONCRETE RESEARCH INSTITUTE

S-100 44 STOCKHOLM, SWEDEN

ABSTRACT

The compressive strength of ordinary Portland cement concrete and cement paste is generally related to its porosity and/or its water to cement ratio. Specimens of cement paste with different water to cement ratios and densities have been prepared and their compressive and flexural strength were determined. The results of the test show that the strength of the cement paste is not significantly influenced by the water to binder ratio. The development of macro and micro cracks govern the mechanical behaviour of the cement paste. Specimens made of gypsum-free Portland cement paste also develop macro and micro cracks.

INTRODUCTION

During the last decades examples of extremely high strength cement pastes and concretes have been presented. In the laboratories it was possible to produce hot pressed pastes (150°C) with a compressive strength of 644 MPa /1/, so called macro-defect free cements (MDF-NIMS) with a compressive strength of 300 MPa and a flexural strength of 150-200 MPa /2/, Densified Systems (Densit) with additions of effective superplasticizers and ultra-fine particles of silica fume with a compressive strength of 200-270 MPa /3/ and polymer impregnated concrete with compressive strengths around 280 MPa. Densified systems and to some extent polymer impregnated concrete have found commercial applications while the other types of high strength cement materials seem still to be in a developing phase. At the Swedish Cement and Concrete Research Institute a project titled "High Strength Low Temperature Bound Ceramics (1988-1989) is in progress with the aim of establishing the know-how of today and to present suitable solutions of material compositions for the use of concrete in new applications in the engineering industry /4/.

MATERIALS AND METHODS

General

Specimens with a high volume of steel fibres (>10%) and cement paste with low water to cement ratio (0.20) have been prepared. The compressive strength of these specimens was around 220 MPa and the flexural strength around 85 MPa. During the tests it was obvious that the limiting part of the composite was the cement paste. Further experiments have therefore been concentrated on the cement paste.

Experiments with ordinary Portland cement

The aim of the experiments with ordinary Portland cement (OPC) was to reduce the porosity of the paste and to increase its density by the addition of superplasticizers and silica fume. The cement paste (120 g) was mixed in a mixer made of a drilling machine allowing a high energy input. The cement paste was cast in Teflon moulds with a height of 74 mm and a diameter of 19 mm. The moulds were thereafter sealed with strong tape.

The mould was stripped after one day and immediately after the density of the specimen was determined. The results are presented in Figure 1.

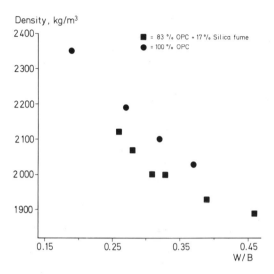

Figure 1. Density of cement paste as a function of the water to binder ratio.

The specimens were then cured in water for six days and stored dry for three weeks before the flexural and compressive strength was determined, see Figure 2 and Figure 3.

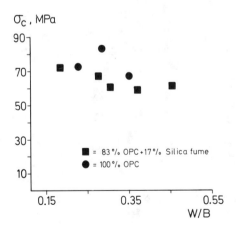

Figure 2. Compressive strength of cement paste as a function of the water to binder ratio.

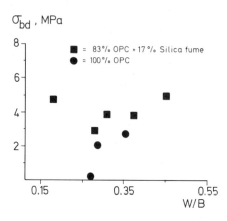

Figure 3. Flexural strength of cement paste as a function of the water to binder ratio.

As can be seen in Figure 2 and Figure 3 the strength of the cement paste is not significantly influenced by the water to binder ratio. The strength of the cement paste is negatively influenced by the development of macro and micro cracks, see Figure 4 and Figure 5.

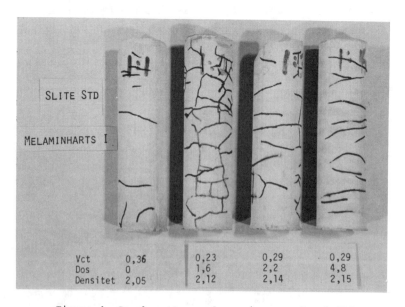

Figure 4. Crack pattern of specimens made of OPC.

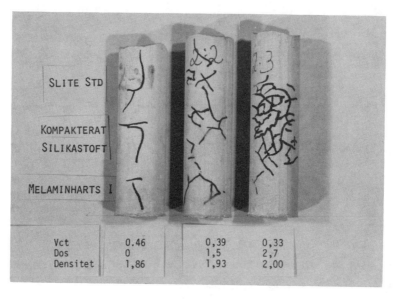

Figure 5. Crack pattern of specimens made of 83% OPC and 17% silica
fume.

Experiments with gypsum-free Portland cement

According to Rio /5/ the sulphur in the gypsum that is used as a binding
regulator should be excluded if you want to get a denser cement paste.
Skvara /6/ and others have shown that it is possible to substitute the
gypsum with alternative additives as for instance polyphenol sulphonate-
carbonate systems.

Experiments with ordinary Portland cement clinker without gypsum
have been made where the clinker was ground together with different
additives. The flexural and compressive strengths of the tested specimens are shown in Table 1.

TABLE 1

Flexural and compressive strength of gypsum-free cement paste specimens.
The setting was regulated by addition of calcium lignosulfonate and
$NaHCO_3$ or K_2CO_3.

	SiO_2 %	W:C	Strength (MPa)	
			flexural	compressive
$NaHCO_3$	-	0.14	8.1	90
$NaHCO_3$	5	0.14	8.8	116
$NaHCO_3$	10	0.15	5.5	156
K_2CO_3	-	0.15	9.0	135

The strength of gypsum-free cement paste specimens is higher than
the strength of ordinary portland cement paste specimens, the water to
cement ratio is lower and the density higher. However, the strength
levels are still low. Even these specimens develop macro and micro
cracks, see Figure 6 and Figure 7.

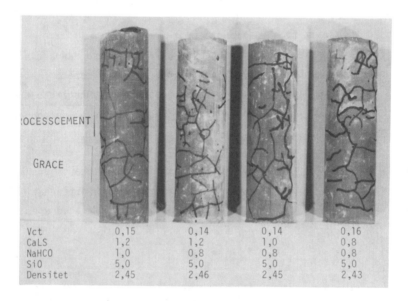

Figure 6. Crack pattern of specimens made of 95% gypsum-free Portland
cement and 5% silica fume set regulated with calcium lignosulphonate and $NaHCO_3$.

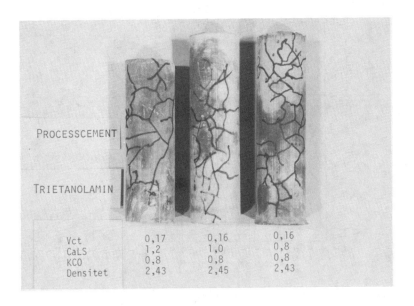

Figure 7. Crack pattern of specimens made of gypsum-free portland cement set regulated with calcium lignosulphonate and K₂CO₃.

DISCUSSION

High strength paste of ordinary Portland cement cracks with time, probably depending on deficiency of water during hydration and shrinkage during drying. According to Grudemo /7/ there is a maximum flexural strength of cement paste stored in water at water cement ratios of 0.5, see Figure 8.

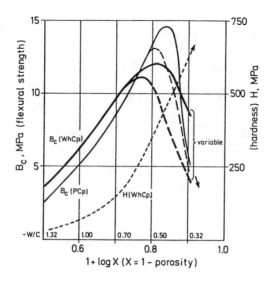

Figure 8. Flexural strength B_c (concentric bending) and surface microhardness (Vickers) for pastes of white cement and Portland
cement, 6 years hydrated, as functions of X = solid-phase
volume according to Grudemo /7/.

The micro and macro crack development cannot be avoided by replacing
gypsum as a setting regulator in the experiments reported herein which
used lignine derivates and alkali carbonates.

CONCLUSIONS

The mechanical properties of Portland cement paste are not governed by
the water to cement ratio and the density as is the case for concrete.
Micro and macro cracks have a large influence on the strength.

REFERENCES

1. Roy, D.M., Gouda R.R.: J Amer. Ceram. Soc. 56, 1973, pp. 549-550.

2. Alford, M Mcn, Birchall J.D.: The Properties and Potential Applications of Macro-Defect-Free Cement. Mat. Res. Soc. Symp. Proc. 1985,
Vol. 42, pp. 265-276.

3. Bache, H.H.: Introduction to Compact Reinforced Composite. Nordic Concrete Research, Publ. No 6, 1987, pp. 19-33.

4. Concrete as Material for the Engineering Industry (Betong som verkstadsmaterial) Swedish Cement and Concrete Research Institute, 1987, pp. 1-76 (in Swedish).

5. Rio, A: Approaching to a Macromolecular Characterization of the C_3S Hydration Process. Paper 67, 6th International Congress Chemical Cement, Moskva 1974.

6. Skvara, F, Rubinova, M: The System Clinker-Polyphenol Sulphonate-Alkalimetal salts. Cement and Concrete Research, Vol 15, 1985, pp. 1013-1021.

7. Grudemo, A: The crypto-crystalline structure of C-S-H gel in cement pastes inferences from X-ray diffraction and dielectric capacity data. Cement and Concrete Research Vol 17, 1987, pp. 673-680.

SOME PROBLEMS CONCERNING THE DETERMINATION
OF THE IMPACT RESISTANCE OF STEEL FIBRE REINFORCED CONCRETE

WOJCIECH RADOMSKI
Warsaw University of Technology, Poland

ABSTRACT

Some chosen and new problems connected with the determination
of the impact resistance of steel fibre reinforced concrete
are presented. They are based on the experiments carried out
by the author and other investigators. The problems discussed
concern the relationship between the impact resistance of
steel fibre reinforced concrete with 2D and 1D orientation of
fibres, the relationship between the energy absorbed during
the static flexure tests and the impact resistance of this material and the effect of the impact velocity as well as the
strain-rate on this resistance.

INTRODUCTION

One of the most important characteristics of steel fibre reinforced concrete /SFRC/ is its considerable impact resistance
determined by experimental studies carried out on specimens of
the material as well as on structural elements of various dimensions and shapes. There are several methods of impact testing of SFRC and depending on the methods applied there are
various versions of the impact resistance definitions. The impact resistance of SFRC cannot therefore be considered as a
single fully-defined physical property of the material. Haw-
ever, it should be emphasized that only in the case of the
swinging pendulum method /Charpy or Izod equipment/ or when
the rotating impact machine is used are the results of investigations physically meaningful. The other methods provide only some approximate information about the impact resistance of
SFRC for practical purposes, but do not allow the determination
of any physical properties of the material. This problem has
been discussed by the author previously [1].

TABLE 1
Factors effecting SFRC properties under impact loads

Matrix		Fibres			Impact load			Response of the member to the impact load	
$\underline{\underline{p}}_m$	$\underline{\underline{\phi}}$	p_f	$\underline{\underline{k}}$	$\underline{\underline{\beta}}$	v	m	N	Ψ_1	Ψ_0
w/c	w/s	d	l	D	α		n		

$\leftarrow \underline{\underline{\tau}} \rightarrow$

p_m = fundamental mechanical properties of a matrix /e.g. static compressive strength, ϕ = size of aggregate used for a matrix, w/c = water/cement ratio, v/s = water/sand ratio, τ = fibre-matrix bond strength, p_f = fundamental mechanical properties of steel used for fibres /e.g. tensile strength/, k, d, l and β = shape, diameter, length and volume content of fibres respectively, D = type of fibre orientation /i.e. 1D, 2D or 3D/, v = velocity of impact, m = mass of missile, N = missile shape factor, α = angle of impact load with respect to the direction of fibres, n = number of blows, Ψ_1 and Ψ_0 = local and overall deformability of SFRC specimen or structural element under impact loads respectively /"hard" or "soft" impact/.

$=$ - relatively often tested,
— - relatively seldom tested,
⋯ - almost unknown.

The impact resistance of SFRC as well as its other dynamic properties depend on many factors characterizing the material / i.e. matrix and fibres/, the impact load and the local and overall deformability of the tested specimens or structural elements. The most important of these factors are listed in Table 1.

It should be emphasized that the effect of some factors /mentioned in Table 1/ on the impact resistance of SFRC is not sufficiently known at present.

It should also be emphasized that the test results concerning the impact resistance are generally of a great scatter due to the heterogeneity of the material. Moreover, the dynamic phenomena produced by the impact loads in the tested specimens or elements are very complex. The interpretation of the test results may therefore be difficult and the theory concerning the dynamic behaviour of SFRC has not been well developed so far.

In this paper some chosen and new problems connected with the determination of the impact resistance of SFRC are presented They are based on the experiments carried out by the author and other investigators.

IMPACT RESISTANCE OF SFRC
WITH 1D AND 2D ORIENTATION OF FIBRES

It is known that depending on the production technology applied, the distribution of steel fibres in the elements may be of different types. This distribution can be theoretically analyzed as three types of the fibre orientation, i.e. three-dimensional /3D/ when the fibres are randomly distributed in the whole volume of the element, two-dimensional /2D/ when the fibres are distributed at random in the parallel planes or surfaces of the element and one-dimensional /1D/ when the fibres are aligned. In practice there are also intermediate cases, i.e. /2D-3D/ or /2D-1D/, because the distribution of the all used fibres in the concrete matrix strictly according to the fibre orientations mentioned above is practically not always possible. The case /2D-3D/ can be observed for instance when the external surfaces of the element have a linearizing effect on the directions of the fibres lying nearby /e.g. in the thin plates/. The case /2D-1D/ occurs more often in practice than the case /1D/ because the linearization of the fibres is never complete.

The linearization of the fibres in the structural elements /or in some parts of the elements only/ may lead to an increase of their resistance. It takes place, for example, when the fibres are aligned according to the principal directions of stresses or strains in the loaded elements.

There are several methods of the linearization of the fibres in the concrete or mortar matrices. In the case of SFRC it is feasible, for example, by the application of a magnetic field [2],[3],[4],[5].

The research on SFRC with the fibres oriented by means of a magnetic field has not been well developed so far, especially when the dynamic properties of the material are concerned. In spite of a very limited number of the tests performed their results indicate that the linearization of the fibres is in some cases justifiable from technical and economical points of view.

The impact resistance of SFRC with 1D and 2D types of fibre orientation has been experimentally compared so far by Skarendahl[4] and by the author[1],[5] only.

Skarendahl performed his experiments on the specimens of 20 x
30 x 120 mm cut out from the plates. Several types of steel
fibres, i.e. plain, deformed and with hooked ends, were used.
The nominal volume content of the fibres /β_n/ in the tested
specimens varied from 0.5 % to 3.0 %. The impact load was produced by the Izod equipment of 40 Nm capacity. In the case of
the specimens with 1D fibre orientation /the fibres magnetically oriented/, the direction of the impact was constant and
perpendicular with respect to the direction of the fibres.
Some of the test results obtained by Skarendahl are shown in
Figure 1.

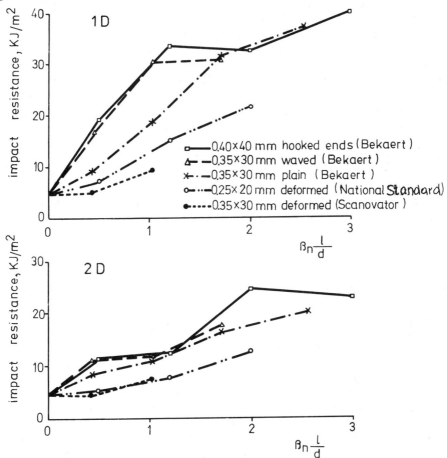

Figure 1. Impact resistance for SFRC with 1D and 2D fibre
orientation.

It is evident that the specimens with the magnetically oriented fibres /1D orientation/ exhibited from 1.07 to 2.78 /app-
roximately twice/ the impact resistance of the specimens with

randomly distributed fibres /2D orientation/.
The author's experiments were to check the effect of the angle
of the impact load with respect to the direction of the fibres
on the impact resistance of the SFRC specimens.
The mix proportions per m^3 of the mortar matrix were as follows:
- portland cement 350   - 550 kg
- sand /grains up to diameter of 2 mm/ - 1635 kg
- water   - 325 kg
- w/c ratio   - 0.59

The plates of 445 x 205 x 25 mm with parallel magnetically o-
riented fibres were cut at different angles with respect to
the direction of the fibres, i.e. at 0°, 30°, 45°, 60° and 90?
The specimens of 25 x 25 x 105 mm were tested with the span
of 50 mm by means of the conventional Charpy equipment of 300
Nm capacity with the impact velocity of 5.2 m/s. The number
of specimens varied from 8 to 14. Plain steel fibres of 0.4 x
40 mm were used. The nominal volume content of the fibres in
the specimens varied from 0.5 % to 3.0 %. Moreover, the specimens with 2D fibre orientation as well as the specimens of
mortar matrix without any fibres were tested. All other details concerning the performed experiments were published elsewhere [5]. The test results are summarized in Figure 2.
The impact resistances of the specimens with 1D and 2D types
of fibre orientation are compared with each other by means of
the ratio $\varkappa$ which can be written as

$$\varkappa = \frac{IR_{1D}}{IR_{2D}} , \qquad /1/$$

where IR_{1D} and IR_{2D} denote the impact resistance of the tested specimens with 1D and 2D fibre orientation, respectively.
Because of the relatively limited number of tests carried out,
the test results should be considered as qualitative only and
their scatter is due to the heterogeneity of the material as
well as to the incomplete linearization of the fibres in the
matrix. In spite of this the test results emphasize the increase in impact resistance of the specimens with increasing
the volume content of the fibres /β_n/ and with the value of
the angle of the impact loads with respect to the direction
of the fibres /α/. The results show that the linearization
of the fibres effects the increase in impact resistance of the
specimens compared to the resistance of the specimens with 2D
fibre orientation when the angle α is equal minimum 60°, e.i.
$\varkappa > 1$ when $\alpha \geqslant 60°$. Specimens with the oriented fibres /1D or-
ientation/ subjected to impact loads perpendicular to the di-
rection of fibres /$\alpha = 90°$/ exhibited from 1.19 to 1.59 the
impact resistance of the specimens with 2D orientation. When

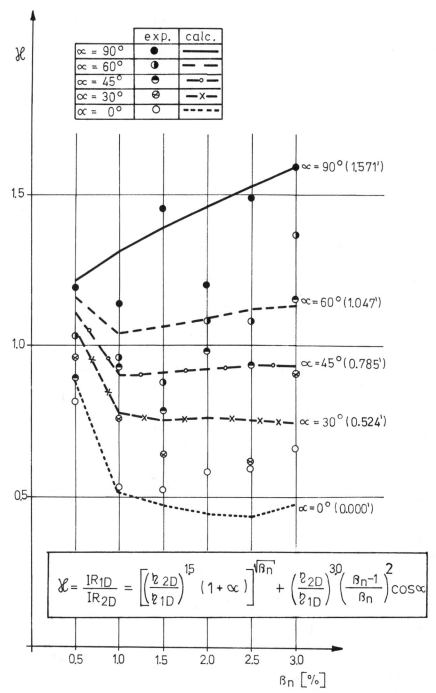

Figure 2. Experimental and calculated values of the ratio $\varkappa$

$\alpha = 60^\circ$ the impact resistance was in general higher than in the case of 2D orientation, whereas when $\alpha = 45^\circ$ or less then, generally, $\varkappa \langle 1$.

On the base of the test results an original empirical formula for the prediction of values of the ratio $\varkappa$ was formulated as follows:

$$\varkappa = \frac{IR_{1D}}{IR_{2D}} = \left[\left(\frac{\eta_{2D}}{\eta_{1D}}\right)^a (1 + \alpha)\right]^{\sqrt{\beta_n}} + \left(\frac{\eta_{2D}}{\eta_{1D}}\right)^b \left(\frac{\beta_n - 1}{\beta_n}\right)^2 \cos\alpha \qquad /2/$$

in which η_{2D} and η_{1D} = the parameters of directional effectiveness of fibres in the cases of 1D and 2D types of fibre orientation respecively, α = the angle of the impact loads with respect to the direction of the fibres in radians, β_n = the nominal volume content of fibres as a percentage, a and b = the exponents determined experimenally.

The comparison between the experimental and calculated values of the ratio $\varkappa$ is presented in Figure 2. The calculated values /$\varkappa_{calc}$/ are determined from the formula /2/. According to the studies performed by Kasperkiewicz [6] it was assumed that $\eta_{2D} = 0.637$ and $\eta_{1D} = 1$. The exponents were assumed as a = 1.5 and b = 3.0. The differences between $\varkappa_{exp}$ and $\varkappa_{calc}$ did not exceed by 28.8 % but the great majority of them varied from 0 % to 20 %. Taking into account the strong heterogeneity of the material it may be concluded that the test results are sufficiently approximated by the calculated values obtained from formula /2/.

Formula /2/ was also used for an approximation of the results of the tests carried out by Skarendahl [4]. The good correlation was found as it is shown in Figure 3.

The form of the formula /2/ seems to be therefore useful for the prediction of the relationship between the impact resistance of SFRC with 1D and 2D types of fibre orientation. This formula requires further experimental verification because of the limited number of tests performed so far.

The test results obtained by the author indicate that when the angle of the impact load with respect to the structural elements made from SFRC is known or may be predicted as varying from 60° to 90°, the linearization of the fibres is justifiable and provide for the increase in the impact resistance of the material. When the angle mentioned above is not known, e.g. the impact loads may occur accidentally with the different angles with respect to the structural elements, it is better to apply SFRC with 2D or 3D fibre orientation. This con-

419

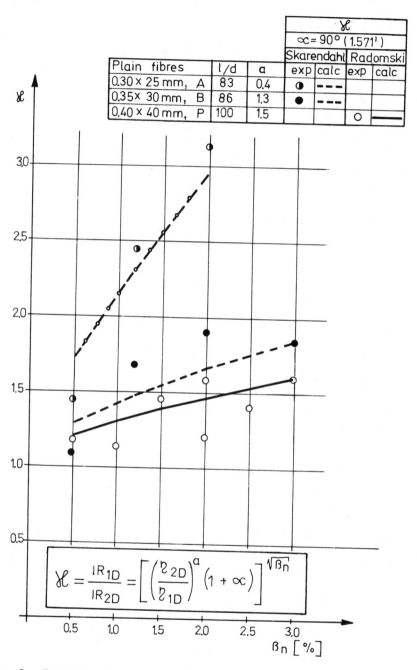

Figure 3. Experimental and calculated values of the ratio ℋ from the tests performed by Skarendahl and by the author.

clusion also requires further investigations.

RELATIONSHIP BETWEEN THE ENERGY ABSORBED
DURING THE SLOW FLEXURE AND THE IMPACT TESTS

It should be emphasized once more that the experimental investigations of SFRC are of fundamental importance, since the im-
pact-load effects on this heterogenous material are very complicated and no theoretical solutions are available as yet.
For this reason there are only some attempts to check the relationship between the static characteristics of SFRC and the
impact resistance of this material. One of such attempts has
been made by Hibbert and Hannant [7] on the base of the analysis performed by Cottrell [8].
Cottrell has shown that the work of fracture due to fibre pull-
out in tension for fibres of circular cross-section is given
to the first approximation by

$$U = \frac{\beta \cdot \tau \cdot l^2}{12d} \qquad\qquad /3/$$

where U is the work of fracture per unit area of specimen
cross-section for a three-dimensional random composite, β is
the volume fraction of fibre in the composite material, τ is
the frictional sliding bond strength /assumed constant/ and l
and d are respectively the fibre length and diameter.
The same relationship /3/ is approximately valid for the static flexure failure provided complete separation of the specimen halves occurs.
Hibbert and Hannant carried out the static and the impact
tests on specimens of SFRC with dimensions 100 x 100 x 500 mm
using several types of fibres. The static tests were performed
to determine the energy absorbed in slow bending in a conventional testing machine. The impact tests were performed by
means of the instrumented Charpy impact machine specially designed and constructed for testing fibre-reinforced concrete.
This machine has been fully described elsewhere [9]. The
loading configuration /centre point loading on a simply supported span of 400 mm/ was the same in the static tests and
the impact tests so that the results are directly comparable.
In Figure 4 the energy obtained from the pendulum amplitude
after the failure and substraction of the kinetic energy of
the broken halves is plotted against $\beta \cdot l^2/d$ for the tested specimens.
The apparently low value of τ for the 0.38 x 38 mm Duoform
fibres is due to the fracture of fibres during the failure of
the specimen. Since extensive fibre pullout did not occur the
analysis is not applicable in this case. The results are shown,

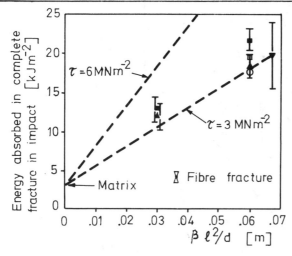

■ Hooked-ends, 0.5 × 50 mm, 2 mths   ∇ Duoform, 0.38×38 mm, 2 mths
▲ Crimped, High C, 0.5×50 mm, 2 mths   △ Duoform, 0.38×38 mm, 2½ years
□ Crimped, High C, 0.5×50 mm, 2½ years   ▼ Duoform, 0.64×60 mm, 2 mths
○ Crimped, Low C, 0.5×50 mm, 2½ years   — — Theoretical ⌶ 90% Confidence intervals for the mean

Figure 4. Energy absorbed in impact of SFRC specimens as a function of fibre volume fraction and fibre size.

however, for comparison with the other fibre types.
It can be seen that the test results correspond to the energy absorbed calculated from the formula /3/ when the fibre-matrix bond strength τ is between 3 MPa and 6 MPa. These values of τ are in good accordance with the experimental data.
The relationship between the work of fracture described by the formula /3/ and the impact resistance of SFRC was also examined by the author. The specimens of 25 x 25 x 105 mm with 2D fibre orientation cut out from plates were tested by means of the conventional Charpy equipment described in the previous paragraph of the paper. The mix proportions of the mortar matrix, type and volume content of the steel fibres were the same as previously reported. Moreover, the specimens with w/c=0.54 were tested. The number of specimens of each type varied from 10 to 14. The test results are shown in Figure 5. The impact resistance of SFRC specimens is plotted against $\beta \cdot l^2/d$. It can be seen that a great majority of the test results correspond to the energy absorbed calculated from the formula /3/ when the fibre-matrix bond strength τ is between 1.5 MPa and 3.0 MPa. These values of τ are in good accordance with the experimental data obtained by Burakiewicz [10].
The results of experiments carried out by Hibbert and Hannant

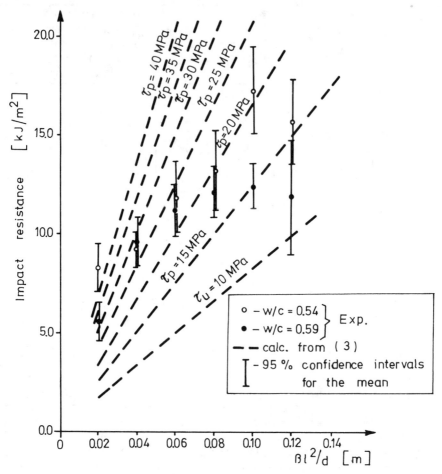

Figure 5. Impact resistance of SFRC specimens as a function
of fibre volume fraction and fibre size

as well as by the author therefore indicate that Cottrell's
formula may be applicable for approximate estimation of the impact resistance of SFRC.
However, Hibbert and Hannant found that for SFRC the energy
absorbed after matrix cracking is not substantially different
under slow flexure and impact conditions. This implies that
the strain rate has little effect on the fibre-matrix bond
properties, particularly the work of fibre pullout. This conclusion is contrary to the newer test results reported in the
next paragraph of the paper.

EFFECT OF THE IMPACT VELOCITY AND THE STRAIN RATE
ON THE IMPACT RESISTANCE OF SFRC

The effect of the impact velocity and the strain rate on the
impact resistance of SFRC has been examined by several investigators, e.g. Shah and Gopalaratnam [11] as well as by the
author [1] .
Shah and Gopalaratnam performed experiments on the SFRC specimens of 76 x 25 x 229 mm. Plain mortar matrix mix used was
proportioned c:s:w=1:2:0.5 /by weight/. Brass coated, smooth
steel fibres of 0.4 x 25 mm were used with volume contents of
0.5 %, 1.0 % and 1.5 %. The static three-point-bend tests
were conducted using a closed-loop universal testing machine
of a capacity of 178 kN. The impact tests were performed by
means of a modified instrumented Charpy equipment with three
different impact velocities, namely 1.30, 1.85 and 2.45 m/s.
Two strain rates: $1 \cdot 10^{-6}$ 1/s and $1 \cdot 10^{-4}$ 1/s were used during
the static tests and three strain rates: 0.017 1/s, 0.09 1/s
and 0.3 1/s were used during the impact tests. The test results are shown in Figure 6.

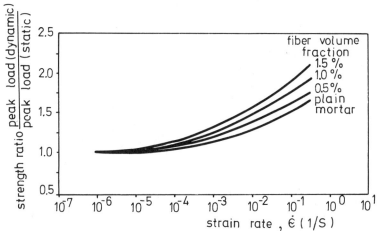

Figure 6. Effect of strain-rate on the flexural strength of
unreinforced and fibre reinforced beams

It can be seen that the effect of strain rate is higher for
SFRC specimens, the more so the higher the volume content of
fibres. The higher strain rate sensitivity of SFRC specimens
than that of mortar specimens is probably due to additional
cracking /both transverse matrix cracking and interfacial cracking or debonding/ generally associated with fibre reinforced concrete specimens and the observation that the strain
rate sensitivity in cement-based composites is related to
crack growth.

On the base of the experiments preformed and the test results
from other investigators, Shah and Gopalaratnam concluded that
SFRC made with weaker matrices, higher fibre contents and larger fibre aspect ratios are more rate sensitive than those made with stronger matrices, lower fibre contents and smaller
fibre aspect ratios. Resulting from the comparison between

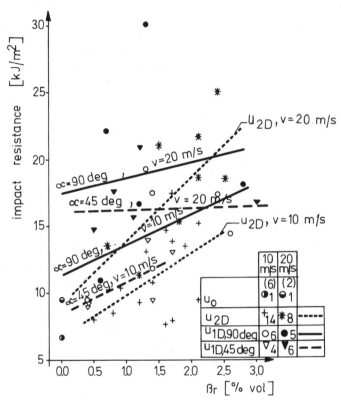

Figure 7. Impact resistance of specimens with 1D and 2D orientation of fibres /U$_{1D}$ and U$_{2D}$/ . Numbers in the table indicate the number of tests.

static and impact tests, Shah and Gopalaratnam observed also
that static flexural toughness tests may be used to estimate
approximately the impact resistance of SFRC, assuming that the
mode of fracture does not change with the rate of loading.
This last conclusion seems to be questionable when the impact
velocity is higher than 5 m/s. It is reinforced by the test
results obtained by the author.
The SFRC specimens of 15 x 15 x 105 mm cut out from plates
were tested by means of the rotating impact machine with the
impact velocities: 10 m/s, 20 m/s and 30 m/s. These test results have been published elsewhere [1] . Some of them are

shown in Figure 7. In spite of a scatter due to the strong heterogeneity of the material, the directions of the straight
lines shown in the figure and determined by the least squares
method, evidently indicate that the impact resistance of SFRC
increases with the increasing velocity of impact load. The author's experiments also indicate that the applications of the
Charpy or Izod equipment /conventional or instrumented/ for
impact testing of SFRC may be insufficient for practical purposes due to the impact velocities produced by this equipment
are relatively small /generally up to 5 m/s only/. For that
reason the applications of the equipment with higher impact
velocities are advisable, e.g. the rotating impact machines.

GENERAL CONCLUSIONS

Some chosen problems concerning the determination of the impact resistance of SFRC discussed in the paper indicate that
this dynamic characteristics requires further investigations
because the knowledge on it is incomplete at present.
First of all it is necessary to provide a standard test method
of testing the material with different impact velocieties more
than 5 m/s only. It is grounded by the development of applications of SFRC in the dynamically loaded structures of various
types.
The test results obtained so far do not allow us to formulate
any general theoretical solutions related to the impact resistance of SFRC.

REFERENCES

1. Radomski, W., Application of the rotating impact machine
   for testing fibre-reinforced concrete, The International
   Journal of Cement Composited and Lightweight Concrete,
   1981, 1, 3-12.

2. Sikorski, C., Patent no 58128, kl. 80a, 51, 1969,/in Polish/.

3. Björklund, F.R., Miller, A.J., Patent: Sverige B /11/, Ut-
   läggningsskrift 7405693-8, 1974, /in Swedish/.

4. Skarendahl, Å., Stålfiberbetong slagseghet vid 1- och 2-di-
   mensionell fierorientering, Fibrobetong, Nordforsks pro-
   jektkommitté for FRC-material, Delrapporter, Sub-Report S,
   Stockholm 1977, S1-S10, /in Swedish/.

5. Radomski, W., Properties of Fibre Reinforced Concrete under Impact Loads, Institute of Roads and Bridges, Warsaw
   University of Technology, Warsaw, 1982, pp. 287, / in Po-
   lish/.

6. Kasperkiewicz, J., Theoretical formulae concerning fibre reinforced concrete like coposites. Report FRC/R/CBI-6, CBI, Stockholm 1974, pp. 64.

7. Hibbert, A.P. and Hannant, D.J., Impact resistance of fibre concrete, Transport and Road Research Laboratory, Supplementary Report 654, Berkshire, 1981, pp. 25.

8. Cottrell, A.H., Strong solids, Proc. Royal Society, A282, 1964, **2**.

9. Hibbert, A.P. and Hannant, D.J., The design of an instrumented impact test machine for fibre concretes. In Testing and Test Methods of Fibre Cement Composites. RILEM Symposium, Sheffield, 1978. The Construction Press, pp.1o7-12o.

10. Burakiewicz, A., Bond strenght between fibres and matrix in fibre reinforced cement composites, Institute of Fundamental Technological Research, Polish Academy of Sciences, Warsaw, 1979, pp. 140, /in Polish/.

11. Shah, S.P. and Gopalaratnam, V.S., Impact resistance measurements for fibre cement composites. In Developments in Fibre Reinforced Cement and Concrete. RILEM Symposium, Sheffield, 1986, vol. 1, pp. 267-276.

APPLICATION OF COMPUTATIONAL MECHANICS RELATED TO FRACTURE CONCEPTS AND RELIABILITY OF COMPOSITES

BERND MICHEL, JOHANN-PETER SOMMER, WERNER TOTZAUER,
KLAUS BRAEMER
Academy of Sciences of G.D.R., The Institute of Mechanics,
PSF 408, Karl-Marx-Stadt, 9010, G.D.R.

ABSTRACT

The development of 'high tech' materials such as ceramics or high strength composites requires the application of new powerful tools of theoretical and experimental mechanics in combination with modern computational aids. The application of so-called 'hybride methods' of solid mechanics has been one characteristic trend of development above all for high strength materials. The main points of this paper deal with the utilization of finite element method (FEM) and boundary element method (BEM) for evaluation of ceramic components and carbon fibre composites. Finally the authors intend to give a short survey of new results combining computer mechanics and experiments carried out by means of laser and speckle interferometry and the micro moire method.

INTRODUCTION

The development of 'high tech' materials such as ceramics or high strength composites requires the application of new powerful tools of theoretical and experimental mechanics. In the field of 'materials mechanics' both modern methods of computational mechanics and progressive experimental methods play an important role for the study of ceramics and composites. One of the reasons for that is the complexity of mechanisms characterizing material properties and material response. Besides the phenomenological modeling more and more theoretically orientated concepts and ideas of micromechanics have to be taken into account too (1), (2). The application of so-called 'hybride methods' of solid mechanics has been one characteristic trend of development above all for high strength materials (3). The authors report on their experience in this field. The main points deal with the application of finite element method (FEM) and boundary element method (BEM) for evaluation of ceramics components and carbon fibre composites (CFP). At the Institute of Mechanics Karl-Marx-Stadt of the Academy of Sciences of G.D.R. the computer code ATOLL has been developed. This software tool allows to optimize the strength and reliability taking into consideration the mechanical and thermal stress analysis of the components. Both deterministic and stochastic (probabilistic) concepts (4) of evaluation have been taken into account (e. g. for the stress-probability-time-SPT diagram) to get information about the strength and failure problems as well. The computer code also allows to evaluate mechanical and thermal field quantities on the basis of modern fracture concepts (5) - (7).

The results do not only take into consideration stress intensity factor or J-integral but also Sih's strain energy density criterion and more generalized integral concepts based on thermodynamics of deformation and fracture. The software may also be used for an advanced study of some current topics on the 'mesoscopic' level of deformation and strength behaviour (matrix-fibre transition regions etc.) when combining computational mechanics and physical modeling.

Finally the authors are going to give a short survey of new results regarding the combination of computer mechanics with experiments carried out by means of laser and speckle interferometry and the micro moire method (determination of microdeformations using scanning electron microscopy) which have been used to investigate special kinds of ceramics and high strength composites. These 'hybrid techniques' can further be improved applying latest tools of digital image processing, being the aim of current research activities (8), (9).

THE ATOLL SOFTWARE SYSTEM

The computer concept ATOLL (10) has been developed as a software tool for numerical modeling of a wide range of scientific problems in solid mechanics, especially in fracture and micromechanics.

ATOLL is suitable as a programming base for

- pre-processing,
- finite element computer codes,
- boundary element computer codes,
- graphic aided post-processing

as well.

In the field of F.E.M. there are some fundamental versions in two and three dimensions for the treatment of following classes of tasks:

- linear elastic problems (isotropic, orthotropic)
- linear stationary heat transfer
- linear transient heat transfer
- non-linear elastic (elastic-plastic problems)
- problems concerning with elastic-viscoplastic materials
- stochastic F.E.M. for linear elastic materials

Beside these versions there are computer codes which take into account coupling effects (first solving the heat transfer problem, then the mechanical one) and software for special investigations and applications (e.g. to determine field quantities that allow to compute related quantities to get some information about strength or failure assessment or characteristics in fracture mechanics like K, J-integral, strain energy density criterion, e.t.c.).

Beside the F.E.M. software there are some powerful B.E.M. codes developed at the Institute of Mechanics: SUNUP (11), (12) and BOAS-3D (13) allow to investigate linear elastic material response in two resp. three dimensions as well based on a so-called direct approach and to compute related quantities in fracture mechanics.

Both groups of numerical methods take advantage of an extensive graphic aided post-processing. This software can be used as the basis for representations of networks, displacements, functions, isolines of computed field quantities for problems in 2D or 3D, respectively (14). Refering to 3D problems, isoline representations can be outlined either at surfaces or at arbitrary planes across the discretized domain. For some more extensive mechanical modeling versions (3D, non-linear material behavior, transient problems) it is desirable to compute characteristical values in fracture mechanics separately as a post-processing step. Therefore, ATOLL includes the needed software components.

NUMERICAL MODELING OF COMPOSITE MATERIALS

In the field of composite materials there are many specimens or construction elements that are large in size compared with the length of 'structural elements' (grains, fibres, conclusions, voids etc.). Therfore, following a purely phenomenological concept, the micromechanical structure can be neglected in a first approach. The well-known homogenization techniques will be a useful tool to take into account some deterministic or stochastic ideas of the mesoscopic and microscopic levels.

However, for a large scale of technological problems linear elastic or elastic-plastic F.E.M. codes can be applied, combined with special routines for calculating material-relevant field quantities for strength and failure analysis as well.

As an example, in order to analyse ceramic structures in thermally and mechanically high loaded diesel engines during the most important running states it is necessary to combine experimental techniques with numerical methods. Numerical simulations of the behaviour of construction elements consisting particularly of ceramics are of special interest, because the state after damage does not allow an effective search for damage reasons or crack initiation locations. Important influences on strength and failure of a diesel engine piston with an embedded ring of ceramics have been discussed on the base of maximal mean stress distributions and the Weibull-Stanley theory (15) - (17).

In many cases composite materials cannot be characterized by deterministic material properties. Often stochastically distributed geometrical parameters (e.g. characteristic length of fibres or cracks) of the specimens or the construction elements have to be taken into account. Here the so-called stochastic F.E.M. is a well suited numerical tool to compute stochastic distributions of displacements, stresses and other related quantities instead of the deterministic ones. Actually a linear elastic version including stochastic material parameters and geometries as well can be used. It is based on the perturbation theory (18).

As an example, performed to a CT-specimen the distribution function of J-integral can be determined in only one computational step. The amount of hardware resources depends on the approximation order and the number of stochastic variables. In many cases it will be only two- or three-times higher compared with a single deterministic computation.

Also at the mesoscopoic level numerical simulations are an universal tool for thermal and mechanical investigations, especially in the field of fracture mechanics. This is one of the most interesting actual topics in research at the Institute of Mechanics. Both F.E.M. and B.E.M. can be used. In the authors opinion the arising main difficulties concerning with the modeling phase are the following:

i.   The optimal suited thermal and mechanical models must be chosen carefully.

ii. The results cannot be better than the quality of the input data, especially of the material parameters.

iii. Sufficiently exact modeling of the boundary conditions is complicated or not possible, in much cases.

This is valid especially for such phenomena like heat transfer conditions between gases and solids or fluids and solids, for the mechanical conditions at inner surfaces between distinct material components (adhere, sliding without or within friction, arising of new surfaces by debonding) or at at crack tip under crack propagation. Therefore, comparisons between numerical results and experimental investigations play an important role.

Some attempts in this field are discussed in (12) using B.E.M. and in (19) using F.E.M.

HYBRID METHODS

The combination of mechanical modeling, experiments and numerical methods leads to the so-called hybrid methods. Some basic ideas are discussed in (3), (20).

Analyzing surface cracks in a fracture mechanics specimen by means of holographic interferometry the displacement field at the surface has to be determined. A main problem is the real fringe pattern interpretation which can be assisted by numerical methods like F.E.M. Uncertainties prevail because real boundary conditions cannot always be described in a correct manner. Rigid body translations or rotations as well as loading point behaviour cannot be completely determined and can contribute to differences between a theoretical prediction and the actual interferogram. Comparing the experimental and the synthetically computed fringe patterns is an efficient way to simulate the effects of several influences to the fringe pattern and to determine the location of the zero order interference line. Modern image analyzers can be used to improve the level of experimental automation (21), (22).

Concerning the scanning micro moire method (6) the analysis of a fringe pattern often performed by use of an image processing system also leads to displacements and related fields at the

surface. In a similar way numerical simulations allow to estimate the influence of systematical errors compared with variations of boundary and other conditions that are essential for
the experimental fringe pattern. This combination of two different methods has been tested for an aluminium alloy specimen at
the Institute of Mechanics in Karl-Marx-Stadt. Because of the
yielding behaviour of aluminium a time dependent computer code
for elastic-viscoplastic material was used (23). In the authors
opinion, this hybrid method is well suited for investigations
of such composite materials that have non-elasic components,
e.g. for ceramics under temperatures higher than 1000 K (high
temperature yielding).

REFERENCES

1. Michel, B., Gruendemann, H., Kaempfe,B., Kuehnert, R.,
   Auersperg, J., Micromechanics – relations to the foundation
   of deformation and fracture. In: Series Fractures Mecha-
   nics, Micromechanics, Coupled Fields (FMC), Academy of
   Scienes of GDR, The Institute of Mechanics, No. 14, Karl-
   Marx-Stadt, 1984.

2. Michel, B., Fortschritte der Physik, 30, 1982, 233-310.

3. Michel, B. and Will, P., Hybrid methods in fracture mechanics and micromechanics. 1st Conference of Mechanics of
   Academies of Socialist Countries, Prague, 1987.

4. Winkler, T. and Michel, B., Progress in mechanics, Moscow
   (in print).

5. Will, P., Michel, B., and Zerbst, U., Engineering fracture
   mechanics, 28, 2, 1987, 197-201.

6. Michel, B., Will, P., and Kuehnert, R., Evaluation of fracture concepts by means of laser experiments. Proc. International Conference on Fracture and Fracture Mechanics
   (ICFFM), Shanghai, 1987, 465-466.

7. Michel, B. and Will, P., Micromechanics and crack spectra.
   Proc. 6th European Conference on Fracture (ECF 6), Amsterdam, 1986, 1, 111-23.

8. Michel, B. and Will, P., Proc. 26th Polish Solid Mechanics
   Conference, Sobieszewo, 1986, 115.

434

9. Michel, B., Will, P., Hoefling, R., and Kuehnert, R., Non-destructive testing of fracture quantities by means of laser interferometry and the scanning micro-Moire-method". 4th European Conference on Non-Destructive Testing, London, 1987.

10. Auersperg, J., Sommer, J.-P., and Hussack, J., Das Programmsystem ATOLL in der Bruchmechanik und Mikromechanik. FMC-Series, Academy of Sciences of G.D.R., The Institute of Mechanics, No. S1, Karl-Marx-Stadt, 1988 (in print).

11. Sommer, J.-P., Programmbeschreibung zum Randintegralprogramm SUNUP. FMC-Series, Academy of Sciences of G.D.R., The Institute of Mechanics, No. 21, Karl-Marx-Stadt, 1986.

12. Sommer, J.-P., Gruendemann, H., and Michel, B., Applications of boundary integral method to computation of displacement and stress fields and related quantities. Proc. Symp. Mechanics of Polymer Composites, Prague, April 1986, 19-23.

13. Gruendemann, H., and Paessler, A., Programmdokumentation und Nutzerhandbuch zum BEM-Programmsystem BOAS. FMC-Series, Academy of Sciences of G.D.R., The Institute of Mechanics, No. 44, Karl-Marx-Stadt, 1989.

14. Auersperg, J., and Hussack, J., Grafische Modell- und Ergebnisaufbereitung als Voraussetzung fuer den Einsatz von FEM- und BEM-Programmen im Rahmen von CAD-Programmen, in Applikation und Forschung. Proc. XI. IKM Weimar 1987, 2, 28-31.

15. Sommer, J.-P., Scherzer, M., Totzauer, W., and Kampmann, J., Rechnergestuetztes Konstruieren im Dieselmotorenbau. Kraftfahrzeugtechnik, 38,5,1988,138-41.

16. Sommer, J.-P., Michel, B., Auersperg, J., and Scherzer, M., Mechankforschung fuer den einsatz von Keramikwerkstoffen unter Nutzung von FEM-Software. In: Proc. Symp. FEM 5, Johanngeorgenstadt, April 1988, ed. by Technical University Dresden (in print).

17. Sommer, J.-P., Totzauer, W., and Scherzer, M., Unterstuetzung des Konstruktionsprozesses fuer Metall-Keramik-Bauteile im Motorenbau. FMC-Series, Academy of Sciences of G.D.R., The Institute of Mechanics, No. 4, Karl-Marx-Stadt, 1987, 85-90.

18. Skurt, L., and Michel, B., A stochastic finite element method applied to fracture problems. FMC-Series, Academy of Sciences of G.D.R., The Institute of Mechanics, No. 23, Karl-Marx-Stadt, 1988, 17-28.

19. Hennig, K, Michel, B. and Sommer, J.-P., Crack tip temperature fields in visco-plastic materials. Proc. 6th Int. Conf. on Fracture (ICF 6), New Delhi, 1984.

20. Sommer, J.-P., Will, P., and Totzauer, W., Verbindung zwischen numerischen Simulationsmethoden und Experiment - Entwicklung hybrider Methoden. FMC-Series, Academy of Sciences of G.D.R., The Institute of Mechanics, No. 14, Karl-Marx-Stadt, 1985, 100-14.

21. Will, P., Totzauer, W., and Michel, B., Analysis of surface cracks by holography. Theor. and Appl. Fracture Mechanics, 9, 1988, 33-8.

22. Will, P., Verallgemeinerte Integralkriterien und ihre Anwendung in der modernen Bruchmechanik. DSc Thesis, Martin-Luther-University Halle, 1988, in: FMC-Series, Academy of Sciences of G.D.R., The Institute of Mechanics, No. 39, Karl-Marx-Stadt, 1988 (in print)

23. Auersperg, J., Kuehnert, R., and Michel, B., Numerische Simulation zur Bewertung von Deformationsvorgaengen der Mikromechanik im Elektronenmikroskop. 9. Tagung uber Probleme und Methoden der mathematischen Physik (TMP), Karl-Marx-Stadt, 1988.

ON THE COMPATIBILITY OF HASSELMAN'S THERMAL SHOCK DAMAGE CRITERIA WITH FRACTURE MECHANICS

H.-J. WEISS, H.-A. BAHR
Zentralinstitut für Festkörperphysik und Werkstofforschung
der AdW der DDR, Dresden 8027

ABSTRACT

By means of a fracture mechanical approach it is shown that si-multaneous unstable crack propagation is apparently absent in thermal shock cracking. Nevertheless, Hasselman's useful re-sults, which were derived from that notion, are reproduced by this approach.

INTRODUCTION

Various aspects of thermal shock damage of brittle materials have been investigated by Hasselman and his co-workers [1,2]. One of the most striking features is the sudden drop of the residual strength as soon as the temperature difference in quenching, ΔT, exceeds a critical value, ΔT_c, as seen in Fig.1. The experimental facts gave rise to the question as to how initial strength, retained strength, and ΔT_c are mutually related. A preliminary answer was provided by Hasselman's model, accor-ding to which the elastic energy built up within the thermally shocked body is set free and then used up in unstable crack propagation. The cracks were supposed to run simultaneously, sharing the available energy among themselves. The greater the number of cracks, the smaller their energy shares. Less energy means less propagation and smaller final crack lengths. Smaller cracks imply higher retained strength. By this kind of ingene-ously simple reasoning, Hasselman was able to explain several essential features of the material behaviour under thermal shock. However, his model was incomplete in the sense that it needed the crack density for an input parameter. Also there was the suspicion that simultaneous unstable crack propagation is not possible under thermal shock. So we were confronted with the fact that there were apparently right predictions de-rived from a dubious model. It should be possible to find a proper fracture-mechanical model that allows to reproduce the facts and to predict the phenomena better than Hasselman's model did.

In the following we show how simultaneous unstable crack propagation, which is a central feature of Hasselman's model, can be ruled out. In doing so we explain our approach, without deriving those results which were published earlier.

STABLE AND UNSTABLE CRACK PROPAGATION

Let us consider the energy release rate of a crack in a thermally quenched body, plotted versus crack length, Fig. 2. At zero crack length, of course, it is always zero. It increases with crack length if the cracks are small compared with the depth of the cooled layer. It is small for a large crack whose tip is located far beyond the quenched layer. Thus the overall shape of the energy release rate versus crack length curve, $\mathcal{G}(1)$, is fixed. In the course of time, with the cooling penetrating further into the body, the $\mathcal{G}$ curves get inflated at first. Later on, as the temperature distribution within the body gradually levels out, the $\mathcal{G}$ curve slowly collapses onto the abszissa axis. The whole family of $\mathcal{G}$ curves generated in the course of time is enveloped by another curve, called the envelope, which is also shown in Fig. 2.

Now let us consider crack propagation starting from some preexisting or initial crack of length l_o. The crack begins to propagate at the instant when its $\mathcal{G}$ has reached a critical level $\mathcal{G}_c$, in the simplest version of linear fracture mechanics. For simplicity, $\mathcal{G}_c$ is assumed to be a constant here, represented by a horizontal line in Fig. 2. (The reasoning could be easily extended to non-constant $\mathcal{G}_c$.) Let us first neglect the existence of 4 horizontal lines in Fig. 2 and focus the attention on the second line. It is seen that at the (normalized) time $\tau = 0.06$ the energy release rate of the initial crack of length l_o has reached $\mathcal{G}_c$. When propagating, the crack releases more energy than required, so it accelerates, which means unstable propagation. It keeps propagating unstably at least as long as its $\mathcal{G}$ is above the $\mathcal{G}_c$ level, and even further because of inertia involved. Then it stops until it is taken along by the inflating $\mathcal{G}$ curve, riding on its downward slope, which means stable propagation. So we have got combined unstable and stable propagation, which already differs from Hasselman's model. It can be deduced from Fig. 2 that for lower ΔT the stable part of the crack path is suppressed in favour of the unstable part such that for certain ranges of the parameters the crack propagation is purely unstable. Let us see whether it can be simultaneous then. For an answer we consider the delay of the second crack with respect to the first one.

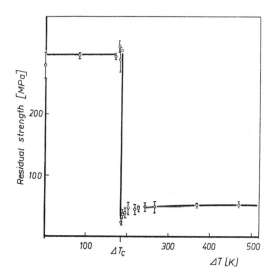

Figure 1. Thermal shock behaviour of polycrystalline Al_2O_3 ceramics from data in [3].

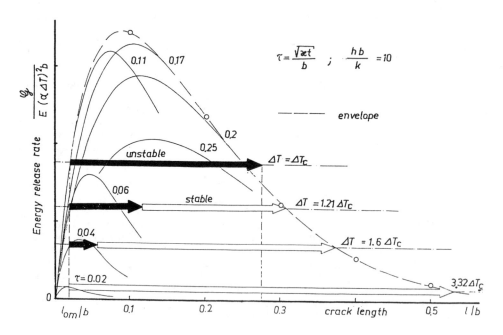

Figure 2. Energy release rate (normalized) versus crack length, for a crack in a slab quenched from one side, after [4]. The change with time is seen from the family of curves. (See the numbers attached to the curves.) The horizontal lines mark the normalized $\mathcal{G}_c$ for some arbitrarily chosen ΔT. One ordinate unit = 0.005.

DELAY IN SUCCESSIVE CRACKING

The initial length of the second crack may have been shorter than that of the first one by an incremental length δl_0. The two cracks of lengths l_0 and $l_0 - \delta l_0$ start to propagate when their energy release rates reach $\mathcal{G}_c$:

$$\mathcal{G}(l_0, t) = \mathcal{G}_c , \qquad \mathcal{G}(l_0 - \delta l_0, t - \delta t) = \mathcal{G}_c \qquad (1)$$

By expanding the latter condition into a Taylor series, one finds the delay δt:

$$\delta t = \delta l_0 \cdot \frac{\partial \mathcal{G}}{\partial l} \Big/ \frac{\partial \mathcal{G}}{\partial t} \qquad at \quad l = l_0 \qquad (2)$$

In order to obtain an estimate of δt from (2), we consider a simple case where, at the instant of crack start, the penetration depth of the cooling, $\sqrt{\varkappa t}$, is small compared with the size of the body, b, but large compared with the crack length:

$$l \ll \sqrt{\varkappa t} \ll b \qquad \varkappa \equiv k / \varrho c \qquad (3)$$

Then, $\mathcal{G}$ can be estimated by

$$\mathcal{G} \approx \sigma^2 l / E \qquad \sigma \approx \alpha E(T - T_0) \qquad (4)$$

where $T - T_0$ is the temperature decrease in quenching. As long as $T - T_0$ is small compared with the temperature difference between body and quenching medium, ΔT,

$$T - T_0 \ll \Delta T \qquad (5)$$

its time dependence can be approximated by

$$T - T_0 \approx \frac{h}{k} \cdot \Delta T \cdot \sqrt{\varkappa t} \qquad (6)$$

From (4) and (6) follows that $\mathcal{G}$, under the above conditions, is proportional to crack length and time:

$$\mathcal{G} \sim l t \qquad \partial \mathcal{G}/\partial l = \mathcal{G}/l \qquad \partial \mathcal{G}/\partial t = \mathcal{G}/t \qquad (7)$$

This reduces (2) to

$$\delta t = \delta l \cdot t / l \qquad (8)$$

The final length of the unstably grown crack is roughly equal to the penetration depth of cooling at the moment of cracking, $\sqrt{\varkappa t}$. The asymptotic velocity of an unstable crack is known to be comparable with the wave velocity, $\sqrt{E/\varrho}$. In realistic situations involving tiny initial cracks, the unstable crack path is at least several times the initial crack length. Then one need not consider the early stage of accelerated propagation because the crack approaches its asymptotic speed after traversing a path of the order of magnitude of its initial length. Thus its average speed will be near $\sqrt{E/\rho}$, and its travel time will be about

$$t_t \approx \sqrt{\varkappa t \rho / E} \tag{9}$$

CONDITION FOR SIMULTANEITY

If the propagation is to be simultaneous, t_t of (9) has to be much longer than the δt calculated above, which leads to

$$\sqrt{\varkappa \rho / E t} \gg \delta / \ell \tag{10}$$

In order to apply this to our original problem, we identify l with the length of the largest initial flaw, l_{om}. Then, δl is the amount δl_o by which the second largest is shorter. t follows from the condition $\mathcal{G}(\ell_{om}) = \mathcal{G}_c$ together with (4) and (6):

$$t = \frac{\mathcal{G}_c}{E \varkappa \ell_{om}} \left(\frac{k}{h \, \alpha \Delta T} \right)^2 \tag{11}$$

With these specifications, (10) becomes

$$\frac{\varkappa h}{k} \alpha \Delta T \sqrt{\frac{\ell_{om} \rho}{\mathcal{G}_c}} \gg \frac{\delta \ell_o}{\ell_{om}} \tag{12}$$

With reasonable assumptions for the numerical values, $l_{om} = 0.1$mm, ρ = 3g/cm , $\mathcal{G}_c$ = 0.1N/cm, α = $3 \cdot 10^{-6}$/K, c = 0.2cal/gK, h =0.05cal/cm^{2}sK, ΔT = 300K, one obtains

$$\frac{\delta \ell_o}{\ell_{om}} \ll 10^{-6} \tag{13}$$

CONCLUSIONS

Condition (13) means that if any two initial flaws were to give rise to simultaneously running cracks, their lengths would have to coincide with this accuracy. This could never be reached with artificially cut notches, let alone natural flaws. Thus one may conclude that simultaneous unstable thermal shock cracking is impossible, provided that l_{om} meets the condition (3). $l_{om} \ll \sqrt{\varkappa t}$ may be rewritten with (11) as

$$\frac{h}{k} \alpha \Delta T \sqrt{\frac{E \ell_{om}^3}{\mathcal{G}_c}} \ll 1 \tag{14}$$

It is fairly well satisfied with the above data and E= 70GPa, k=0.004 cal/sKcm. Because of the very restrictive condition (13), one feels that simultaneous unstable propagation will be absent even if (14) is poorly met.

Since the maximum of the $\mathfrak{G}(l)$ curves is located approximately at $l \approx \sqrt{\varkappa t}$, initial cracks larger than $\sqrt{\varkappa t}$ find themselves on the downward slope of the $\mathfrak{G}(l)$ curves and therefore cannot propagate unstably. It seems that thermal shock cracking is either unstable and not simultaneous or else it is stable and simultaneous. Thus it appears that simultaneous unstable propagation can be ruled out in large parameter ranges and is probably absent in any thermal shock cracking. (It is known to be present in ballistic impact, for instance.)

RELATION TO OTHER WORK

The subject of this paper is a selected piece from a wider research program concerning thermal shock damage. The activities include the fracture analysis of multiply notched strips under thermal shock as well as experiments to produce thermal shock crack patterns. The calculations involved analytical, finite element and boundary element methods. The results have been published recently [4,5]. Their immplications for the thermal shock damage criteria, including the reproduction of Hasselman's results, have been discussed earlier [6,7].

Related experiments yielded the crack patterns shown in Fig. 3. They exhibit a hierarchical order of cracks with even a tendency to regularity though they were grown without making notches into the samples [7].

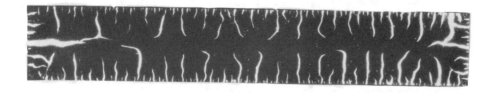

Figure 3. Crack patterns obtained by quenching in water a Al_2O_3/glass ceramic slab on its circumferential faces with ΔT =340K. Cracks were made visible by fluorescent dye penetration. Curtesy by G. Fischer.

The formation of a hierarchical order in the final crack distribution was first investigated by Bazant [8] and Nemat-Nasser [9]. They assumed that a regular array of growing cracks showed a special type of instability where every second crack is left behind while the others advance. With this assumption they were able to calculate the instants of instability or bifurcation of the solution. We adopted that view in our earlier papers [5]. Our latest numerical results, however, show that this is not necessarily so: Regular crack arrays evolve in a more complex way [10], and the bifurcation mode proposed by [8,9] is definitely ruled out in some cases. It has been shown that bifurcation modes with every third or fourth crack propagating will be present then. It is not yet clear whether there are parameter ranges where the bifurcation mode with every second crack advancing is realized or serves as a good approximation in the presence of randomness.

The latter investigations have been done not only with respect to the strength of ceramics. It is hoped that the theory of crack pattern formation will contribute to a deeper understanding of structure formation in general.

Summarizing we can state that Hasselman's thermal shock damage criteria have been given a fracture-mechanical support which also carries additional results reaching beyond. The scheme developed for this purpose provides insight into a variety of thermal shock damage phenomena, and it seems that its explanatory power has not yet been exhausted.

REFERENCES

[1]   Hasselman, D.P.H., Unified theory of thermal shock fracture initiation and crack propagation of brittle ceramics, J. Amer. Ceram. Soc., 1969, 52, 600-4.

[2]   Hasselman, D.P.H. and Singh, J.P., Role of mixed-mode crack propagation in thermally shocked brittle materials, Theoret. Appl. Fracture Mech., 1984, 2, 59-65.

[3]   Bertsch, B.E., Larson, D.R. and Hasselman, D.P.H., Effect of crack density on strength loss of polycrystalline Al_2O_3 subjected to severe thermal shock, J. Amer.Ceram. Soc., 1974, 57, 235-6.

[4]   Bahr, H.-A., Balke, H., Kuna, M. and Liesk, H., Fracture analysis of a single edge cracked strip under thermal shock, Theoret. Appl. Fracture Mech., 1987, 8, 33-9.

[5]   Bahr, H.-A., Weiss, H.-J., Maschke, H.G. and Meissner, F., Fracture analysis of multiple crack propagation in a strip under thermal shock, Theoret. Appl. Fracture Mech., in print.

[6]   Bahr, H.-A., Weiss, H.-J., Heuristic approach to thermal shock damage due to single and multiple crack growth, Theoret. Appl. Fracture Mech., 1986, 6, 57-62.

[7]   Bahr, H.-A., Fischer, G. and Weiss, H.-J., Thermal shock crack patterns explained by single and multiple crack propagation, J. Mater. Sci., 1986, 21, 2716-20.

[8]   Bazant, Z.P., Ohtsubo, H. and Ach, K., Stability and post-critical growth of a system of cooling or shrinkage cracks, Internat. J. Fracture, 1979, 15, 443-56.

[9]   Nemat-Nasser, S., Keer, L.M. and Parihar, K.S., Unstable growth of thermally induced interaction cracks in brittle solids, Internat. J. Solids Structures, 1978, 14, 409-30.

[10] Meissner, F., Diploma report, Technical University Karl-Marx-Stadt, 1988.

FE-CALCULATION DEMONSTRATING THE INFLUENCE OF THE MATRIX INTERFACE LAYER ON THE FREE EDGE STRESSES IN A LAMINATED CFRC COMPOSITE

M. LINDNER, K. KROMP

Max-Planck-Institut für Metallforschung
Institut für Werkstoffwissenschaften
Seestr. 92, D-7000 Stuttgart 1, FRG

ABSTRACT

Finite element caluculations were made for a $(0/90)_s$ CFRC composite and were compared with the results of a CFRP composite. The influences of a thin interface matrix layer upon the interlaminar stresses σ_{zz} and σ_{yz} were examined by modelling the interface layer using three different thicknesses h_o.

NOMENCLATURE

b................Half width of laminate cross section
h_m ....Ply thickness
h_o ....Thickness of interface layer
E_{ij} ....Elastic moduli of a 0° ply
G_{ij} ....Shear moduli of a 0° ply
ν_{ij} ....Poisson's ratio
Q_{ij} ....Elements of stiffness matrix
x, y, z ....Cartesian coordinates
u, v, w ....Displacements in x, y, z directions
U, V, W .....Displacement functions in x, y, z directions
σ_{ij} ....Stress components
ϵ_{ij} ....Strain components
ϵ_o ....Uniform axial strain

1. INTRODUCTION

The laminate composites of Carbon Fibre Reinforced Plastic (CFRP) and Carbon Fiber Reinforced Carbon (CFRC) show a very high strength and a "ductile" behaviour compared to brittle materials and have a low density. CFRC composites are especially suitable for high temperature applications. Therefore they are very important to aerospace technology. A single lamina of CFRP or CFRC may be considered to be homogeneous and anisotropic. The composite consists of a stack of such plies. Since each ply may have a different orientation and/or elastic moduli, the composite as a whole must be treated as an inhomogeneous anisotropic material.

Delamination, particularly between plies, has been observed as a major failure source, [1],[2],[3]. It is caused by interlaminar stresses in the region of the free edge [4],[5], arising by the mismatch of material properties [6].

The free edge stress problem has been investigated by several authors using different methods: finite difference method [4], perturbation theory [7], [8] boundary layer theory [9], [10], and finite elements method [11], [12], [14],[20],[21]. These authors mainly examined graphite - epoxy laminates, consisting of transversal isotropic plies with the following elastic properties: $E_{11} = 138 GPa$, $E_{22} = E_{33} = 15 GPa$, $G_{12} = G_{13} = G_{23} = 5.9 GPa$, and $\nu_{12} = \nu_{13} = \nu_{23} = 0.21$.

During the fabrication process, thin matrix rich layers (interface layers) form between the plies. These interface layers were not considered by the above authors. Kim and Hong [15] and Wu [16], however, considered these layers in their investigations, considering also that delamination occurs mainly between plies. Both modelled this isotropic interface layer with a thickness of $h_o = 0.1 * h_m$. While Kim and Hong avoided the oscillating behaviour of the stresses ahead of an interface crack by using the isotropic interface layer, Wu, among others, investigated its influence on the interlaminar stress distribution.

The material properties used in this investigation simulate a $(0/90)_s$ CFRC composite. The plies consist of a woven mat with warp und weft and may therefore not be considered to be transversal isotropic. The influence of the interface layer is shown by modelling it with $h_o = 0.05 * h_m$, $h_o = 0.1 * h_m$ and $h_o = 0.2 * h_m$.

2. THE FREE-EDGE PROBLEM

The $(0/90)_s$ CFRC - composite has a width of 2b and a height of $4h_m$, as modelled in figure 1, consisting of four homogeneous anisotropic plies. Due to the symmetry, one eighth of the whole structure will be enough for the examination, figure 1. The displacement field is assumed as follows [4]:

$$u(x, y, z) = \epsilon_o x + U(y, z)$$
$$v(x, y, z) = V(y, z)$$
$$w(x, y, z) = W(y, z);$$

where ϵ_o is a uniform axial strain and U, V, W are functions of coordinates y and z only.

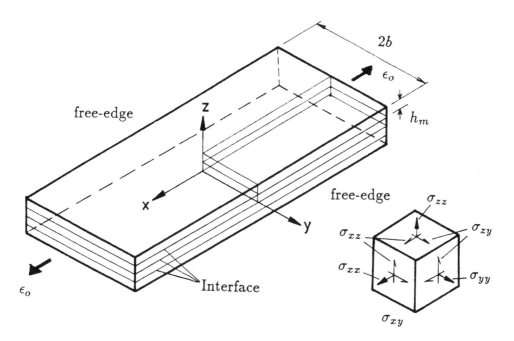

Fig. 1. Laminate configuration and stresses.

As elastic moduli for a 0° ply are used:

$E_{11} = 70 GPa$ $\qquad\qquad (10.15 * 10^6 psi)$

$E_{22} = 55 GPa$ $\qquad\qquad (7.98 * 10^6 psi)$

$E_{33} = 9 GPa$ $\qquad\qquad (1.31 * 10^6 psi)$

$G_{12} = G_{13} = G_{23} = 4.5 GPa$ $\qquad (0.65 * 10^6 psi)$

$\nu_{12} = \nu_{13} = \nu_{23} = 0.21$

For the isotropic carbon interface layer the values of $E = 4 GPa$, $\nu = 0.36$ were assumed. The relation of the elastic moduli reflects the woven structure of the plies [3]. From the above data one can calculate the stiffness matrix for the finite element calculation as follows [17], [18]:

$$
\begin{pmatrix} \sigma_1 \\ \sigma_2 \\ \sigma_3 \\ \sigma_4 \\ \sigma_5 \\ \sigma_6 \end{pmatrix} =
\begin{pmatrix}
Q_{11} & Q_{12} & Q_{13} & 0 & 0 & 0 \\
Q_{21} & Q_{22} & Q_{23} & 0 & 0 & 0 \\
Q_{31} & Q_{32} & Q_{33} & 0 & 0 & 0 \\
0 & 0 & 0 & Q_{44} & 0 & 0 \\
0 & 0 & 0 & 0 & Q_{55} & 0 \\
0 & 0 & 0 & 0 & 0 & Q_{66}
\end{pmatrix} *
\begin{pmatrix} \epsilon_1 \\ \epsilon_2 \\ \epsilon_3 \\ \epsilon_4 \\ \epsilon_5 \\ \epsilon_6 \end{pmatrix}
$$

$$Q_{11} = E_{11}(1 - \nu_{23}\nu_{32})/\Delta \qquad Q_{44} = 2G_{23}$$
$$Q_{22} = E_{22}(1 - \nu_{31}\nu_{13})/\Delta \qquad Q_{55} = 2G_{13}$$
$$Q_{33} = E_{33}(1 - \nu_{12}\nu_{21})/\Delta \qquad Q_{66} = 2G_{12}$$

$$Q_{12} = (\nu_{21} + \nu_{31}\nu_{23}E_{11})/\Delta = (\nu_{12} + \nu_{32}\nu_{13}E_{22})/\Delta = Q_{21}$$
$$Q_{13} = (\nu_{31} + \nu_{21}\nu_{32}E_{11})/\Delta = (\nu_{13} + \nu_{12}\nu_{23}E_{33})/\Delta = Q_{31}$$
$$Q_{23} = (\nu_{32} + \nu_{12}\nu_{32}E_{22})/\Delta = (\nu_{23} + \nu_{21}\nu_{13}E_{33})/\Delta = Q_{32}$$
$$\Delta = 1 - \nu_{12}\nu_{21} - \nu_{23}\nu_{32} - \nu_{31}\nu_{13} - 2\nu_{21}\nu_{32}\nu_{13}$$

3. FINITE ELEMENTS MODEL

a. Without interface layer

The program PERMAS [19] is used for the computations. The three- dimensional modelling was performed with HEXEC20 (20 nodal points per element) and HEXEC15 (15 nodal points per element) elements of Serendipity type.

A convergence study with a coarse, a medium and a fine mesh were carried out. The structure of the fine mesh near the free edge is shown in figure 2a. For the expected stress singularity near the free edge a finer partition in this region is used. The coarse mesh has 44 elements and 326 nodes, the medium mesh 224 elements and 1401 nodes and the fine mesh 432 elements and 2757 nodes. The ratio width to height is equal to eight. The lamina is subject to uniform axial strain $\epsilon_o = 1 * 10^{-2}$.

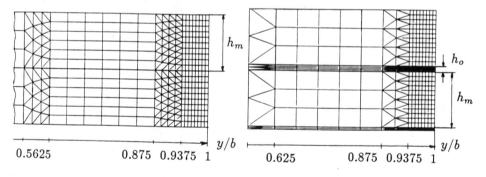

Fig. 2a. Fine FE-mesh without interface layer.

Fig. 2b. Fine FE mesh with interface layer.

b. With interface layer

The finite element model for $h_o = 0.1 * h_m$ is shown in figure 2b. Convergence studies for the interlaminar stresses were also performed. 470 elements with 3233 nodal points for the fine mesh proved to be sufficient. The uniform axial strain is $\epsilon_o = 1 * 10^{-2}$.

4. ANALYSIS UND DISCUSSION

In order to check the finite element calculations, the elastic moduli from Raju-Crews [21] and Wang-Crossman [20] were additionally used. The results agreed very well, demonstrating the sufficiency of the fine partition. At first the influence of the elastic moduli is discussed, and then the effects of the isotropic interface layer are analysed. Only the results for the fine mesh are shown.

a. Analysis without interface layer

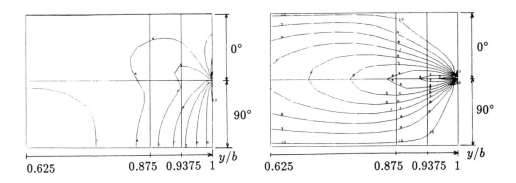

Fig. 3a. Isocontours for σ_{zz}
near the free edge.

Fig. 3b. Isocontours for σ_{yz}
near the free edge.

Figure 3 a. and b. show the isocontours of the interlaminar stresses σ_{zz} und σ_{yz}. The high stress gradients at the interface are clearly visible. The stress distribution for σ_{zz} and σ_{yz} along the interfaces is shown in figure 4 and 5. The maximum stress is obtained near the 0/90 interface, figure 6. Comparing figure 6 and 7 one can see, that the stress gradient is localized to the free edge region.

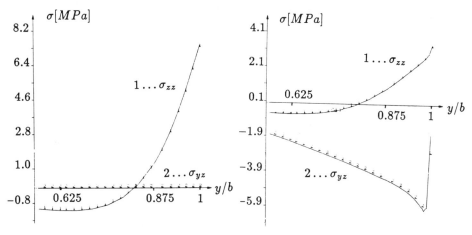

Fig. 4. σ_{zz} and σ_{yz} along y,
$z = 0$.

Fig. 5. σ_{zz} and σ_{yz} along y,
$z = h_m$.

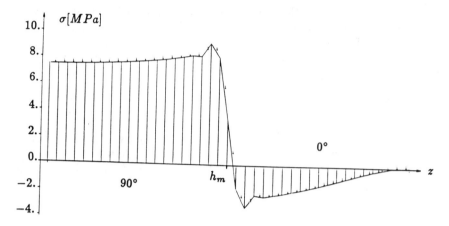

Fig. 6. σ_{zz} along the free edge, $y = b$.

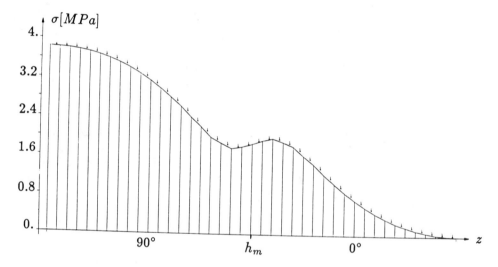

Fig. 7. σ_{zz} along z for $y = 0.9375 * b$.

The principle pattern of the interlaminar stresses are comparable with those of Wang-Crossman [20], Raju-Crews[21] and Wu[16]. With the CFRC data lower stress values are obtained. For the normalized maximum stresses near the 0/90 interface different results are reported:

$$\text{Raju[21]} : \sigma_{zz}/(10^6 \epsilon_o) = \quad 3 * 10^{-3} MPa$$
$$\text{Wu[16]} : \sigma_{zz}/(10^6 \epsilon_o) = 3.7 * 10^{-3} MPa$$
$$\text{present} : \sigma_{zz}/(10^6 \epsilon_o) = 0.9 * 10^{-3} MPa$$

Wu used elastic moduli which are slightly different from those of Raju-Crews.

b. With interface layer included

The isocontours in figure 8 are seen to be very similar to those in figure 3. The only difference is, that stress gradients also arise in the middle of the laminate, that is, in the 90/90 interface, figure 8.

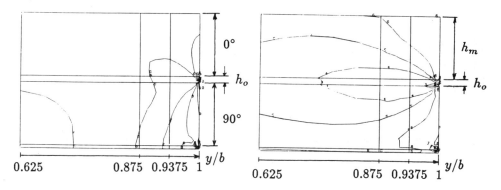

Fig. 8a. Isocontours for σ_{zz}
near the free edge.

Fig. 8b. Isocontours for σ_{yz}
near the free edge.

Figures 9 - 12 give the interlaminar stresses for the different interfaces for $h_o = 0.1 * h_m$. When comparing these results, the form of the distribution of σ_{yz} in the matrix-90° interface is of interest. Of importance is also σ_{zz} in the matrix interfaces, where one can observe compressive stresses in the matrix-0 interface and very high tensile stresses in the 90-matrix interface, figure 13. The stress gradient is taken up by the matrix layer. Examining the interlaminar stress distribution with interface layer and comparing it with the results without an interface layer, it can be seen, that the stresses near the interface are higher, whereas they are lower within the plies, see figure 6 and 13.

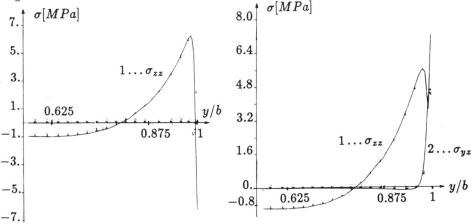

Fig. 9. σ_{zz} and σ_{yz} along y,
$z = 0$.

Fig. 10. σ_{zz} and σ_{yz} along y, for matrix-90° interface, $(z = 0.05 * h_m)$.

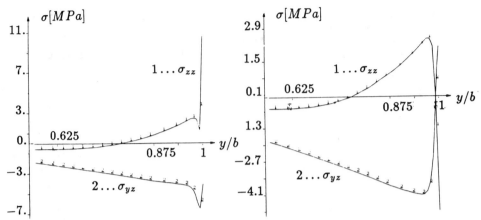

Fig. 11. σ_{zz} and σ_{yz} along y, for 90°-
matrix interface, $(z = 1.05 * h_m)$.

Fig. 12. σ_{zz} and σ_{yz} along y, for matrix-
0° interface, $(z = 1.15 * h_m)$.

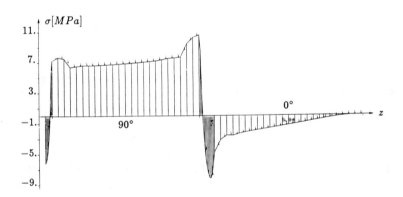

Fig. 13. σ_{zz} along the free edge, $y = b, h_o = 0.1 * h_m$.

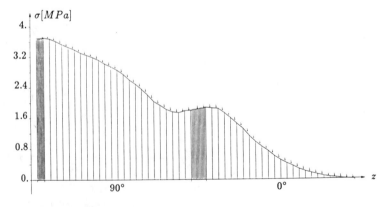

Fig. 14. σ_{zz} along z for $y = 0.9375 * b, h_o = 0.1 * h_m$.

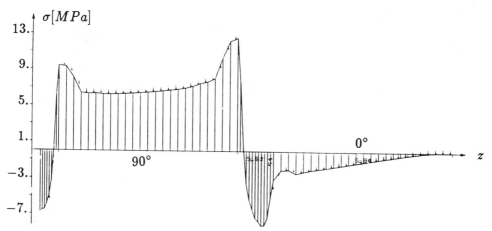

Fig. 15. σ_{zz} along the free edge, $y = b, h_o = 0, 2 * h_m$.

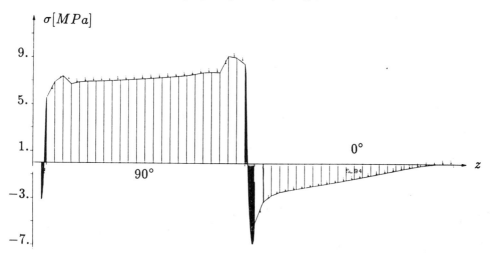

Fig. 16. σ_{zz} along the free edge, $y = b, h_o = 0.05 * h_m$.

The calculations with $h_o = 0.05 h_m$, $h_o = 0.1 h_m$ and $h_o = 0.2 h_m$ confirm this tendency, figure 14 - 16; thicker interface layers create higher interlaminar stresses near the interface and lower the gradient inside the interface layer.

CONCLUSION

The interlaminar stresses computed for the CFRC composite show peak stresses lower than the calculation for the CRFP composite. This may be an effect of the woven structure which leads to a lower elastic moduli mismatch. Taking into account the

452

thin matrix interface layer (an isotropic layer consisting of nearly pure carbon), the peak stresses near the interface at the free edge increase with increasing thickness h_o, while the stress gradients in the interface layer decrease. At a distance h_m from the free edge, the stresses were only about $10 - 20\%$ of the free edge peak stresses. The matrix interface layer is the weakest link of the composite and is exposed to the highest stress gradient. It will therefore be the most important failure source of the laminate.

Acknowledgement:
The authors wish to thank the Deutsche Forschungsgemeinschaft for its financial support under Nr. Pa 241/9.

REFERENCES

1. Garg A. C., Delamination - a damage mode in composite structures, Engng. Fracture Mech., 1988, **29**, pp. 557 - 584.

2. O'Brien T. K., Characterisation of delamination onset and growth in a composite laminate, ASTM STP 775, 1982, pp. 140 - 167.

3. Vogel W., Haug T., Kromp K., Popp, Schädigungsmechanismen in multidirectionalen CFC-Werkstoffen, Fortschrittsbericht der Deutschen Keramischen Gesellschaft (ceramic forum international), Vol. 2, No. 3, 1986/87, pp. 29 - 38.

4. Pipes R. B., Pagano N. J., Interlaminar stresses in composite laminates under uniform axial extension, J. Compos. Mater., 1970, **4** , pp. 538 - 548.

5. Pagano N. J., On the calculation of interlaminar normal stress in composite laminates, J. Compos. Mater., 1974, **8**, pp. 65 - 81.

6. Herakovich C. T., On the relationship between engineering properties and delamination of composite materials, J. Compos. Materials, 1981, **15**, pp. 336 - 348.

7. Hsu P. W., Herakovich C. T., Edge effects in angle-ply composite laminates, J. Compos. Mater., 1977, **11**, pp. 422 - 428.

8. Hsu P. W., Herakovich C. T., A perturbation solution for interlaminar stresses in bidirectional laminates, ASTM STP 617, 1977, pp. 296 - 316.

9. Tang S., A boundary layer theory - Part I: Laminated composites in plane stress, J. Compos. Mater., 1975, **9**, pp. 33 - 41.

10. Tang S., Levy A., A boundary layer theory - Part II: Extension of laminated finite strip, J. Compos. Mater., 1975, **9**, pp. 42 - 52,

11. Whitcomb J. D., Raju I. S., Goree J. G., Reliability of the finite element method for calculating free edge stresses in composite laminates, Computers & Structures, 1982, **15**, pp. 23 - 37.

453

12. Herakovich C. T., Influence of the layer thickness on the strength of angle-ply laminates, J. Compos. Mater., 1982, **16**, pp. 216 - 227.

13. Rohwers K., Stresses and deformations in laminated test specimens of carbon fiber reinforced composites, DFVLR - FB 82-15, Institut für Strukturmechanik, Braunschweig, FR Germany, 1982.

14. Rybicki E. F., Schmueser D. W., Fox J., An energy relaease rate approach for stable crack growth in the free-edge delamination problem, J. Compos. Mater., 1977, **11**, pp. 470 - 487.

15. Kim K. S., Hong C. S., Delamination growth in angle-ply laminated composites, J. Compos. Mater., 1986, **20**, pp. 423 - 438.

16. Wu C. M. L., Nonlinear analysis of edge effects in angle ply laminates, Computer & Structures, 1987, **25**, pp. 787 - 798.

17. Vinson J. R., Sierakowski R. L., The Behavior of Structures Composed of Composite Materials, Martinus Nijhoff Publishers, 1986.

18. Lekhnitskii S. G., Theory of Elasticity of an Anisotropic Body, Mir Publishers, Moscow, 1981.

19. Schrem E., PERMAS Handbook for Linear Static Analysis, INTES GmbH, Stuttgart, 1985.

20. Wang A. S. D., Crossman F. W., Some new results on edge effects in symmetric composite laminates, J. Compos. Mater., 1977, **11**, pp. 92 - 106.

21. Raju I. S., Crews J. H. jr., Interlaminar stress singularities at a straight free edge in composite laminates, Computers & Structures, 1981, **14**, pp. 21 - 28.

AN ATTEMPT TO EVALUATE THE INTRINSIC
STRENGTH AND STRAIN OF CARBON FIBRES

WITOLD ŻUREK, IZABELLA KRUCIŃSKA
Technical University Łódź
Żwirki 36, 90-924 Łódź, Poland

ABSTRACT

The method of evaluation of strength and strain of carbon
fibres basing upon stress-strain relationships of straight
fibres and dimensions of loops at break is presented.

INTRODUCTION

Present theories of mechanical performance of composite
materials usually require the knowledge of the intrinsic
strength behaviour of fibre in a region of the critical lo-
ad-transfer, i.e. of dimensions from 0,1 mm to 0,3 mm. Reliable strength measurements of fibres at such extremely short
distances are beyond the capabilities of classical methods [1].
Moreover, carbon fibres are brittle with mechanical characteristics dependent on the defects randomly distributed along
the fibre volume. A direct consequence of this fact is that
the fibre strength is dependent on gauge length. This dependence precludes the evaluation of intrinsic strength and
failure strain of carbon fibres from the measurements completed under a longer gauge length than the transfer length. The
gauge length influence on the evaluation of fibre intrinsic
strength was extensively discussed in [2] - [4]. However, it
is not known how to determine the intrinsic breaking strength
and strain of very short specimens of fibres. In this paper
the investigation of the breaking force of looped fibres is
presented. The method of computing the strength and strain
at break of the fibres is proposed.

MODEL OF THE INFLUENCE OF FIBRE CROSS-SECTION UNEVENNESS ON
THE FAILURE STRAIN AND BREAKING FORCE

Let us assume that the distribution of the specimen cross-section area along the fibre length could be described by a rectangular distribution, i.e.

$$\varphi(A) = 1/(A_2 - A_1), \qquad (1)$$

where A_1 is the minimum and A_2 the maximum cross-section area of the specimen. Then, for the mean strain of the specimen the following relation obtained from a simplified model is derived in [5]:

$$\varepsilon_s = \int_{A_1}^{A_2} f(F/A) \, \varphi(A) \, dA. \qquad (2)$$

Here, F is the loading force, A the local variable specimen cross-section area and $f(F/A)$ the relation between strain and stress,

$$\varepsilon = f(F/A) \qquad (3)$$

The experimental results indicate that in the case of carbon fibres the relation (3) is linear

$$\varepsilon = \frac{1}{E} \frac{F}{A} = b \frac{F}{A}, \qquad (4)$$

hence using (4) and (1) in (2) the following relation is obtained

$$\varepsilon_s = \int_{A_1}^{A_2} b \frac{F}{A} \frac{1}{A_2 - A_1} \, dA = b \frac{F}{A_2 - A_1} \ln \frac{A_2}{A_1}. \qquad (5)$$

The maximum values of strain occur at the points of the smallest diameter of the fibre,

$$\varepsilon_{max} = b \, F/A_{min}. \qquad (6)$$

Introducing new constant X

$$X = A_2/A_1 , \qquad\qquad (7)$$

after a simple transformation of eqn. (5) the following equation is obtained:

$$\varepsilon_s = b \frac{F}{A_1} \frac{1}{X - 1} \ln X = \varepsilon_{max} \frac{1}{X - 1} \ln X , \quad (8)$$

where:

ε_s - strain of uneven specimen,
ε_{max} - max strain in the specimen.

The numerical value of X can be found by assuming that the relation between stress and strain is the same in each segment along the length of the fibre. Consequently, the breaking load for the longer fibres is the minimum force within the fibre length, proportional to the minimum cross-section area

$$F_{rs} = \sigma A_1 , \qquad\qquad (9)$$

where σ is the breaking stress of the fibre.
For the tested fibres it was assumed that

$$F_r = \sigma A_m , \qquad\qquad (10)$$

where A_m is the mean cross-section area of the fibres.
For rectangular distribution:

$$A_m = (A_2 + A_1) / 2 , \qquad\qquad (11)$$

hence

$$A_1 = 2 A_m / (1 + X) . \qquad\qquad (12)$$

Therefore,

$$F_{rs} = 2 F_r / (1 + X) , \qquad\qquad (13)$$

where F_{rs} is breaking force of uneven specimen and F_r is breaking force of even specimen. Hence,

$$X = (2 F_r/F_{rs}) - 1 \ . \qquad (14)$$

The X value can be estimated from the axial tension test and bending test, which give F_{rs} and F_r values.

When the looped fibre is loaded, its curvature is increased and the rupture strain in the outer zone at the moment of the break of fibre is

$$\mathcal{E} = \frac{d}{2\varrho} \ , \qquad (15)$$

where d – diameter of the fibre cross-section,
  ϱ – radius of loop curvature.

The Young modulus can be found out from equation

$$E = \frac{T}{2} \cdot \frac{64}{\pi d^4} R\varrho \ , \qquad (16)$$

where T is the force recorded in the Instron tester and 2R is the maximum width of the loop.

MATERIALS AND METHODS

Two kinds of carbon fibres made in Poland were analyzed: 38/III and WS/213, and the following properties were measured:

  - diameter with the use of a projecting microscope,
  - strength and strain of straight fibres,
  - loop diameter and loading force at break.

Mechanical properties were determined with the use of an Instron tester. Conditions of analysis of straight fibres were as follows:

  - gauge length - 10, 25 mm,
  - cross-head speed - 2 mm/min,
  - temperature of air - 20°C,
  - relative humidity of air - 65%.

The tested fibres were glued in the paper frames as shown in Fig.1 in such a way, that from the same filament two tandem samples were prepared of various length: first - 10 mm and second - 25 mm.

In the case of bending tests the samples were prepared as previously, the only difference being the length of "window" in the frame was 70 mm. The framed fibre was drawn through the wire loop fastened in the upper jaws of Instron tester (Fig.2).

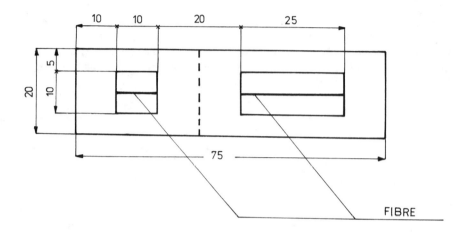

Fig.1. Test fibres glued to paper frames for stress-strain tests

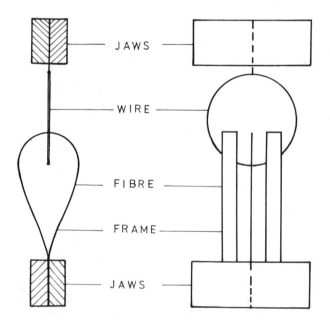

Fig.2. Fibre and wire loops fastened in a Instron tester

The diameter of wire cross-section was equal to 56 μm. The
fibre loop was formed by clamping together both ends of the
framed fibre in the bottom jaws of Instron tester. The plane
of fibre loop was perpendicular to the wire loop plane.
After cutting the paper frame and drawing down the bottom
jaws of Instron tester at the constant speed of 2 mm/min
the changes of the loop shape and load were observed. When
the loading force bending the fibre T reached the value
50 μN the loop shape was recorded using Fastax camera. The
speed of camera was equal to 500 pictures per second.
The analysis of the loop shape was carried out using a projection microscope. Last picture just before and after failure of a fibre is presented in Figure 3.

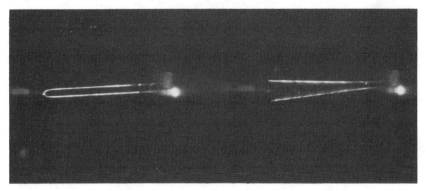

Fig.3. Fibre loop just before (left) and after the failure
(right)

RESULTS AND DISCUSSION

Starting from the top point of fibre loop just before its failure the width of loop was measured at 39 μm intervals along
its length with the use of projection microscope, (magnification 130 x). The diagram representing mean values of half
width of loop based on results from 19 fibres is presented in
Fig.4 a and b . The numerical values are given in the Table 1.

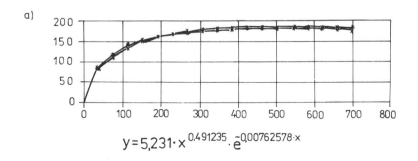

$$y = 5,231 \cdot x^{0,491235} \cdot e^{-0,00762578 \cdot x}$$

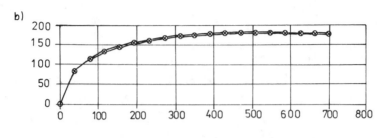

$$y = 5{,}042 \cdot x^{0{,}475514} \cdot e^{-0{,}0070588 \cdot x}$$

Fig.4. Half width of fibre loop at the failure of fibre,
a) for fibre 38/III, b) for fibre WS/213.

TABLE 1
Mean values of width of loops $[\mu m]$

Distance from the loop top	39	78	117	156	195	273	429	702
Fibre 38/III	178	246	286	312	330	350	372	384
Fibre WS/213	166	226	264	290	308	328	350	362

Mean values and standard deviations of force T were:
- for the fibre 38/III 3.57 mN and 0.69 mN,
- for the fibre WS/213 3.43 mN and 0.404 mN.

After these results the tensile strength, Young modulus and breaking strain were calculated (Table 2).

TABLE 2
Mean values of tensile strength, Young modulus and breaking strain of fibres

Kind of fibre	Looped fibre (mm)		Gauge length			
			10 (mm)		25 (mm)	
	38/III	WS/213	38/III	WS/213	38/III	WS/213
Tensile strength (GPa)	6.22	6,65	2.87	4.68	2.68	4.62
Young modulus (GPa)	182.5	185.0	134.7	173.3	197.1	226.5
Breaking strain (%)	3.41	3.60	2.13	2.70	1.36	2.04
Fibre diameter (μm)	8.25	7.78				

Comparison of these values shows that the intrinsic value of breaking strain of fibres without faults is much greater than found out from the measurement of breaking strain of straight fibres, when the distance between clamps is equal to 25 mm. Similar differences can be observed also for the strenght measurements. Only Young modulus has similar values when based on results found out from loading of straight or looped fibres.

CONCLUSIONS

Comparison of mechanical properties found out from the test of carbon fibres in straight and looped forms shows that it is possible to determine intrinsic material characteristics based on values of critical load, shape and the loop dimensions and fibre diameter.

REFERENCES

1. Żurek, W., Kocik, M., Całka, W., Jakubczyk, J., Tensile properties of carbon fibres. Fibre Sc. and Tech., 1981, 15, pp. 223-234.

2. Beetz, Ch., P.Jr., The analysis of carbon fibre strength distributions exhibiting multiple modes of failure. Fibre Sc. and Techn., 1982, 16, pp. 45-59.

3. Beetz, Ch., P.Jr., A self consistent Weibull analysis of carbon fibre strength distributions. Fibre Sc. and Techn., 1982, 16, pp. 81-84.

4. Hitchon, J.W., Phillips, D.C., The dependence of the strength of carbon fibres on length. Fibre Sc. and Tech., 1979, 12, 217-233.

5. Żurek, W., Yarn Structure (in Polish). Wydawnictwa Naukowo-Techniczne, Warszawa, 1971, pp. 156-174.

DELAMINATION OF POLYMER COMPOSITE SANDWICH STRUCTURES THE INTERACTION OF THE CONTROLLING PARAMETERS

BARRY W STAYNES
Department of Civil Engineering
Brighton Polytechnic
Brighton, U.K.

ABSTRACT

The composite structures formed when Portland cement concrete is surfaced or repaired with Polymer Concrete or Polymer Modified Cement Concrete are subject to complex states of time dependent stress which can result in cracking and delamination failures. The engineering properties of the substrate and surfacing materials which influence the performance of such composite structurers are identified and methods that have been developed to measure shrinkage at all stages of cure, thermal effects and tensile properties are discussed. A finite element computer model is being developed to enable the sensitivity of composite structures to the controlling parameters will enable engineers to make well founded engineering decisions on the selection of surfacing and repair materials.

INTRODUCTION

The use of Polymers in Concrete is now well established and the development of advanced materials for a range of applications has increased rapidly in recent years. This trend is likely to continue [1] and as repair and renovation now constitutes a very high proportion of construction output in developed countries, understanding of the nature of the composite behaviour of polymers with the more conventional construction materials is becoming increasingly necessary. The importance being attached to this aspect of civil engineering is borne out by the following examples : In the United States the Army Corp of Engineers has promoted a special research programme on the maintenance, repair and rehabilitation of large hydraulic structures as they are responsible for well over 150 dams and similar structures which are at least 50 years old. Many are showing evidence of the need for restoration [2]. Similarly in the United Kingdom, central government through the Science and Engineering Research Council and the Department of the Environment has made repair and maintenance a "preferred area for research." This is well justified as repair and maintenance is 40% of construction output in an extremely buoyant market [3], and in West Germany new investment is expected to account for only 20% of future highway work [4].

The factors responsible for creating the need for repair and renovation can be sub-divided into :

Natural Catastrophes (earthquakes, floods); Climate/Environmental Factors (high soil chloride concentrations, rapid microbiological digestion, heavy use of de-icing salts); Conflicts (physical damage); Design/Construction Errors (chloride attack, reinforcement corrosion; alkali silica reaction) and of course Age Effects (carbonation, long term erosion).

As structures made of concrete and allied materials are involved in approximately half of the repair and renovation work, it is important to understand and evaluate the performance of the materials to be used to make good damaged/eroded concrete,and their composite behaviour with the parent structures. While there are many cases where replacement of material in depth is required these are likely to be individual rather than general cases. However, a high proportion of repairs involve the application of surface layers or patches varying in thickness from a few mm to 50 or 100 mm.

It is the behaviour and performance of the laminated structures resulting from the application of surface layers that is the subject of this paper. However, there are wider implications,as similar laminated structures are formed when chemically resistant linings are applied to the inside of tanks and pipes, and as specialised floor/highway/bridge deck finishes.

PERFORMANCE IN SURFACING AND REPAIR APPLICATIONS

Two materials being widely used for the repair and renovation of damaged concrete are Polymer Concretes (PCs) and Polymer Modified Cement Concretes (PMCCs). These materials are very different in both composition and behaviour. PCs are highly filled polymers (e.g.epoxide resins) and contain no hydraulic cement-water phase. PMCCs on the other hand contain up to about 20% of latex (e.g. Styrene/Butadiene resin;

acrylic) dispersed through the water phase of Portland Cement Concrete.
The relevant properties of these materials will be discussed.

There are also two basic types of repair and these have been designated
Structural Composite Surfacings (SCS) and Protective Composite Surfacings
(PCS). In the case of SCS the repair material is required to act comp-
ositely with the parent structural material to resist significant stress
due to the primary loading of the building. PCSs on the other hand
provide a protective and/or cosmetic surface which may well be subjected
to stresses, but the building is not reliant on it for its basic
structural integrity.

The aim of UK research programmes is to enable engineers to make well
founded engineering judgements in the selection of materials for specific
repairs and/or surfacing applications. To do this it is necessary to
establish the interaction of the parameters which govern peformance.
Failures usually take the form of delamination of the composite structure
or cracking of the surfacing or substrate.

As it is necessary to measure the controlling parameters of the potential
materials, a secondary aim of the research is to ensure that there are
suitable standard test procedures available to enable the full range
of parameters to be measured.

Controlling Parameters

Fundamental engineering properties and their interactions that need to be
evaluated include :

(i)   Coefficient of thermal expansion

(ii)   Shrinkage at all stages of normal cure

(iii)   Post cure shrinkage

(iv)   Limiting tensile strain

(v)   Adhesion properties of the surfacing material

(vi)   Creep/relaxation properties

(vii)   Tensile stiffness and strength of the substrate

(viii) Tensile stiffness, strength and Poisson's ratio of the surfacing
        material.

The sensitivity of the above parameters to practical conditions such as
inaccurate proportioning and/or mixing of the surfacing material
ingredients, temperature effects and the influence of flexure of the

structure due to service loading during the surfacing operation need to be evaluated.

Various International Standards and other methods are available to measure many of the above parameters [7 to 34],but conspicuous by their absence are methods of determining the "fundamental" as opposed to "emperical" values of (ii),(iii), (iv) and (vi) above.

Industrial Experience

Data relating to the failure of in particular PCs, was obtained from experienced engineers and contractors by means of questionnaires and interviews.

The main findings and reasons for failure of surfacings were :

(i)   Short Term Failure

- poor mixing, poor preparation, lack of temperature conditioning, inaccurate batching.

(ii)   Medium/Long Term Failures

- mismatch of engineering properties of surfacing and substrate,

- lack of appreciation of strain-related rather than stress related effects,

- inadequate wetting of the bond interface,

- lack of, or erroneous information on selection and application of materials.

(ii)   General

- Poor site procedures due to inadequate training and/or supervision,

- temptation to accept cheap materials due to either over or poor specification.

Investigations [5] have shown that poor performance of Structural Composite Surfacings occurs due to :

(i)   Thermal cracking on curing at the repair substrate interface particularly for polyester resin mortars.

(ii)   High deflection of beams as a result of using repair material with too low a stiffness in the compression zone.

(iii)   Localised cracking of substrate in members repaired on the tensile face with polymer resin mortar, with long term durability implications.

(iv)   Failure of tensile face repairs in regions of high shear force.
Practical experience has shown [6] that general material selection can be
based on the following :

CONDITION	REPAIR MATERIAL
1. Large areas thicker than 25mm	Sprayed or trowelled concrete with or without admixtures or bonding aids.
2. Surface repairs 12-25mm thick	Polymer Modified Cement Concrete.
3. Large surface repairs less than 12mm.	Epoxide Resin Mortar.
4. Small surface repairs less than 12 mm.	Epoxide or Polyester Resin Mortar.

Measurement of Shrinkage and Coefficient of Thermal Expansion

The following apparatus and procedure has been developed to enable
shrinkage of the polymer mortars to be measured at all stages of cure.
It also facilitates the measurement of coefficient of thermal expansion.

The apparatus shown in Figure 1 consists of a pressed steel trough
shaped mould 10 mm deep and 400 mm long. The trough is lined with PTFE
film to prevent bonding of the PC or PMCC sample cast in it, and is
mounted on an insulated casing. A temperature sensor is attached to the
underside of the steel trough to facilitate measurement of temperature
changes in the steel and, indirectly, the sample   Electro-
magnetic movement transducers with very low frictional resistance are
rigidly attached to either end of the trough to measure the movement of
the stop-ends of the mould. Bonding of the specimens to the stop-ends
is ensured by means of a keying strip.

As the stop-ends are free to move and the treated PTFE prevents bonding
to the mould, shrinkage of PC/PMCC specimens can be measured from the
moment they cease to flow. Thus shrinkage from a very early age through
all stages of normal cure, and post cure shrinkage can be measured. The
coefficient of thermal expansion of specimens cast into the apparatus
can be readily measured by measuring the movement of the stop-ends
following changes of temperature of the sample by conditioning to

stability in an incubator. Samples cast separately can also be assessed
for thermal expansion characteristics by cutting them to suit, laying
them in the trough, and bonding their ends to the stop-ends of the
mould. It is necessary to correct direct readings of the transducers
for expansion/contraction of the steel trough - hence the need for the
temperature sensor mentioned above. Typical shrinkage characteristics
are shown in Figures 2 and 3. The coefficient of thermal expansion of
a typical epoxide resin PC is 80×10^{-6} m/m$^{\circ}$C.

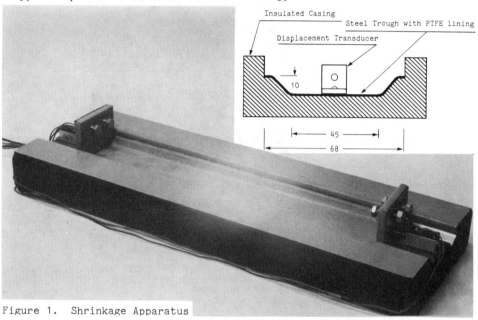

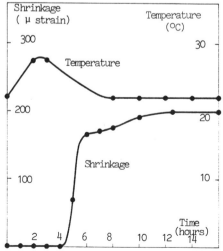

Figure 1. Shrinkage Apparatus

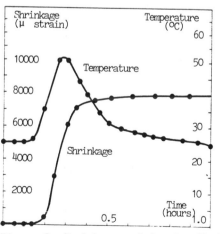

Figure 2. Shrinkage - Epoxide Mortar   Figure 3. Shrinkage - Polyester Mortar

Tensile Tests

Figure 4 shows the form of the tensile specimens for both short-term and creep testing. A special anchorage system has been devised consisting of metal plates and machined steel cross bars - Figure 5. Single specimens are tested for short-term strength, stiffness and Poissons ratio.

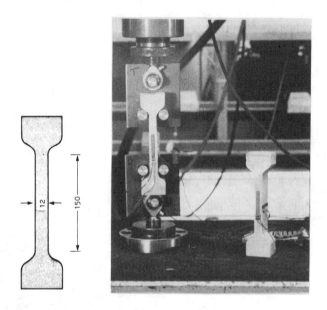

Figures 4a & b   Tensile Test Specimen   (40 mm thick)

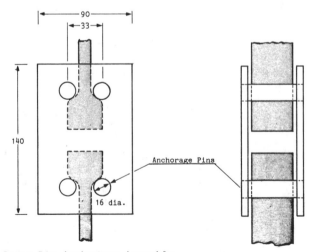

Figure 5.   Creep Rig Anchorage Assembly

Consistent results and ultimate failure in the middle third of the neck of the specimens indicates that the apparatus is effective at inducing uniaxial tension. Three specimens are mounted in series for creep testing enabling spread of values between specimens to be measured for identical load and environmental conditions. Five creep lines are operated in ambient conditions while a further five are mounted in an incubator cabinet - see Figure 6.

Figure 6.   Creep Apparatus.

To determine the influence of temperature on tensile performance under short-term and sustained loading, specimens were cast/cured at 5, 20 and 30°C. The following results are typical for epoxide resin mortars formulated for optimum casting at 20°C and 5°C respectively, cast at a range of temperatures but all tested at 20°C.

	Casting Temperature ($^{\circ}$C)	Ultimate Tensile Strength (MN/m^2)	Modulus of Elasticity (GN/m^2)	Tensile Strain at Failure (μ strain)
EP 20	5	15.3	10.5	1970
	20	16.4	9.3	2200
	30	17.2	10.7	2180
EP 05	5	14.3	19.3	830
	20	12.4	16.4	820
	30	10.8	15.2	790

Similar tests on two polyester resin mortars, one flexible the other stiff, gave the following results :

PF$_F$	5	11.8	20.4	6300
	20	9.5	9.8	7500
PR$_S$	5	12.1	20.0	790
	20	8.2	7.6	1100

Delamination Failure

Observation of failures of thin polymer mortar surfacings indicates that tensile stresses are induced in the surface causing cracking at regular intervals (approximately one metre centres) and this is followed by a delamination/peeling effect at the edges of the cracks. Such failures can be readily explained as once good bond to the substrate has been achieved any subsequent differential movement between the surfacing and the underlying concrete will induce stresses. Since any substrate concrete in a dry condition will have stabilised so far as shrinkage is concerned, these induced stresses will be tensile if the surfacing material undergoes cure shrinkage and/or post-cure shrinkage. Normal cure shrinkage of 2 to 8 mm can be expected over a length of 1 metre and such effects are likely to be enhanced by cooling due to environmental temperature change. A reduction from say 20°C to -10°C will induce differential movements in excess of 1 mm at either side of a 1 m strip.

Finite element analysis indicates that movements of the above magnitude will induce stresses likely to cause delamination failure either of the surfacing or the bond interface, particularly during periods of partial cure, or in concrete substrates. The above analysis is complicated by reductions in the induced stresses due to relaxation effects. As the PC modulus of elasticity and creep resistance is increasing during cure, and it is at this stage that the greatest amount of shrinkage is taking place, complex time dependent finite element analysis is needed. Such analysis is "in hand" the aim of which it to determine the relative sensitivity of delamination failure to the various controlling parameters listed earlier.

CONCLUSION

The development of methods of measuring the fundamental properties of polymer mortar surfacing materials from initial stages of cure and of a finite element computer model which can take into account time dependent effects, will enable stresses and strains in a surfacing material, concrete substrate and their interface to be determined. The sensitivity of these parameters will be assessed and engineering judgement can then be used to select PC systems which are unlikely to induce delamination failures.

REFERENCES

1. Nutt, W.O. and Staynes, B.W., Polymers in Concrete - The Next 25 Years. Proceedings of 5th International Congress Polymers in Concrete, Brighton Polytechnic, September 1987, p413-416.

2. Scanlon, J.M., REMR Research Programme Development Report. Final Report, U.S. Army Corps of Engineers Waterways Experimental Station, Vicksburg, MS., Feb.1983.

3. Department of the Environment - Construction Forecast 1987, Her Majesty's Stationery Office.

4. Rabe, O., Die Unterhaltung von Stahlbeton und Spannbetonbrucken, Bauingenieur 56, (1981), p431-437.

5. Emberson, N.K., and Mays, G.C., Polymer Mortars for the Repair of Structural Concrete : The Significance of Property Mismatch., Proceedings of 5th International Congress Polymers in Concrete, Brighton Polytechnic, September 1987, p335-341.

6. Shaw, J.N.D., Concrete Decay : Causes and Remedies, 11th Conference on Our World in Concretes and Structures, Singapore, August, 1986.

7. BS6319 : Parts 1 to 10 : 1987, Testing of Resin Compositions for use in Construction - various tests.

8. RILEM - Technical Committee 52 - RAC, Publications 1-9.

9. Agrément, European Union, MOAT No.21 1982, Directives for the assessment of ceramic tile adhesive.

10. France NF.T30-602 - Determination of adhesion strength by pull-off.

11. Germany DIN: 53.504, Testing of elastomers tensile properties.

12. Japan JIS: A6203 - Polymer dispersions for cement modifiers.

13. JIS: A1181-1186, Methods of test for polyester resin concrete.

14. UK BS.5270: 1976, Polyvinyl acetate emulsion bonding agents for internal use with gysum building plasters.

15. UK. Ryder, JF., Construction and Industry, Aug.10th 1957, pp 1090-1092. Methods for testing the adhesion of plaster to concrete.

16. UK BS.3900(E10): 1979 and ISO 4624: 1978. Paints and varnishes. Pull-off test for adhesion.

17. USA ASTM:C882-1978, Bond strength for epoxy-resin system used with concrete (to hardened or freshly mixed).

18. USA ACI Standard 503.1-79, Standard specification for bonding hardened concrete, steel, wood, brick and other materials to hardened concrete with a multi-component epoxy adhesive.

19. USA Proposed ACI Standard 503.2-79, Standard specification for bonding plastic concrete to hardened concrete adhesive.

20. USA Proposed ACI Standard 503.4-79, Standard specification for repairing concrete with epoxy mortars.

21. Canada 71-GP-30M, Epoxy and modified mortar systems for installation of quarry tiles.

22. France NF.P 85-507 to 518, Sealants - various tests.

23. France NF.T.30 - 062 and 700 to 708, Paints - various tests.

24. France NF.T.76-122, Characterisation of structural adhesives. Performance test under permanent shear stress.

25. France NF.T 76-107, Adhesives. Determination of tensile lap-shear strength of high-strength adhesive bonds.

26. Germany DIN: 52:455 pts 1-4, Sealants - various tests.

27. Germany DIN: 52:265, Testing adhesives for bonding ceramics.

28. Japan JIS: 1612/3, Bond strength tests for board adhesives.

29. Japan JIS: A6024, Epoxy injection adhesives for building repair.

30. JIS:K6848-6862, Rules for testing and methods for tensile stength of adhesives and other properties.

31. USA ASTM D2197, 2833, 3359, 3730, 4541, Paints and varnishes - methods of testing adhesion.

32. USA ASTM STP 649/1978, Adhesion measurements of thin films, thick films and bulk coatings.

33. Long, A.E., and MacMurray, A., The pull-off partially destructive test for concrete, Proceedings of Conference on Insitu/ non-destructive testing of concrete, Oct. 1984, p327-350.

34. Naderi, M., Cleland, D.J., and Long, A.E., Polymer Modified Repair Materials - Strength and Durability, Proceedings of 5th International Congress, Polymers in Concrete, Brighton Polytechnic, 1987, p309-313.

FAILURE ANALYSIS OF COMPOSITE LAMINATES WITH FREE EDGE

S. G. ZHOU* and C. T. SUN
School of Aeronautics and Astronautics
Purdue University
West Lafayette, Indiana 47907
U.S.A.

ABSTRACT

A failure theory which combines classical lamination theory with free-edge interlaminar stress was employed to predict the strength of laminates with free edges. Laminates studied were $[\pm\theta]_{2s}$ angle-ply laminates, $[\pm\theta/90]_s$ laminates, and fiber-dominated $\pi/4$ laminates. The present theory was capable of predicting failure loads for $[\pm\theta]_{2s}$ and $[\pm\theta/90]_s$ laminates as well as delamination on-set loads for $\pi/4$ laminates.

INTRODUCTION

Flat coupon specimens are often used to measure the strength of composite laminates. Because of high stresses near the free edges, failure in a laminate often occurs first along the free edges. Thus, one must realize that the data obtained in this manner could be the free edge interlaminar strength of the laminate rather than the true laminate strength. Because classical failure criteria using classical lamination theory fail to account for the free edge effect, their predictions of strength usually are higher than the actual data. More realistic laminate failure criteria should be based on free edge stresses.

A previous study [1] showed that, under off-axis loading, coupon specimens of quasi-isotropic laminates ($\pi/3$ and $\pi/4$) exhibit a staircase-like-crack-delamination pattern at the free edge. Since this initial failure at the free edge triggers unstable failure progression into the interior of the laminate, the strength of these specimens can be estimated by the load which causes free-edge failure. An average-stress-criterion based on stresses near the free edge was developed in [1] for laminate strength prediction. Analytical results showed good agreement with experimental data.

---

* Currently, School of Aeronautics, Northwestern Polytechnical University, Xian, Shaanxi, People's Republic of China

In this paper, the failure criterion developed in [1] was employed to study failure in a number of matrix-dominated laminates, such as $[\pm\theta]_{2s}$, $[\pm\theta/90]_s$ and some fiber-dominated $\pi/4$ laminates. For some $\pi/4$ laminates, test results showed that open-mode delamination could occur along interfaces adjacent to the $90°$ layer. Such failure of delamination depends on the stacking sequence of the laminate. Results indicated that the present method can also be used to predict the onset of open-mode delamination.

MATERIAL PROPERTIES

The mechanical properties of AS4/3501-6 graphite/epoxy composite are

$E_1 = 20.16$ msi $\quad X = 320$ ksi $\quad X' = -292$ksi

$E_2 = 1.43$ msi $\quad Y = 8.2$ ksi $\quad Y' = -30$ ksi

$G_{12} = 0.76$ msi $\quad S = 16$ ksi $\quad v_{12} = 0.30$

where E_1 is the longitudinal modulus of elasticity, E_2 the transverse modulus, G_{12} the in-plane shear modulus, v_{12} the longitudinal Poisson's ratio, and X (X'), Y (Y'), and S are the longitudinal, transverse, and in-plane shear strengths, respectively. The in-plane shear strength was determined by using a $[\pm 45]_{2s}$ specimen. Thus, this value contains lamination effects. This value is higher than that of 14.4 ksi reported in the literature [2].

The mechanical properties of T300/5208 graphite/epoxy composite are [3]

$E_1 = 19.85$ msi $\quad X = 219$ksi $\quad X' = -X$

$E_2 = 1.39$ msi $\quad Y = 6.24$ksi $\quad Y' = -25$ksi

$G_{12} = 0.65$ msi $\quad S = 11.3$ksi

Five additional constants are given by

$$E_3 = E_2 \quad , \quad G_{13} = G_{23} = G_{12} \quad , \quad v_{13} = v_{23} = v_{12}$$

The following interlaminar strengths were found suitable for graphite/epoxy composite.

$$Z_i = Y; \quad S_i \leq S.$$

The values of interlaminar strengths will be discussed later.

FAILURE CRITERION FOR LAMINATES

Failure in the laminate is predicted by three failure criteria. Three modes of failure are included, i.e., in-plane, interlaminar, and surface ply free edge failures.

In-plane Failure in a Lamina

For in-plane failure in a lamina, the Tsai-Hill criterion was used with classical lamination theory to predict failure load. The Tsai-Hill criterion is given by

$$\left[\frac{\sigma_{11}}{X}\right]^2 - \left[\frac{\sigma_{11}}{X}\right]\left[\frac{\sigma_{22}}{X}\right] + \left[\frac{\sigma_{22}}{Y}\right]^2 + \left[\frac{\sigma_{12}}{S}\right]^2 = 1 \tag{1}$$

Laminate failure analysis was performed using a step-by-step stiffness reduction procedure. The mode of failure was determined by examining the ratios of $\frac{\sigma_{11}}{X}$, $\frac{\sigma_{22}}{Y}$, and $\frac{\sigma_{12}}{S}$; the maximum ratio would distinguish between a fiber failure mode or matrix failure mode. In the subsequent failure analysis, reduced moduli should be used to account for the failure. The reduction procedure was as follows: E_1 reduced to zero for fiber breakage failure; E_2 and G_{12} reduced to zero for shear matrix failure; and only E_2 reduced to zero for transverse matrix failure.

Interlaminar Failure at Free Edge

At each interface, the interlaminar failure criterion used in this study is given by

$$\left[\frac{\sigma_z}{Z_i}\right]^2 + \left[\frac{\sigma_{zy}}{S_i}\right]^2 + \left[\frac{\sigma_{zx}}{S_i}\right]^2 = 1 \tag{2}$$

where σ_z, σ_{zy}, and σ_{zx} are the average interlaminar stresses over a critical distance of $2t$ ($t = $ ply thickness), Z_i is the peel strength, and S_i is the interlaminar shear strength. For AS4/3501-6 graphite/epoxy composite, the following values are recommended.

$$Z_i = Y = 8.2 \text{ ksi} \quad , \quad S_i = S = 16.0 \text{ ksi}$$

For T300/5208 graphite/epoxy composite, $Z_i = Y = 6.2$ ksi, and $S_i = 10.0$ ksi were used.

The above interlaminar strengths do not account for the effect of in-plane matrix cracks. When matrix cracking exists in adjacent laminas, the interlaminar strengths were assumed to be reduced to

$$Z_i^* = rfxZ_i \quad , \quad S_i^* = rfxS_i$$

In this study, rf = 0.9 was found to be suitable for graphite/epoxy composite.

As discussed in [1], we assume that once an interface fails, total laminate failure results, then the lowest interlaminar failure load should be regarded as the failure load of the laminate.

Surface Ply Failure at Free Edge

Since the in-plane stresses (σ_{11}, σ_{22}, σ_{12}) in the surface ply near the free edge are not constant, the average stresses over a distance of 2t from the free edge were used in the Tsai-Hill criterion for failure prediction. Further, since the surface plies are partially free from constraints (the lamination effect), the in-plane shear strength should be lower than that measured with [±45]$_{2s}$ specimen. Thus, we took the value S' = 14.4 ksi (for AS4/3501-6) reported in most literature. For T300/5208 graphite/epoxy composite we found S' = 8.2 ksi was suitable.

Here, we modify the free edge failure criterion by including the surface ply failure mode. The modified criterion states

$$\sigma_{ult} = \text{Min} \,(\sigma_{ult}^{int}, \sigma_{ult}^{SP})$$

where σ_{ult} is laminate failure stress, σ_{ult}^{int} is the smallest interface failure stress, and σ_{ult}^{SP} is the failure stress for the surface ply.

Free Edge Stress Analysis

The stress state near the free edge of a laminate is three-dimensional in nature. However, for long specimens, the state of stress can be regarded as independent of x-axis in the longitudinal direction. With this assumption, the problem was reduced to a two-dimensional problem with y and z as independent variables.

Because the deformation was independent of the loading axis, a pseudo three-dimensional displacement field was written in the form,

$$u = e_o \, x + U(y,z) \quad , \quad v = V(y,z) \quad , \quad w = W(y,z)$$

where u is the displacement in the x-direction, v the y-direction, w the z-direction, and e_o is the applied uniform axial strain. Symmetry allowed the investigation to be limited to a quadrant of the cross-section.

The finite element model used here was a nine-node isoparametric element which is much more efficient in handling high stress gradients than the constant strain elements used by other authors. The generalized plane strain analysis used here was formulated to consider orthotropic properties.

In this study, the problem to be analyzed was a laminate under in-plane uniaxial tension. For stress analysis, the region of interest was subdivided into a number of elements with a very dense, and uniform mesh used near the free edge.

Stress distribution in the neighborhood of the free edge was calculated using the pseudo 3-D finite element program. In-plane stresses were σ_{11}, σ_{22}, σ_{12}, and interlaminar stresses were σ_{zz}, σ_{zy}, and σ_{zx}. For ease of comparison, a tensile in-plane load (N_x) of 1000 lb/in was used in the analysis.

Accompanying the free-edge interlaminar stress analysis was the use of reduced lamina moduli to account for the lamina matrix cracking predicted by the Tsai-Hill criterion. Caution must be exercised in this reduction of elastic constants. When a certain elastic constant is reduced, in order to satisfy constraint conditions on the elastic constants, the remaining constants should also be adjusted. For example, due to transverse matrix failure, E_2 was reduced to a small value (e.g. 100 psi). In order to meet the constraint condition like $v_{23} < E_2/E_3$, v_{23} should also be reduced.

Average stresses over a distance of 2t (t= ply thickness) from the free edge were used in the criterion for failure prediction. For comparison, the interlaminar stresses listed in Tables were assumed to be loaded under a tensile load of 1000 lb/in.

ANGLE-PLY LAMINATES

Angle-ply $[\pm\theta]_{2s}$ laminates of T300/5208 graphite/epoxy composite were analyzed using the proposed failure criterion. In ensence, Tsai-Hill, interfacial failure, and surface-ply failure loads were obtained first and the laminate strength was taken to be the lowest failure load among the three.

In Table 1, average interlaminar stresses in the $[\pm\theta]_{2s}$ laminate with critical length 2t are shown. No reduced moduli were used in this calculation. From Table 1, we note that shear stress σ_{zx} dominates interface failure in these three cases. Results also show that the shear stress in interface 2 is much smaller than that in interface 1 and interface 3. This is because layer 1 and layer 2 balance each other through interface 1, and layers 3 and 4 balance through interface 3.

These average interlaminar stresses were used in Eq. (2) to predict interlaminar failure at the free edge. The results are presented in Table 2.

Table 3 lists the predicted failure stresses for the three modes of failure, i.e., in-plane (Tsai-Hill), interlaminar, and free-edge surface-ply failures. The test data were reported by Rotem and Hashin [3] and by Oplinger et al [4]. Note that Rotem and Hashin used T300/5208 graphite/epoxy while Oplinger et al used MOD II/5206 graphite/epoxy. However their data appeared similar, and thus were treated equally.

The results in Table 3 are also presented in graph form in Fig. 1. From these results, the following observations were made.

- For the fiber-dominated laminate ($\theta = 0°$) and highly matrix- dominated laminates ($\theta > 30°$), failure initiates from the interior of the laminate and the classical Tsai-Hill criterion is accurate.

- For small θ values ($\theta = 5°$ - $10°$), failure is dominated by the surface-ply failure at the free edge. Both Tsai-Hill and interlaminar failure criteria predict much higher strengths in this region.

- In the region $\theta = 15°$ - $30°$, free edge (interlaminar or surface-ply) failure dominates.

TABLE 1

Average interlaminar stresses in $[\pm\theta]_{2s}$ laminates under in-plane load $N_x = 1000$ lb/in

θ	Interface	σ_{zz} (psi)	σ_{zy} (psi)	σ_{zx} (psi)
10°	1	0	-13	1755
	2	-1	6	-434
	3	2	23	1755
15°	1	-2	-36	2507
	2	-6	20	-764
	3	5	71	2504
22.5°	1	-13	-103	3541
	2	-26	58	-1301
	3	13	203	3524
30°	1	-33	-205	4441
	2	-58	114	-1771
	3	26	394	4392
45°	1	-56	-351	4257
	2	-90	168	-1648
	3	45	614	4180

TABLE 2
Predicted interlaminar failure load of $[\pm\theta]_{2s}$ laminate with T300/5208 composite

$[\pm\theta]_{2s}$ Laminate	7.5°	10°	15°	22.5	30°	45°
Interface 1	7418	5698	3989	2823	2249	2341
Interface 2	34008	23041	13089	7679	5634	6035
Interface 3	7419	5698	3992	2833	2268	2367
Pred. Load (lb/in)	7418	5698	3989	2823	2249	2341
Strength (ksi)	185	142	100	71	56	58

TABLE 3
Predicted and experimental laminate strength (ksi) for $[\pm\theta]_{2s}$ T300/5208 composite

Criterion θ	Tsai-Hill	Interlaminar Failure	Surface-Ply with $S'=8.2$	Experimental Failure
0°	219.0		219	219
7.5°	210.0	185	136	
10°		142	120	107*
15°	167.0	100	99	100
22.5°	115.0	71	68	
30°	53.0	57	57	59.0 (57*)
45°	21.7	59	20.4	21.5 (21*)
60°	13.0			11.6
75°	6.6			6.2
90°	6.24			6.24

* Data taken from [4].

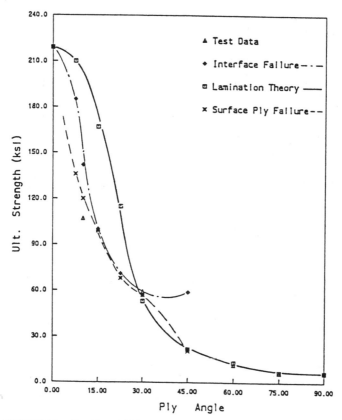

FIGURE 1. Predicted failure loads and test data of $[\pm\theta]_{2s}$ laminates

$[\pm\theta/90]_s$ LAMINATES WITH DIFFERENT STACKING SEQUENCES

The second group of matrix-dominated composite laminates was $[\pm\theta/90]_s$ type. They included $[90/\pm30]_s$, $[30/90/-30]_s$, $[90/\pm45]_s$, and $[\pm45/90]_s$ laminates (seven specimens each). The composite was AS4/3501-6 graphite/epoxy.

The failure loads predicted using classical Tsai-Hill criterion are 93.4 ksi for $[\pm30/90]_s$ laminate and 48 ksi for $[\pm45/90]_s$ laminate. These predicted failure loads are much higher than the test data. Thus, failure in these laminates must have initiated from the free edges.

Following the same procedure presented in the previous sections, predicted failure loads can be determined; these results are listed in Table 4. The corresponding interlaminar strengths used in this prediction are also listed in this table. At interface 2, because of the matrix crack, the critical strengths were reduced to $S_i^* = 14.4$ ksi, and $Z_i^* = 7.4$ ksi. However, at interface 3 (in the middle plane of the $90°$ layer), because of vanishing interface shear stress, $Z_i = 8.2$ ksi was retained. From the results presented in Table 4, it is obvious that free edge interlaminar strength agrees with the laminate strength.

TABLE 4
Laminate failure stress (ksi) for $[\pm\theta/90]_s$ type laminates (Material AS4/3501-6)

Failure Mode	Laminates		
	$[90/30/-30]_s$	$[30/90/-30]_s$	$[30/-30/90]_s$
Tsai-Hill Prediction	93.4	93.4	93.4
Surface Ply Failure	456.4	71.0	69.4
Interface 1 Failure	91.8 [1]	56.9 [1]	64.5 [2]
Interface 2 Failure	70.6 [2]	55.8 [1]	61.7 [1]
Interface 3 Failure			66.3 [3]
Failure Prediction	70.6	55.8	61.7
Experimental Data	71.6	58.4	

Failure Mode	Laminates		
	$[90/45/-45]_s$	$[45/90/-45]_s$	$[45/-45/90]_s$
Tsai-Hill Prediction	48.0	48.0	48.0
Surface Ply Failure	218.8	28.9	32.3
Interface 1 Failure	50.1 [1]	33.5 [1]	55.7 [2]
Interface 2 Failure	40.2 [2]	33.0 [1]	32.2 [1]
Interface 3 Failure			34.3 [3]
Failure Prediction	40.2	28.9	32.2
Experimental Data	36.4		32.0

where [1] $Z_i^* = 7.4$ ksi, $S_i^* = 14.4$ ksi

[2] $Z_i = 8.2$ ksi, $S_i = 16.0$ ksi

[3] $Z_i = 8.2$ ksi

Calculation also showed open mode delamination would occur when tensile loads increase to 61.7 ksi ([±30/90]$_s$) and 32.2 ksi ([±45/90]$_s$). Because both of these delamination loads were very close to the predicted final failure loads, no significant open mode delamination should be expected. This expectation was confirmed by observation during testing.

QUASI-ISOTROPIC LAMINATES

Open Mode Delamination

Many reports have shown that for a $\pi/4$ quasi-isotropic laminate, when the interlaminar normal stress σ_{zz} is tensile and is the dominant component among the free edge stresses, then open mode delamination may occur. Examples of such laminates include [±45/90/0]$_s$, [±45/0/90]$_s$, and [0/±45/90]$_s$ laminates. In these laminates, the open-mode delamination crack may propagate into the laminate before final failure of the laminate. For these laminates, classical failure criteria are not suitable for laminate strength prediction.

Table 5 lists delamination and failure loads for four $\pi/4$ laminates with different stacking sequences.

TABLE 5
Experimental and predicted failure stresses using Tsai-Hill criterion for different cases of $\pi/4$ laminate

Laminate	Del. Stress (ksi)	Ult. Stress (ksi)	Tsai-Hill (ksi)
[0/90/±45]$_s$		115.5	115.3
[±45/90/0]$_s$	84.4	95.4	115.3
[±45/0/90]$_s$	65.5	84.7	115.3
[0/±45/90]$_s$	84.6	97.1	115.3

These data are the average values of at least four specimens tested for each laminate. Except for the [0/90/45/−45]$_s$ laminate, the other three laminates showed open mode delamination at loads significantly lower than the respective ultimate loads. Except for [0/90/45/−45]$_s$ laminate, the Tsai-Hill criterion fails to predict the ultimate stress.

Test results showed that open-mode delamination occurred along the interfaces adjacent to the 90° layer. From the analytical results of interlaminar stress distributions along interface 3 of [45/−45/0/90]$_s$ laminate, at interface 3, high normal stress σ_{zz} along the free edge is evident. The interlaminar failure criterion given by Equation (2) was employed to predict open mode delamination load. The delaminate on-set stress for each interface adjacent to the 90° layer was calculated and the results presented in Table 6. The lowest interface failure

TABLE 6
Open mode delamination on-set stresses (ksi) for π/4 laminates

Laminate	[±45/90/0]ₛ	[±45/0/90]ₛ	[0/±45/90]ₛ
Interface 2	88.1		
Interface 3	85.7	68.9	84.3
Interface 4		67.1	91.5
Prediction	85.7	67.1	84.3
Exp. Data	84.4	65.5	84.6

* Interface 4 is at the middle plane

stress is the delamination stress for the laminate. The predictions agree very well with the experimental results. Note that the free edge of the specimens tested were well polished. It is possible to obtain lower failure loads with unpolished specimens.

If open mode delamination is regarded as total loss of load carrying function of the laminate, then the present failure criterion based on interface strength is adequate for laminate failure prediction for fiber-dominated π/4 laminates that are prone to open mode delamination.

Ultimate Laminate Stress

For laminates which are prone to open-mode delamination, progression of delamination as the applied load increases must be modeled first. In addition, the effect of delamination on ultimate strength must be estimated. This failure analysis is not an easy task.

From the experimental results for π/4 laminates, we note that if delamination does not occur (e.g. [0/90/45/−45]ₛ) then Tsai-Hill criterion can predict laminate strength. If delamination occurs before total laminate failure, then the laminate strength is lower than that predicted by Tsai-Hill criterion.

In view of the foregoing, we propose the following simple estimate of laminate strength.

$$\sigma_{ult} = A\,\sigma_{del} + (1-A)\,\sigma^*_{ult} \tag{3}$$

where σ_{ult} is the laminate strength, σ_{del} the delamination stress, σ^*_{ult} the laminate strength predicted by Tsai-Hill criterion, and A is a parameter to be determined experimentally. Note that if $\sigma_{del} = \sigma^*_{ult}$, then $\sigma_{ult} = \sigma^*_{ult}$.

Table 7 presents the laminate strengths obtained from Equation (3) with $A = 0.6$ for the delamination-prone π/4 laminates. It is evident that the predictions agree with the experimental results very well.

TABLE 7
Ultimate laminate stresses for delamination-prone $\pi/4$ laminates

Laminate	$[\pm45/90/0]_s$	$[\pm45/0/90]_s$	$[0/\pm45/90]_s$
σ_{del} (ksi)	85.7	67.1	84.3
σ^*_{ult} (ksi)	115.3	115.3	115.3
σ_{ult} (ksi)	97.5	86.4	96.7
Exp. Data	95.4	84.7	97.1

CONCLUSION

Failure of matrix-dominated ($[\pm\theta]_{2s}$, $[\pm\theta/90]_s$) laminates and fiber- dominated $\pi/4$ laminates has been studied. A new multiple-mode failure criterion was proposed. This new criterion, which took the free-edge effect into account, was found capable of predicting laminate strengths for the matrix-dominated laminates and delamination on-set stress for $\pi/4$ laminates.

Acknowledgement

This work was supported by Naval Research Laboratory under Contract No. N00014-84-K-2006.

REFERENCES

[1] Sun, C.T. and Zhou, S.G., "Failure of Quasi-Isotropic Laminates with Free Edges," submitted for publication.

[2] Qualification Tests on Hercules Magnamite AS4/3501-6, May, 1984, Delsen Testing Laboratories, Inc.

[3] Rotem, A., and Hashin, Z., "Fatigue of Angle-Ply Laminates," *AIAA Journal,* Vol. 14, 1976, pp. 868-872.

[4] Oplinger, D.W., Parker, B.S. and Grenis, A., "On the Three- Dimensionality of Failure Modes in Angle-Ply Strips under Tension," AMMRC MS 76-2, Sept. 1976, pp. 263-286.

MACRO-STRUCTURAL EFFECTS IN A UNIAXIAL TENSILE TEST ON CONCRETE

DIRK A. HORDIJK[*], HANS W. REINHARDT[**]
[*]Stevin Laboratory, Delft University of Technology, Stevinweg 4,
NL-2628 CN Delft, The Netherlands
[**]Darmstadt University of Technology, Alexanderstrasse 5,
D-6100 Darmstadt, West Germany

ABSTRACT

The complete tensile stress-deformation relation for concrete as obtained in a deformation controlled uniaxial tensile test cannot unreservedly be regarded as a material property. A simple model is given to demonstrate how a macro-structural behaviour of the specimen affects the obtained $\sigma-\delta$ relation. For an accurate simulation of a tensile test a finite element analysis should be performed. With the model, however, it is possible to investigate the way the results of such a test are influenced by, for instance, specimen dimensions, boundary conditions and deficiencies in test performance, relative easily.

INTRODUCTION

The tensile strength of concrete is about 5 to 10 per cent of its compressive strength. Therefore, concrete is a material most suitable to withstand compressive stresses rather than tensile stresses. Nevertheless, interest in the tensile behaviour of concrete increases. The most important reason for this increasing interest in the tensile properties of concrete is the introduction of fracture mechanics in the field of concrete structures and the increasing importance of numerical analyses.

The tensile behaviour of concrete is determined by the complete stress-deformation relation under tensile loading. Previously, it has been assumed that the $\sigma-\delta$ relation obtained from a deformation controlled tensile test directly yields the material property. Recently, however, it has been demonstrated by experiments [1] and by finite element analyses [2] that the post-peak behaviour can be affected by a (macro-)structural behaviour of such a specimen. In this paper it will be demonstrated that this phenomenon can also be studied by means of a simple numerical model.

MOTIVE AND SCOPE

Concrete is an elastic-softening material. When straining a concrete bar in tension, it displays a linear relation almost up to the peak load. After peak load the stress decreases with increasing deformation. This decreasing stress transferring capacity is due to the development of a process zone in the tensile bar. The width of this process zone is not exactly known, but it is assumed to be in the order of several times the maximum aggregate size. Outside this process zone the material unloads.

Since the last decade, deformation controlled uniaxial tensile tests are being performed in the Stevin Laboratory in order to determine the tensile properties of concrete. In these tests, it is intended to create a crack (process zone), while the crack surfaces remain parallel to each other, because only in those circumstances the obtained $\sigma-\delta$ relation can directly be regarded as a material property. In the beginning, the tensile bar was placed between two hinges. Then strong non-uniform crack openings were observed in the post-peak region. Therefore, the loading rig was adapted in a way that rotation of the end-platens was prevented, in order to force the crack to open more uniformly. After this adaption, however, still non-uniform crack openings were observed, which were, as contrasted with the earlier experiments, confined to only a small part of the loading path, just beyond peak load, whereafter the deformations over the process zone were again uniformly distributed. A peculiar result from these tests was that the descending branch displayed irregularities, which were not observed in the experiments with the hinges. A typical result obtained on a lightweight concrete is shown in Fig. 1.

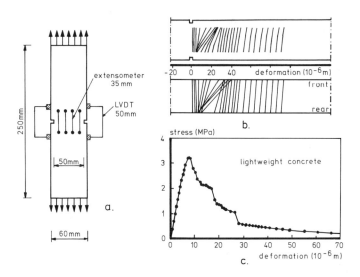

Fig. 1. Experimental result; specimen (a), deformation distributions (b) and $\sigma-\delta$ relation (c).

Based on the idea that for a softening material and an average deformation equal to the deformation at peak load, the tensile force will be lower in case of non-uniform crack opening than in case of a uniform one, a model was introduced [1]. The model explains why there are non-uniform crack openings in a deformation controlled tensile test on a concrete tensile bar. According to this model the non-uniform crack opening can be limited by an increasing rotational stiffness of the boundary of the process zone. This means that the specimen itself also contributes to this rotational stiffness. The model was verified by experiments in which the specimen length was varied. Since this phenomenon can be regarded as a macro-structural behaviour, it should also be obtained with a finite element analysis of a tensile test. Indeed, the results of such a calculation (see Rots et al. [2]) appeared to be in agreement with the model. With the FE-programme, a parameter study can be performed. The influence of, for instance, boundary conditions (hinges or non-rotatable loading platens), specimen length, specimen stiffness, eccentric loading, initial stresses and assumed stress-crack opening relation on the tensile behaviour, can be studied. Such a study, however, will be rather time consuming and demands a powerful computer. Therefore, a relative simple computer programme was written for the model as described in [1], which enables to study several phenomena with a personal computer.

The phenomena mentioned have also been reported by other investigators. For example, non-uniform crack openings were reported in [3,4,5]. As far as the discontinuities in the descending branch are concerned, the results reported in [5,6,7,8] display comparable features. Van Mier [5] also pointed to a connection between the two phenomena and the importance of the boundary conditions. The typical shape of the descending branch, a plateau followed by a sudden jump, was explained by Van Mier and Nooru-Mohamed [8] with a two stage fracture mechanism. In their approach, first perimeter cracks develop along the circumference of the specimen, which is followed by bending of intact ligaments between the perimeter cracks.

MODEL AND SOLUTION PROCEDURE

The model is based on the theory of bending in combination with a process zone model. The tensile specimen with its boundary conditions is replaced by a small zone (in the following denoted as fracture zone) of the specimen, that encompasses the process zone (see Fig. 2). The surroundings of this zone is replaced by translational and rotational springs, that are connected to this zone through a rigid interface. For the stiffness of the specimen against rotation the beam theory is used, $M = (EI/\ell)\phi$. The model does not account for the influence of horizontal displacements. For the sake of clarity, there is no direct relation between the depth of the fracture zone, which will

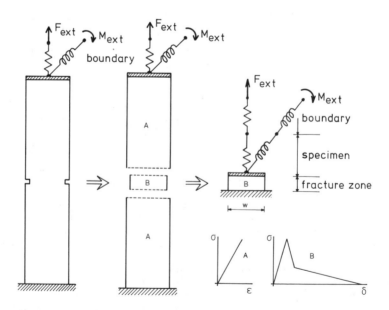

Fig. 2. Model for a tensile experiment on a concrete bar.

be arbitrarily chosen, and the width of the process zone in an actual experiment. In fact, the depth of the fracture zone links the deformation distribution in this part of the cross-section where a process zone is active, with the strain distribution in that part where the material still behaves linear elastically (see also Fig. 3). At this instant it is not clear which value should be taken for the depth of this fracture zone in order to get the best approximation of a real experiment. This shortcoming, however, is less important, since it is intended to study with the model some features of a uniaxial tensile test, rather than to simulate such a test as good as possible. For the latter objective a FE-analysis must be performed.

Input for the material behaviour is Young's modulus, tensile strength, fracture energy and stress-crack opening relation. These parameters result for a certain depth of the fracture zone in a σ-δ relation for this zone.

The solution procedure for this model is as follows. For an arbitrarily chosen combination of mean deformation δ_m and rotation ϕ of the fracture zone, the deformation distribution of this zone is defined (Fig. 3). With the σ-δ relation, the corresponding stress distribution can then be obtained. This stress distribution can be replaced by an internal force F_{int} and an internal moment M_{int}. This combination of δ_m and ϕ will only be a solution for the model in case where equilibrium and compatability exists at the boundary. For the boundary of the fracture zone in this tensile experiment, it means that the resulting internal moment M_{int} due to the imposed rotation ϕ should make equilibrium with the external moment M_{ext} that belongs to the same rotation. It may be obvious, that for every δ_m there does at least one solution exist, which belongs to a rotation equal to zero. These solutions belong to the uniform crack openings and yield as a result the σ-δ relation used as input. We are interested, however, in other possible solutions than those that belong to a uniform crack opening. In order to find these solutions, the force and moment in the fracture zone will be calculated for a great number of rotation angles. The procedure can then be repeated for different mean deformations.

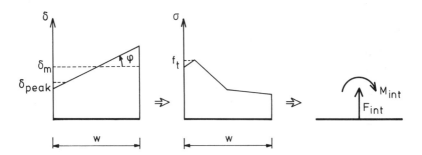

Fig. 3. Deformation and stress distribution for an assumed δ_m and ϕ and the resulting M_{int} and F_{int}.

COMPARISON WITH FE-CALCULATION

In this Section it will be demonstrated that solutions with non-uniform deformations can also be found with the described model. For reasons of comparison, the input as used in the FE-calculation of the tensile experiment [2] is also taken as input for the model. The specimen dimensions are $250*60*50$ mm^3, while two saw cuts reduced the critical cross-sectional area to $50*50$ mm^2. In the first instance, 35 mm was taken for the depth of the fracture zone. This length is equal to the measuring length of the extensometers in the experiments and will therefore also be taken as reference length for the σ-δ relations. The tensile strength was 3.4 MPa and for Young's modulus a value of 18000 MPa was taken, which was obtained in the experiments on a lightweight concrete [1]. The rest of the fracture mechanics parameters can be found in Fig. 4. The rotational stiffness of the boundary of the specimen was specified as 10^6 Nm/rad. The translational stiffness plays no role in this model for the tensile test.

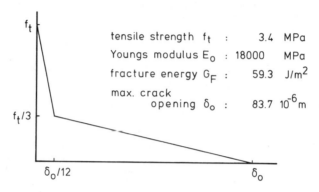

Fig. 4. Input parameters for the material behaviour of the concrete.

Moment-Rotation Relations for different mean Deformations
For values of the mean deformation δ_m ranging from $5 \cdot 10^{-6}$ m to $15 \cdot 10^{-6}$ m with steps of $0.1 \cdot 10^{-6}$ m and rotation angles varying between 0 and $400 \cdot 10^{-6}$ rad with steps of $4 \cdot 10^{-6}$ rad, the corresponding internal moment and internal force were calculated. The relation between M_{int} and φ for a number of δ_m-values is given in Fig. 5a and 5b. The deformation at peak load, δ_{peak} is equal to $6.611 \cdot 10^{-6}$ m. It appears that M first becomes negative for rotations starting from zero (signs as indicated in Fig. 5) in case $\delta_m < \delta_{peak}$ and first becomes positive in case $\delta_m > \delta_{peak}$. In Fig. 5 the linear relation between the external moment at the boundary of the fracture zone, M_{ext}, and an imposed rotation, φ, is included in a way that combinations of δ_m and φ that fulfil the requirement of equilibrium, can directly be taken from the figure, as being the points of intersection of the straight line and the curved lines. It can be seen that there exists equilibrium for a number of δ_m-values, while the rotations are not equal to zero. For a number of M_{int}-φ curves there are two points of intersection besides the origin. The meaning of this will become clear from the σ-δ relations. From the curves in Fig. 5 it can be suggested that there are no non-uniform solutions outside the range of δ_m, that was investigated.

Stress-Deformation Relations
The average stress for the points of equilibrium is determined by the corresponding internal force divided by the critical cross-sectional area. The equilibrium points with φ not equal to zero result in the curved line of Fig. 6. The σ-δ relation that was used as input and which is equal to the solution with the uniform deformations is also plotted in this figure. It appears that there exists a second equilibrium path for a part of the descending branch, which is accompanied with non-uniform crack openings. With the information of Fig. 5, where it was observed that M_{int} has another sign for a very small rotation and $\delta_m < \delta_{peak}$ compared with a very small rotation and $\delta_m > \delta_{peak}$ it can be assumed that the equilibrium path with the non-uniform deformations starts at peak load. In literature, such a point where one equilibrium path splits into two or more equilibrium paths is known as a bifurcation point (see also [9]).

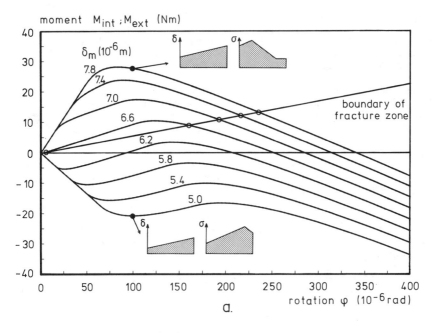

a.

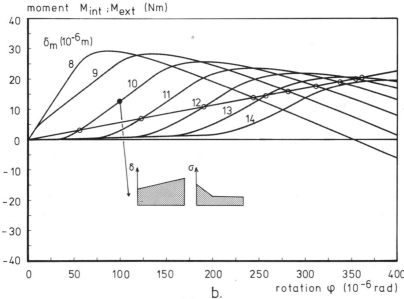

b.

Fig. 5. Relation between δ_m, ϕ and M_{int} for the fracture zone and the boundary condition for the tensile experiment (linear M_{ext}-ϕ relation).

In Fig. 6 it can be seen that in the equilibrium path with the non-uniform deformations two snap-backs (stress as well as deformation decreases) occur. This explains why for certain values of δ_m two solutions with non-uniform deformations were found. It may be obvious that in a deformation controlled experiment with the deformation of a 35 mm base as control parameter, it is not possible to obtain these snap-backs. Then a steep drop will occur as indicated by the dashed lines in Fig. 6.

Fig. 7 shows the σ-δ relations for the 'uniform' and 'non-uniform' crack openings as obtained with the FE-analysis [2]. For this calculation the average crack opening was used as control parameter. As far as the two equilibrium paths are concerned, the results of Figs. 6 and 7 display the same features. Nevertheless, also some differences between these figures can be observed. For instance, the shape of the curves near peak load differ. The reduced middle cross-sectional area results for the FE-calculation in stress-concentrations near the notches. Therefore in Fig. 7 the word symmetric is used instead of uniform. These stress-concentrations cause the different areas in the middle cross-section not to reach the tensile strength at the same instant, but the small areas near the notches reach f_t first. Thereafter, the centre part of the middle cross-section reaches f_t, while the areas near the notches unload. As a result, the peak of the stress-deformation diagram is less sharp. The second difference can be observed in that part of the curve where the strong snap-back occurs. This is due to a difference in solution procedure. In the FE-calculation the solutions are obtained with an incremental method, which means that unloading as indicated in Fig. 7 can be incorporated. In the model as described in this paper a solution for an arbitrarily chosen deformation is calculated directly. This demands an unambiguous relation between deformation and stress, which means that an unloading path as used in the FE-calculation cannot be implemented. In fact, the unloading path in this model is equal to the loading path (see Fig. 6).

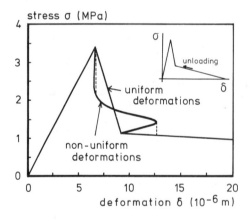

Fig. 6. Stress-deformation relations predicted by the model.

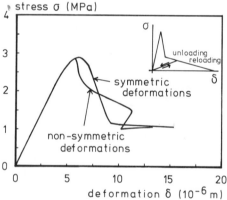

Fig. 7. Stress-deformation relations predicted by the FE-analysis [2].

Deformation Distributions

The deformation distribution for a number of δ_m-values, while the equilibrium path with the non-uniform deformations is followed and δ_m continuously increases like in an experiment, are plotted in Fig. 8. The resemblance with results obtained in an actual experiment (see [1]) is rather good. It should be noted, however, that in an experiment the phenomenon occurs three-dimensionally, while it is studied with a two-dimensional model.

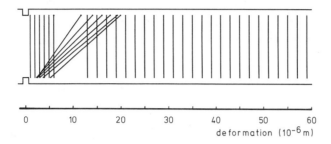

$$\begin{array}{cccccccc} 0 & 10 & 20 & 30 & 40 & 50 & 60 \end{array}$$
deformation $(10^{-6}\,\text{m})$

Fig. 8. Deformation distributions obtained with the model.

Influence of Depth of the Fracture Zone
In order to investigate the significance of the depth of the fracture zone, which was arbitrarily chosen as 35 mm, for the obtained results, the calculation has been repeated with 2.5 mm and 50 mm, respectively, for this parameter. The $\sigma-\delta$ relations for the 35 mm base, are plotted in Fig. 9. It can be seen that there is an influence of this parameter. The special features, however, are not significantly affected by varying this parameter between 2.5 mm and 50 mm.

INFLUENCE OF SPECIMEN SIZE AND BOUNDARY CONDITIONS

Specimen Length
Previously, the influence of specimen length was investigated experimentally and showed the non-uniform crack openings and discontinuities in the descending branch to become less important for decreasing specimen length [1]. This phenomenon can now easily be studied with the model. A shorter specimen length results in an increasing rotational stiffness of the specimen and therefore also of the boundary of the fracture zone. This means that the straight line in Fig. 5, that represents the $M_{ext}-\phi$ relation, will have a steeper slope. As a result, the maximum rotation for the equilibrium path with the non-uniform deformations, will decrease. The $\sigma-\delta$ relation for a specimen length of 125 mm, while other parameters were kept the same, is given in Fig. 10. It can be concluded, that the experimentally obtained results are confirmed by the model.

Hinges instead of Non-Rotatable End-Platens
For the situation of a tensile bar placed between two rotatable end-platens (hinges) the combination of δ_m and ϕ has to fulfil the requirement that the resulting internal moment is zero. For this boundary condition, the model also predicts the non-uniform deformations to start at peak load. In this case, however, the rotation keeps on increasing for an increasing δ_m, and the strong snap-back as was observed in case of non-rotatable end-platens is not obtained (see Fig. 10). This result looks to be in accordance with experimental observations as discussed before.

DISCUSSION

Besides the examples that have been given, more phenomena can be studied with the model. For instance, the influence of imperfections in test performance, like eccentric loading in case of hinges or the existence of an initial rotation in case of non-rotatable end-platens, can be subject of such a study. In the latter case, the line for the M_{ext}-ϕ relation in Fig. 5 is shifted to the right. Then, only one equilibrium path (non-uniform deformations) will be found, because the origin in Fig. 5 does no longer lead to a solution. This result is in accordance with the theory that bifurcation points mainly occur in perfect structures [9]. Since concrete is an inhomogeneous material, the equilibrium path with the non-uniform deformations, will always be found in experiments.

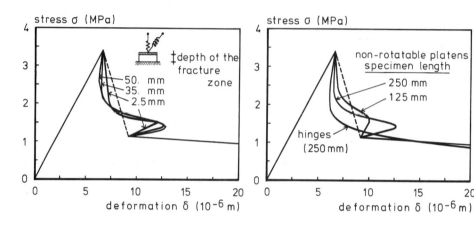

Fig. 9. Stress-deformation relations
obtained with different values for
the depth of the fracture zone.

Fig. 10. Stress-deformation relations
for varying specimen length
and boundary condition.

For all the calculations presented in this paper, the input for the material was kept the same. It may be clear that a parameter study with regard to these input parameters is also possible. It is surmised that this macro-structural effect can also play a role in the post-peak behaviour of other softening materials than normalweight or lightweight concrete. Besides for tensile tests, it is supposed that the model can also be used for studying features in bending tests on plain beams. This, however, has not been checked so far.

Another macro-structural effect that may occur in a uniaxial tensile test, is the effect of initial eigenstresses in the specimen, due to differential shrinkage [10] or temperature. For this problem, the model should be adapted in a way that the width of the specimen can be split up in a number of elements, so that each element can have its own initial deformation.

CONCLUDING REMARKS

It can be concluded that the model for a uniaxial tensile test, as described in this paper, is suitable for studying several phenomena that occur in uniaxial tensile tests on softening materials like concrete. For simulating an experiment as good as possible, FE-calculations should be performed. With the model and a small computer, however, it is possible to get a better understanding of the behaviour of a specimen in a uniaxial tensile test and of the influence of several parameters on this behaviour, rather quickly.

As far as the typical shape of the descending branch, a plateau followed by a steep drop, in a uniaxial tensile test on concrete is concerned, it is demonstrated that this shape can be the result of the macro-structural effect as discussed in this paper. The results do not necessarily support the explanation for the typical shape of the descending branch, as proposed by Van Mier and Nooru-Mohamed [8]. The influence of specimen dimensions and boundary conditions on this behaviour, as observed in experiments, was confirmed by calculations with the model.

ACKNOWLEDGEMENT

This investigation was partly supported by the Netherlands Technology Foundation (STW).

REFERENCES

1. Hordijk, D.A, Reinhardt, H.W. and Cornelissen, H.A.W., Fracture mechanics parameters of concrete from uniaxial tensile tests as influenced by specimen length. In Fracture of Concrete and Rock, RILEM-SEM Int. Conf., eds. S.P. Shah and S.E. Swartz, Houston, 1987, pp. 138-49.

2. Rots, J.G., Hordijk, D.A., and De Borst, R., Numerical simulation of concrete fracture in 'direct' tension. In Numerical Methods in Fracture Mechanics, eds. A.R. Luxmoore et al., Pineridge Press, Swansea, 1987, pp. 457-71.

3. Notter, R., Schallemissionsanalyse für Beton im dehnungsgesteuerten Zugversuch. Dissertation, Zürich, 1982.

4. Scheidler, D., Experimentelle und analytische Untersuchungen zur wirklichkeitsnahen Bestimmung der Bruchschnittgrössen unbewehrter Betonbauteile unter Zugbeanspruchung. DAfStb, Heft 379, 1987, 94 pp.

5. Van Mier, J.G.M., Fracture of concrete under complex stress. HERON, No. 3, 1986, 90 pp.

6. Willam, K., Hurlbut, B. and Sture, S., Experimental and constitutive aspects of concrete failure. In Finite Element Analysis of Reinforced Concrete Structures, eds. C. Meyer and H. Okamura, Tokyo, 1985, pp. 226-54.

7. Budnik, J., Bruch- und Verformungsverhalten harzmodifizierter und faserverstärkter Betone bei einachsiger Zugbeanspruchung. Dissertation, Ruhr-Universität, Bochum, 1985.

8. Van Mier, J.G.M. and Nooru-Mohamed, M.B., Geometrical and structural aspects of concrete fracture. Int. Conf. on Fracture and Damage of Concrete and Rock, Vienna, July 1988.

9. De Borst, R., Non-linear analysis of frictional materials. Dissertation, Delft University of Technology, Delft, 1986, 140 pp.

10. Hordijk, D.A. and Reinhardt, H.W., Fracture of concrete in uniaxial tensile experiments as influenced by curing conditions. Int. Conf. on Fracture and Damage of Concrete and Rock, Vienna, July 1988.

INFLUENCE OF AGGREGATE SIZE ON THE POST-PEAK TENSILE BEHAVIOUR OF CONCRETE IN CYCLIC TESTS

SZCZEPAN WOLIŃSKI
Technical University Rzeszow
Department of Building Structures
ul. Pow. Warszawy 6, 35-040 Rzeszów, Poland

ABSTRACT

Two types of deformation-controlled uniaxial tensile tests under the post-peak cyclic loading were performed on five types of the ordinary concrete. The influence of the maximum aggregate sizes and the loading type upon the stiffness degradation during unloading and reloading, stress drop from the tensile envelope curves and residual compressive deformations were studied. The results indicate that these features are not affected by the aggregate particle size but are influenced by the type of loading. An analytical expression to describe the post-peak cyclic response of concrete has been found and a model suitable for numerical analysis has been developed.

INTRODUCTION

The knowledge of the tensile proporties of concrete including the strain softening behaviour is necessary for rational calculations of the crack width, deflection, the bond and ahear transfer phenomena of concrete and reinforced concrete structures and for better understanding of their failure. This information is also of great importance for the fracture mechanics application in the design and numerical analysis of concrete structures. Among the numerous variables which influence the softening behaviour of concrete under the tensile loading there are there main categories: first, mix compositions and properties of the constituent materials, second, the age, moisture and curing conditions and third, the loading types. The available information of the influence of these variables on the tensile proporties of concrete is scarce and conflicting [1,2,3]. The most convincing way of determining parameters for describing a complete stress-deformation curve in tension is by means of a uniaxial, stable tensile test [4]. The purpose of the research reported herein was to study the influence of the maximum aggregate size on the tensile softening behaviour of concrete in the post-peak cyclic tests.

EXPERIMENTAL

Two types of deformation-controlled uniaxial tensile tests under the post-peak cyclic loading were performed on five types of ordinary concrete. Five concretes of similar mechanical properties but with different maximum aggregate particles (D_{max} = 2 mm, 4 mm, 8 mm, 16 mm and 32 mm) and different gradiation were used. The mix compositions and results of the standard 28-day tests are given in Table 1.

TABLE 1
Composition of concrete mixes and their 28-day properties

Concrete No./D_{max}	Mix proportions cement:water: sand 0-2:2-4: gravel 4-8:8-16: 16-32 /by weight/	f_c (N/mm^2)	f_{ts} (N/mm^2)
No.1 / 2 mm	1:0.5:4.29	41.8	3.0
No.2 / 4 mm	1:0.5:2.47:2.02	46.8	2.59
No.3 / 8 mm	1:0.5:2.42:0.97:1.44	47.2	3.14
No.4 /16 mm	1:0.5:2.11:0.72:1.03:1.28	51.7	3.15
No.5 /32 mm	1:0.5:1.85:0.64:0.74:0.90:1.17	48.3	3.12

f_c = cube compressive strength (mean of 3 specimens)
f_{ts}^c = cube splitting strength (mean of 3 specimens)

The tests were carried out on prismatic specimens 150 mm long, 60 mm wide and 50 mm thick, with saw-cuts 5 mm deep and 5 mm wide on both sides. The geometry of the specimens was selected in order to guarantee a uniform distribution of the deformation within the specimen cross-sections. They were sawed out of 50 mm thick panels casted in a battery mould. Curying conditions were as follows: after 2 days the mould was stripped and the panels were stored under water for 13 days, then cut to the proper dimensions and dried in the laboratory up to the

time of testing at 28-35 days. The closed-loop electro-hydrau-
lic loading equipment described in [3] was used. The constant
rate of deformation was about 0.08 μm/s with all the tests.
The longitudinal deformations of the specimens were measured
with two LVDTs of 50 mm long base mounted on the sides, three
extensometers of the gauge length of 110 mm and eight of 35 mm
mounted over the reduced cross-section in the front and the
rear of the specimens.

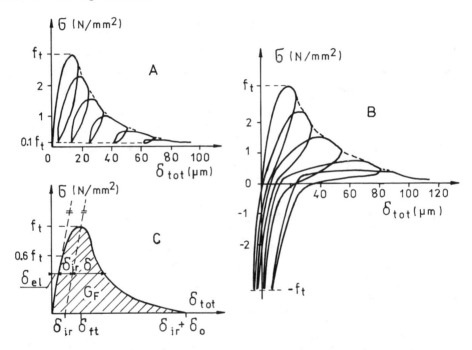

Figure 1. Average stress-deformation curves from the tensile-
tensile (A) the tensile-compressive (B) cyclic tests
and an envelope curve (C).

Two types of the post-peak cyclic tests: tension-tension and
tension-compressive (Fig.1) were carried out. During the first
loading the peak stress was reached. Then the stress drop of
about 0.16 N/mm^2 was allowed and the deformation direction was
reversed. For the tensile-tensile loading the unloading proceeded until about 0.1 f_t stress level and to about $-f_t$ for
the tensile-compressive loading. Then the deformation direction was reversed again. For both loading types the loading
went up to the tensile envelope curve and the procedure was

repeated at least seven times.

At least three specimens of each concrete type were tested
under both types of loading.

TEST RESULTS

Some more important material properties obtained from the test
results for both loading types of concretes tested are given
in Table 2.

TABLE 2
Average material properties from A and B type cyclic test
results

Concrete No./D_{max}	f_t (N/mm^2)		E_o (N/mm^2)		δ_{ft} (µm)		G_F (N/m)	
	A	B	A	B	A	B	A	B
No.1 / 2 mm	2.23	2.29	36843	32165	4.94	5.07	72.2	76.5
No.2 / 4 mm	2.42	2.39	32997	33404	4.63	5.81	88.1	85.0
No.3 / 8 mm	2.67	2.87	34305	29905	5.27	6.23	107.8	100.2
No.4 /16 mm	2.65	3.36	35883	32989	5.12	6.84	105.2	128.8
No.5 /32 mm	2.71	3.07	36139	30402	5.21	6.85	113.1	122.2

f_t = direct tensile strength, E_o = Young´s modulus (tension),
δ_{ft} = deformation at peak loading, G_F = fracture energy.

The three main behaviour characteristics of concrete in cyclic
tests: stiffness degradation, stress drop from the envelope
curve and residual compressive deformations were studied.

Stiffness degradation
The stiffness of a specimen was defined as a slope of a tangent or a secant line corresponding to a given part of the
stress-deformation curve. The initial stiffness C_o was referred to the tangent of the ascending branch. The unloading and
the reloading values of stiffness were estimated as the slope
of the secant line through an appriopriate point on the stress-
deformation curve and the minimum of a cycle (for A type loading) or as the intersection point of the curve with δ-axis
(for B type loading). The localization of these points is
shown in Figs. 2 and 3.
The stiffness ratio during unloading or reloading to the initial stiffness of a specimen was assumed to be a measure of

the stiffness degradation during succesive loading cycles.
This relative stiffness was plotted against the crack opening
deformation at the start of the unloading in each cycle for
each concrete tested and for both types of loading Figs. 2
and 3 .

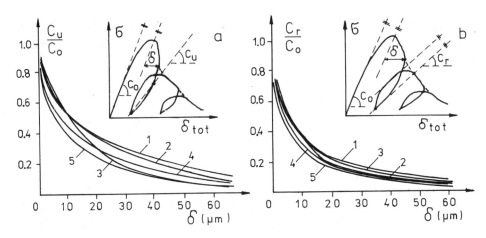

Figure 2. Stiffness degradation for A type tests: unloading (a)
and reloading (b).

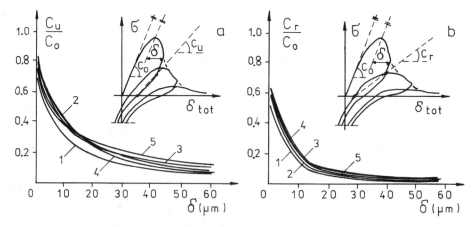

Figure 3. Stiffness degradation for B type tests: unloading (a)
and reloading (b).

As it can be noticed from Figs. 2 and 3 the reloading stiffness shows a steeper decrease with the crack opening deformation than the unloading one for all types of concrete and both
loading types. However, the scatter of experimental results

observed was greater for the unloading.

In none of the cases any distinct correlations were found between the stiffness degradation and the maximum aggregate size. Exponential functions were fitted by means of the least squares method to obtain analytical descriptions of $C/C_0 - \delta$ relations:

$$C_u/C_0 = \exp\left(-0.171\delta^{0.648}\right), \quad s = 0.072 \quad (1)$$

for all A type tests during the unloading,

$$C_r/C_0 = \exp\left(-0.375\delta^{0.478}\right), \quad s = 0.035 \quad (2)$$

for all A type tests during the reloading,

$$C_u/C_0 = \exp\left(-0.272\delta^{0.549}\right), \quad s = 0.056 \quad (3)$$

for all B type tests during the unloading,

$$C_r/C_0 = \exp\left(-0.523\delta^{0.513}\right), \quad s = 0.019 \quad (4)$$

for all B type tests during the reloading,
s is a standard deviation of the error in the direction of
C/C_0.

Stress drop

The stress drop from the envelope curve was defined as the fall of stress at the start of the unloading corresponding to the constant crack opening deformation. The ratio of the stress drop to the stress at the start of the unloading was found to be linearly correlated with the deformation at the start of the loading cycle. In the case of the tension-tension loading a rise in stress was also observed during the reloading from minimum of a cycle. Moreover, the stress drop and rise in the same cycles are equal. For both types of loading the stress drop is not significantely influenced by the maximum aggregate sizes.

A close fit of the data for all concrete types and the corresponding values of the linear correlation coefficient r are given by:

$$\Delta\bar{\sigma}/\bar{\sigma} = 0.254 - 0.00215\delta, \quad r = -0.983 \quad (5)$$

for all A type tests and

$$\Delta \bar{\sigma}/\bar{\sigma} = 0.337 - 0.00438\,\delta\,, \qquad r = -0.920 \qquad (6)$$

for all B type tests.

In Fig. 4 the average stress drop-crack opening deformation lines for all A and B type tests are shown. A slower decrease of stress drop with the crack opening deformation can be observed in the case of the tensile-tensile loading.

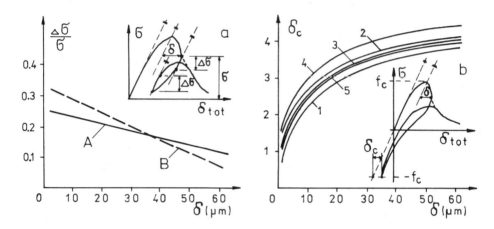

Figure 4. Relative stress drop versus the crack opening deformation (a) and the residual compressive deformation versus the crack opening deformation (b).

Residual compressive deformation

The residual compressive deformations of the specimens were measured for the tensile-compressive loading. Their values were assumed to be equal to a distance from the maximum compressive stress point in a cycle to the tangent line corresponding to the ascending tensile branch of the stress-deformation curve in the direction of δ_{tot}- axis (Fig. 4). A slow increase in the residual deformations with an increase in the crack opening deformations can be observed. No significant correlations were found between the residual compressive deformations and the maximum aggregate size but a scatter of test results decreases considerably with an increase in the particle size. A power function was used for fitting the test results for all the concretes:

$$\delta_c = 1.257\,\delta^{\,0.298}\,, \qquad s = 0.158 \qquad (7)$$

THE POST-PEAK TENSILE CYCLIC BEHAVIOUR MODELLING

A model of the post-peak cyclic behaviour of the concretes investigated for the tensile-tensile and the tensile-compressive loading is proposed. The average stress-total deformation curves are split into two parts: the ascending parts in which a unique relation exists between the stress and the strain that consists of an elastic component and an irreversible one: $\varepsilon = \varepsilon_{el} + \varepsilon_{ir}$, the descending parts in which a unique relation exists between the stress and the crack opening deformation: $\delta = \delta_{tot} - \delta_{el} - \delta_{ir}$. The tensile envelope curve conception is used [2,5]. The model is split into elastic and strain softening behaviour (Fig. 5). The unloading and reloading steps are modelled as follows:

1. The unloading starts at $\sigma = \sigma_1$ on the envelope curve and takes place at constant crack opening deformation δ_1 down to the value of $\sigma_1 - \triangle\sigma$ (segment AB). The stress drop $\triangle\sigma$ can be obtained from formula (5) or (6).

2. Then the unloading follows the line BC with the slope C_u which can be calculated from (1) or (3), until the minimum of the cycle $\sigma = \sigma_2$ (for A type tests) or $\sigma = 0$ point (for B type tests) is reached.

3. For A type tests the reloading in tension starts from σ_2 and at constant δ_2 progresses up to the value of $\sigma_2 + \triangle\sigma$ until the reloading stiffness C_r is reached (segment CD). This stiffness is maintained to the intersection point E with segment AB. From point E the crack opening deformation increases at a constant stress value until the envelope curve is reached at point F.

4. For B type tests the loading in compression follows the line CD and then DE down to the maximum compressive stress. For a known α angle having a constant value of about 45° point D may be found and point E is determined by the values of δ_1 and δ_c (7). The unloading in compression starts at a constant δ_c value to reach the reloading stiffness at point F. Then the unloading in compression and the reloading in tension follows line FG with the slope C_r up to the intersection point G with line AB. From this point the crack opening deformation incre-

ases at a constant stress value until the envelope curve is reached at point H.

5. Then the stress-deformation curve follows the envelope curve till the next unloading.

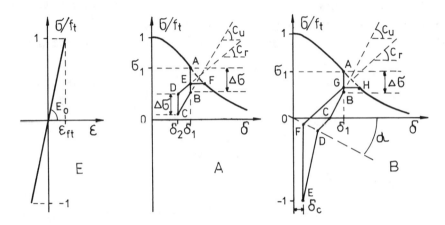

Figure 5. Model for the cyclic loading: the pre-peak range (E) the post-peak range for the tensile-tensile (A) and for the tensile-compressive loading (B).

According to the envelope curve conception the crack opening deformation is assumed to be unaffected by the type of cyclic loading [5]. This implies that the energy dissipated during cyclic loadings does not affect the fracture energy. Nevertheless, the model takes this energy into account.

CONCLUSIONS

The stiffness degradation, the stress drop from the tensile envelope curve and the residual compressive deformations for the concretes investigated are not affected by the maximum aggregate size and the composition of concrete. The rates of concrete stiffness degradation during the tensile-compressive loading are considerably greater than in the case of the tensile-tensile type of loading.

The stress drop from the tensile envelope curves was found to be inversely proportional to the crack opening deformation at the start of a loading cycle.

The residual deformations at the compressive loading increase

with an increase in the crack opening deformation.
The results of the cyclic tests have been used to establish
the model suitable for numerical analysis. The model comprises all the features observed in the experiments including
energy dissipation during the cyclic loading.

ACKNOWLEDGMENTS

These investigations were carried out at the Stevin Laboratory, Delft University of Technology, during the author´s stay
there. The author is greatly indebted to Prof. H.W., Reinhardt for his valuable suggestions in the subject of the experiments, to Ir. D.A., Hordijk for his substantial assistance
and to Mr. A.S., Elgersma for performing the tests.

REFERENCES

1. Swamy, R.N., Fracture mechanics applied to concrete. In
   Developments in Concrete Technology - 1, ed., F.D. Lydon,
   Applied Science Publishers LTD, London, 1979, pp. 221-81.
2. Woliński, Sz., Hordijk, D.A., Reinhardt, H.W. and Cornelissen, H.A.W., Influence of aggregate size on fracture
   mechanics parameters of concrete. The International Journal of Cement Composites and Lightweight Concrete, 1987,
   2, 95-103.
3. Woliński, Sz., Influence of aggregate size and loading type on the tensile behaviour of concrete. Report, Stevin
   Laboratory, Delft University of Technology, April 1986.
4. Hillerborg, A., The theoretical basis of a method to determine the fracture energy G_F of concrete. Matériaux et
   Constructions, 1985, 106, 291-96.
5. Reinhardt, H.W., Cornelissen, H.A.W. and Hordijk, D.A.,
   Tensile tests and failure analysis of concrete. Journal of
   Structural Engineering, 1986, 11, 2462-77.

FRACTURE BEHAVIOUR OF PLAIN CONCRETE IN BENDING

Janusz KASPERKIEWICZ
Institute of Fundamental Technological Research
Polish Academy of Sciences
00-049 - Warszawa, Świętokrzyska 21, POLAND
Piet STROEVEN and Dik DALHUISEN
Delft University of Technology, Department of Civil Engineering
2628 CN - Delft, Stevinweg 1, The NETHERLANDS

ABSTRACT

Several series of notched concrete specimens have been tested in pure bending to observe fracture behaviour. Particular attention was given to the effects of structural inhomogeneity and of the age. Due to the rigidity of the loading system it was possible to control continuously the crack opening and deflection over the complete loading–unloading–reloading cycles. Various parameters of the fracture process were recorded, such as the crack tip position and the surface strains ahead of the notch.

The fracture surface energy γ_f of the concrete was found not to depend on the direction of testing. Neither any effects of the position during casting have been observed. Average values of γ_f were about 65 N/m, with a rather large scatter. A weak correlation has been observed between the crack mouth opening at ultimate loading and γ_f. Additionally, γ_f was found to increase in time.

AIM AND SCOPE OF THE TESTS

The experiments were planned taking into account earlier fracture mechanics tests in this series [1], [2]. Principal goal was to observe the crack development in pure bending in a concrete typical for building practice. Material composition has therefore been chosen roughly in accordance with a Reference Concrete proposed by Dutron in 1974 [3], that has also been employed in other fracture mechanics tests [4]. Two mixes have been designed with a slightly differing grading, thus yielding two different material structures. The four-point bending test was given preference over the three-point one, proposed by a RILEM Recommendation [5], since it allows the crack to develop undisturbed by stresses induced by the loading system.

The same RILEM Recommendation proposes to take the loading direction perpendicular to the direction of casting. In practice crack propagation may however occur in the direction of casting. The question therefore arrises whether anisotropy and inhomogeneity effects are negligible. The present tests aimed at evaluating possible differences in the mechanical behaviour between "horizontally" and "vertically" loaded specimens, and between "horizontally" loaded ones obtained from top and bottom sides, respectively, of the original beams.

The effect of the stiffness of the loading machine on the mechanical characteristics was studied by testing part of the specimens in a soft testing machine (Amsler), while the remaining part was tested in a specially-designed deformation-controlled one. Specimens from one of the series have been tested at 100, 174, and 405 days, to be able to observe time effects. The other specimens have been tested at about 128 days. Some of the tests have been prematurely terminated after passing the maximum bending load. These specimens have been restored in the fog room. Finally, they have been loaded again and tested in the same position at the age of more than one year, allowing to study the self healing capacity of the concrete.

MATERIALS AND SPECIMEN PREPARATION

A cement content of 300 kg/m^3, a maximum grain size 31.5 mm, and a water-cement ratio of 0.58 are specified for the Reference Concrete [3]. These data were also accepted in the experiments, except for the grading characteristics. Instead, two mixes with a maximum grain size of 16 and 4 mm, respectively, have been designed. The two grading curves have been derived from the curve of the Reference Concrete by geometric transformation [6], reducing, as a consequence, the fineness moduli to below the RILEM proposal.

The composition of a batch of the first mix (RefC) amounted to: cement – 20.18 kg, water – 11.70 litres, aggregates – 3.48, 4.75, 10.43, 9.16, 10.14, 17.68, 30.60 and 29.68 kg, for the respective sieve fractions: <0.125, 0.125-0.25, 0.25-0.50, 0.5-1, 1-2, 2-4, 4-8 and 8-16 mm. The composition of the second mix (7DD) was similar except for the aggregate grading. The relevant data amounted to: 7.25, 9.90, 21.73, 19.08, 21.13, 36.83, 0.0 and 0.0 kg for the heretofore given sieve openings

The specimens were casted in steel moulds 100x100x500 mm. Next, they were stored in the fog room, except for a short period in which the sawing operations were accomplished (including a 4-5 mm wide and 20 mm deep notch at the bottom side of the specimens). Apart from the as-casted specimens, referred to as plain, and indicated by the code letter P, the sawing in vertical direction (=direction of casting) yielded specimens of about half the original size; they are referred to as "sawn vertically", indicated by the code SV . Alternatively, by sawing in a horizontal direction, two half-size specimens were obtained, referred to as "sawn horizontally top" and "sawn horizontally bottom", indicated by the respective codes SHT and SHB.

LOADING SYSTEM, PROGRAM OF MEASUREMENTS AND OTHER OBSERVATIONS

The centrally-notched beams spanned 450 mm, with a distance between the concentrated loads of 150 mm. A monitoring system of 14 gauges was used. Values of loading, crack mouth opening, deflection under the load, position of the crack tip, and six surface strains ahead of the notch were simultaneously recorded and thereupon processed by an EPSON QX-10 computer (Fig.1).

The loading process was controlled by means of the crack opening displacement, recorded on channel "1". Loading was generally performed in a non-monotonic way; i.e. testing encompassed three to ten unloading-reloading cycles. The following loading procedures have been adopted:

- 36 specimens (RefC) have been tested at about 128 days in a soft testing machine, yielding a value of equivalent bending strength f'_f;
- 48 specimens (RefC) have been tested at an approximate age of 128 days in a deformation-controlled mode, providing information on f'_f, fracture energy, surface strains, crack openings, crack tip positions and deflections. Of this group of specimens:
   - 26 were loaded to beyond ultimate, whereupon the load was released and the specimens replaced in the fog room until final test-loading at an age of about 400 days;
   - 22 were test-loaded to fracture.
- 16 specimens (7DD) have been test-loaded at the respective ages of 100, 174 and 405 days to fracture, in a deformation-controlled way.

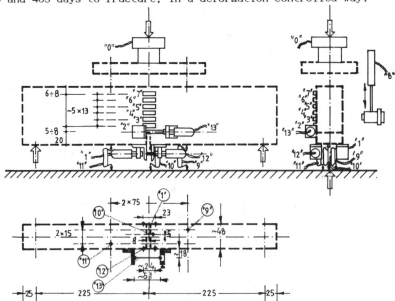

Figure 1. Position of the gauges used for monitoring the behaviour of the beams in flexure. Channels "9" to "11" yield information on deflections, "1", "12" and "13" on the crack mouth opening, "2" to "7" on surface strains ahead of the notch, "8" on the position of the crack tip, and "0" and "15" on the load intensity.

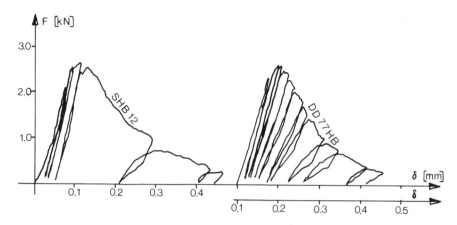

Figure 2. Two examples of load-deflection curves of similar specimens from the SH-category tested at the approximate age of 128 days.

Fig. 2 presents representative examples of the obtained diagrams. The fracture surface energy γ_f is calculated as the area under the envelope curve of the load-displacement diagram divided by twice the ligament cross-sectional area of the beam.

The beam specimens were additionally used to determine the dynamic value of Young's modulus by pulse velocity measurements (on base 100 mm), and the density of the material. The compressive strength was determined on the two broken parts of the beams. Some average data pertaining to the properties of the various categories of specimens are presented in Table 1.

After termination of the tests, the central portion was cut from the specimen and sectioned; a fluorescent spray was used to visualize the microcracks and the macrocrack. Observations were performed under illumination by UV-light. A qualitative assessment of the images showed

TABLE 1

Group averages of various mechanical and other properties of the investigated categories of concrete specimens

		ρ [g/cm^3] avg. (CoV)	E_d' [GPa] avg. (CoV)	$[CMOD]_{ult}$ [mm] avg. (CoV)	f_f' [MPa] avg. (CoV)	f_c' [MPa] avg. (CoV)
P	RefC	2.316 (1%)	49.4 (3%)	0.037 (---)	4.05 (8%)	43.6 (7%)
SV	RefC	2.303 (2%)		0.024 (42%)	3.94 (15%)	27.6 (11%)
SHT	RefC	2.272 (1%)	47.0 (3%)	0.033 (40%)	3.78 (14%)	35.0 (10%)
	7DD	2.201 (1%)	39.1 (3%)	0.019 (38%)	3.28 (12%)	37.5 (3%)
SHB	RefC	2.297 (1%)	48.7 (4%)	0.021 (27%)	4.03 (15%)	38.8 (5%)
	7DD	2.189 (1%)	39.2 (4%)	0.020 (30%)	3.49 (13%)	38.4 (5%)

similar features as found in direct tensile tests [7]. The nonuniform distribution of the stresses over the depth of the beam, however, strongly influences the crack coalescence and succeeding crack concentration processes. The latter leads to the creation of a somewhat meandering and sometimes branching macrocrack. As in the tensile tests, crack growth and crack coalescence proceed in a quasi-stochastic way, also outside the path of the macrocrack. We will report on these aspects of the investigations elsewhere.

FRACTURE SURFACE ENERGY AND OTHER MECHANICAL DATA

Inherent in mechanical testing is the variation in observations. Scatter decreases for specimens with higher ratio of minimum linear dimensions over maximum grain size; for the RefC mix this ratio is about 3, which is low (too low!) for structure-insensitive properties. Structure sensitive properties require even considerably larger values to suppress experimental scatter to acceptable proportions. This is experimentally confirmed.

TABLE 2

Crack mouth opening displacement and fracture surface energy for RefC specimens tested at about 128 days

Specimen No.	Crack mouth opening displacement [mm]				Fracture surface energy γ_f [N/m]			
	(SHT)	(SHB)	(SV)	(P)	(SHT)	(SHB)	(SV)	(P)
PH011				0.03687				95.87
SHT12	0.04763				80.16			
SHB12		0.02468				83.38		
SHT13	0.01696				45.36			
SHB13		0.01899				67.15		
SVA18			0.02519				81.80	
SVB18			0.04519				80.78	
SHB21		0.02468				70.06		
SVA22			0.02326				59.15	
SVB22			0.01635				44.37	
SVA25			0.01848				68.22	
SHB27		0.02488				68.45		
SHB36		0.01280				37.58		
SHB38		0.02173				71.84		
SHT38	0.02864				65.62			
SHB44		0.02773				48.35		
SHB48		0.01777				55.62		
SHT48	0.03940				59.73			
SVB51			0.01727				48.47	
SHB53		0.01016				67.64		
SVA55			0.02153				56.91	
SHB67		0.02163				66.75		
Average:	0.03316	0.02051	0.02390		62.72	63.68	62.81	95.87
C.o.Var.:	40.1%	27.5%	41.5%		23.0%	20.5%	23.5%	----

Tables 2 and 3 present a survey of the fracture energy data of 22 'RefC' and 16 '7DD' specimens loaded as shown in Fig.1. Structural variation is obvious. Nevertheless, average values for crack mouth opening displacement for SHT and SHB specimens (RefC type), reveal segregation of the structural components. This effect is not manifested by the more homogeneous, fine-grained mix (Table 3), however. Fracture behaviour of SV-beams will be governed predominantly by the bottom side, so that only a modest increase in crack mouth opening displacement in comparison to the SHB-average value can be expected. The experiments display such characteristics, although the differences are not significant.

TABLE 3

Crack mouth opening displacements and fracture surface energy for 7DD specimens tested at 100, 174 and 405 days

Specimen No.	Crack mouth opening displacement [mm]		Fracture surface energy γ_f [N/m]	
	(SHT)	(SHB)	(SHT)	(SHB)
DD71HT	0.01208		48.96	
DD71HB		0.02072		46.34
DD72HT	0.01131		49.94	
DD72HB		0.02026		49.46
DD73HT	0.02742		61.05	
DD73HB		0.01767		49.31
DD74HT	0.02529		54.76	
DD74HB		0.00930		57.67
DD75HT	0.00934		50.90	
DD75HB		0.01981		50.93
DD76HT	0.01828		62.34	
DD76HB		0.02153		69.35
DD77HT	0.02509		65.92	
DD77HB		0.01767		53.97
DD78HT	0.02255		54.36	
DD78HB		0.03098		54.69
Average:	0.01892	0.01974	56.03	53.97
C.o.Var.:	38.0%	30.2%	11.3%	13.3%

Differences between the average values of the fracture surface energy of the P, SHT, SHB, and SV categories are insignificant. Due to segregation in coarse-grained mixes, the quality of concrete is lowest at the top part of P-specimens (see Table 1). As a result, the resistance to crack opening will be relatively low in SHT specimens (Table 2). The higher deformability is however accompanied by a lower stress transfer capacity, so that energy consumption in the two types of SH-specimens turned out to be quite similar. Obviously, the fine-grained material manifests no segregation. It should be noted here, that such trends cannot be demonstrated, of course, on the basis of individual results (or specimens), since mechanical data and fracture mechanical properties are determined on material volumes having a different material structure.

In spite of the scatter a non-linear correlation can be observed in a graph relating the crack mouth opening displacement at ultimate load to the fracture surface energy, as shown in Fig. 3. This tendency could justify the application to concrete of the COD concept. More tests have to be performed on different types of concrete to check this statement, however.

The changes in the fracture surface energy with the age of concrete are displayed in Fig. 4. These changes resemble those in the compressive strength due to age of the material, an effect well-known in concrete technology.

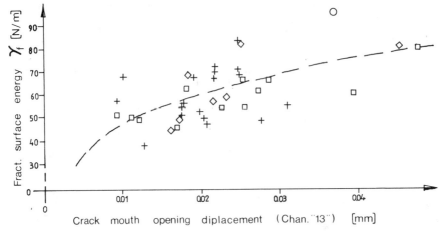

Figure 3. Dependence of the fracture surface energy γ_f on the crack mouth opening displacement at ultimate loading

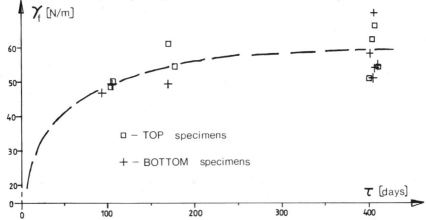

Figure 4. Changes of the fracture surface energy in time. Experimental results obtained on 16 specimens of the 7DD series are presented.

EFFECT OF CRACK PROPAGATION ON COMPLIANCE

The unloading-reloading loops in the load-deflection diagrams allowed to calculate the compliance of the beam during crack propagation. These results were compared with the outcomes of a FEM analysis with the STRUDL program, in which material behaviour is assumed to be fully elastic. A value of 46 GPa was adopted for Young's modulus of concrete, which is slightly more than the experimental value of 43 GPa.

The agreement between the experimental data and the results of the FEM approach is quite satisfactorily, as can be seen in Fig. 5. The latter overestimates the compliance for small crack extension and underestimates it for larger cracks. The first effect is due to growth and coalescence of cracks outside the major crack path. The second effect is the results of load transfer due to crack meandering and branching (particle interlock).

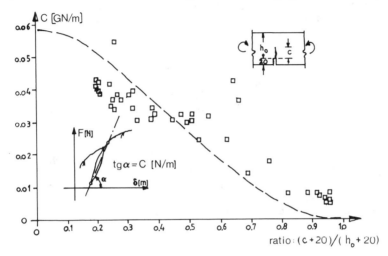

Figure 5. Relationship between the cracked fraction of the beam and its compliance. The definition of the compliance is given in the insert on the lower left hand side, showing part of the load-displacement (F-δ) diagram, while linear dimensions in the cracked area are defined in the insert at the top right. The dashed curve is obtained by a FEM analysis with the STRUDL program (E_d = 46 GPa).

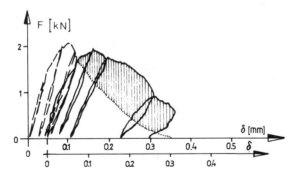

Figure 6. Example of the self-healing effect. The load-deflection diagrams are from specimen SHT21 (RefC). The specimen was first loaded beyond ultimate at 123 days (dashed line at the left). It was reloaded at an age of about 400 days after storage in the fog room (continuous line at the right). The dotted line corresponds to the predicted behaviour of the specimen at 123 days, should the test not have been prematurely terminated.

SELF-HEALING OF CONCRETE

Fig. 6 displays the load-deflection diagrams of a single specimen, loaded and unloaded after 123 days, stored in the fog room, and reloaded after more than one year. The second curve is shifted to the right to simulate a continuous experiment of loading, unloading and reloading. The continuous line represent the observed behaviour at the second loading, the dotted part - the expected behaviour of the beam when loaded to its ultimate deformational capacity at the first trial. The shadowed area between both curves corresponds to the self-healing effect of the concrete. The energy gain due to the self-healing effect was shown to be of the same order of magnitude as γ_f at about three months of age (!).

CONCLUSIONS

From the obtained test results, of which only a minor portion could be presented, it is hardly possible to derive meaningful correlations between fracture mechanics parameters such as γ_f, but supposedly also for K_{Ic}, and mechanical properties, like E_d, f'_f, f'_c, and ρ of single specimens. This is due to the natural structural variation of the material and the size limitations of the specimens. Since such correlations can not be obtained for similar material volumes, because of the destructive character of the approach, it is only sensible to correlate such properties of "representative" material volumes. For structure-sensitive properties this requires the handling and testing of very large specimens. This is, at least, unpractical. Therefore large numbers of "identical" - smaller - specimens have to be used to yield reliable group averages, as in the present tests.

Segregation and anisotropy effects on γ_f could not be observed, although segregation effects were detected in crack mouth opening displacements at ultimate loading for the coarse-grained mix. The observed values of γ_f were on the average 65 N/m for these mixes, ranging from 38 to 95 N/m, with a 22.6% coefficient of variation. The values for the fine-grained mixes were slightly lower (with an average value of about 55 N/m). f'_f values were nevertheless quite similar in both test setups (Table 1), so that such differences should be attributed to differences in the loading procedure (e.g. loading rate).

It was shown that γ_f and the compressive strength of concrete show a similar improvement with age. A significant contribution of self healing to fracture resistance could be detected by continuing the test after restoring for more than one year in the fog room.

The unloading-reloading cycles allowed calculation of the compliance as a function of crack extent in the bent beam. A comparison of such data with the outcomes of a FEM analysis revealed the cracked section of the beam to underestimate the actual extent of the damage. It is nowadays indeed recognized that the formation of a macrocrack is proceeded and accompanied by the growth and coalescence of microcracks (see, eg. [7]). Moreover, the origin of a macrocrack from coalescing microcracks, that are predominantly situated in interfacial areas, leads to a meandering path of the macrocrack. In a more advanced state of cracking it is also possible to observe branching in the macrocrack and/or a significant deviation of the crack path from the assumed cracked cross-section. Such features in crack development favour, on the other hand, the transfer of stresses in the cracked cross-section. As a result the FEM approach underestimates the load bearing capacity of the yielding beam.

ACKNOWLEDGEMENTS

The tests have been performed at the Stevin Laboratory of the Civil Engineering Department, Section of Material Science, Delft University of Technology (DUT). The tests were executed during a stay of the first author in Delft, made possible by a research fellowship granted by DUT. The tests form part of a five-year's co-operation program between DUT and IPPT PAN, the home institute of the first author.

REFERENCES

1. Kasperkiewicz, J., Dalhuisen, D. and Stroeven P., Structural effects in the fracture of concrete. In Brittle Matrix Composites 1, ed-s A.M. Brandt and I.H. Marshall, Elsevier Applied Science Publishers, London 1986, pp. 537–48.

2. Kasperkiewicz, J. and Dalhuisen D., Crack propagation in plain concrete beams subjected to bending. Report Stevin Laboratory 1-86-10, Delft University of Technology, Delft, May 1986.

3. Dutron P., Mise au point d'une composition de béton de référence pour recherches et essais en laboratoire. Mat. et Constr., 1974, 7, 207–24.

4. Hillerborg A., Results of three comparative tests series for determining the fracture energy Gf of concrete. Mat. et Constr., 1986, 19, 407–13.

5. RILEM Draft Recommendation, Determination of the fracture energy of mortar and concrete by means of three-point bend tests on notched beams. Mat. et Constr., 1985, 18, 285–90.

6. Kasperkiewicz, J., Dalhuisen, D. and Stroeven, P., Fracture mechanics tests of reference concretes in pure bending. Report Stevin Laboratory, Delft University of Technology, Delft (to be published).

7. Stroeven, P., Characterization of microracking in concrete. In: RILEM Conference: Cracking and durability of concrete, Saint Rémy lès Chevreuse, Aug/Sept, 1988 (to be published).

MIX PROPORTIONING AND PROPERTIES OF EPOXY-MODIFIED MORTARS

YOSHIHIKO OHAMA*, KATSUNORI DEMURA* AND TAKAYUKI OGI**
* College of Engineering, Nihon University,
Koriyama, Fukushima-ken, 963 Japan
** Hasegawa Komuten Co., Ltd., Tokyo, 105 Japan

ABSTRACT

This paper deals with the selection of the optimum mix proportions of epoxy-modified mortars prepared using a bisphenol A-type epoxy resin in the same method as conventional cement mortar, and the properties of the epoxy-modified mortars with the optimum mix proportions. For the purpose of examining the mix proportioning, the epoxy-modified mortars using three hardener systems, a high range water-reducing agent and four nonionic surfactants as dispersants and an antifoamer are prepared with various hardener contents, dispersant contents and polymer-cement ratios, and tested for flexural and compressive strengths. The optimum mix proportions of the epoxy-modified mortars which are obtained from the test results are as follows: modified polyamide-amine hardener content 55% (of epoxy resin), water-reducing agent content 3.0% (of epoxy resin), nonionic surfactant content 7.5% (of epoxy resin), antifoamer content 1.0 % (of epoxy resin), ratio of cement to sand 1:3 and polymer-cement ratios 30 to 60 %. The epoxy-modified mortars with the optimum mix proportions were tested for adhesion, waterproofness, carbonation, chloride ion penetration, drying shrinkage and chemical resistance. The effect of the polymer-cement ratio on the properties of the epoxy-modified mortars is discussed. In conclusion, the properties of the epoxy-modified mortars are greatly improved with an increase in the polymer-cement ratio except for the drying shrinkage.

INTRODUCTION

Epoxy resin has superior properties such as high adhesion and anticorrosion, and has widely been used as adhesives and anticorrosives in the construction industry in the world. Provided the incorporation of the epoxy resin into cement mortar can give its superior properties to the mortar, it is possible to produce a highly polymer-modified cement mortar. The first patent of an epoxy-modified cement system was taken by Donnelly in 1965 [1]. Since the patent, 30 or more papers on the epoxy resin modification of the cement mortar and concrete have already been published [2]. Most epoxy resin-based cement modifiers dealt with in the papers are specially compounded by the manufacturers, and the procedures for mixing them to fresh mortar and concrete are considerably complicated. The objective of this study is to solve such problems. Accordingly, in the study all the materials used are commercially available, and mixed in the same method as conventional cement mortar or latex-modified mortar without any special or complicated mixing procedures.

This paper consists of two parts as follows:(1) The selection of the optimum mix proportions of the epoxy-modified mortars prepared without the use of specially compounded epoxy resin-based cement modifiers in the same manner as conventional cement mortar, and (2) The investigation concerning the properties of the epoxy-modified mortars with the optimum mix proportions.

MATERIALS

Cement and Aggregate
The ordinary portland cement and Toyoura standard sand as specified in JIS(Japanese Industrial Standard) were used in all the mix proportions of mortars.

Cement Modifiers
Commercial cement modifiers used were based on a bisphenol A-type epoxy resin, diglycidyl ether of bisphenol A (DGEBA), which was mixed with three types of hardeners, modified polyamide-amine (MPAA),modified aliphatic polyamine (MAPA) and modified amine (MA) at the hardener contents recommended by the respective manufacturers. The hardener content was expressed as follows:

Hardener content (wt%) = [hardener/ (DGEBA + hardener)] x100

The chemical structure of the epoxy resin is shown in Figure 1.

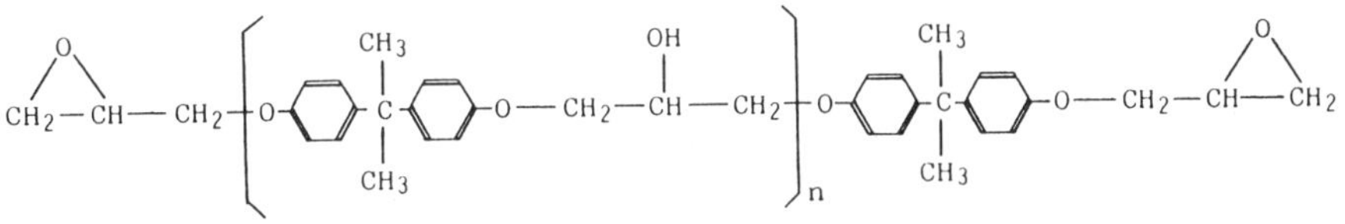

Figure 1. Chemical structure of epoxy resin.

Admixtures
Commercial admixtures employed were a polyalkyl aryl sulfonate-type high range water-reducing agent, four polyoxyethylene nonylphenol ether-type nonionic surfactants with HLB (hydrophile-liophile balance) values of 9.5,10.9,12.6 and 13.3, and a silicone emulsion-type antifoamer. The water-reducing agent and nonionic surfactants were used as dispersants. The content of the admixtures was expressed as wt% of DGEBA.

SELECTION OF OPTIMUM MIX PROPORTIONS

Testing Procedures
Preparation of specimens :In accordance with JIS A 1171(Method of Making Test Sample of Polymer-Modified Mortar in the Laboratory), epoxy

modified mortars were prepared with a cement : sand ratio of 1 : 3 (by weight) and polymer-cement ratios [calculated as (DGEBA +hardener)/cement ratios by weight] of 0,30,40 and 50%, and their flows were adjusted to be constant at 160±5. In this case, the premixtures of DGEBA, hardeners and desired admixtures were mixed with the required amount of the mixing water to a dry blend of the cement and sand. The mix proportions of the mortars are given in Table 1 . Mortar specimens 40x40x160mm were molded, and then subjected to a 2-day-20°C-80%R.H.-moist plus 5-day-20°C-50%R.H.-dry cure.

Strength test: The cured mortar specimens were tested for flexural and compressive strengths according to JIS A 1171.

TABLE 1
Mix proportions of epoxy-modified mortars

Cement :Sand (By Weight)	Type of Hardener	Hardener Content (%)	Water-Reducing Agent Content (%)	HLB of Nonionic Surfactant	Nonionic Surfactant Content (%)	Antifoamer Content (%)	Polymer-Cement Ratio (%)	Water-Cement Ratio (%)	Flow
1:3	-	-	0				0	65.5	160
							40	56.0	155
				-	0	0	30	56.2	160
							40	52.8	156
							50	49.2	156
					1.0			55.2	160
						1.0		56.0	162
				9.5	3.0	0		51.0	160
						1.0		56.0	160
					7.5	0		45.0	156
						1.0		54.0	162
					15.0	0		41.5	160
						1.0		51.5	164
	MPAA	55	3.0		1.0		40	55.2	163
				10.9	3.0			51.0	160
					7.5			45.0	162
					15.0			41.0	164
					1.0			55.0	161
				12.6	3.0			51.0	159
					7.5			45.0	158
					15.0			41.0	160
					1.0			52.0	156
				13.3	3.0			47.0	164
					7.5			44.0	162
					15.0			40.5	159
			5.0			0	30	51.0	156
							40	49.3	160
							50	48.0	155
			8.0				30	50.0	165
				-	0		40	48.0	162
							50	43.0	160
	MAPA	60	3.0				30	60.0	163
							40	53.6	162
							50	51.4	163
	MA	50					30	65.4	165
							40	61.0	165
							50	55.8	160

Test Results and Discussion

Figure 2 represents the effects of type of hardener and polymer-cement ratio on the flexural and compressive strengths of epoxy-modified mortars. The flexural and compressive strengths of the epoxy-modified mortars are considerably affected by the type of hardener and polymer-cement ratio. Generally the epoxy-modified mortars with a modified polyamide-amine hardener provide superior flexural and compressive strengths, which are about 65 kgf/cm^2 and 300 kgf/cm^{2}respectively at a polymer-cement ratio of 50 %. Therefore, the modified polyamide-amine hardener is recommended as a hardener for the epoxy-modified mortars. This agrees with the results of Popovics's study [3].

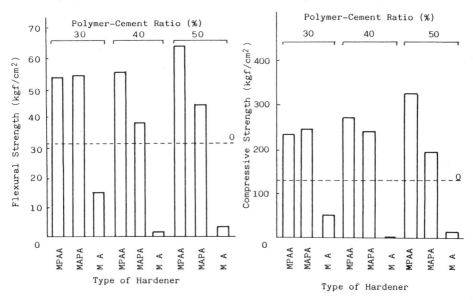

Figure 2. Effects of type of hardener and polymer-cement ratio on flexural and compressive strengths of epoxy-modified mortars.

Figure 3 illustrates the effects of water-reducing agent content and polymer-cement ratio on the flexural and compressive strengths of epoxy-modified mortars with the modified polyamide-amine hardener(hardener content, 55%) recommended from the above test results. Irrespective of the polymer-cement ratio, the flexural and compressive strengths of the epoxy-modified mortars increase with raising water-reducing agent content, and reach the maximum at a water-reducing agent content of 3.0 %. In general, the flexural and compressive strengths increase with an increase in the polymer-cement ratio. The highest flexural and compressive strengths are obtained at a water-reducing agent content of 3.0 % and a polymer-cement ratio of 50%, and are about 60 kgf/cm^2 and 300 kgf/cm^2 respectively. Such effectiveness of the water-reducing agent also agrees with Popovics's results[3]. Accordingly a water-reducing agent content of 3.0 % is recommended for the mix proportions of the epoxy-modified mortars.

Figure 4 exhibits the effects of nonionic surfactant content, HLB and antifoamer on the flexural and compressive strengths of epoxy-modified mortars with a modified polyamide-amine hardener content of 55%, a polymer-cement ratio of 40% and a water-reducing agent content of 3.0 %, which are the optimum values recommended from the above test results. The flexural strength of the epoxy-modified mortars with a nonionic surfactant of HLB 9.5

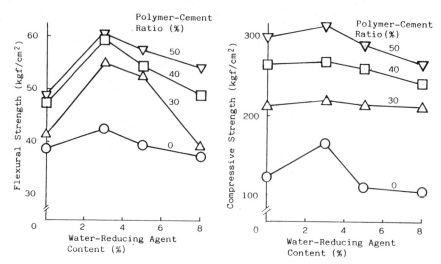

Figure 3. Effects of water-reducing agent content and polymer-cement ratio on flexural and compressive strengths of epoxy-modified mortars.

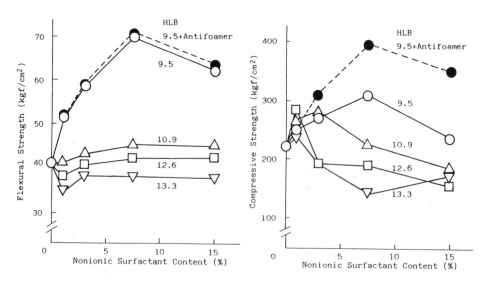

Figure 4. Effects of nonionic surfactant content, HLB and antifoamer on flexural and compressive strengths of epoxy-modified mortars.

increases with an increase in the nonionic surfactant content, and attains to the maximum of about 70 kgf/cm^2 at a nonionic surfactant content of 7.5 %. The effect of the antifoamer on the flexural strength is not remarkable. The epoxy-modified mortars with nonionic surfactants of HLB 10.9 to 13.3

provide nearly constant flexural strength regardless of the nonionic surfactant content. Generally the compressive strength of the epoxy-modified mortars increases with a raise in the nonionic surfactant content, and reaches the maximum, depending on their HLB. Decreasing HLB tends to provide the maximum compressive strength at higher nonionic surfactant content, and the maximum at smaller HLB is higher than that at larger HLB. In particular, the epoxy-modified mortar with an HLB-9.5-nonionic surfactant content of 7.5% gives the highest compressive strength of about 300 kgf/cm^2. In addition, the effect of the antifoamer on the compressive strength is markedly recognized, and the addition of the antifoamer causes a 33 % increase in the compressive strength. Appropriate nonionic surfactant content and HLB of the surfactant for the mix proportions of the epoxy-modified mortars are 7.5 % and 9.5 respectively.

From the above test results, the optimum mix proportions of epoxy-modified mortars are recommended as shown in Table 2 .

TABLE 2
Optimum mix proportions of epoxy-modified mortars

Cement: Sand (By Weight)	Type of Hardener	Hardener Content (%)	Water-Reducing Agent Content (%)	HLB of Nonionic Surfactant	Nonionic Surfactant Content (%)	Antifoamer Content (%)	Polymer-Cement Ratio (%)	Water-Cement Ratio (%)	Flow
	–	–	0	–	0	0	0	65.5	160
							30	55.0	165
1 : 3	MPAA	55	3.0	9.5	7.5	1.0	40	54.0	162
							50	50.0	164
							60	47.5	161

PROPERTIES

Testing Procedures

Preparation of specimens: According to JIS A 1171, epoxy-modified mortars were prepared with the mix proportions as shown in Table 2. Mortar specimens with the geometry indicated in Table 3 were molded, and then given a 2-day-20°C-80%R.H.-moist plus 5-day-20°C-water plus 21-day-20°C-50%R.H.-dry cure.

TABLE 3
Geometry of specimens

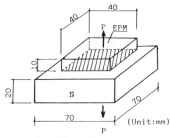

Type of Test	Shape	Geometry of Specimen (mm)
Compressive Strength	Cylinder	50x100
Adhesion in Tension	–	(See left figure)
Water Absorption	Beam	40x40x160
Water Permeation	Disk	150x40
Carbonation	Beam	100x100x400
Chloride Ion Penetration	Cube	100x100x100
Drying Shrinkage	Beam	40x40x160
Chemical Resistance	Cube	40x40x40

S : Substrate
EPM : Epoxy-Modified Mortar
▨ : Bonding Joint

Various tests: Cylindrical mortar specimens 50x100mm were tested for compressive strength according to JIS A 1108 (Method of Test for Compressive Strength of Concrete), and the compressive stress-strain curves were automatically recorded. Mortar specimens, specially shaped ones, beams 40x40x160mm and disks 150x40mm were tested for adhesion in tension, water absorption and drying shrinkage, and water permeation, respectively, in accordance with JIS A 6203 (Polymer Dispersions for Cement Modifiers). Beam mortar specimens 100x100x400mm were tested for carbonation through exposure in air with a CO_2 concentration of 5% at 20°C and 60%R.H. for 182 days. Cube mortar specimens 100x100x100mm were tested for chloride ion (Cl^-) penetration through immersion in saturated NaCl solution at 20 °C for 91 days. Cube specimens 40x40x40mm were tested for chemical resistance through immersion in various test solutions at 20°C for 28 days.

Test Results and Discussion

Figure 5 represents the compressive stress-strain curves for epoxy-modified mortars. The compressive strength of the epoxy-modified mortars and their extensibility estimated from the longitudinal and lateral strains increase remarkably with an increase in the polymer-cement ratio. Both longitudinal and lateral strains of the epoxy-modified mortar with a polymer-cement ratio of 60% attain to about 2.5 times those of unmodified mortar. Such large extensibility in compression is reported by Nawy et al[4]. The Poisson's ratio of the epoxy-modified mortars increases gradually with raising polymer-cement ratio.

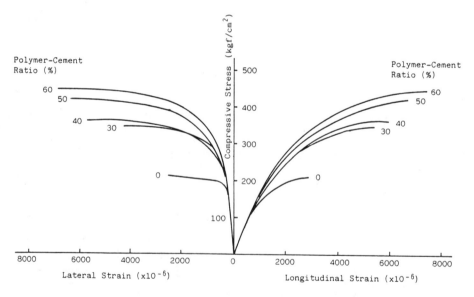

Figure 5. Compressive stress-strain curves for epoxy-modified mortars.

Figure 6 indicates the adhesion in tension of epoxy-modified mortars to ordinary cement mortar before and after warm-cool cyclings. When the polymer-cement ratio increases to 30 %, the adhesion of the epoxy-modified mortars before warm-cool cycling is about twice that of unmodified mortar, and becomes nearly constant, depending on the failure mode. Like the adhesion behavior before warm-cool cycling, when the polymer-cement ratio rises to 30 %, the adhesion of the epoxy-modified mortars after warm-cool cycling is about 3 times that of the unmodified mortar, and becomes nearly constant, depending on the failure mode.   Regardless of the polymer-cement ratio, the adhesion after warm-cool cycling is reduced by about 25 % of that before warm-cool cycling.

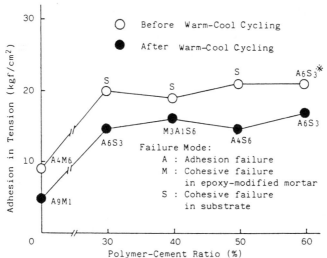

Note: ※ ; The respective approximate rates of A,M and S areas
in the total area of 10 on the failed crosssections
are expressed as suffixes for A,M and S.

Figure 6. Adhesion of epoxy-modified mortars before and after warm-cool
cyclings.

Figure 7 shows the effect of polymer-cement ratio on the water absorption and permeation, carbonation and chloride ion penetration of epoxy-modified mortars. The water absorption of the epoxy-modified mortars decreases with increasing polymer-cement ratio, and becomes about one-third of that of unmodified mortar at polymer-cement ratios of 50 and 60%. As the polymer-cement ratio increases to 30%, the water permeation of the epoxy-modified mortars provides about one-tenth of that of the unmodified mortar, and then becomes nearly constant.   From Figure 7 , it is concluded that their waterproofness is excellent. As the polymer-cement ratio rises to 30 %, the carbonation depth of the epoxy-modified mortars decreases sharply to about one-fourth of that of the unmodified mortar, and then becomes nearly constant. When the polymer-cement ratio increases to 30 %, the chloride ion penetration depth of the epoxy-modified mortars is reduced to about a half of that of the unmodified mortar, and then becomes nearly constant.

Figure 8 gives the relation between the dry curing period and drying shrinkage of epoxy-modified mortars.   The drying shrinkage of the epoxy-modified mortars increases with additional dry curing period, and becomes nearly constant at an age of 28 days.   The drying shrinkage tends to increase with an increase in the polymer-cement ratio, and there is little difference in the drying shrinkage between the polymer-cement ratios of 40, 50 and 60 % at an age of 28 days.

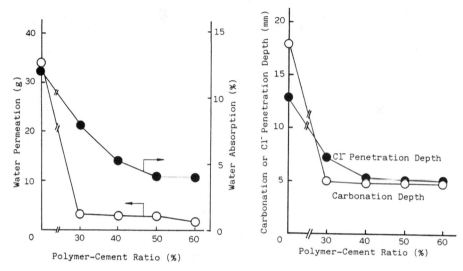

Figure 7. Effect of polymer-cement ratio on water absorption and permeation, carbonation and chloride ion penetration of epoxy-modified mortars.

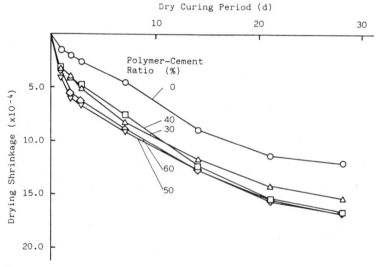

Figure 8. Dry curing period vs. drying shrinkage of epoxy-modified mortars.

Figure 9 exhibits the weight change of epoxy-modified mortars immersed in various test solutions for 28 days. It appears that the chemical resistance of the epoxy-modified mortars is superior to that of unmodified mortar because of the excellent chemical resistance of the epoxy resin used.

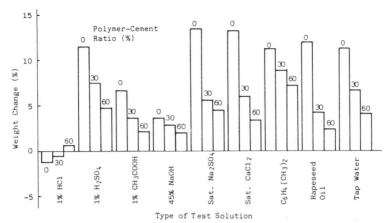

Figure 9. Weight change of epoxy-modified mortars immersed in various test solutions for 28 days.

CONCLUSIONS

(1)Epoxy-modified mortars which are prepared by mixing most popular bisphenol A-type epoxy resin with commercial modified polyamide-amine hardener, polyalkyl aryl sulfonate-type water-reducing agent, polyoxyethylene nonylphenol ether-type nonionic surfactant and silicone emulsion-type antifoamer into cement mortar have excellent properties comparable to ordinary polymer-modified mortars using latex-type cement modifiers[5]. The optimum mix proportions of the epoxy-modified mortars are shown in Table 2. Their disadvantage is a need of much higher polymer-cement ratio than the ordinary polymer-modified mortars. Therefore, the development of low-cost, effective dispersants is expected in the near future.
(2)As polymer-cement ratio increases from 30 to 60%, generally the extensibility, strength and chemical resistance of epoxy-modified mortars are considerably improved, but their adhesion, waterproofness, carbonation depth, chloride ion penetration depth and drying shrinkage are nearly constant. Of the improved properties, the large extensibility is very interesting.

REFERENCES

1. Donnelly, J.H., Inorganic cement-epoxy resin composition containing animal glue. USP 3198758, 1965.
2. Ohama,Y., Bibliography on Polymers in Concrete, International Congress on Polymers in Concrete,1987.
3. Popovics, S., Modification of portland cement concrete with epoxy as admixture. In Polymer Concrete:Uses, Materials, and Properties SP-89, ed., J.T. Dikeou and D.W. Fowler, American Concrete Institute, Detroit, 1985, pp.207-229.
4. Nawy, E.G., Ukadike, M.M. and Sauer, J.A., High-strength field polymer modified concretes. Journal of the Structural Division, Proceedings of the American Society of Civil Engineers, 1977,103(ST12), 2307-2322.
5. Ohama, Y.,Chapter 7. Polymer-modified mortars and concretes. In Concrete Admixtures Handbook: Properties, Science and Technology,ed.,V.S. Ramachandran, Noyes Publications, Park Ridge, New Jersey, 1984, pp.337-429.

MEANING OF SYNERGY EFFECTS
IN COMPOSITE MATERIALS AND STRUCTURES

LECH CZARNECKI

Civil Engineering Dept.

Warsaw Technical University

Al. Armii Ludowej 16, 00-637 Warszawa, Poland

VLADIMIR WEISS

Civil Engineering Faculty

Czech Technical University

Thakurova 7; 166 29 Prague, Czechoslovakia

ABSTRACT

Synergism - in the lexicon meaning - the simultaneous action of agencies (or components) which, together, have greater total effect than the sum of their individual effects. In the engineering - the synergism means very often some type of "puzzle". There are "uncertainty"in manufacturing and exploitation as well as something "unexpected" in engineering design. Synergistic effects as the research and technical problem have been generally characterized in the paper. The development of the synergy meaning and understanding from the historical point of view has been discussed. Several examples of synergy phenomena in material engineering have been described.

INTRODUCTION

Any new synergy phenomena discovered by experience or test is unexpected and means initially always a puzzle or may be evaluated as an error. The confusion is usually even greater for the reason that the understanding of the synergy is having more than just one meaning.

The aim of presented contribution is to describe the use of this word from the historical viewpoint in various fields and then to give an answer to some fundamental questions:

- how to define the synergic effects, what is a synergy phenomenon and what is not,
- what kind of effects can be expected in composite materials and structures,

and

- how to use synergic effects (possibilities and limitations).

THE STATE OF ART

The word "synergy" is of greek origin and was already used e.g. by Demosthenes (350 BC) in three senses, namely "help in the work", "cooperation in proceeding" or "participation" [1]. Later on, the word "synergy" (or "synergism") was borrowed by Latin and, subsequently, by many other languages and was then used under various, mostly very specific meanings, as stated in encyklopaediae and lexicons.

Probably the oldest of the mentioned meanings is that in the sphere of theology and denotes a doctrine that in salvation is needed a co-operation of divine grace and human activity (Webster's New Dictionary 1981). In a similar sense, these terms were rather later applied as the main principle of the happening to metaphysics, ethics and sociology: co-operation of the man with the universe, with the world and with the society. The notion of synergy is presented also for the region of natural sciences in majority of lexicons on two, exceptionally on three examples. First, in physiology, it is a muscle functioning in co-operation with another one to produce a specific movement. Second in pharmacy and medicine, synergy is defined as follows: A remedy acting similarly to another remedy and increasing its efficiency when combined with it. The third discovered example refers to biochemistry: The increased effect of antioxidants, due to the simultaneous action of so-called synergists.

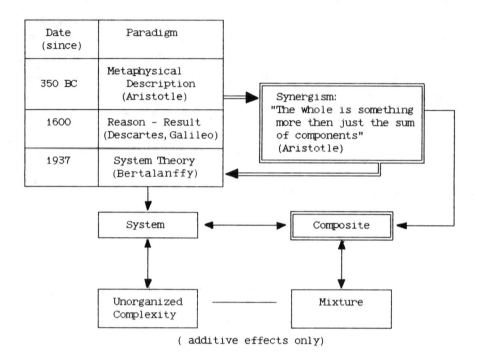

Figure 1. The synergism as a "bridge" between two paradigms in scientific workshops. "System" and "Composite" as the derivatives of the synergism. "Mixture" and "Unorganized complexity as the antonymus

The synergy can be derived from Aristotle's dictum: "The whole is more then just the sum of components".This definition could be treated as the "bridge" between the metaphysical description - the old paradigm dominating in the scientific workshops since Aristotle (350 BC) and the system theory as the contemporary paradigm (Fig. 1). The Aristotle's dictum can be also written as the definition for a composite:

$$Y(A, B) = Y(A) + Y(B) + Y(A) * Y(B) + \ldots,$$

mechanism additive   synergistic effect

where

Y(A,B), Y(A), Y(B) - given property of the composite and its components, respectively,

A, B - contents (vol.) of suitable components.

In contrary to the composite the properties of "mixture" are just the sum of the properties of their components, according to the"rule of mixtures":

$$Y(A;B) = Y(A) + Y(B),$$

where

Y(A;B) - given property of mixture.

DEFINITION OF SYNERGY ENGINEERING SCIENCES

The general definition of synergy for engineering science could run as follows: co-operative simultaneous action of discrete agencies such that their total effect is greater than the sum of individual effects taken independently, under the condition that all these agencies lead to a final effect of the same kind.

Nevertheless, one limitation of this general definition is necessary mostly in the field of structural engineering, for co-operation of components. Higher effects are here achieved very often only by their complementary functions: e.g. co-operation of reinforcement with concrete in reinforced or prestressed elements, stiffening of frames by inserted panels against effects of horizontal loads, joining of components in composite structures and in sandwich elements or even co-operation of struts in lattice structures. To avoid confusion in this field, it is necessary to classify all these and similar examples as trivial applications of the term synergy and to narrow this notion only for such cases where the higher final effect shown by a system of components is achieved by changes in properties or in behaviour at least of one component due to interaction with at least one an other of them.

It is important for all acting agencies to have the final effect of the same kind; for the opposite case, i.e. if one agency itself cannot produce such an effect but can intensify the effect of other agencies, the pertinent expression is stimulation.

The difference between the notions of synergy and stimulation is more clarified on several examples taken from the field of material engineering. The well-known term "stress-corrosion" where mechanical stress accelerates the progress of corrosion but it does not excite this chemical process by itself is a good example for stimulation. Nevertheless, if the fracture of the body is chosen as criterion, then both chemical agents and also stress itself can lead to destruction of the integrity of the body, and, therefore, in cases of overpassing the"rule of mixtures", it is a good example of synergy.

Accelerated processes of ageing or degradation of materials where the deterioration of the material represents the measure of their effects are mostly synergic. It is possible to find many examples of the combined action of aggressive liquids together with y-rays, that of heat with humidity or with stress [2], that of aggressive environments with temperature [3], etc. The earlier loss of passivation of steel by carbonating concrete in the presence of chloride ions has synergic character too (Fig. 2).

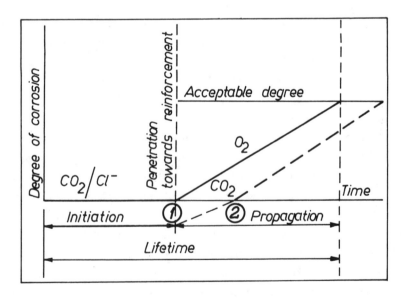

Figure 2. Model of the corrosion process of steel in concrete: (1) - in the presence of chloride (critical value pH=9), (2) - no chlorides (ph=11); pH ≥ 13.5 metastate state.

Two examples are presented for phenomena of stimulation: first, accelerating influence of air streaming on the destruction of plastics due to the action of oxygen [2], second, increase in velocity of diffusion and sorption of liquids in porous solids subjected to stress [4].

BACKGROUND OF SYNERGY PHENOMENA

Investigation of observed synergies only in a phenomenological way is usually unsatisfactory because it does not allow reliably to forecast new results and therefore less to make extrapolations beyond the hitherto existing limits. This, it is always necessary to apply a heuristic approach to the explanation of each synergy the cause of which consists in other processes or on other level then usually followed for its description.

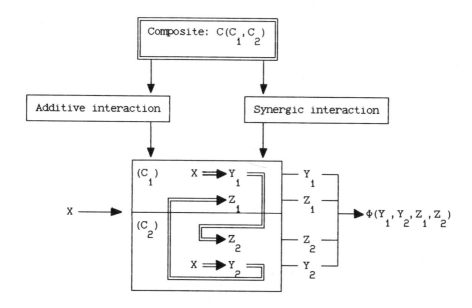

Figure 3. Schematic presentation of the composite (C) properties as the superposition of additive - and synergic ($C_1 \rightleftarrows C_2$) interaction between two components (C_1, C_2): X-external loads, Y_1, Y_2, Z_1, Z_2 -intrinsic material response, Y, Z, Φ - exterior material response - material properties.

The heuristic approach is now demonstrated by several examples. The properties of the given composite materials (Fig. 3) depend on fractional volume and bulk properties according to the additive mechanism as well as on component surface and interfacial interaction - synergic mechanism [5]. In this meaning a microfiller is acting almost solely according to synergy way. Synergic effects of UV-radiation with mechanical stress, if need be with aggressive chemical agents, on the durability of polymers are based on the process of rupture and recombination of chemical bonds when recombinations under stress are not possible [6]. Thermal fluctuaction movements of particles and chains mean a further source of synergy; increase in temperatures causes increase in mobility and desorientation of these chains and therefore, non-linearities in strength relations. Another process facilitating development of cracks and deformations in thermoplastics under a simultaneous action of stress and liquids is the adsorption of these liquids on arising new surfaces of cracks and defects [7]. Overpassing the percolation threshold can be the cause of sudden changes of properties of concretes or other composites as influenced by the production of their structure of individual components [8].

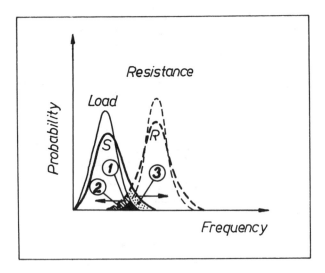

Figure 4. The structural failure as the result of Load - Resistance synergism [12]: 1 - construction development, 2 - Inadequate process technology, 3 - Aggressivity of envizonements.

The significant increase in strength and stiffness of polymer composites with increasing inner surface consists mainly in the improvement of the orientation of molecular chains of the binder on the interface binder - filler [9]. The observed increase in strength and extensibility of brittle or quasi-brittle materials efficiently cooperating with tougher materials is caused by rather great energy absorption in the tougher materials spanning critical structure cracks in the more brittle material, despite of rather low induced interactive forces which could not markedly improve neither the static nor the dynamic equilibrium of these cracks [10,11]. Another example of synergy is that of increased long-term deformations of cement concrete due to simultaneous action of shrinkage and creep. The background of this phenomenon consists predominantly in complex sorption processes of water on submicroscopical structure elements due to changes in relative local humidity influenced both by the action of stress and of environment. Further example the synergy effect on the durability of cement concretes is represented by alkali-silica reaction [12].

In some meaning, the structural failure of constructions can be also considered as the result of synergism - the result of exterior load and material resistance cooperation (Fig. 4).

CONCLUDING REMARKS

Civil Engineers can meet the synergy effects during:
- manufacturing and on-site using of composite materials and managing it by proper material designing,
- making a prognose of material behaviour under given conditions; the synergy as the source of uncertainty,
- exploatation of construction - mostly as the negative processes from the durability point of view which need suitable prevention means.

The limited sources of raw materials and energies together with increasing demands of the mankind need to use advanced materials and structures; the most rational solutions consist in the wide application of composite materials and structures. Nevertheless, a simple use of various mixture rules for composites offers no great advantages in the field; much more effective ways are those discovering and exploiting possible synergy effects: to promote the positive processes and to avoid the negative ones.

534

REFERENCES

1. Pape, W., Griechisch-deutsches Handwörterbuch., Vieweg u. Sohns, Braunschweig, 1958.

2. Benlian, D., Comment est combattu le vieillissement des plastiques., L'usine nouvelle, 1969, No 10, pp. 133-139.

3. Menges, G., Riess, R., Taprogge, R., Verformungsverhalten thermoplastischer Kunststoffe in flüssigen Medien., Materialprüfung, 1972, No 5, pp. 141-146.

4. Pancevskij, D.P., Popova, M.B., Stepanov, P.D., Lenskij, O.P., Influence of the stress - state on diffusion kinetics of inert liquids into polyformaldehyd. (in Russian), Fiziko - Chemicheskaya mechanika materialov, 1973, No 5, pp. 37-40.

5. Czarnecki, L., Introduction of material model of polymer concrete., Proc. of Fourth Int. Congress on Polymers in Concrete., ed., H.Schulz, Darmstadt, 1984, pp. 59-64.

6. Bober, T.B., Regel, V.R., Sanfirova, T.P., Cernyj, N.B., Instigating the influence of UV-radiation on durability of polymers subjected to loading in vacuum and air.(in Russian), Mechanika polimerov, 1968, No 1, pp. 661-664.

7. Tynnyj, A.N., Gural, V.M., Kalinin, N.G., Sosh, A.I., About the stress relaxation of polymethylmetacrylate in liquid media.(in Russian), Fiziko - Chemicheskaya mechanika materialov, 1969, No 3, pp. 376-377.

8. Solomatov, V.I., Byrovoj, V.N., Abbashanov, N.A., Concrete like a composite material., VzNIINTI, Tashkent, 1985.

9. Czarnecki, L., Lach, V., Structure and fracture in polymer concretes: some phenomenological approaches.,Proc. of European Mechanics Colloquium on Brittle Composites -1., ed., A.M. Brandt, I.H. Marshall, Elsevier Appl. Sci. Publishers, London, 1985, pp. 241- 261.

10. Weiss, V., Cracking and failure in materials with closely spaced reiforcement and similar materials.Failure of brittle materials with tougher surface coating. Własności mechaniczne i struktura kompozytów betonowych., ed., A.M. Brandt, Ossolineum, Wrocław - Warszawa - Kraków - Gdańsk, 1974, pp. 501-516(A), pp. 517-525(B).

11. Weiss, V., Cooperation of tougher coating with concrete under different types loading., Proceeding ISAP '86 - RILEM (Aix-en-Provence), Chapman and Hall, London, 1985, pp. 289-296.

12. Idorn, G.M., The concrete future., Danish Concrete Institute, DK 2840 Holte, 1980.

ESTIMATION OF TYPE AND MAGNITUDE OF MACROSTRUCTURE
DEFECTS OF POROUS CONCRETES

V.A. Titov, T.P. Vaschenko, V.P. Varlamov, N.K. Sudina
All-Union Research Institute of Building Materials & Structures
Karl Marx Str. 117, Kraskovo, 140080
Moscow Reg. USSR

ABSTRACT

Herein is contained a proposed method for determining the strength
parameters of porous materials using a flat ended punch. Its relative
efficiency is highlighted.

The mechanical properties of porous concrete depends on macro
structural defects and composition of calcium hydrosilicates. An exact
estimation of these parameters is very important in porous concrete
technology.

Existing methods for measuring macrostructured defectivity usually
consists of visual examination of the geometry and size of porosity.
These methods, along with analysis methods for determining the
composition of calcium hydrosilicates (X-ray analysis, differential
thermal analysis, etc.) do not demonstrate a direct relationship between
macrostructure defectivity and mechanical properties. Thus, all of
these methods can not be used to control technological processes. These
parameters can not be taken into consideration as methods of analysis for
solid phases calls for sophisticated instrumentation and equipment.

Alternatively, existing methods for measurement of the stress and
strain parameters of concrete do not allow determination of the causes of
material failure. Thus it is impossible to outline logical waysof
eliminating the defects of macrostructure and solid phase.

Considerable research is directed towards investigating the
relationship between macrostructure defectivity and mechanical
properties of porous concrete. The authors have carried out an
efficient method for measuring the value and type of macrostructure
defectivity.

The proposed method consists of determining the strength parameters of porous materials under local loading by a punch with a flat working surface. These parameters were measured at different places. The advantages of the proposed methods are :

(i)   it does not require sophisticated instrumentation equipment

(ii)   it allows determination of both real and maximum (which can be reached with a given composition of calcium hydrosilicates) strength of the concrete

(iii)   it allows an estimated value and type of macrostructured defectivity.

The proposed method allows detection of any macrostructure defects of porous concrete, non-homogeneous structure(e.g. some pores of greater diameter and chains of pores), non-optimal porous composition, cracks, etc.

The prototype idea for the present method was used to determine the strength of concrete with the help of a ball press tool. Pressure to the test piece transmitted by the ball and along with the diameter of the imprint gives the strength of the concrete.

Ball press tools produce complex loading of the contact layer since the area of the working zone changes during the loading process. Thus it is difficult to analyze the limit strength of the porous contact layer and it is also impossible to estimate the macrostructure defectivity.

A full description of the present method is presented herein. The test pieces of porous concrete must be cubes of edge length 15-20 cm, the average density of the concrete must be between 500 and 1100 kg/m^3 and the general compressive strength may be 2.5 - 17.5 MPa.

Loading of the test piece is made incrementally at several areas of surface. Local loading is produced by a punch with a flat working surface of 3-9 cm^2. Between 8-12 points must be uniformly distributed along the surface of the test piece.

Concrete behaviour under local loading (produced by a small punch) differs from that of a standard compressive test. There is a thin layer of concrete being worked and thus a small number of structural elements in it. Thus maximum stresses appear primarily in the layer nearest to the punch. At every loading point the value of ultimate stress (σ max) and maximum deformation (Δ max) were determined. The maximum deformation in this case is the punch displacement in the loading direction relative to the unloaded surface. This displacement must be measured when the maximum stress occurs. The value of Δ max is closely linked to the pore size and the strength of between-pore partitions. The authors assumed that the value of Δ max is the thickness of the layer of destructed between-pores partitions. This result was obtained using statistical analysis of experimental data and also data obtained using laser and visual methods. The value of the layer of destructed between-pores partitions is between .1 and .8 mm.

As previously mentioned, the present method did not require sophisticated equipment. It only required :

a)   an appropriate press tool
b)   a cylindrical punch with a flat working surface
c)   any appropriate indicator for measuring displacement (Fig. 1).

The loading was applied incrementally using equal increments (ΔN). The stress in the concrete under the punch at the current stage of loading is N_i/F, where F is the square size of the punch. Displacement of the punch at the current stage is defined as Δi.

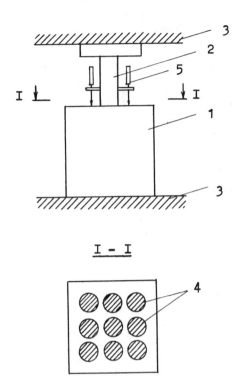

Figure 1. Scheme of the testing process.
1 - test piece of the concrete
2 - cylindrical punch
3 - foundation of the press tool
4 - points of the loading
5 - indicator

It is important to measure the punch displacement relative to the unloaded surface of the test piece. This gives very reliable data for the limit strength of the contact layer. When the stress under the punch reaches its maximum value, it produces the "fluidity status" of the concrete. At this status there exists constant stress under the punch, with increasing punch displacement.

Existing press tools only allow measurement of the load magnitude. Consequently the "status of fluidity" can only be measured using deformation measurement techniques. During testing no destruction of

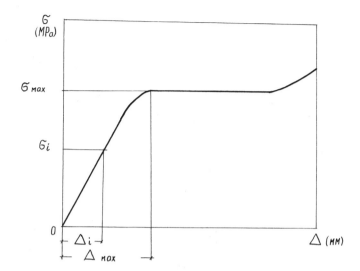

Figure 2. The stresses in the concrete under the punch depending on the punch displacement

test piece takes place, but there is a consolidation of concrete under the punch. The latter takes place, due to the destruction of the between-pores partitions in the concrete under the punch. This can be seen in Fig. 2, as after the horizontal part of the curve an increase is apparent.

It can be stated that the maximum stresses in the concrete under the punch equal the limit strength of the contact layer of the between-pores partitions Rc.

After the previously described testing, the surface of the test piece is levelled out, e.g. by grinding, and then the general compressive strength Rb is determined.

Generally speaking, it can be stated that the local limit strength of the defect less contact layer with homogenous spherical porosity exceeds the general compressive strength by 2-3 times. Such differences appear to be due to various macrostructure defects which tend to decrease the general compressive strength. The probability of these defects increases as the specimen size increases. When the local strength is determined according to the method described herein, the macrostructure defects at the local points of loading are reduced to a minimum or eliminated. Thus, the local strength of the defectless contact layer defines the maximum strength of porous concrete with given composition of hydrosilicates which can be reached (R_c^{max}).

The value of the macrostructure defectivity can be defined by the following parameters :

a) The ratio of limit strength of the contact layer of the between-pores partitions (R_c^{max}) to the general compressive strength Rb. In general, this ratio defines the value of macrostructure defectivity. The ratio R_c^{max}/Rb for defectless macrostructure lies near to unity. For a macrostructure with a large number of defects the ratio is between 1.5 and 3. Thus, it can be concluded that a greater value of this ratio equates to a greater value of macro-

TABLE 1

Strength parameters and features of macrostucture of
the pieces of concrete.

No	Rb (MPa)	Rc (MPa)/Δ(MM)				(Rc/Δ)a	R_c^{max}/Rb	R_c^{max}/R_c^{min}
1	2.39	$\frac{7.8}{0.12}$	$\frac{9.1}{0.14}$	$\frac{4.3}{0.07}$	$\frac{6.8}{0.11}$	64	3.8	2.1
2.	3.30	$\frac{4.6}{0.19}$	$\frac{4.2}{0.18}$	$\frac{4.3}{0.19}$	$\frac{3.9}{0.17}$	23	1.4	1.2
3	5.53	$\frac{7.4}{0.26}$	$\frac{7.2}{0.25}$	$\frac{7.1}{0.25}$	$\frac{7.7}{0.27}$	29	1.4	1.1
4	6.08	$\frac{18.6}{0.29}$	$\frac{20.1}{0.32}$	$\frac{12.8}{0.21}$	$\frac{22.4}{0.35}$	62	3.3	1.76
5	17.5	$\frac{43.2}{0.50}$	$\frac{46.8}{0.56}$	$\frac{40.2}{0.46}$	$\frac{38.8}{0.45}$	86	2.7	1.19

structure defectivity.

b)   The value of the strength of the contact layer of between-pores
partitions R_c and ratio of maximum strength of the contact layer
R_c^{max} to minimum R_c^{min}.   This data allows the influence of
defectivity on the strength of the concrete to be defined.

c)   The linear dimensions of the destructed layer and average ratio of
the strength of the contact layer to the punch displacement (Rc/).
The former allows the optimum pore size to be designed, the latter
allows estimation of the influence of the features of the
composition of calcium hydrosilicates on the strength of the
concrete.   From the experimental data it can be concluded that the
perfect composition of calcium hydrosilicates gives the value of the
previously mentioned ratio near 100.

Hence the proposed method allows the following parameters to be
directly defined :
- strength of the between-pores partitions
- value of macrostructure defectivity
- optimum pore size for any porous composite.

A large amount of experimental data and subsequent analysis allows
the porous concretes defined in Table 1 to be classified.

The macrostructure of test piece No. 1 shows a large amount of
defects, with the ratio R_c^{max}/Rb equal to 3.8.   The test pieces Nos. 4
and 5 also have defects of macrostructure with ratios of 3.2 and 2.7
correspondingly.   The ratio R_c^{max}/Rb for test pieces No. 2 and 3 equal
1.4.   Hence it can be concluded that the macrostructure of those samples
is sufficiently good.

As an example the macrostructure defectivity which can be defined
using the ratio R_c^{max}/R_c^{min} for test pieces 1 and 4 equals 2 and for test
piece No. 5 is 1.2.   On the other hand, the defectivity for test pieces
Nos. 1, 4 and 5 were generally the same.   This means that there were
defects of porosity (non-optimal pore sizes, etc) in the macrostructure
of test pieces Nos. 1 and 4.   When high values of Rc are apparent it can

be concluded that the porous composition of these concretes has little or no defects, but the defects of macrostructure are cracks. Neighbouring values of the strength of the contact layer shows that there is homogenous porosity of the concrete (No. 1.2) ratio $(R^c/\ )a$, for the test pieces Nos 3 and 2 equals 29 and 23 (conventional units) correspondingly and for test piece No. 5, 86 units. It can be concluded that essential improvements of the composition of calcium hydrosilicates are required.

The present conclusions are confirmed with the help of traditional methods. The results obtained make possible effective steps to eliminate technology shortcomings and to improve mechanical properties of concrete. The method may be used for quick testing of the quality of concrete in mass production environments.

INVESTIGATIONS OF THE INFLUENCE OF A SORT OF COARSE AGGREGATE ON FRACTURE TOUGHNESS OF CONCRETE

ANDRZEJ BOCHENEK
Department of Metallurgy
Technical University of Czestochowa
GRZEGORZ PROKOPSKI
Department of Civil Engineering
Technical University of Czestochowa

ABSTRACT

The fracture toughness in Mode II of concretes made of different coarse aggregates (granite, basalt, gravel and burned shale) has been investigated. Stress intensity factor K_{IIc} and fracture energy J_{IIc} have been plotted against type of coarse aggregates. Structural analysis on Quantimet 720 and fractographical investigations on SEM Cambridge Scientific S4-10 Microscope, have shown influence of the sort of coarse aggregate used for concrete making, on fracture micromechanism of concrete.

INTRODUCTION

It has been stated in tests of the influence of coarse aggregate on concrete strength, that the morphological characteristics of aggregate and the strength of contact layers between aggregate and mortar play predominant part. The adhesiveness between aggregate and mortar is to be one of the most important factors of concrete strength. Decrease of the adhesiveness between aggregate and mortar (e.g. by aggregate coating with paraffine [1]) makes concrete weak, whereas adhesiveness increase causes growth of the concrete strength.

Under the influence of external loads the cracks occuring in a concrete, among other things as a result of shrinkage, are to be joined and extended, being propagated through the matrix (mortar) or through the grains. There is the minor probability of cracks going through the grains in high strength concretes, whereas in concretes having grains weaker of matrix, than the fractures are going just through the grains [2]. Comparison between calcareous and gravel concretes had shown that after 28

days maturing, the calcareous concrete had the greater fracture
toughness than gravel concrete [3].

The tests presented in the paper [4] have shown that
reinforced concrete beams made of concretes of B20 class,
containing the following aggregates : basalt, calcareous and
rounded coarse, proved distinct differences in the values of
cracking moments, in the morphology of cracks and in the
quickness of cracks propagation according to growing load.

The tests results of the stress intensity factor K_{IIc} and
fracture energy J_{IIc} in connection with the fracture
micromechanism, have been presented in this paper. The tests
have been carried out in Mode II (shear) on concretes made of
different aggregates with the same size of coarse grains.

The test method based on Mode II is to be especially
essential with regard to cement pastes, mortars and concretes
considering their low shear strength [5, 6, 7].

OBJECT AND RANGE OF TESTS

A trial has been undertaken in this paper to establish the
influence of the sort of coarse aggregate on the fracture
toughness of concretes. Four sorts of concretes have been
tested with constant water/cement ratio w/c = 0.55 and
containing the following aggregates : natural gravel, basalt,
granite and burned shale. Each of these concretes consisted of
coarse aggregate with grains of 4-8 mm in diameter, quarz
sand with grains of 0-2 mm and Portland cement of quality "35".

The concrete cubes size of 0.15 m had been used, in groups
of three for compressive strength tests and also the concrete
cubes size of 0.15 m with two primary fissures, in groups of
six for fracture toughness tests. The above mentioned groups of
specimens were made of each sort of concrete mixes (Table 1).

TABLE 1
Compositions of concrete mixes

Kind of concrete	Component of mixe kg/m^3			
	aggregate 4-8 mm	sand 0-2 mm	cement "35"	water
basalt	1300	740	320	175
gravel	1200	710	310	170
granite	1300	700	310	170
burned shale	600	400	415	230

The specimens were kept for 7 days in water bath and for the
next 21 days in natural laboratory conditions. After 28 days of
maturing, the specimens were subjected to the compressive
strength (Table 2) and fracture toughness tests and also to the
microscopic observations.

TABLE 2
Compressive strength of tested concretes

Kind of concrete	basalt	gravel	granite	burned shale
Compressive strength [MPa]	26.8	27.1	36.6	18.2
standard deviation	2.2	2.5	2.8	2.1

TESTS OF FRACTURE TOUGHNESS

The tests of fracture toughness has been carried out on the post shown in Fig 1 . For such specimens a wide range of independence of the compliance function Y(a/W) of the geometrical value a/W is existing [8].

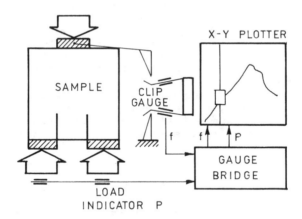

Figure 1. Scheme of the test stand.

The formula of the stress intensity factor K_{IIc}, given in the paper [6], has the following shape :

$$K_{IIc} = \frac{5.11\ P_Q}{2\ B\ b}\ \sqrt{\pi\ a}$$

where : P_Q — value of the critical force initiating a fracture development, identified in graphs as a slight refraction or extreme of the curve (Fig. 2), b — height of the specimen above the fissure (size of the uncut part of a specimen), a — length of the fissure, B — specimen thickness.

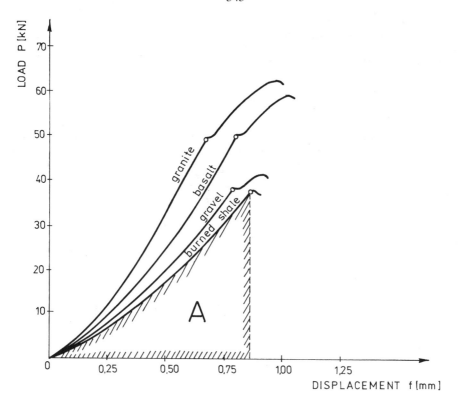

Figure 2. Exemplary load-displacement diagrams.

Calculations of the fracture energy J_{IIc} have been made according to formula given in the paper [9]:

$$J_{IIc} = \frac{A}{2\,B\,b}$$

where : A — area under graphs (Fig. 2) i.e. the energy accumulated in a specimen till the moment of the primary fissure initiation.

RESULTS OF TESTS OF FRACTURE TOUGHNESS

It has been stated, as a result of tests of fracture toughness, that the maximum value of the stress intensity factor K_{IIc} = 5.139 MN/m$^{3/2}$ had possessed concrete made of granite aggregate, next K_{IIc} value amounted 4.448 MN/m$^{3/2}$ — for the basalt concrete, 3.969 MN/m$^{3/2}$ — for the gravel concrete and 3.538 MN/m$^{3/2}$ had shown concrete made of burned shale aggregate (Table 3, Fig. 3). The concrete made of burned shale showed the greatest dispersion of results, which can be

explained   by heterogeneity of this   aggregate.   The   greatest
convergence of results has been found to be critical for   concrete of
gravel aggregate with the greatest homogeneity of grains.

TABLE 3
Stress intensity factor K_{IIc}

Kind of concrete	granite	basalt	gravel	burned shale
K_{IIc} [MN/m$^{3/2}$]	4.434	3.860	4.229	4.106
	4.106	5.297	4.024	3.613
	5.749	4.517	3.613	2.012
	5.666	5.050	3.490	3.613
	5.092	3.900	4.476	5.133
	5.789	4.065	3.983	2.751
Mean value	5.139	4.448	3.969	3.538
Standard deviation	0.726	0.613	0.369	1.079

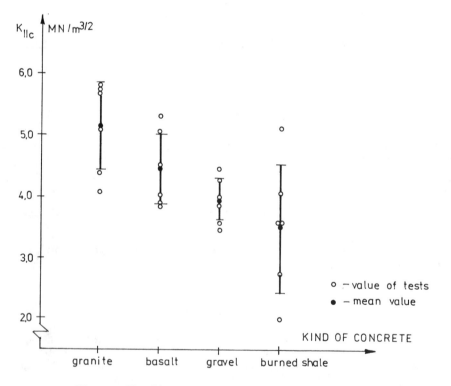

Figure 3. Stress intensity factors K_{IIc}

The tests have shown   that the   greatest   fracture   energy   was

1880 N/m — for the granite concrete, 1388 N/m — for the basalt
concrete, 1114 N/m — for the gravel concrete, and 835 N/m for
concrete made of burned shale aggregate (Table 4, Fig.4).

TABLE 4
Fracture energy J_{IIc}

Kind of concrete	granite	basalt	gravel	burned shale
J_{IIc} [N/m]	2143	1035	1618	973
	1187	1581	335	1251
	1991	2053	1069	375
	1420	1123	718	707
	2119	1554	1395	827
	2420	987	1549	881
Mean value	1880	1388	1114	835
Standard deviation	474	415	508	291

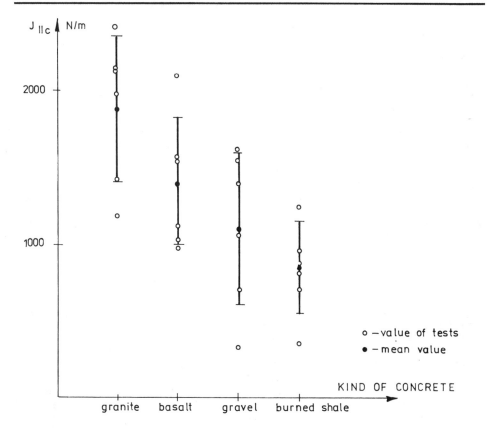

Figure 4. Fracture energies J_{IIc}

Macroscopic observations of the obtained specimens fractures have shown that amount of the split right across grains of coarse aggregate was dependent on a sort of aggregate used for specimens making. Computation of the average total surface of grains split right across was made with help of the computerical picture analyser the Quantimet 720. The tests have shown, that the specimens made of burned shale concrete had the greatest average total surface of grains split right across, amounting about 3100 mm^2. The next surface value of about 1830 mm^2 has been stated for specimens made of granite concrete, the value of about 1150 mm^2 — for specimens of gravel concrete and the lowest value of about 960 mm^2 — for specimens of basalt concrete (Table 5).

TABLE 5

Results of tests Quantimet 720

Kind of concrete	granite	basalt	gravel	burned shale
	1579.8	945.0	1110.5	3190.7
	1610.8	1093.5	1265.9	3037.3
Intergranular	1951.6	958.5	1066.0	2546.4
crack area	1858.0	945.0	999.4	3144.7
mm^2	1982.5	837.0	1177.1	3969.9
	2013.5	958.5	1265.9	2975.9
Mean value	1832.8	956.3	1147.4	3098.6
Standard deviation	191.0	81.6	108.5	372.0

The tests made with help of the Quantimet 720 have proved a connection of the used sort of coarse aggregate with obtained values of the K_{IIc} and J_{IIc}. The relatively great total surface of grains split right across in the granite concrete also great parameters K_{IIc} and J_{IIc}, had been caused by the strongly developed surface of grains of the granite aggregate.

MICROFRACTOGRAPHIC TESTS

Surfaces of the specimens fractures, obtained in the tests of fracture toughness, were subjected to microfractographic observations. The areas were observed laying in the vicinity of the top of primary fissure. The areas of about 600 mm^2 in each sort of concrete were subjected to observations. The tests were made on the scaning microscope type Cambridge Scientific S4-10 using magnifications of 50 to 2000 times. The pictures presented in the paper are being characteristic for the observed character of fracturing occuring in the particular sorts of concretes. It has been stated, that the cohesion forces between aggregate and mortar, being an effect of physical and chemical processes during maturing of concrete, had the predominant influence on decohesion on the boundary between aggregate grains and mortar. The concrete made of burned shale aggregate showed the greatest cohesion forces. The burned shale grains had their surface topography very strongly developed. The greatest average (of particular specimens) surface of grains split right across occured as a result of strong setting between burned shale grains and mortar. Surfaces of the natural gravel grains had the lowest cohesion properties. It was observed in this case, that the greatest amount of fractures was running along the border line between gravel grains and mortar. Only inconsiderable amount of gravel grains split right across.

The cleavable fractures right across the mortar had been observed in case of gravel concrete, causing in some fragments of specimens exposure of primary cracks, originated in maturing process of concrete (Fig. 5) and disclosure of pores existing in concrete. In the basalt and granit concretes fracture were running in cleavable way across grains (Fig. 6), or in case the other configuration of fracture, along the border line between grain and mortar. The fractures of granite grains had a shape of characteristic terraces with clefts (Fig. 6). The terraces in basalt concrete had their surface more developed (Fig. 7) than in case of granite concrete.

The quasi-plasticity phenomenon in basalt concrete occured most strongly in the grains alone during their disintegration in the failuring process. The fractures at the contact surface of basalt grains and mortar was rarely observed.

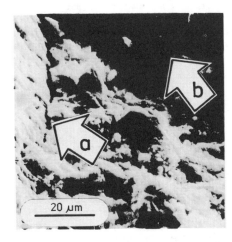

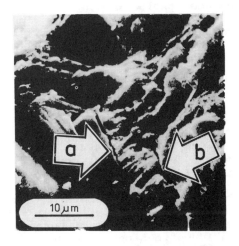

Figure 5. Microfracture on the border line between grain and cement paste (a) and running across the cement paste (b). In contact between grain and cement paste visible the pillar-shaped contact layer perpendicularly situated.

Figure 6. Cleavable fracture of a specimen of granite concrete. There are visible microfractures on the boundary of granite grain (a) and pillar-shaped contact layer and between contact layer and cement paste (b).

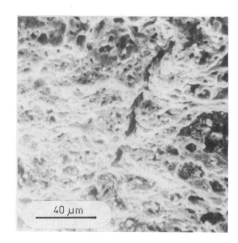

Figure 7. Terraced cleavable microfracture running across basalt grain (a).

Figure 8. Transcrystalline fracture across burned shale grain with ramifications.

The fracture toughness of grains in concrete of burned shale aggregate was lower than that of mortar and contact layers of mortar with strongly developed surfaces of burned shale grains. Such a fact caused that the fracturing occured mainly across burned shale grains , with not great amount of side ramifications (Fig. 8). The firmly porous structure was observed inside grains of burned shale aggregate. Such a structure is characteristic for the way of production of this sort of aggregate.

CONCLUSIONS

The stress intensity factor K_{IIc} amounted 5.139 MN/m$^{3/2}$ — for granite concrete, 4.448 MN/m$^{3/2}$ — for basalt concrete, 3.969 MN/m$^{3/2}$ — for gravel concrete and 3.538 MN/m$^{3/2}$ — for concrete made of burned shale aggregate (Fig. 3).

The fracture energy J_{IIc} showed similar tendencies. Namely it had the value 1880 N/m — for granite concrete, 1388 N/m — for basalt concrete, 1114 N/m — for gravel concrete and 835 N/m — for concrete made of burned shale aggregate (Fig. 4).

Investigations have proved that the average total surface of grains split right across, in case of concrete made of burned shale aggregate was greater (much greater than the surface of gravel grains split right across) than in case of gravel concrete, at the little difference between values of the stress intensity factor K_{IIc}. It bears witness to essential influence of surface topography and shape of aggregate grains on the fracture toughness of concrete. These two factors are shaping the strength parameters of the contact layer between aggregate and mortar.

The observed quasi-plasticity phenomenon in case of granite and basalt concretes, was mainly caused by cleavable character of splits of the gravel grains in failuring process and also by arising of characteristic terraces with clefts.

Near-by position of the point of initiation of primary fissure and the point of maximum loading, in case of gravel concrete, was caused by a feeble juncture between aggregate grains and mortar in consequence of a poor developed and unvarying surface of the spherical grains of gravel. It generated in maturing concrete the loosening at the contact of gravel grains and mortar, leading to an easy development of the primary fissure in failuring process.

The coincidence of the point of initiation of primary fissure and the point of maximum loading, in case of concrete made of burned shale aggregate, was caused by the low mechanical properties of this aggregate.

The primary fissure initiation caused immediate failure of a specimen in the plane of development of primary fissure. The fracture was going through all grains of the burned shale aggregate laying within this plane.

REFERENCES

1. Kuczynski, W., Wplyw kruszywa grubego na wytrzymalosc betonu. Arch. Inz. Lad., 1958, 4, 181-209, (in Polish).
2. Brandt, A.M., Zastosowanie doswiadczalnej mechaniki zniszczenia do kompozytow o matrycach cementowych. In Mechanika kompozytow betonopodobnych. PAN-Ossolineum, 1983, pp. 449-501, (in Polish).
3. Naus, D.J. and Lott, J.L. , Fracture toughness of Portland cement concretes. ACI Journal, June 1969, 481-489.
4. Flaga, K. and Furtak, K., Wplyw rodzaju kruszywa na poziomy naprezen krytycznych w betonie sciskanym. Arch. Inz. Lad., 1981, 27, 655-665, (in Polish).
5. Davies, J., Morgan, T.G. and Yim, A., The finite element analysis of a punch-trough shear specimen in Mode II. Int. J. Frac., 1985, 28, R. 3-10.
6. Watkins, J., Fracture toughness test for soil-cement samples in Mode II. Int. J. Frac., 1983, 23 ,R. 135-138.
7. Davies, J. and So, K.W. Further development of fracture test in Mode II. Int. J. Frac., 1986, 31, R. 19-21.
8. Dixon, J.R. and Strannigan, J.S., Determination of energy release rates and stress-intensity factors by the finite-element method. J. Strain Analysis, 1972, 7, 125-131.
9. ASTM E 813-81, JIc, A measure of fracture toughness.

FRACTURE MECHANICS OF MICRO-CONCRETES

RYSZARD FRACKIEWICZ
DANUTA WALA
Technical University of Wroclaw
Institute of Building
Wybrzeze Wyspianskiego 27
50-370 Wroclaw, Poland

ABSTRACT

Research has been carried out on cement matrix mortars and on fibre-reinforced mortars.The materials tested were of very different microstructures. Both standard and fracture mechanics parameters have been measured. The strong relationship between the microstructure and fracture behaviour of the materials has been observed. Lifetime prediction procedure and proof testing have been carried out.

INTRODUCTION

Brittle fracture mechanics applied to glass and ceramics has proved to be a valuable research tool. In particular, the $v(K_I)$ relationship enables the quantitative description of the mechanism of a delayed fracture.The development of linear elastic fracture mechanics have lead to the possibility of predicting time to failure of brittle materials subjected to stresses lower than their impact strengths.

Numerous efforts were undertaken in order to transform brittle fracture mechanics methods developed and tested for glass, ceramics and metals to the area of brittle concrete-like materials [1], though the complexity of their microstructures caused many difficulties in the reliable interpretation of the results.

This paper reports on the results of our own research on the possibility of predicting the lifetime of cement pastes and mortars as well as on the establishing the relationship between the microstructural and brittle fracture mechanics parameters.

MATERIALS

The research has been carried out on cement matrix composites of maximum aggregate size 2.5 mm and on fibre reinforced mortars. The materials were of different amount and size of aggregate, kind of stone (quartz, basalt), porosity, water-cement ratio and kind of fibres (steel, plastic; 10-15 mm long, diameter .2 mm). Table 1 containes mix proportions and the other microstructural data of the materials tested.

TABLE 1

Mix proportions and microstructural data of mortars

No	Mix proportions aggr.	w/c	Density g/cm^3	Porosity %	E 10^{10} J/m^2		fibre admix
1	–	.3	1.91	8.7	3.1	.33	–
2	–	.4	1.67	15.0	2.1	.20	–
3	–	.5	1.60	16.7	1.6	.26	–
4	sq	.5	2.20	8.4	4.3	.34	–
5	fq12	.4	2.01	11.5	2.9	.27	–
6	fq30	.4	2.06	15.4	3.1	.30	–
7	fq60	.5	2.11	16.8	3.2	.32	–
8	cq12	.4	2.02	13.6	3.8	.33	–
9	cq30	.4	2.09	12.6	4.0	.36	–
10	cq60	.5	2.20	10.6	4.3	.37	–
11	sb	.5	2.39	12.3	4.6	.36	–
12	fb12	.4	2.20	15.1	3.0	.26	–
13	fb30	.4	2.03	18.5	3.1	.27	–
14	fb60	.5	2.09	22.5	3.4	.28	–
15	cb12	.4	1.96	14.8	4.0	.34	–
16	cb30	.4	2.15	13.6	4.3	.35	–
17	cb60	.5	2.32	12.9	4.8	.36	–
18	sq	.53	2.27	10.9	.4.3	.33	steel
19	sq	.53	2.24	10.1	4.3	.33	plast.
20	sb	.53	2.49	11.2	4.3	.34	steel

Abbr.: f-fine(0-.5 mm), c-coarse(1-2.5 mm), s-standard,
b-basalt, q-quartz, 12,30,60-average volume amount
of sand (%)

TESTING TECHNIQUES

Standard and fracture mechanics parameters of materials tested have been obtained by the following means:

-flexural strength, R_g: cuboids .04x.04x.16 m, .1 m span;

-compressive strength, R_s: halves of the broken cuboids;

-Young's modulus, E, and Poisson's ratio, ν,: ultrasonic method;

-critical stress intensity factor, K_{IC},: 3-point bending of notched beams [2], Instron 1126 testing machine;

-fracture surface energy, γ_f,: Davidge-Tappin method [3];

-subcritical crack growth velocity, v,: double torsion method, .01x.09x.16 m plates with initial notch and groove; the $v=AK_I^n$ relationship has been assumed (A and n are established experimentally).

Fracture mechanics parameters of materials tested are given in Table 2.

TABLE 2

Standard and fracture mechanics parameters of mortars.

No	R_g	R_s	K_{IC}	γ_f	n	S_n	$-\log A$	S_A
	MPa	MPa	$MNm^{-3/2}$	Jm^{-2}				
1	11.9	52.1	.52	6.3	9	6	56	4
2	9.1	45.0	.44	2.2	15	6	87	5
3	7.7	36.1	.45	1.8	19	7	110	5
4	8.6	35.2	.64	12.8	9	12	57	8
5	11.9	53.0	.66	8.4	11	8	68	7
6	10.2	42.6	.58	11.0	6	10	39	9
7	5.5	31.1	.44	13.1	8	15	49	13
8	9.8	50.9	.54	11.9	20	10	119	8
9	9.0	56.2	.56	23.5	17	13	102	11
10	6.6	42.5	.48	30.0	10	20	62	15
11	8.8	48.5	.71	25.3	16	16	105	13
12	11.5	53.0	.72	15.6	18	11	110	8
13	10.2	50.3	.69	18.2	6	15	40	13
14	5.7	42.0	.28	21.1	27	22	159	18
15	10.4	55.7	.66	19.4	30	11	177	9
16	9.8	68.1	.62	31.5	11	18	68	15
17	7.6	43.7	.69	73.8	6	24	40	22
18	14.5	38.7	1.44	670.0	–	–	–	–
19	7.5	33.3	.63	70.0	–	–	–	–
20	12.5	52.0	1.36	540.0	–	–	–	–
S	5.5	9.3	10.0	15.3				

Abbr.:S-mean standard deviation ,%

LIFETIME PREDICTION

The basis for the lifetime prediction is a relationship between the stress intensity factor, K_I, and the faw length, a,:

$$K_I = 6 Y a^{1/2} \qquad (1)$$

where: 6 - apparent tensile stress,
a - apparent flaw length,
Y - dimensionless geometric parameter.

As a result of assuming the $v(K_I)$ relationship as $v=AK_I^n$ one can obtain the formula for time to failure:

$$t = \frac{2}{A \; 6^2 \; Y^2 \; (n-2) \; K_{IC}^{n-2}} \qquad (1)$$

Fracture mechanics parameters of brittle materials enable working out diagrams of their service lives and designing proof tests that can eliminate, before the real application in service, all the specimens that contain flaws of dimension too large to let the material work under the service stress for a desired period.

The specimens are subjected to the proof stress, 6_p, higher than the service stress, 6_a. The specimens for which

$$6_p \; a_p^{1/2} \; Y > K_{IC} \qquad (2)$$

(where a_p-critical flaw length for 6_p) fail. The survival of the proof test by the specimen quaranties that K_I in the tip of the longest flaw does not reach K_{IC}.

The minimum lifetime of the specimen is given by the formula:

$$t_{min} = 2 \left(\frac{6_p}{6_a}\right)^{1/2} \Big/ \; [(n-2) \; A \; 6_a^2 \; Y^2 \; K_{IC}^{n-2}] \qquad (3)$$

So, the minimum lifetime is expressed by means of subcritical crack growth parameters (n, A, K_{IC}) and $6_p/6_a$ ratio. The reliability of the proposed method strongly depends on the accuracy of determining subcritical crack growth parameters (K_{IC}, n, A). The lifetime and proof test diagrams for the selected mortars are given in fig.fig.1-4.

556

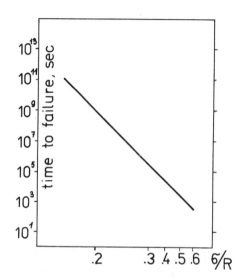

Figure 1. Lifetime prediction
diagram for mortar
No 8

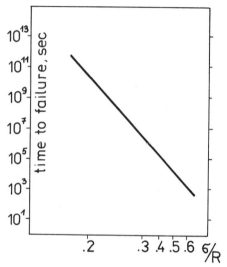

Figure 2. Lifetime prediction
diagram for mortar
No 15

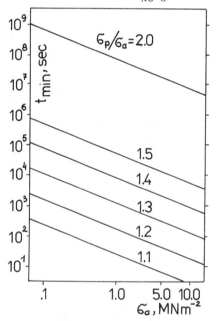

Figure 3. Proof test diagram
for mortar No 8

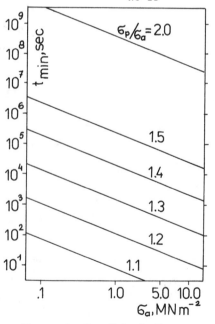

Figure 4. Proof test diagram
for mortar No 15

RESULTS

Flexural strength

Within the range of materials tested it is noticeable that the higher the porosity and an amount of aggregate the lower the flexural strength. This decrease is more intensive for the mortars of uniform sizes of sand. For 60% amount of sand the highest flexural strength is observed for a standard mortar containing quartz aggregate of 0-2.5mm size. Fibres substantially increase the flexural strengths of mortars.

Compressive strength

Compressive strength of the mortars decrease with the increase of porosity and a volume share of fine aggregate. Materials containing coarse aggregate exhibit their maximum compressive strength at about 30%vol. amount of sand. The highest compressive strength (68.1 MPa) has been obtained for a mean size coarse basalt aggregate mortar (No 16). Addition of fibres does not substantially increase the compressive strengths of the mortars.

Young's modulus

Young's modulus of the mortars decrease with the increase of porosity and slightly increase with the increase of the volume share of sand.

Fracture mechanics parameters

K_{IC} decrease with the increase of porosity of the materials tested. The higher the amount of fine quartz sand the lower the K_{IC}, while for the basalt aggregate the influence of the volume changes of sand seem negligible. It is clearly visible that in all cases basalt aggregate effects in higher K_{IC} values than the quartz sand. Admixture of steel fibres (1.5% vol.) increase the K_{IC} values of the mortars of 110% (see No 18, No 20). Fracture toughness of the materials (assesed by means of n and -logA) increase with the porosity. The highest toughness exhibit the mortars of low w/c ratio, high percentage of microporosity and of coarse basalt aggregate at about 10% vol.(No 15). Effective fracture surface energy, γ_f, decrease for the lower values of porosity and considerably increase for the higher porosities (No 1 - No 3). The increase of volume amount of all the kinds of aggregates increase γ_f.

This effect is clearly visible for the coarse basalt grains. Admixture of steel fibres (1.5% vol.) effects in the increase of γ_f from 12.8 J/m^2 to 670 J/m^2 (No 4, No 18).

CONCLUSIONS

On the basis of the results obtained it is possible to establish a relationship between the mortars' microstrctures and their standard mechanical properties. For the micro-concretes K_{IC}, n, -logA and γ_f can be considered the materials' constants describing their toughness to subcritical crack growth and to impact damage. The best way to increase the fracture toughness of the cement matrix matrials is to increase the values of K_{IC}, n and -logA. Within the range of materials tested attempts to model the microstructures of the mortars in order to obtain the highest possible values of all the three parameters simultaneously is difficult. For the cement composites subjected to cyclic loads the optimalisation of their microstructures should increase the values of γ_f. In the light of the test results, however, the main microstructural parameters that are responsible for the high γ_f values decrease both R_g and K_{IC}.

REFERENCES

1. Mindess S., Application of fracture mechanics to cement and conrete. Proc.: Mechanics of Concrete-like Composites, IPPT PAN, Warszawa-Jablonna 1979.
2. Evans A.G., Fracture mechanics determinations. In Fracture Mechanics of Ceramics vol.1, pp 17-48, Plenum Press, New York 1974.
3. Davidge R.W., Tappin G., The effective surface energy of brittle materials. J. Mater. Sci.,1968, 3, 165-173.

LOADING RATE SENSITIVITY OF CONCRETE-LIKE COMPOSITES UNDER TENSILE LOADING

MICHAL A. GLINICKI

Institute of Fundamental Technological Research

Polish Academy of Sciences

Swietokrzyska 21, 00-049 Warsaw, Poland

ABSTRACT

The paper presents some experimental results on the loading rate influence upon the tensile fracture behaviour of concrete-like composites. Uniaxial tensile tests on concrete and fibre reinforced concrete specimens were carried out at variable loading rates: from 0.001 MPa/s up to 500 MPa/s. The fibre reinforced concrete comprised a volume fraction of 1.5% and 2.5% of plain steel fibres. The effects of the rate of loading on the tensile strength and the stress-strain relationship were analysed. The tests revealed a significant influence of the applied loading rate upon the tensile strength and the ultimate strain of the composites; for ordinary concrete specimens an improvement of about 70% and 80% over the static values was gained respectively. An insignificant influence of the fibre content on the tensile strength of fibre reinforced concrete was observed.

INTRODUCTION

The material properties of cement-matrix composites under high rates of loading may significantly differ from that under static loading. The loading rate effect of ordinary concrete have been investigated extensively. Most studies, however, have been directed to the case of compressive loading and only few studies are available for the dynamic tensile loads ([1],[2]). The relevant research on fibre reinforced concretes and polymer modified concretes is even less documented, although the recognised advantages of these concretes are evident enough for succesful applications in structures subjected to impact and explosion loads ([3]).

The experimental data concerning the loading rate influence upon the tensile strength of ordinary concrete are summarized in Fig.1 (after [1]).

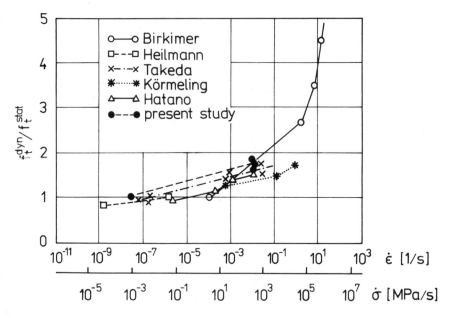

Figure 1. Relative increase of tensile strength of concrete as a function of stress or strain rate.

All investigations show an increase in the strength; the ultimate strain increase has also been observed. A similar increasing trend has been detected for fibre reinforced concrete (FRC) , however it seems to be dependent on the shape and the type of fibres used. The investigations on the fracture toughness of concrete and FRC show even higher loading rate sensitivity of this parameter, even up to twice greater than the tensile strength sensitivity ([4],[5]).An influence of concrete technological parameters (e.g. cement content, w/c ratio) on the loading rate sensitivity has been also found out, but its dependency on the internal structure of material has not been studied in detail. The present research is aimed to fill this gap in understanding the dynamic fracture of cement-based materials. Adopting the meso level approach for structure consideration both porosity and aggregate and/or fibre inclusion content are considered. This paper presents the preliminary test results and further experiments are in progress.

DETAILS OF EXPERIMENTAL PROGRAMME

Materials and Specimens

For preparation of specimens a basic concrete mix of following proportions
was used: cement 1.0, sand 0-2 mm, 2.0, aggregate 2-4 mm,0.54, water, 0.5.
The FRC mixes used the same matrix and comprised fibre volume contents of
1.25 % and 2.5 %. Straight smooth steel fibres of the aspect ratio of
40/0.4=100 were used. Prepared "paddle -shaped" specimens, shown in Fig.2,
had the thickness of 43-46 mm. Some of the specimens were notched in the
middle with a notch of 20 mm at botch sides.

Testing Procedure and Measurements

For the uniaxial tensile tests the steel grips, shown in Fig.3, were
applied. The tests were carried out on an Instron testing machine at a
constant rate of the cross-head displacement. For static testing the rate
of 0.2 mm/min was applied. To obtain higher rates of loading the testing
machine was switched to dynamic mode and a single impulse of ramp or
rectangular shape was used to control the displacement speed. Selected
amplitude and frequency parameters resulted in the loading rates up to 10^6
times higher than in static tests.

During testing the applied load, the cross-head displacement and
longitudinal deformation of the specimen were recorded. For the strain
measurement two sets of devices were applied.

In the case of static tension two extensometers with a measuring
length of 245 mm attached to both faces of the specimen were used. The
static test data, i.e. the load-deformation relationship and the load-
cross-head displacement relationship were recorded on X-Y plotters.

Dynamic deformation measurements were conducted using a LVDT
transducer (measuring length of 60 mm) and an Instron extensometer with a
measuring length of 55 mm (Fig.3). The load output from the load cell, the
cross-head displacement signal and two deformation signals were
simultaneously monitored using two storage oscilloscopes. The triggering of
the scope sweep was acomplished using the cross-head displacement signal.
To have a permanent record of the test data the stored signals were played
back on an X-Y plotter and photographs of the screen display were also
taken. Due to unexpected oscillations of dynamic deformation signal
recorded by the extensometer these data were rejected and are not shown
further on.

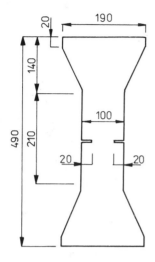

Figure 2. Dimensions of the
specimen tested under tensile
loads.

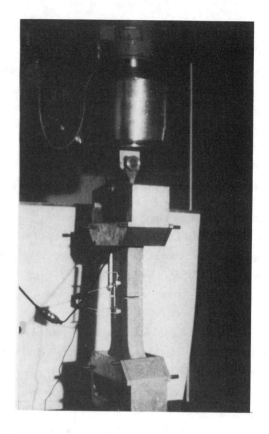

Figure 3. Specimen in the grips and devices
for deformation measurements.

TEST RESULTS AND DISCUSSION

Static tensile tests resulted in the tensile strength of concrete of 3.0
MPa (an average of four specimens). The applied loading rate was about 10^{-3}
MPa/s. The values of the strain at peak stress obtained for three specimens
were within the range 0.07-0.092 %. (average values of two readings at both
faces of each specimen. The fourth specimen failed outside the mesuring
length of the extensometers.

Typical results obtained from dynamic tensile tests on concrete and
fibre reinforced concrete are shown in Fig.4. The values of stress and
strain are plotted with respect to time. The stress-time curves for
specimens for which records of the deformation failed, are shown in Fig.5.

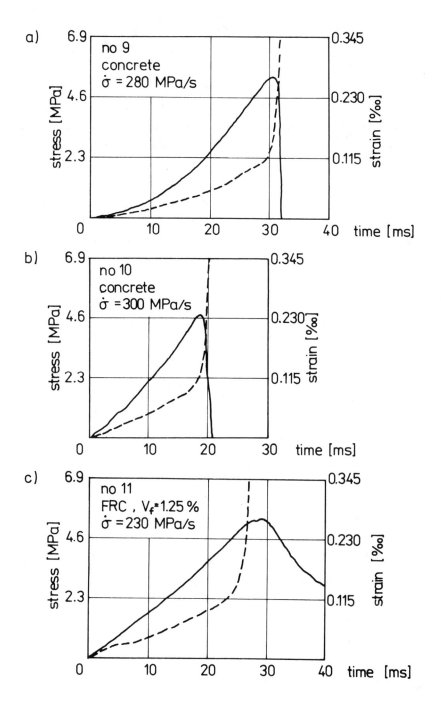

Figure 4. Recorded stress-time curves (continuous lines) and strain-time curves (dashed lines) for concrete and fibre reinforced concrete specimens.

The attained values for the tensile strength of concrete were within the range 4.8-5.6 MPa. The recorded deformation at peak stress was 0.13-0.16 %.. These results were obtained for the loading rates $2-5*10^2$ MPa/s that corresponded to the loading rise time of 16-33 milliseconds.

It can be concluded that the tensile strength of concrete was significantly influenced by the applied rate of loading. An average increase of 70 % over the static value was obtained due to an increase of the loading rate from 10^{-3} MPa/s to $2-5*10^2$ MPa/s. This is close to the values reported in the literature (see Fig.1).

The loading rate efect on the strain at peak stress was also found to be significant. The applied increase of the rate of loading resulted in increase in the ultimate strain of about 80 % over the static value. The attained values of the strain of concrete are slightly lower in comparison to the test data given elsewhere (e.g. Tinic and Brühwiler [6], Zielinski [7]) but the relative increase agrees fairly well.

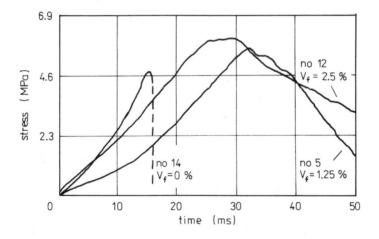

Figure 5. Recorded stress-time curves for concrete and fibre reinforced concrete specimens.

The tests on FRC specimens at the loading rate $2-3*10^2$ MPa/s resulted in following values of the tensile strength: 5.25 MPa and 5.7 MPa for fibre volume content of 1.25 %, 5.5 MPa and 5.93 MPa for fibre volume content of 2.5% . The results are evidently higher that the tesile strength of concrete at the same loading rate. However the influence of fibre volume content on the loading rate sensitivity is not clear. In fact, the differences in the strength lie within the scatter of results so it can be said that no significant influence of fibre content has been found. In view of possible considerable scatter in the number of fibres within one batch containing the same percentage by volume of fibres the preceeding statement should be checked by counting the actual number of fibres in fractured cross-section (not done yet). However the similar conclusion was proposed by Körmeling [5] who also used straight smooth fibres. The evidence of fibre shape influence upon the loading rate sensitivity can be found in Naaman tests [8]; i.e. it is much more pronounced in the case of deformed fibres than for straight smooth fibres.

As it can be seen from Figs.4 and 5 the decrease of the load after its maximum value ocurred in about 1 milisecond for concrete specimens. In the case of FRC the decrease was gradual and it seemed to be sensitive to fibre content; i.e. for V_f=1.25 % it was about 0.3 MPa/s and for V_f=2.5 % it was about 0.1-0.2 MPa/s. This implies an increase of the fracture energy of FRC for increasing fibre content. Refering again to Körmeling tests [5] a confirmation of such effect can be found.

The records of stress-time curves and strain-time curves shown in Fig.4 provided a possibility to obtain stress-strain diagrams which are given in Fig.6. Presented relationships for high loading rates were not obtained by manual transformation but simply by an automatic plot of stored data in both channels of an oscilloscope versus each other by an X-Y recorder. The data are shown only up to the maximum stress. The average values of strain measured on both faces of a specimen were plotted against the stress values to give static stress-strain curves that are also shown in Fig.6. The shapes of the static curves are linear up till failure. For dynamic curves a deviation from linear can be noticed just before maximum stress. According to Reinhardt [9] such deviations can be attributed to higher amount of microcracking in the whole volume of a specimen.

A comparison of the dynamic test results obtained for notched and unnotched specimens gave essentialy no difference due to the presence of a notch. To justify this fact results of Eibl and Curbach [10] can be recalled. The explanation given by the authors is that in the case of high rate load the incoming stress reacts less to the changing boundary condition at the notch than in the case of static loading.

The examination of fractured surfaces of concrete specimens revealed that at high loading rates failure ocurred through aggregate inclusions while this phenomenon was not apparent at low loading rates. The described

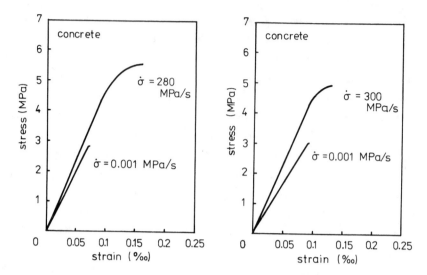

Figure 6. Relationship between tensile stress and strain for concrete specimens tested at different loading rates.

phenomenon has been already found out by several authors, e.g. [6],[7], neverthless it can not explain the full range of the dynamic strength increase. The other reasons for the tensile strength improvement due to increasing of loading rate are supposed to have an origin in the dynamics of the deformation process of the heterogenuous material, which features include inertial effects, a limited crack velocity and stress-vave reflections ([10],[11]).

CONCLUSIONS

From the test results the following conclusions can be formulated:

The rate of tensile loading of about 1000 MPa/s, that roughly corresponds to the upper range of an earthquake loading rate, is found to be the highest rate available using the described testing equipment.

The influence of the loading rate on the tensile strength and the ultimate strain of concrete is significant. An increase of the loading rate from 10^{-3} MPa/s up to $5*10^2$ MPa/s resulted in the increase of the strength of about 70% and of the ultimate strain of about 80% over the respective static values.

The tensile loading sensitivity of fibre reinforced concrete is similar to that of ordinary concrete, i.e. the effect of a double change of fibre volume content has not been found, possibly because the straight smooth steel fibres were used.

REFERENCES

1.Suaris, W. and Shah, S.P., Mechanical properties of concrete subjected to impact. Intern. Symp "Concrete Structures under Impact and Impulsive Loading", Berlin, 1982, 33-62.

2. Sierakowski, R.L., Dynamic effects in concrete materials. NATO-ARW "Application of Fracture Mechanics to Cementitious Materials", ed. S.P.Shah, Evanston, 1984,391-411.

3. Radomski, W., Properties of fibre reinforced concrete under impact loads (in Polish). Warsaw University of Technology, Warsaw, 1983.

4. Brühwiler, E., Wittmann, F.H., Rokugo, K., Influence of rate of loading on the fracture energy and strain softening of concrete. Trans. 9th Intern. Conf. on SMiRT, Lausanne, 1987, vol.H, 25-33.

5. Körmeling, H.A., SFRC in uniaxial impact tensile loading at 20°C and at -170°C. Third Intern. Symp. on Developments in Fibre Reinforced Cement and Concrete, Sheffield, 1986, Vol.1, 259-266.

6. Tinic, C., Brühwiler, E., Effect of compressive loads on the tensile strength of concrete at high strain rates. Intern. J. Cement Comp., 7,no.2,1985, 103-108.

7. Zielinski, A.J., Fracture of concrete and mortar under uniaxial impact tensile loading. Delft University Press, Delft, 1982.

8. Naaman, A.E., Fiber reinforced concrete under dynamic loading. Intern. Symp. on Fiber Reinforced Concrete, ed. G.C.Hoff, ACI SP-81, Detroit, 1984.

9. Reinhardt, H.W. Tensile fracture of concrete at high rates of loading. NATO-ARW "Application of Fracture Mechanics to Cementitious Materials", ed. S.P.Shah, Evanston, 1984, 413-442.

10. Eibl,J., Curbach, M., Behaviour of concrete under high tensile loading rates. Trans. 9 th Intern. Conf. on SMiRT, Lausanne, 1987, vol.H, 245-250.

11. Mianowski,K., Dynamic aspects in fracture mechanisms. In Brittle Matrix Composites-1, eds. A.M.Brandt & I.H.Marshall, Elsevier, 1986, 81-91.

CRITERION OF CRACK PROPAGATION ALONG THE INTERFACE
BETWEEN DISSIMILAR MEDIA

Mieczysław JARONIEK
Institute of Applied Mechanics
Technical University of Łódź, Poland

ABSTRACT

A central crack between two homogeneous linearly elastic and isotropic media of different properties is considered.
The elastic properties of the two media are characterized by their Poisson's ratios ν_i and their Young's moduli E_i, (i=1,2). The numerical solution is obtained using the finite element method. The characteristic parameter of a debonding criterion is obtained from the definition for the bi-material stress intensity factors $K_I^{(i)}$ and $K_{II}^{(i)}$ for a crack between two dissimilar materials in the case of CT-specimens and DCB-specimens. The debonding of the interface can be investigated using a formulation based on the strain energy relase rate G. This paper reviews the experimental procedure used to determine the state of stresses along the interface between the concrete and polymer concrete. Using a load-displacement curve and the photoelastic measurement results the stress intensity factors $K_I^{(i)}$ and strain energy relase rate G were determined for the interface between the fibre concrete and polymer concrete.

INTRODUCTION

For an infinite bi-material plate with a central crack along the interface subjected to biaxial loading (Fig.1) the analytical solution was given in references [1] to [6].
The solution to the problem is obtained using the complex variable approach. The critical stress intensity factors $K_{IC}^{(i)}$ can be determined from the following conditions:

$$
\begin{aligned}
K_{IC}^{(1)} &= \sqrt{2\pi r_o} \ \sigma_{\theta\theta}^{(1)}(\theta_o^{(1)}, r_o) & 0 &\leq \theta_o^{(1)} \leq \pi \\
K_{IC}^{(2)} &= \sqrt{2\pi r_o} \ \sigma_{\theta\theta}^{(2)}(\theta_o^{(2)}, r_o) & -\pi &\leq \theta_o^{(2)} \leq 0 \\
K_{IC}^{(b)} &= \sqrt{2\pi r_o} \ \sigma_{\theta\theta}^{(b)}(0, r_o) & \theta_o^{(1)} &= \theta_o^{(2)} = 0
\end{aligned}
\tag{1}
$$

Upper indices (1),(2) and (b) correspond to a crack propagating in materials (1),(2) or along the interface between them
respectively. Angles $\theta_o^{(1)}$, $\theta_o^{(2)}$ correspond to the maximum
values of $G_{\theta\theta}$ and $K_{IC}^{(1)}$, $K_{IIC}^{(2)}$ are the critical stress intensity
factors of these two media while $K_{IC}^{(b)}$ denotes the critical
bonding factor.

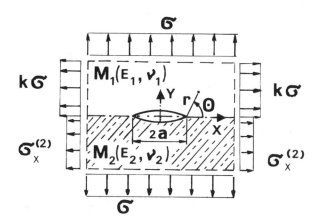

Figure 1. General geometry and central crack between two
isotropic media of different properties E_i , ν_i

The numerical and experimental procedures were applied to
determine the stress intensity factors $K_{IC}^{(i)}$ and the strain
energy relase rate G in the case of the compact tension (CT)
specimen and the double cantilever beam (DCB) specimen.

NUMERICAL ANALYSIS OF STRESS AND STRAIN DISTRIBUTION

The distribution of stresses and displacements has been calculated for the compact tension - CT specimen and for the double
cantilever beam - DCB specimen. The configuration of these
specimens is given in Fig.2 and 3, respectively. The geometry
and materials of models were chosen to reflect the actual
specimens used in the experiments.
     It is assumed that the crack exsists between two homoge-
neous and isotropic media, M_1 (E_1, ν_1) and M_2 (E_2, ν_2) for
example between the polymer concrete and the concrete rein-
forced with steel fibres (SFRC). In fact these media M_1 and
M_2 should be considered as macroscopically homogeneous ones.
They are characterized by their Young's moduli E_i and their
Poisson's ratios ν_i : E_1 = 29600MPa, ν_1 = 0,15
                       E_2 = 20800 MPa, ν_2 = 0,3
Since an analytical solution for this problem is not available, the distribution of stresses and displacements has been
calculated using the finite element method. The strain energy
relase rate G is equal in this case to the Rice J - integral

which can be found from the following expression:

$$J = \int_s \left(\tfrac{1}{2}\, \sigma_{ij}\, \varepsilon_{ij}\, dx_2 - T_i^{(n)}\, \frac{\partial u_i}{\partial x_1}\, ds \right) \tag{2}$$

or from numerical calculations using the relation:

$$J = \sum_{i=1}^{k} \left\{ \frac{1}{2}\left[\frac{1}{E_i}\left(\sigma_{yi}^2 - \sigma_{xi}^2\right) - \frac{\tilde{\tau}_{xyi}^2}{\mu_i} \right] n_{1i} - \right.$$
$$\left. - \left[\frac{\tau_{xyi}}{E_i}\left(\sigma_{xi} - \nu_i \sigma_{yi}\right) - \sigma_{yi}\frac{\Delta v_i}{\Delta x_i} \right] n_{2i} \right\} \Delta S_i \tag{3}$$

The J – integral can be also found out from the results of calculation on the basis of the force displacement curve P-Δ.

$$J = \frac{2A}{bB} \tag{4}$$

where: A – area under load-displacement curve,
B – thickness of the specimen,
b – ligament depth.

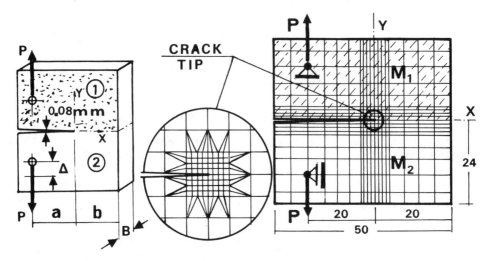

Figure 2. General view of the CT – specimen and mesh for a numerical calculation.

The stress intensity factors $K_I^{(i)}$, $K_{II}^{(i)}$ in the medium i and the value of $A(o)$ can be determined from approximate expressions for the local stress distributions corresponding to the maximum values of $\sigma_{\theta\theta}^{(i)}$

$$\sigma_y^{(i)} = \frac{K_I^{(i)}}{\sqrt{2\pi r}}\cos\frac{\theta}{2}\left(1+\sin\frac{\theta}{2}\sin\frac{3\theta}{2}\right) + \frac{K_{II}^{(i)}}{\sqrt{2\pi r}}\sin\frac{\theta}{2}\cos\frac{\theta}{2}\cos\frac{3\theta}{2}$$

$$\sigma_x^{(i)} = \frac{K_I^{(i)}}{\sqrt{2\pi r}}\cos\frac{\theta}{2}\left(1-\sin\frac{\theta}{2}\sin\frac{3\theta}{2}\right) - \frac{K_{II}^{(i)}}{\sqrt{2\pi r}}\sin\frac{\theta}{2}\left(2+\cos\frac{\theta}{2}\cos\frac{3\theta}{2}\right) + A(o) \tag{5}$$

$$\tau_{xy}^{(i)} = \frac{K_I^{(i)}}{\sqrt{2\pi r}}\sin\frac{\theta}{2}\cos\frac{\theta}{2}\cos\frac{3\theta}{2} + \frac{K_{II}^{(i)}}{\sqrt{2\pi r}}\sin\frac{\theta}{2}\left(1-\sin\frac{\theta}{2}\sin\frac{3\theta}{2}\right)$$

where: A(o) - far field stress component.

The cartesian components of stress $G_x^{(i)}$, $G_y^{(i)}$ and $\tau_{xy}^{(i)}$ obtained by finite element method allow to determine the values of stress intensity factors $K_I^{(i)}$, $K_{II}^{(i)}$ and A(o) using eq (5).

NUMERICAL RESULTS

The stress and displacement distribution has been calculated for CT-specimen and DCB-specimen using the finite element method [11]. The stress intensity factors of the two media $K_I^{(i)}$ and $K_{II}^{(i)}$ were determined from eq.(5) using the cartesian components $G_x^{(i)}$, $G_y^{(j)}$ and $\tau_{xy}^{(i)}$ obtained from FEM for the tearing forces P = 100 N. The critical values of the stress intensity factors K_{IC}, K_{IIC} and G_C were evaluated using critical values of tearing forces P_C obtained experimentally.

where:
$$K_{IC}^{(i)} = \beta K_I^{(i)} \qquad\qquad G_C = \beta^2 G \qquad\qquad (6)$$
$$\beta = P_C/P \qquad\qquad i = 1,2$$

A finite element mesh of a CT-specimen and DCB-specimen used for numerical simulation of the stress distribution was as shown in Figs 2 and 3 and the numerical results are given in Table 1. Using the FEM the isochromatic fringe pattern has been also calculated. The pattern obtained from the numerical calculations for CT-specimen and the distribution of the iso-chromatics fringes obtained experimentally in the photoelastic coating is shown in Fig.4. The numerical and experimental results for DCB-specimen are presented in Fig.5.

TABLE 1

Critical values of stress intensity factors, results of calculations

Specimens	Media	$K_{Ic}^{(i)}$	$K_{IIc}^{(i)}$	$K_{Ic}^{(b)}$	J_c
		(MN/m$^{3/2}$)			(N/m)
CT		P_C = 375 N			
CT	M_1 - SFRC	1.62	1.73	1.16	45
	M_2 - polymer concrete	1.37	0.914		
DCB		P_C = 3200 N			
DCB	M_1 - SFRC	1.47	1.28	0.90	205
	M_2 - polymer concrete	1.104	0.634		

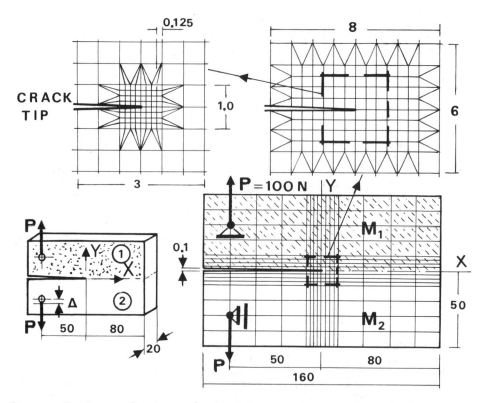

Figure 3. General view of the DCB-specimen and mesh for
a numerical simulation.

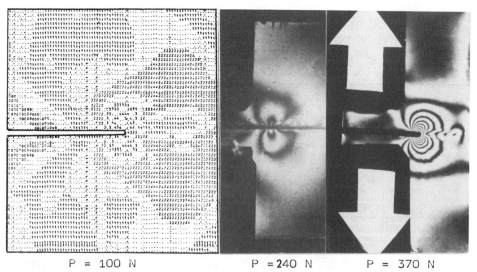

P = 100 N   P = 240 N   P = 370 N

Figure 4. Isochromatic fringe pattern in the CT-specimen,
obtained from FEM and experimentally in the
photoelastic coating.

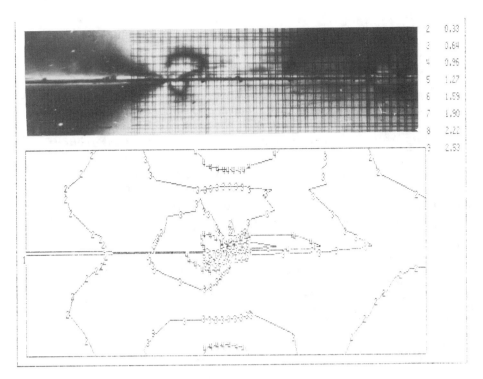

Figure 5. DCB-specimen. Isochromatic in photoelastic coating
and from FEM.

EXPERIMENTAL RESULTS

In order to analyse the influence of specimen geometry on the
crack development the experimental tests have been carried
out using following specimens composed of two media:
- type 1 - CT specimens Fig.2 made according to the ASTM
  E399 and E813,
- type 2 - DCB-specimens Fig.3.
The $K_{IC}^{(i)}$ value can be determined from Irwin method [10] using
isochromatic fringe loops which occur in the region adjacent
to the crack tip by employing the following relation:

$$K_{IC}^{(i)} = 2\tau_m^{(i)} \sqrt{2\pi r_i} \; \varphi_i(\theta^{(i)}) \tag{7}$$

where:

$$\tau_m^{(i)} = 0.5 f_i n_i$$

$\theta^{(i)}$ - angle corresponding to the point in the isochromatic fringe loop which occurs in the
region adjacent to the crack tip where
$\partial\tau_m^{(i)}/\partial\theta^{(i)} = 0$

$$\varphi_i(\theta^{(i)}) = \frac{1}{\sin\theta^{(i)}} \left[1 + \left(\frac{2}{3\operatorname{tg}\theta^{(i)}}\right)^2 \right]^{-\frac{1}{2}} \left(1 + \frac{2\operatorname{tg}\frac{3}{2}\theta^{(i)}}{3\operatorname{tg}\theta^{(i)}}\right)$$

$$f_i = f \frac{E_i(1+v)}{E(1+v_i)}$$ – photoelastic material fringe value on the medium surface i

$\quad\quad f$ – material fringe value of the photoelastic coating

E_i , E, v_i, v – the Young's moduli and Poisson's ratios of the medium i and photoelstic coating, respectively

r_i, $\Theta^{(i)}$ – polar coordinates at single measurement point in the medium i

$\quad\quad n_i$ – isochromatic fringe order obtained experimentally or from numerical calculation

The critical value of the J-integral (the strain energy release rate G) was determined according to expression (4) from load-displacement curve obtained using X-Y plotter. The stress intensity factor $K_{IC}^{(i)}$ were determined by applying the Irwin method from (7) and photoelastic measurement results. The displacements and the stresses in the interface between concrete and the polymer concrete was determined by applying the strain gauges.
The experimental and calculated values of $K_I^{(i)}$ factors and critical values of the J_C – integrals for three specimens of each type are given in Table 2.
As the crack propagated through medium 1 in all cases, the values of $K_I^{(2)}$ are smaller than the critical ones.

TABLE 2

Stress intensity factors of two media. Experimental results

	Specimens	$K_{IC}^{(1)}$	$K_I^{(2)}$	J_C
		(MN/m3/2)		(N/m)
CT	M₁ CRACK M₂	1.703 1.33 1.46	1.39 0.981 1.05	135 77 123
DCB	M₁ CRACK M₂	1.28 1.49 1.62	0.92 1.06 1.15	94 141 186

CONCLUSIONS

A generally good agreement has been obtained between numerical and experimental values of K_{IC} shown in Tables 1 and 2. In the models build up of two media the critical values of factors K_{IC} satisfied the inequalities:

$$K_{Ic}^2 > K_{Ic}^1 > K_{Ic}^b$$

and accordingly the crack initiation occurred always in the

interface. Next, it was found out that the crack propagated
in the fibre concrete which was less homogeneous, more porous
and more brittle $(E_1 > E_2)$ than the polymer concrete.
In order to analyse the influence of the configuration of a
specimen on the crack development and its propagation tests
have been carried out using two types of specimens CT and DCB.
It was found out that lower rigidity of cantilevers of DCB
specimens enables one to investigate the crack propagation in
the initial phase of fracture.
In the case of CT specimens made according to ASTM E 399
the brittle fracture occurs rapidly when critical load is
reached.

ACKNOWLEDGEMENT

This research was supported by CPBP PROBLEM 02.21 in coopera-
tion with the Institute of Fundamental Technological Research,
Polish Academy of Sciences.
The author is most grateful to Professor A.M.Brandt for many
valuable remarks with regard to the content of this paper.
Preparation of the polymer concrete specimens by Professor
L.Czarnecki, Warsaw Technical University, is gratefully
acknowledged.

REFERENCES

1. Sneddon, I.N., The Use of Integral Transforms. McGraw-Hill,
   New York, 1972.

2. Williams, M.L., The stresses around a fault or crack in
   dissimilar media. Bull.Seismol.Soc.Amer. 1959, 49,
   pp.199-204.

3. Rice, J.R. and Sih, G.C., Plane problem of crack in
   dissimilar media. J.Appl.Mech. 1965, E32, 418-432.

4. Clements, D.L., A crack between dissimilar anisotropic
   media. Int.J.Eng. 1971, 9, 257-265.

5. Willis, J.R., Fracture mechanics of interfacial cracks.
   J.Mech.Phys.Solids, 1971, 19, 353-368.

6. Piva, A. and Viola, E., Biaxial load effects on a crack
   between dissimilar media. Nota tecnica n.35.1979, Istituto
   di Scienza delle costruzioni,University of Bolonia, Italy.

7. Szmelter, J., The finite element method programs
   (in Polish), Arkady, Warszawa 1973.

8. Brandt, A.M., Influence of the fibre orientation on the
   energy absorption at fracture of SFRC specimens.
   In Brittle Matrix Composites 1, eds. A.M.Brandt,
   I.H.Marshall, Elsevier Applied Science Publishers, London
   1986, pp.403-420.

9. Irwin, G.R., Dally, J.W., Kobayashi, T., Fourney, W.L.,
   Etheridge, M.J. and Rossmanith, H.P., On the determination
   of the a-K relationship for birefringent polymers.
   Exp.Mech. 19, 4, 1979, pp.121-128.

10. Irwin, G.R., The dynamic stress distribution surrounding a running crack - a photoelastic analysis. Proc.of SESA, 16, 1, 1958, pp.92-96.

11. Zienkiewicz, O.C., The Finite Element Method in Engineering Science, McGraw-Hill, London, 1971.

12. Sih, G.C. and Liebowitz, H., Mathematical theories of brittle fracture in Fracture, vol.II, Academic Press, New York and London, 1968, pp.67-190.

13. Czarnecki, L., Lach, V., Structure and fracture in polymer concretes: some phenomenological approaches. as [8] pp.241-261.

14. Jaroniek, M., Niezgodziński, T., Studies of fracture and the crack propagation in the concrete and the polymer concrete, as [8], pp.355-370.

15. Czarnecki, L., The Status of Polymer Concrete. Concrete International, vol.17, no.7, pp.47-53, 1985.

16. Czarnecki, L., Grabowski, L., Criterion of cracking resistance of glass fiber reinforced resins. Proceedings of the International Symposium, ISAP'86, Adhesion between Polymers and Concrete, RILEM TC-52, Aix-en-Provence, France, Chapman and Hall, London, New York, 1986, pp.152-165.

17. Jaroniek, M., Niezgodziński, T., Fracture interfaciale entre les matériaux polymères et le béton armé, as [16], pp.32-40.

18. Jaroniek, M., Mesures photoélastiques de facteur d'intensité de tension de matériaux armés. Proc.of the 1-st International Congress RILEM, Paris-Versailles 1987, Chapman and Hall, London, New York, 1987, pp.671-677.

A COMPARISON OF THEORETICAL MODELS FOR THE MIXED MODE FRACTURE BEHAVIOUR OF THERMOSETTING RESINS

CARLOS A.C.C. REBELO
Secção Autónoma de Engenharia Mecânica - FCTUC
3000 Coimbra, Portugal

PAULO M. S. T. DE CASTRO
ANTÓNIO TORRES MARQUES
Departamento de Engenharia Mecânica - FEUP
Rua dos Bragas, 4099 Porto Codex, Portugal

ABSTRACT

A study on the mixed mode I and II crack propagation in a thermosetting polyester resin is described. Experimental measurements of the direction of initial crack extension, fracture stress and critical interaction of the stress intensity factors K_I and K_{II} are compared with theoretical predictions based on the maximum circumferential stress criterion and the minimum strain energy density criterion. It is shown that the minimum strain density criterion provides a better interpretation of the experimentally observed behaviour.

INTRODUCTION

Thermosetting resins are being widely used as matrix materials of fibrous composites in many engineering applications. From a structural point of view, the matrix of these systems is responsible for transferring the load to the reinforcing fibres and thus it plays an important role in the fracture behaviour of the composite.

In a series of previous papers [1-3], the authors have already analysed the influence of cure conditions on the tensile properties and fracture toughness of thermosetting polyester resins commonly used in glass-fibre reinforced plastics, as well as the mixed mode fracture behaviour of one of these resins. The experimental results obtained from fracture tests under combined modes I and II of loading are now used to carry out a detailed research on the performance of two theoretical models for the prediction of the crack growth direction and the critical stress in these circumstances.

FRACTURE CRITERIA FOR MIXED MODE LOADING

The problem of fracture under modes I and II of loading is usually studied using as a basic model the situation sketched in Figure 1, in which a straight crack of length 2a contained in a plate of infinite dimensions is angled to the loading direction. The determination of the critical remote stress σ_c necessary to propagate the crack and the fracture angle θ_0 which defines the direction of initial crack extension, are the main questions to be solved. Several fracture criteria have been developed to answer correctly these questions. Among those criteria, the maximum circumferential stress criterion, proposed by Erdogan and Sih [4], and the minimum strain energy density criterion, introduced by Sih [5], are the ones which have been discussed in greater detail.

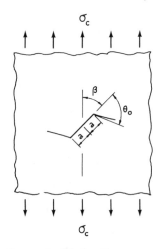

Figure 1. Angled crack under uniaxial loading.

Erdogan and Sih [4], basing their analysis on the singular term of the series expansions for the polar components of the crack tip stress field [6], postulated that fracture initiates following a radial direction to which corresponds a maximum circumferential stress and occurs when this stress reaches a critical value. Under these conditions, the equations which relate the fracture angle θ_0 and the critical stress σ_c with the crack angle ß, are:

$$\cotan \beta = \frac{\sin\theta_0}{1-3\cos\theta_0} \quad , \quad \beta \neq 0 \tag{1}$$

$$\frac{\sigma_{Ic}}{\sigma_c} = \sin \beta \cos(\theta_0/2)[\sin \beta \cos^2(\theta_0/2) - \frac{3}{2}\cos \beta \sin \theta_0] \tag{2}$$

where σ_{Ic} is the critical remote stress for ß = 90°.

Later on, Williams and Ewing [7] have shown that the introduction of non-singular terms on the series expansions for the stresses gives a better correspondance between the theoretical and experimental results. Including the second term of the asymptotic solutions for the stresses, Williams and Ewing modified the maximum circumferential stress criterion postulating that the fracture initiates following the direction on which the circumferential stress has a maximum value but at a small critical distance from the crack tip. Fracture occurs when the maximum circumferential stress, at that distance, reaches a critical value. According to this modification and taking into account a correction pointed out by Finnie and Saith [8] to the Williams and Ewing analysis, the equations to determine θ_0 and σ_c as a function of ß are as follows:

$$[1+(\frac{2r_0}{a})^{1/2}\frac{16\sin(\theta_0/2)}{3\tan\theta_0}]\tan^2\beta - [\frac{1-3\cos\theta_0}{\sin\theta_0}]\tan\beta -$$

$$-(\frac{2r_0}{a})^{1/2}\frac{16\sin(\theta_0/2)}{3\tan\theta_0} = 0 \tag{3}$$

$$\frac{\sigma_{Ic}}{\sigma_c} = \sin\beta\cos(\theta_0/2)[\sin\beta\cos^2(\theta_0/2) - \frac{3}{2}\cos\beta\sin\theta_0] +$$

$$+(\frac{2r_0}{a})^{1/2}(\cos^2\beta - \sin^2\beta)\sin^2\theta_0 \tag{4}$$

where r_0 is the mentioned critical distance.

Using a different approach, Sih [5] has developed a fracture criterion in which the intensity of the strain energy density field is the parameter of interest. According to this criterion, often referred to as the S - criterion, the fracture initiates on a direction corresponding to a minimum of the strain energy density, and the propagation occurs when this minimum reaches a critical value. If the strain energy density is computed using the singular solution for the crack tip stresses, the equations which relate θ_0 and σ_c with ß, for plane strain conditions, are:

$$2(1-2\upsilon)\sin(\theta_0-2\beta) - 2\sin[2(\theta_0-\beta)] - \sin 2\theta_0 = 0 \quad , \quad \beta \neq 0 \tag{5}$$

$$\frac{\sigma_{Ic}}{\sigma_c} = [\frac{4\pi\mu(a_{11}\sin^2\beta + 2a_{12}\sin\beta\cos\beta + a_{22}\cos^2\beta)\sin^2\beta}{1-2\upsilon}]^{1/2} \tag{6}$$

where υ is the Poisson ratio, μ is the shear modulus and a_{11}, a_{12}, a_{22} are coefficients depending on the material elastic constants [5].

It must be noted that for each value of ß, the equations (1), (3) and (5) give a negative and a positive value for θ_0, which correspond, respectively, to the application of a tensile or compressive load.

Theoretical curves of θ_0 versus ß and σ_{Ic}/σ_c versus ß based on exact solutions for the stresses, have been obtained by Sih and Kipp [9] for both the maximum circumferential stress and minimum strain energy density criteria. In the case of the maximum circumferential stress criterion, it is interesting to notice that for small values of r_0/a the use of equations (3) and (4) gives very similar results to the ones calculated with exact solutions.

The combination of the mode I and mode II stress intensity factors, K_I and K_{II}, at the onset of fracture, is sometimes used to define the critical propagation conditions of a crack under a general two-dimensions stress system. Taking into account that for the considered geometry $K_I = \sigma \sqrt{\pi a} \sin^2\beta$ and $K_{II} = \sigma \sqrt{\pi a} \sin\beta\cos\beta$, the K_I and K_{II} values related to the fracture initiation can be easily obtained using the equations:

$$\frac{K_I}{K_{Ic}} = \frac{\sigma_c}{\sigma_{Ic}} \sin^2 \beta$$

$$\frac{K_{II}}{K_{Ic}} = \frac{\sigma_c}{\sigma_{Ic}} \sin\beta \cos \beta \qquad (7)$$

where K_{Ic} is the critical value of K_I for ß = 90°

EXPERIMENTAL

The material tested was the Crystic 272 polyester isophtalic resin, supplied by Quimigal(Portugal). The cure system consisted a peroxide methyl ethyl ketone (Butanox M50) and a cobalt octoate accelerator (TP 395 VZ). Catalyst and accelerator were used in weight percentage of 2% and 0.1% respectively. Mixtures were always subjected to a vacuum degassing, in order to eliminate possible air bubbles.

The experimental characterization of the resin's fracture behaviour under mixed mode loading was carried out performing uniaxial tensile tests on specimens of dimensions 350mmx160mmx4mm containing central cracks 16mm long, with different inclinations. Specimens were machined from 4 mm thick plates, cast between two flat glass plates clamped together against a silicone rubber gasket. Before machining, all the plates were subjected to a 24 hours cure at room temperature, followed by a 3 hours post cure at 80°C. In order to obtain the cracks, a 12mm long notch was produced with a 0.3mm thick saw, starting from a 2 mm diameter central hole. Nominal notch inclinations were 0°, 15°, 30°, 45°, 60°, 75° and 90°, measured from the load direction. In both ends of the notch, a natural crack approximately 2mm long was obtained by a light impact of a razor blade. For each value of notch inclination, 3 specimens were prepared and tested at 21 ± 2^oC and $60+10\%$ relative humidity.

Specimens with central crack parallel to the load direction were tested in a Tinius Olsen universal testing machine, with a 300 kN load cell. All the other tests were carried out using an Hounsfield Type "W" tensometer, with a 20 kN load cell. The displacement rate for both machines was 2mm. min-1. Specimens were tested using grips fastened by three equally spaced screws. All screws were identically tightened using a torque wrench, up to an equal pre load such that during testing the load was entirely transmitted by friction between grips and specimen surface. The grips were assembled with the help of a set of special tools designed to ensure their proper alignment. For each test, a load versus time record was obtained.

After testing, the length 2a of the precrack, inclination ß and fracture angle θ_0 were measured. The practical impossibility of producing natural cracks with exactly the same orientation as the initial notches, lead to the appearance in the same specimen, of natural cracks with different inclinations. Since fracture is a localized phenomenon, the natural cracks inclinations were used to correlate with the fracture angle. Each plate specimen gave, therefore, two pairs of ß and θ_0 values. The crack length was measured as the projection in the notch direction of the distance between the natural crack tips. All values are an average of measurements carried out using a Mitutoyo microscope, with 30 x magnification.

RESULTS

Fracture angle θ_0, as a function of initial inclination ß, is represented in Figure 2.

Figure 3 presents the corrected values of the ratio between remote critical stress for ß = 90°, σ_{Ic}, and remote critical stress for different initial crack inclination, σ_c, as a function of ß. The correction consisted of multiplying σ_{Ic} and σ_c respectively by $\sqrt{a_I}$ and $\sqrt{a}$, where a_I is semicrack length for $ß_{nom}$ = 90°, in order to make it possible the comparison of the critical stresses. The normalizing factor used was $\sigma_{Ic}\sqrt{a_I}$ = 0,381 MNm-3/2, which is the average of $\sigma_{Ic}\sqrt{a}$ values for all the $ß_{nom}$ = 90° specimens. Following the procedure proposed by Williams and Ewing [7], in their pioneering work on the inclined crack problem, each σ_c value was correlated with the lower of both ß values. This supposes that fracture is controlled by the crack tip corresponding to the lower inclination.

Figure 4 presents the critical combination of K_I and K_{II}, normalized with the critical value of the stress intensity factor in pure mode I, K_{Ic} = 0,675 MNm-3/2.

In all these figures, a comparison of experimental data and theoretical predictions given by the maximum circumferencial stress criterion is shown. The theoretical curves presented were obtained using singular solutions for the crack tip stresses or the two-term asymptotic approximation based on a crack length 2a=16mm and critical distances r_0 = 0,04 mm and r_0 = 0.16 mm.

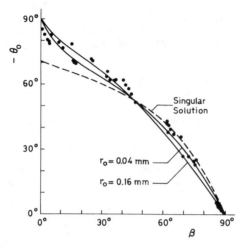

Figure.2. Fracture angle versus crack orientation - Experimental results and theoretical predictions based on the maximum circumferential stress criterion.

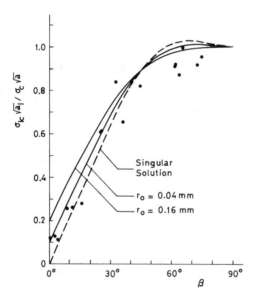

Figure 3. Normalized critical stress versus crack orientation - Experimental results and teoretical predictions based on the maximum circumferential stress criterion.

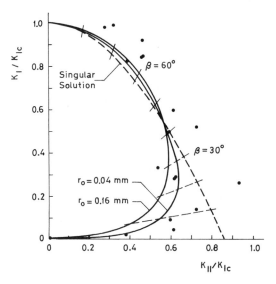

Figure 4. Critical interaction of K_I and K_{II}, normalized with K_{Ic} - Experimental results and theoretical predictions based on the maximum circumferential stress criterion.

Figures 5, 6 and 7 present similar comparisons, using now the predictions provided by the minimum strain energy density criterion when singular or exact solution for the stresses are considered. These predictions assume plane strain conditions and a Poisson ratio of $\upsilon = 0.33$ which is thought to be a good estimate for the studied material. The full curves in Figures 5 and 6 were obtained by Sih and Kipp [9] from an exact stress solution, derived for a degenerated elliptic hole.

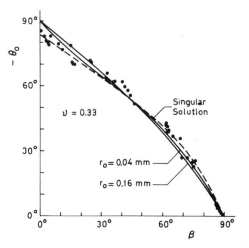

Figure 5. Fracture angle versus crack orientation - Experimental results and theoretical predictions based on the minimum strain energy density criterion.

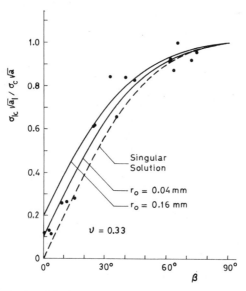

Figure 6. Normalized critical stress versus crack orientation - Experimental results and theoretical predictions based on the minimum strain energy density criterion.

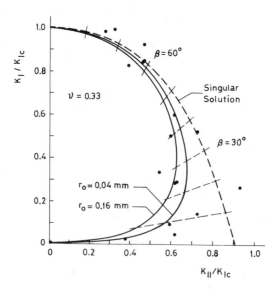

Figure 7. Critical interaction of K_I and K_{II}, normalized with K_{Ic} - Experimental results and theoretical predictions based on the minimum strain energy density criterion.

DISCUSSION

The effectiveness of the analysed criteria to predict the variation of the fracture angle with respect to the crack orientation can be assessed in Figures 2 and 5. It can be seen that, for the maximum circumferential stress criterion, the theoretical curves based on non-singular solutions for the stresses are the ones in best agreement with the experimental data, although it is not possible to define the value of the critical distance which gives more precise predictions. For the minimum strain energy density criterion, anyone of the curves describe satisfactorily the obtained results.

As far as the variation of σ_{Ic}/σ_c versus ß is concerned, the observation of figures 3 and 6 suggests that the minimum strain energy density criterion is the one that provides a better interpretation of the experimental results.

The critical interaction curves of K_I and K_{II}, that in the literature reviewed were only found for singular solutions, have been here worked out for the non-singular ones. Figures 4 and 7 show again that the minimum strain energy density criterion gives a better description of the experimental observations.

CONCLUSIONS

For the Crystic 272 resin, the direction of the initial extension of a crack loaded under mixed modes I and II, is predicted with a reasonable approximation by the maximum circumferential stress criterion, considering non-singular solutions for the stresses and critical distances between 0.04 and 0.16mm.

Equally approximated are the predictions based on the minimum strain energy density criterion, regardless of the use of singular or exact solutions.

For the Crystic 272 resin, the critical stress at which the propagation of an angled crack occurs is better predicted by the minimum strain energy density criterion than by the maximum circumferencial stress criterion.

REFERENCES

1.   C. A. C. C. Rebelo, A. T. Marques, P. M. S. T. de Castro, The influence of cure conditions on the fracture of non-reinforced thermosetting resins, in: Brittle Matrix Composites,A. M. Brandt, I. H. Marshall, eds., Elsevier Applied Science, 1988, pp.305-309.

2.   C. A.C. C. Rebelo, A. J. M. Ferreira, A. T. Marques, P. M. S. T. de Castro, The influence of processing conditions on mechanical and fracture properties of GRP plates, in: Mechanical Behaviour of Composites and Laminates, W. A. Green, M. Micunovic, eds., Elsevier Applied Science, 1987, pp. 54-63.

3.   C. A. C. C. Rebelo, A. T. Marques, P. M. S. T. de Castro, Fracture characterization of composites in mixed mode loading, 6th European Conference on Fracture, ECF 6, Amsterdam, 15-20 June 1986.

4. Erdogan, J. G. and Sih, G. C., On the crack extension in plates under plane loading and transverse shear, J. Basic Eng., 1963, **85**, 519-527.

5. Sih, G. C., Mechanics of Fracture, Vol I, Noordhof International Publishing, Leyden, 1973, XXI-XLV.

6. Williams, M. L., On the stress distribution at the base of a stationary crack, J. Appl. Mech., 1957, **24**, 109-114.

7. Williams, J. G., and Ewing, P. D., Fracture under complex stress - the angled crack problem, Int. Journ. Fract. Mech., 1972, **8**, 441-446.

8. Finnie, I. and Saith, A., A note on the angled crack problem and the directional stability of cracks, Int. Journ. of Fracture, 1973, **9**, 484-486.

9. Sih, G. C. and Kipp, M. E., Discussion on "Fracture under complex stress - the angled crack problem" by J. G. Williams and P.D.Ewing, Int. Journ. of Fracture, 1974, **10**, 261-265.

MINERAL POLYMER MATRIX COMPOSITES

G.PATFOORT, J.WASTIELS, P.BRUGGEMAN, L.STUYCK

Department of Constructions
Vrije Universiteit Brussel (VUB)
Pleinlaan 2, 1050 Brussels, Belgium

ABSTRACT

The intrinsic properties of mineral polymers are compared to the properties of organic polymers and cementitious materials. It follows that the toughness of mineral polymers can be increased by the addition of fibres, and that it is possible to incorporate a higher volume content of long fibres than is possible in cement. An application of this is presented under the form of tensile and bending characteristics of a jute reinforced laminate with mineral polymer matrix.
Discussion of the results reveals that a tensile and bending strengthening and toughening is possible, and that the NPL-theory for continuous fibres with frictional bond is valid, if one takes into account the different behaviour of the fibres before and after matrix cracking.

MINERAL POLYMERS (MIP)

Mineral polymers (MIP) are materials that have similar properties as ceramics, porcelain, pottery or glasses. They are however not obtained by fusion or firing at high temperature. MIP are inorganic materials that are obtained by chemical polymerisation or polycondensation of different oxydes from H, Li, Na, K, Al, Si, P,...
At present, the research efforts at Brussels University (V.U.B.) are concentrated on formation of MIP at atmospheric pressure and low temperature : between room temperature and 100°C. A series of polymers have been developed and tested for different properties. MIP can be prepared from powders, grains, pastes or liquids, and can thus be manufactured with processes as pressing, extrusion and casting. Composite materials technologies, such as lamination and filament winding, can also be used.
Due to their chemical nature, MIP present some attractive properties, compared to metals and organic polymers : chemical inertia, temperature resistance, low energy input, incombustibility. On the other hand they present, just as other ceramics, an important lack of toughness and thus a brittle behaviour and limited tensile strength.

MIP IN FIBRE REINFORCED COMPOSITES

Tough and strong materials can be obtained with organic polymers reinforced
with strong fibres. In table 1, some properties of organic polymers are
compared to the corresponding ones of materials of ceramic type, like
cement, MIP or ceramics. It follows that fibre addition to organic polymers
is used for stiffening and strengthening of the matrix

TABLE 1
Typical Properties of Matrix Materials

ORGANIC POLYMERS		CEMENT CERAMICS MIP

Low Stiffness	↔	High Stiffness
High Strain Capacity	↔	Low Strain Capacity and Brittle
Limited Temperature Use	↔	Moderate to High Temperature Use
Not Fire Resistant	↔	Fire Resistant
Easy Processing for Composites	↔	Generally difficult Processing for Composites

(although the point of view that some matrix is needed for protecting the
stiff and strong fibres is also correct), while for the ceramic materials
fibre addition is mainly aimed to improve the toughness and tensile
strength. Whether this is technically and economically possible for brittle
matrices depends on a number of factors, some of which are represented in
table 2 in qualitative form.
Taking into account that for increasing the tensile cracking stress essentially stiff fibres are needed, and that for improving toughness and strain
capacity high volume contents of long (or continuous) fibres are needed,
the picture of tables 1 and 2 is not so positive for the brittle matrices.
Indeed, the fabrication at high temperature of ceramics limits the choice
of fibres and causes thermal expansion problems, while the rheology of a
hydraulic cement makes it very difficult, if not impossible, to incorporate
high volume contents of long fibres due to the poor penetration of the
bundles. Some mineral polymers however consist,after mixing,the raw materials,
of a liquid with low viscosity, comparable to an organic resin. With these
MIP it is possible to fabricate at low temperature fibre reinforced composites with an amount of continuous fibres that exceeds the critical fibre
content, needed for the obtention of the strengthening.
In this case, the existence of large flaws can be avoided, and the formation of a large amount of microcracks during straining of the composite
leads to a high fracture energy.

TABLE 2
General Characteristics of Brittle Matrices

	CEMENT	CERAMICS	MIP
Fabrication Temperature	Low	High	Low
Possible Temperature of Use	Limited	High	High
Cost	Low	High	Low
Processing for Composites	Difficult	Very Difficult	Easy

APPLICATION

Jute Reinforced Laminate

To assess the possibilities of reinforcing MIP with long fibres, a series of jute reinforced laminates with MIP matrix was fabricated. The used jute consisted of an orthogonal woven mat with a surface density of approximately 210 g/m^2. Two parallel laminae of this mat were used as reinforcement for the MIP, resulting in an orthotropic ($E_1 = E_2$) laminate with a thickness of 2.2 mm, and a volume content of fibres of $V_f = 0.17$.

The laminate was made with the hand lay-up technique, in a very similar way to the fabrication of a resin matrix composite. After lamination, a curing was performed during 10 hours at 60°C and atmospheric pressure. The laminate was then cut with a diamond blade parallel to one of the fibre orientations, to obtain tensile and bending coupons of a width of 25 mm.

The relevant characteristics of the matrix were obtained on a bulk MIP :

E_m = 8000 MPa   as Young's modulus

ε_{mu} = 0.03 %   as ultimate tensile strain

σ_{mu} = 2.4 MPa   as ultimate tensile stress

The characteristics of the jute fibre were determined in two ways :

- by a tensile test on an individual bundle of fibres with a length of 300 mm :

E_f = 8030 MPa   (standard dev. : 1150 MPa)

ε_{fu} = 3.65 %   (s.d. : 0.75 %)

σ_{fu} = 123 MPa   (s.d. : 19 MPa)

It must be emphasized that the large standard deviations result from irregularities in bundle geometry and composition due to the natural origin of the fibre, and that the stiffness of the bundle is very low at low strains, and increases gradually to the value of 8030 MPa at and beyond a strain of 1.5 % : the "rope" effect is obvious.

- by a tensile test of a polyester-jute laminate. Young's modulus, computed from the law of mixtures, and knowing the volume contents of fibres and matrix, and Young's modulus of the matrix and the composite, is the following : E_f = 28.400 MPa.

Ultimate tensile stresses and strains could not be obtained because it was not possible to incorporate the critical volume content of fibres.

The striking difference between the stiffness values of jute, obtained in two different ways, is mainly attributed to the different loading conditions : in the bundle test, the "rope" effect can fully develop and bundle characteristics are measured, while in the composite test the adherence between fibres and matrix leads to the measurement of fibre characteristics rather than bundle characteristics.

Tensile Characteristics

Tensile testing was performed on the laminate coupons in an INSTRON test machine at a displacement rate of 0.2 mm/min (strain rate of approx. 0.2 %/min). Longitudinal strain was measured by means of an extensometer, attached to the specimen. A typical stress-strain curve is given in figure 1 (mean results of 4 tests are used for the calculations).

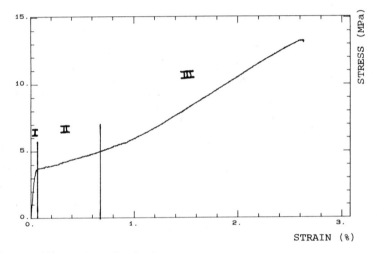

Figure 1. Tensile test on laminate.

Three stages of behaviour can be distinguished, corresponding to the NPL-theory for continuous aligned fibres with frictional bond between the fibres and matrix [2] :

- an initial stage before matrix cracking (I). The measured stiffness of the composite is 8900 MPa. Application of the law of mixtures, with an effective volume content of fibres $V_f' = 1/2 \ V_f = 0.085$ (half of the fibres is directed perpendicular to the applied stress, half is aligned with it) leads to a fibre stiffness of 26.600 MPa. This corresponds to the value obtained in the polyester composite test, where a perfect bond exists between fibres and matrix.
  The mean cracking strain of the matrix ε_{mu} equals 0.03 %, the same as for the plain matrix.

- a stage of crack formation (II) during which load is transferred from the matrix to the fibres in the vicinity of the cracks. Two deviations are observed from the theoretical curve :

. the behaviour does not occur at constant composite stress, but a slightly increasing one. This deviation was found to be more pronounced at
higher fibre contents

. the strain increase during this stage is higher than predicted by the
theory. Experimentally a mean value of 0.67 % is observed. Theoretically
one gets $1/2\ \alpha\ \varepsilon_{mu}$, with $\alpha = E_m V_m / E_f V'_f$ equal to the relative stiffness
of the matrix to the fibres. Using the "fibre" modulus of 26.600 MPa,
the theoretical value is only 0.044 %, while using the "bundle" modulus
of 8030 MPa one gets 0.15 %.

These deviations are believed to be caused by the gradual transfer of the
behaviour conditions of the fibres from continuity with the matrix before
cracking, to bundle behaviour after cracking.
The effective modulus of the fibre bundle in the strain region of interest is experimentally much lower than the measured value of 8030 MPa at
higher strains, leading to higher straining in the crack formation stage.
than is predicted by the theory, which does not take into account this
effect.

- a stage of crack opening (III) during which the fibres are stressed up
to failure. The composite modulus at higher strains indicates that the
effective bundle modulus is 5800 MPa, a lower value than the free bundle
modulus. Calculated fibre failure stress is $\sigma_{fu} = 160$ MPa, and the corresponding strain is $\varepsilon_{fu} = 3.7$ %. The ultimate composite strength is
$\sigma_{cu} = 13.5$ MPa (st. dev. 0.4 MPa) which corresponds to 560 % of the
matrix strength.
The energy absorbed until failure of the composite is about 580 times
the energy absorbed by the matrix alone.

Bending Characteristics

Three point bending tests with a span of 120 mm were performed on some of
the tensile coupons at a displacement rate of 2 mm/min (maximum strain rate
of approx. 0.2 %/min for elastic behaviour). A typical composite stress-
strain curve is given in figure 2.

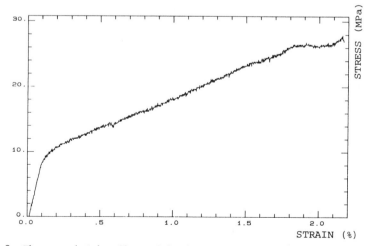

Figure 2. Three point bending of laminate.

Again a marked increase in strength and energy absorption capacity is noted : the mean MOR equals 25.5 MPa (1.9 times the ultimate tensile strength, which corresponds very well with the NPL-theory [2]).

The strain at maximum stress is lower in bending than in tension. This is attributed to the fact that the strain value is calculated from the central deflection by elastic beam theory, which leads to an underestimation of the real value.

Calculation of the compressive stress at the ultimate strength leads to the conclusion that the composite fails nearly simultaneously in compression and in tension.

CONCLUSIONS

It is possible to manufacture long fibre reinforced composites, using mineral polymers as matrix material. The fibre content can be relatively high, much higher than possible with cement, due to the favourable rheological properties of MIP. As such a tough composite material with incombustible matrix can be obtained. The behaviour of this composite material can be described by the NPL-theory for continuous fibres with frictional bond, taking into account the different fibre bundle behaviour before and after matrix cracking.

REFERENCES

1. Davidovits, J., Lavau, J., Solid Phase Synthesis of a mineral Block-polymer by low Temperature Polycondensation of Alumino-Silicate Polymers, IUPAC International Symposium on Macromolecules, Royal Inst. of Technology, Stockholm, Sweden. 1976.

2. Aveston, J., Mercer, R.A., Sillwood, J.M., The Mechanism of Fibre Reinforcement of Cement and Concrete (Part I), National Physical Laboratory, publ. n°SI 90/11/98, January 1975.

ANALYSIS OF MATERIAL EFFICIENCY OF RESIN CONCRETES

ZDZISLAW PIASTA
Technical University of Kielce
25-314 Kielce,Al.1000-lecia 5,Poland

LECH CZARNECKI
Warsaw Technical University
00-637 Warszawa,Al.Armii Ludowej 16,Poland

ABSTRACT

For building applications, material design and general useability under given conditions is an important engineering consideration. In the present paper, a procedure based on overall desirability is postulated. The concept of overall desirability is a means of estimating composites useability taking into account all of the relevant properties.

Analysis of the overall desirability behaviour for two polyester concretes and two epoxy concretes is presented. Results of the statistical and graphical analysis can be treated as a quantitative representation of the forecasts based on the theoretical considerations of the material model for resin concretes.

INTRODUCTION

Resin concretes are used as a rule in rugged conditions. Material costs of resin concretes are very high.The material requirements dictated by various service conditions could be fulfilled by resin concretes only when simultaneously several composite properties are considered [1].Engineering intuition is usually insufficient in the design of resin concretes because of complex mechanism of the forming of concrete properties.

Combining the response surface methodology and multicriteria optimisation based on an overall desirability function is an efficient method of the solution of that problem.

MATERIAL AND METHODS

Assumptions

Composite properties $y^{(1)}, \ldots, y^{(m)}$ can be treated as the effect of the action of controllable factors $z_1, \ldots, z_n$ related to the composition and technology of the composite and of uncontrollable noise factors .Occurrence of a phenomenon of internal interactions between components, structure ,and energy state is a characteristic feature of the composite [1].

Resin concretes considered in the paper have been treated as mixtures of four components:

M—microfiller,

S—sand,

G—gravel,

B—resin binder.

Selection of the most desirable fractions of the particular components has been formulated as an optimisation problem. Three factors:

$z_1 = (M+S+G)/B$ — ratio of aggregate to resin binder,

$z_2 = M/(M+S+G)$ — fraction of microfiller in aggregate,

$z_3 = S/(M+S+G)$ — fraction of sand in aggregate

have been taken as decision variables.

Three responses:

$y^{(1)}$ — compressive strength,

$y^{(2)}$ — flexural strength,

$y^{(3)}$ — apparent density

have been chosen as criteria variables.

Laboratory Tests

Laboratory tests have been performed according to four experimental programs due to a three-factor rotatable second order design with two central points [2] (Table 1).

TABLE 1

Experimental design

No. of experiment	Code values of factors		
	x_1	x_2	x_3
1	-0.58	-0.58	-0.58
2	0.58	-0.58	-0.58
3	-0.58	0.58	-0.58
4	0.58	0.58	-0.58
5	-0.58	-0.58	0.58
6	0.58	-0.58	0.58
7	-0.58	0.58	0.58
8	0.58	0.58	0.58
9	-0.97	0	0
10	0.97	0	0
11	0	-0.97	0
12	0	0.97	0
13	0	0	-0.97
14	0	0	0.97
15	0	0	0
16	0	0	0

The programs varied due to the kind of resin binder (epoxy E
or two kinds of polyester P), the maximal diameter of gravel
grain (4 or 10 mm) and the dimension of the experimental
domain (Table 2).

TABLE 2

Experimental domain

z_i	Base level z_{i0}				Unit range of variability Δz_i			
	P108(4)	P109(4)	E5(4)	E5(10)	P108(4)	P109(4)	E5(4)	E5(10)
z_1	8.0	8.0	6.75	6.75	1.72	1.72	1.72	1.72
z_2	0.2	0.2	0.2	0.3	0.1	0.1	0.2	0.2
z_3	0.3	0.3	0.3	0.2	0.1	0.1	0.2	0.2

In the framework of each program 16 various compositions have
been prepared in accordance with the experimental design.
Moreover 4 additional replications have been done in the
central point of the experimental domain to estimate the
experimental error variances.
The levels of factors z_i in the compositions have been
obtained from the relation:

$$z_i = z_{i0} + x_i \Delta z_i, (1)$$

where :
z_i is the value of the i-th factor,
z_{i0} is the base level of the i-th factor,
Δz_i is the unit range of variability of the i-th factor,
x_i is the code value of the i-th factor in the experimental
        design.

Each experimental result of compressive and flexural
strength as well as density of resin concretes has been
obtained as a mean of 7 to 10 measurements on different
specimens.

Overall Desirability
The question which arises is how to find such a point in the
experimental domain which leads to the "best" overall set of
response values. The approach applied in this paper to solve
that problem is connected with an overall criterion of
desirability [3]. For each feature $y^{(i)}$ $(i=1,...,m)$ one should
choose such two levels $y_1^{(i)}$ and $y_h^{(i)}$ of its values so that the
material properties are "satisfactory" from the practical
point of view if $y^{(i)} \in < y_1^{(i)}, y_h^{(i)} >$.
Desirability d_i of the particular value $y^{(i)}$ of the feature
is defined then by the formulae:

$$d_i = \exp\left[-\exp\left[-\frac{y^{(i)} - y_1^{(i)}}{y_h^{(i)} - y_1^{(i)}}\right]\right] \qquad (2)$$

It follows from formulae (2) that values of $y^{(i)}$ are transformed into $Y^{(i)} = \left[y^{(i)} - y_1^{(i)}\right] / \left[y_h^{(i)} - y_1^{(i)}\right]$ at first and next into $d_i = \exp\left[-\exp\left(-Y^{(i)}\right)\right]$ (Fig.1).

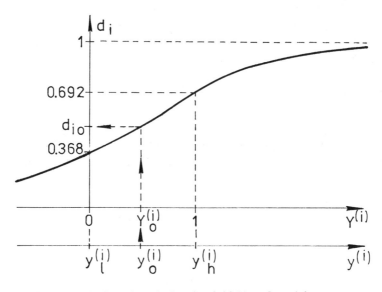

Figure 1.Graph of desirability function.

Desirability d_i is an increasing function of $y^{(i)}$,which takes values from the interval $(0,1)$.Levels of d_i close to 0 are related to the unacceptable values of feature $y^{(i)}$;levels close to 1 are related to the best values.The set of the "satisfactory" values of the feature is transformed by function (2) into the interval $<0.368,0.692>$.Taking into account the form of function (2),an overall desirability D has been defined as the weighted exponential mean of desirabilities d_i $(i=1,\ldots,m)$, i.e.

$$D = \exp\left[\exp\left[-\sum_{i=1}^{m} w_i \frac{y^{(i)} - y_1^{(i)}}{y_h^{(i)} - y_1^{(i)}}\right]\right], \qquad (3)$$

where weights $w_i \in \langle 0,1 \rangle$ and $\sum_{i=1}^{m} w_i = 1$.

Values of D range in the interval $(0,1)$. Higher values of D point to a better overall desirability of the considered composite material in the sense of its predicted application.

RESULTS

Experiments have been performed according to the program presented in the preceding section. Values of the overall desirability have been computed for all 80 resin concretes. The weights and intervals of satisfactory levels for the three considered criteria variables of the concretes have been chosen as in Table 3.

TABLE 3

Weights and satisfactory values for criteria variables of resin concretes

Variable $y^{(i)}$	Unit	Weight w_i	Interval $\langle y_1^{(i)}, y_h^{(i)} \rangle$
Compressive strength $y^{(1)}$	MPa	0.5	$\langle 50,70 \rangle$
Flexural strength $y^{(2)}$	MPa	0.3	$\langle 25,30 \rangle$
Apparent density $y^{(3)}$	kg/m^3	0.2	$\langle 2000,2150 \rangle$

It follows from Table 3 that the mechanical strength requirements predominate over tightness requirements represented by density in the predicted application of concrete.

Second order regression functions have been employed to

approximate the relationships between the overall desirability
D and the material factors over a region of interest [4].
Values of D corresponding to the points of the experimental
design have been used to the estimation of regression
coefficients.The following functions with significant ($\alpha=0.1$)
regression coefficients only have been obtained :
—for the P108(4) concrete:

$$D=0.934+0.028x_2-0.047x_1^2-0.0566x_2^2-0.0164x_3^2+0.018x_1x_3, \quad (6)$$

—for the P109(4) concrete:

$$D=0.934+0.021x_1+0.025x_2-0.053x_1^2-0.027x_2^2+0.028x_1x_3, \quad (7)$$

—for the E5(4) concrete:

$$D=0.822-0.031x_1+0.066x_2+0.047x_3^2 \quad (8)$$

For the E5(10) concrete a model with all the coefficients has
been used:

$$D=0.874-0.024x_1+0.084x_2+0.081x_3-0.065x_1^2-0.104x_2^2-0.091x_3^2+$$
$$+0.01x_1x_2+0.077x_1x_3-0.041x_2x_3. \quad (9)$$

Values of the coefficient of multiple correlation are higher
than 0.9 in all cases.That means that the degree of
association existing between experimental data and fitted
values obtained from relations (6)-(9) is high.
Analysis of variance has been another method of checking
whether the fitted equations provide an adequate representation
of the overall desirability.In all four cases the F ratios of
the residual variance to the experimental variance have been
smaller than the critical values of the F statistics for
significance level $\alpha=0.05$.It means that there is no reason to
doubt the adequacy of functions (6)-(9).

DISCUSSION

Practical inferences arising from relationships (6)-(9) have

been formulated on the basis of an analysis of their graphic
representation.To prepare the figures it was necessary to fix
the value of one of the considered material factors at a
certain level.

To illustrate relation (6) factor x_2,i.e. fraction of sand in
gravel has been fixed at the code level 0.247 .This value of
that factor leads to the maximal overall desirability.It
follows from Fig.2a that maximal values of D are reached in
the experimental domain.The composition of the most desirable
P108(4) concrete is following:11% of polyester ,20% of
microfiller, 26.5% of sand and 42.5% of gravel.

Relation (7) is represented graphically in Fig.2b in the form
of lines of equal values of the overall desirability D.

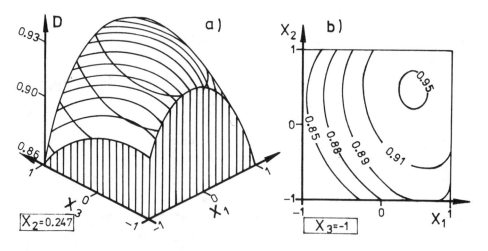

Figure 2.Effect of material factors on overall desirability
of P108(4)[a] and P109(4)[b] concretes.

A value of factor x_3 i.e. fraction of sand in aggregate has
been fixed at the code level -1, which leads to maximal D.A
maximal value of D is related to the following composition of
the P109(4) concrete : 10.5% of polyester ,21.5% microfiller,
36% of sand and 32% of gravel.

Relations (8) and (9) are illustrated graphically in the

triangular coordinate system.This system includes three of
the four components of the considered concretes.Fraction of
resin binder has been fixed at the base level ,i.e. 13%.
Taking into account relation (1) ,the form of function (8)
with microfiller M, sand S and gravel G as arguments is
following:

$$D=0.822-0.031\frac{(M+S+G)/0.13-6.75}{1.72}+0.066\frac{M/(M+S+G)-0.2}{0.2}+$$
$$+0.047\frac{S/(M+S+G)-0.3}{0.2} \quad (10)$$

Fig. 3a is a graphical illustration of relation (10).An
analogic procedure applied to relation (9) leads to Fig.3b.

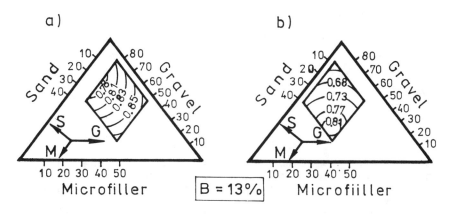

Figure 3.Effect of aggregate components on overall
desirability of E5(4)[a] and E5(10)[b] concretes.

<u>CONCLUSIONS</u>

Combining the response surface methodology and the
multicriteria optimisation based on the overall desirability
function is an effective method of composite material
designing .
Relationships between material factors and the overall

desirability have been obtained on the basis of one series of
experiments .These relations can be used many times to solve
optimisation problems and to predict the effect of changes in
the material composition on material properties.

In three of the four considered cases the experimental domain
includes the point related to the most desirable concrete.For
all the considered concretes an optimal microfiller content
is about twice higher than the resin binder content .This
result corresponds to conclusions obtained on the basis of
theoretical considerations when one assumes obtaining
concrete with resin binder exclusively in "ordering state"[5].
Values of the overall desirability for the polyester
concretes are greater than those for the epoxy concretes.It is
probably connected with lower viscosity of polyester binder
and in consequence with better homogenization of the
composite.

Among the polyester concretes,slightly higher desirability
is observed for the P108(4) concrete .Resin binder P108
includes addition of paraffin ,which can reduce adhesion of
binder to aggregate and in consequence to decrease mechanichal
strength.

REFERENCES

1.Czarnecki L.,The status of polymer concrete.Concrete
  International Design and Construction,1985,77,47-53.
2.Piasta Z.,Rusin Z.and Piasta J.,Examples of applications of
  the second order design in experimental investigations in
  the concrete technology.Archiwum Inżynierii Lądowej,1986,32,
  277-290 (in Polish).
3.Harrington E.C.,The desirability function.Industrial Quality
  Control,1965,21,494-498.
4.Kalinina E.W.,Klimova L.Z. and Lapiga A.G.,Application of
  desirability function in regression analysis.Zavodska
  Laboratoria,1981,47,56-60,(in Russian).
5.Czarnecki L. and Lach V.,Structure and fracture in polymer
  concretes :some phenomenological approaches.In Brittle
  Matrix Composites -1,ed.,A.M.Brandt and I.H.Marshall,
  Elsevier Applied Science Publishers ,London,1986,pp. 241-261.

NEW METHOD FOR DETERMINATION OF ROCK COEFFICIENT OF CRITICAL STRESS INTENSITY

VU CAO MINH
Institute of Earth's Sciences,NCSR,
Nghia Do,Tu Liem,HANOI.
Institute of HEG,WU, Al.Żwirki i Wigury 93,WARSAW.

ABSTRACT

The author considered the energy balance during routine tension test and derived new formulas for the coefficient of critical stress intensity in tensile mode. The experimental verification and reference data comparison had shown preciseness of the method. Determined values of the coefficient of critical stress intensity showed to be the material constant. This method is simple /no notches or precracks are needed/, and can be adopted for most of routine tension tests for brittle materials.

INTRODUCTION

The coefficient of critical stress intensity in tensile mode is an important parameter to investigate and predict the fracture processes /fragmentation, stability, hydraulic fracturing, rockburst, tectonic faulting, earthquakes.../. Up to day methods for its determination are brought up from that used for metal and alloy like Single Edge Notch Beam /SEND/, Double Edge Notch /DEN/, Compact Tension /CT/, Double Torsion /DT/, Short Rod /SR/, Dish-Shape Compresion. In effect, these methods are troublesome for geomaterials in sample preparing /original shapes, notchmaking, etc/ and getting adequate instrumentation. To overcome such difficulties, the routine tension tests had been reconsidered, in which the energy approach had been used to analyse the relationship between the potential fracture energy and energy to form new surfaces. This way occurred to be successful in determination ord investigation of the coefficient of critical stress intensity in tensile mode.

THEORETICAL BASE

During uniaxial compression or tension tests of brittle rocks
when new fissure forms up, we have observed simultaneously
the drop of axial stress, Vu Cao Minh /4/. Such almost
vertical drop of axial stress when the axial strain remains
unchanged, means the **loss** of the elastic potential energy.
Most of this energy loss is used to create a new fissure
surface. In his early works, Griffit supposed that the
surface energy of the crack, 2a long, existing in any body
is equal to the loss of potential energy of the volume around
a crack with radius R = a .

   The above mentioned situation looks like the case when
we compress a cylindrical rod along its long side and cause
the fracture through out its diameter /brasilian tension test/
So if we make the energy analysis of the test, we can find
out relationship linking energetical parameters, then derive
the formula determinating the coefficient of critical stress
intensity /K_{Ic}/. Let consider following cases :

Brasilian Tension Test
First, we suppose that all potential energy is transferred to
surfaces of fracture. In this case, with sample volume
$V = \pi L D exp.2 /4$, the elastic potential energy stored up to
failure /peak tensile stress/ is equal to :

$$W = \frac{\sigma_t^2}{2E} \cdot \frac{\pi D^2 L}{4} \qquad /1/$$

where σ_t -tensile strength, E -elastic modulus, D -sample
diameter, L -sample length. Meanwhile the energy that is
needed to create a fracture surface /the surface with size D
and L/, is :

$$f = 2 \gamma DL \qquad /2/$$

where γ -surface energy density. For brittle materials there
is well known relationship between surface energy density
and coefficient of critical stress intensity K_{Ic} as

$$\gamma = \frac{K_{Ic}^2}{2E} \qquad /3/$$

So, by considering that the energy loss due to other effects
like plastic deformation, acoustic emission, etc is small
and can be omitted in this case, then by combining equations
/1/, /2/, /3/, we obtain

$$K_{Ic} = \sigma_t \sqrt{\frac{\pi D}{8}} \qquad /4/$$

In the brasilian tension test, tensile strength is calculated by:

$$\sigma_t = \frac{2P}{\pi DL} \qquad /5/$$

thus, in other form we get

$$K_{Ic} = \frac{P}{L\sqrt{2\pi D}} \qquad /6/$$

The formulas /4/ and /6/ allow to determine K_{Ic} from the simple brasilian test.

Uniaxial Tension Test

In laboratory practice, uniaxial tension tests are very often in use. We will take into account two cases with cylindrical and rectangular samples. Taking the same rule of energy transference as mentioned at the beginning, and following the same way as for the brasilian test, we obtain :

For a cylindrical sample /V $= \pi D^3/6$ /

$$K_{Ic} = \sigma_t \sqrt{\frac{D}{3/1-\nu^2/}} \qquad /7/$$

For a rectangular sample, where thickness b is smaller than wideness a and $V = \pi a^2 b/4$

$$K_{Ic} = \sigma_t \sqrt{\frac{\pi a}{8/1-\nu^2/}} \qquad /8/$$

EXPERIMENTAL VERIFICATION

The formulas /4/, /6/, /7/, /8/ have been derived in hypothetical-theoretical way. Their preciseness needs to be verificated. To do so, the brasilian tension test with new interpretation method and the toughness test proposed by Shendi-Horvat G. /5/ had been parallelly carried out to make comparison with each other. The last one, called the Notch-Dish-Shape compression test, was chosen because the other ones are more troublesome in sample preparing. The mudstone samples taken from eastearn Poland were used. The coefficient of critical stress intensity for notch-dish-shape case was calculated by the formula :

$$K_{Ic} = 1.264 \, PN^{\frac{1}{2}}/DL \qquad /9/$$

where P –fracture load, N –notch depth, D –dish sample diameter, L –sample thickness. The results are given in Table 1.

TABLE 1
K_{Ic} values for mudstone, determined by the Notch-
Dish-Shape toughness test and New Method

Sampl. No.	D /cm/	L /cm/	N /cm/	P /kN/	K_{Ic} /MPam$^{1/2}$/
MT-1	2.17	0.59	0.2	0.65	0.29
MT-2	2.17	0.95	0.2	1.00	0.27
MT-3	2.17	0.66	0.2	0.63	0.25
MN-2	2.19	2.30		2.50	0.29
MN-9	2.16	2.33		2.85	0.33
MN-10	2.17	2.31		2.38	0.28
MN-11	2.16	2.30		2.20	0.26
MN-12	2.19	2.31		2.40	0.28

The K_{Ic} values for samples MT-1 - MT-3 were calculated by
formula /9/, for samples MN-2 to MN-12 by formula /6/. As can
be seen, the two methods gave nearly the same critical stress
intensity values. The discrepancy lies within the test error
and rock properties variation.

REFERENCE DATA COMPARISON

To get more confirmation, the proposed method had also been
adopted to calculate K_{Ic} for other rocks basing on tensile
strength data published by other authors and then compared
with the K_{Ic} values given by the same sources. The formula
/4/ was used, and results are established in Table 2.

TABLE 2
K_{Ic} values for various rocks, determined by toughness
tests and these from the formula /4/

Rock	Reference	Test	σ_t /MPa/	K_{Ic}^x	K_{Ic}^{xx}
Syenite Wh.	Huang J.A./2/	STPB	11.1	1.1-1.9	1.55
Syenite Dar.	- " -	ATPB	13.2	1.6-2.0	1.85
Basalt	- " -	- " -	21.5	2.2-3.0	3.01
Dolostone F.	Gusallus K.K./1/	SR	13.3	1.66	1.86
Dolostone O.	-" -	SR	13.0	1.78	1.82
Sandstone G.	-" -	SR	10.1	1.47	1.41
Limestone I.	-" -	SR	11.9	1.36	1.66
Limestone R.	-" -	SR	15.0	2.06	2.10
Dolostone K.	-" -	SR	12.1	1.66	1.69

Dolostone M.	- " -	SR	12.1	1.80	1.69
Dolostone R.	- " -	SR	17.0	2.47	2.38
Granite Ch.	Labuz J.F. /3/	DEN	7.56	1.20	1.26
Granite Ro.	- " -	DEN	4.63	0.74	0.78

K_{Ic}^{X} = Toughness test value /MPam$^{\frac{1}{2}}$/.

K_{Ic}^{XX} = Value determined by formula /4/, /MPam$^{\frac{1}{2}}$/.

STPB -Symetrical 3-Point Bending, ATPB -Asymetrical 3-Point
Bending, SR -Short Rod, DEN -Double Edge Notch.

In Table 2 the K_{Ic} values for granites Ch. and Ro. were
obtained for rectangular sample so the formula/8/ was used.
Larger discrepancy exists in the case of the limestone I.,
because of   big variation of its properties /variation
coefficient of tested K_{Ic} comes up to 32%/. The other
difference between the Ic tested and calculated values are
whithin the range of 10%. So the proposed method brings to
the results in more than acceptable agreement with complicated method of fracture toughness tests.

VARIATION OF K_{Ic} WITH SAMPLE SIZE

Many strength parameters of rocks /σ_c, σ_t, I_c/ show size-
dependent tendency, where the tensile and point load
strengths are the most sensitive to size changes. In Fracture
Mechanics there are still different opinions about K_{Ic} nature,
whether it is a material constant or only the indice one.
To give more data for this the author examined the brasilian
test for basalt samples with 3 series of diameters. The
results gained by new method are shown in Table 3.

TABLE 3
K_{Ic} values for bazalts samples of different sizes

Samp. No.	D /cm/	L /cm/	P /kN/	K_{Ic} /MPa m$^{\frac{1}{2}}$/
BwgkI 2.3	2.16	2.12	21.1	2.71
2.5	2.16	2.10	16.5	2.14
2.7	2.16	2.10	17.8	2.31
BwgkII 4.5	3.53	3.64	38.2	2.23
4.6	3.53	3.53	45.9	2.77
BwgkIII 5.1	4.93	5.20	69.0	2.39
5.3	4.92	5.24	69.0	2.37
5.6	4.95	5.11	62.7	2.21

Table 3 shows that the critical stress intensity coefficient in tension mode is not sensitive to sample size, at least for the case considered. It gives more to confirmate that stress intensity coefficient is the material constant and it can be called as given in the title the coefficient /not a factor/.

CONCLUSIONS

The experimental investigation and reference data comparison have showed the preciseness of the proposed method for determination of critical stress intensity coefficient in tension mode. The whole procedure is simple due to use of routine tension tests. K_{Ic} value behaves as the material constant. This method can be also used to determine fracture surface energy. The energy approach is a useful way to analyse the fracture processes.

REFERENCES

1. Gunsallus,K.K.,Kulhavy,F.H., A comparative evaluation of rock strength measures. Int.J.Rock Mech.Min.Sci.Geomech. Abstr., 1984, 5, 233-48.

2. Huang,J.A.,Wang,S., An experimental investigation concerning the comprehensive fracture toughness of some brittle rocks. Int.J.Rock Mech.Min.Sci.Geomech.Abstr.,1985, 2, 99-104.

3. Labuz,J.F.,Shah,S.P.,Dowding,C.H., Experimental analysis of crack propagation in granite. Int.J.Rock Mech.Min.Sci. Geomech.Abstr.,1985, 2, 85-98.

4. Minh,V.C.,Energetical Analysis of Rock Deformation and Failure , Dr.Sc thesis, Warsaw University,1988.

5. Shendi-Horwath,G.,On the fracture toughness of coal. Austral.J.Coal Min.Tech.Res.,1982, 2, 51-57.

CRACKING RESISTANCE OF CONCRETES
WITH DIFFERENT STRUCTURURAL HETEROGENEITY

Yu.V. ZAITSEV*, K.L. KOVLER**, R.O. KRASNOVSKY***,
I.S. KROL**** and M. TACHER*****
*Polytechnical Inst. (VZPI), P. Korchagin Str., 22,
Moscow 129805, USSR

**Research Inst. of Building Materials & Structures
(VNIISTROM), K.Marx Str., 117, Kraskovo, Moscow Reg., 140080, USSR

*** Research Inst. of Physico and Radio Technical
Measurements (VNIIFTRI)

****Research and Project Inst. ORGENERGOSTROY

*****Moscow Civil Engineering Inst. (MISI)

ABSTRACT

Herein is contained details of an experimental technique for bend testing
of concrete specimens which allows the cracking resistance and load-
deflection relationships for each specimen to be obtained.   Specimens of
11 concrete mixes of different structural heterogeneity were tested with
5 initial notch depths.   It is argued that different characteristics of
cracking resistance of concrete at varying stages of crack initiation and
growth must be used.   The dependance of these characteristics on the
content and grain size of aggregates at constant mechanical properties of
hardened cement paste matrix has been shown.

INTRODUCTION

Many mechanical characteristics of concrete are normal in Soviet Building
Codes in relation to the main design parameter - compression strength,
and the structure of concrete is not normally taken into account.
However, it is well known that directional modification of the structure
can enhance the tensile strength, elastic modulus, cracking and frost
resistance, and other properties, thereby allowing the necessary load
carrying capacity and reliability of concrete structures.

It is usual for heavy concretes to have a certain structure.   After
selecting quantitative mixed proportions for producing dense packed
coarse aggregates, the intergranular volume must be filled by cement-sand

mortar.   Such composite material is often called concrete.

With extension of the present applications sphere for concrete technology
and other future developments, a number of new structural requirements
become evident.   The simplest way to create the required material
consists of varying the aggregate content and particle size.   For
instance, when placing concrete by pneumatic or mechanical sprinkling, it
is necessary to reduce grain rebounding by decreasing the relative
content of course aggregates and thus creating a structure with floating
aggregates which are divided by more thick mortar seams.   In armocement
and other structures, fine-grained concrete obtained by the total
exclusion of course aggregates is used.

Many authors have studied the structural influence on compression and
tensile strength, but few have considered cracking resistance.   This is
related to the lack of well approbated standard techniques for
determination of cracking resistance.   One of the first standard
methods was developed recently (1).   This regulates the measurement of
fracture energy G_f using the total area under the complete load-
deflection diagram of beams under bending.   However, G_f does not take
into account different physical peculiarities related to the resistance
of concrete to crack initiation and future stable propagation.

The main thrust of the present paper is that different characteristics of
cracking resistance of concrete must be used at the various stages of
crack initiation and growth.   The authors have tried to answer the
question   - how do these characteristics depend on relative content and
grain size of aggregates at constant mechanical properties of hardened
cement paste (hcp) matrix.

MATERIALS AND METHODS

The concrete structure was varied by a number of components and the
change of relative contents of these components within each structure.
Lots of variation were chosen so that the materials would not show an
essential discontinuity of the structure.   The mechanical properties of
hcp matrix remain constant for all mixes considered.   This was ensured
by the consistency of the effective water-cement ratio (0.236) in the hcp
matrix, bearing in mind the water requirements of the aggregates which
were found experimentally.

The raw materials used were: Portland cement brand M400, quartz sand as
a fine aggregate with a fineness modulus of 2.18 and water requirements
of 7.5%, crushed granite stone as course aggregate with particle sizes of
5-10 mm (20%) and 10-20 mm (80%) and water requirements of 0.7%.

Prismoidal specimens of 70 x 70 x 280 mm were formed in the laboratory.
The vibration time was chosen to ensure maximum compaction factor for
each mix.   If this was less than 0.05%, the specimen was rejected.   The
samples were cured at air moisture of 95-98% and temperature of 18-22°C.
for 28 days, then isolated from environmental water exchange prior to
testing at 1-year age.   Such conditions ensured a degree of consistency
of mechanical properties for all samples tested.

The main mechanical characteristics of all concretes, compressive and
tensile strength (R_b, R_{bt}) and modulus of elasticity E_b were

preliminarily determined - Table 1.

TABLE 1
Concrete mixes

Material	Mix Code	Weight content				Volume content			E_b, GPa	R_{bt}, MPa	R_b, MPa
			aggregates				aggregates				
		cement	fine	coarse	water	hcp	fine	coarse			
hcp	HCP	1	0	0	0.24	1	0	0	23	2.1	66
fine concrete	A	1	0.5	0	0.30	0.77	0.23	0	25	6.4	54
	B	1	1	0	0.35	0.64	0.36	0	24	12.1	50
	C	1	2	0	0.44	0.50	0.50	0	17	11.2	40
	D	1	4	0	0.64	0.39	0.61	0	15	2.8	15
coarse concrete	A1	1	0.5	0.5	0.27	0.62	0.19	0.19	29	6.7	57
	A2	1	0.5	1.1	0.28	0.50	0.16	0.34	30	6.3	54
	A3	1	0.5	2	0.28	0.40	0.12	0.48	32	8.3	57
coarse concrete	C1	1	2	2	0.40	0.33	0.34	0.33	32	10.2	36
	C2	1	2	3	0.41	0.28	0.29	0.43	32	9.4	47
	C3	1	2	4	0.44	0.25	0.25	0.50	30	7.0	36

Prior to bend testing, the specimens were notched using a diamond cutting disk such that initial cracks were formed with relative length $\lambda_0 = 1/8$, 1/4, 3/8 and 1/2 from section depth.

The experimental technique for determining crack resistance was as follows.   Elastic elements were inserted between the press plates and the specially constructed stiff frame.   Thus it is possible to obtain a complete load-displacement curve, including the descending part, by loading with a constant rate.   Load-line deflection gauges were mounted on the special frame in support sections which excluded the influence of irregular movement.   A number of supports were used i.e. a swinging knife-edge instead of a roller in order to exclude horizontal support reactions and a fixed spherical hinge for excluding torsional moments.

Each specimen was tested in bending 3 times.   For this purpose special extenders were mounted on 2 halves of the specimen.   The extenders had counterweights for a more exact load requirement by taking sample dead weights into account.   This bend test method is reported in detail in Ref. (2).

RESULTS AND DISCUSSION

Pairs of curves of load-deflection (F-f) and load-crack mouth opening displacement (F-v) or F-f and v-f were obtained using a tensometric dynamometer displacement gauges and 2 X-Y recorders.   It is noted that as a rule, the ascending branch of the F-f diagram is approximately linear up to approximately the maximum (critical) load F_c.   Descending branches asymptotically tended to the deflection axis (Fig. 1).   This can always be observed when the dead weight influence is absent.

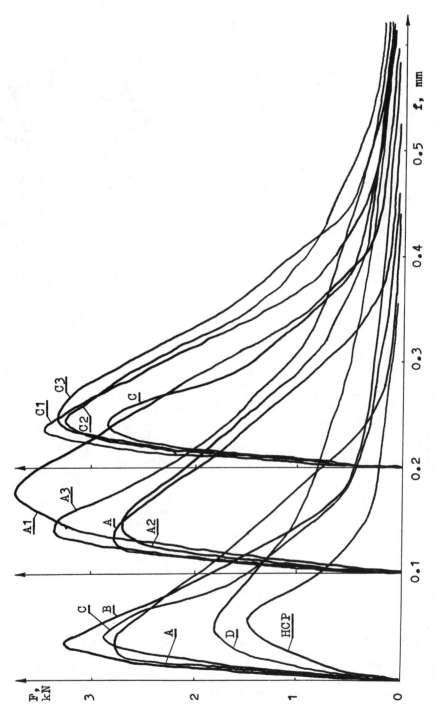

Figure 1. The examples of load-deflection curves at $\lambda_o = 1/4$.

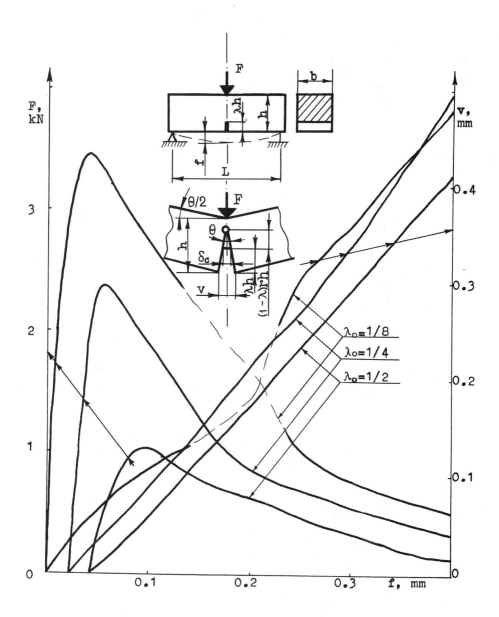

Figure 2. Load-deflection and crack mouth opening displacement-deflection curves of concrete A3 and bend beam scheme.

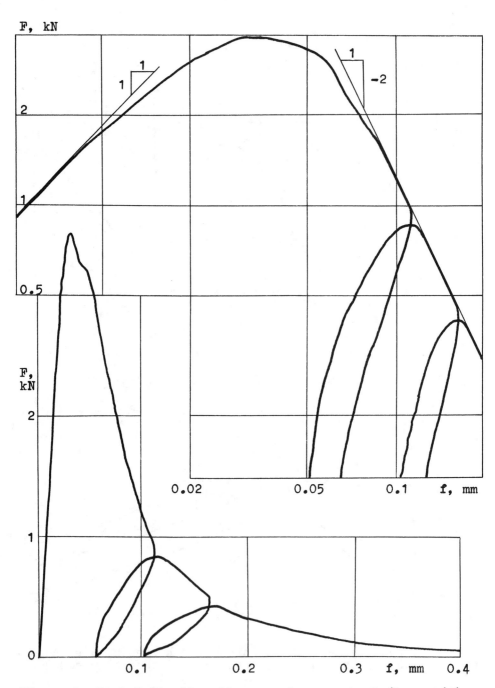

Figure 3. Load-deflection diagram of concrete C (λ_o = 1/8) in ordinary and logarithmic coordinates.

It is known that different parts of the F-f diagram correspond to different stages of deformation. The initial linear part reflects an elastic response. Some nonlinearity can appear near the maximum load F_c owing to micro-cracking and slow extension of pre-existing bond cracks. Finally, the extensive micro-cracks coalesce with bond cracks resulting in the initiation of large macro-cracks and subsequent reduction in the load carrying capacity.

For specimens with short notches ($\lambda_o < 1/8$), after load F_c unstable crack growth was sometimes observed, and the X-Y recorder moved quickly with the discontinuous track shown in Fig. (2). For further stable crack growth, it was necessary for total press load to be increased. Unstable crack growth can be explained by the large deformation energy released kinematically. In the authors' opinion, the specimens with depth/length ratios of 1/4 and short initial notches, should not be used for the exact measurements of stable crack propagation resistance.

It is interesting that both ascending and descending branches of the load-deflection curve in logarithmic co-ordinates may be described by two different straight lines with the slope of 1 and 2 accordingly, and related to each other by some short intermediate part (Fig. 3). Alternatively, the descending branch is a square hyperbola in normal co-ordinates.

The resistance to crack initiation (critical stress intensity factor K_{1c}) was defined from F_c, λ_o, beam span L, thickness b and depth h:

$$K_{1c} = \frac{3 F_c L}{2 b h^{3/2}} (\lambda h)^{1/2} (1.93 - 3.07\lambda + 14.53\lambda^2 - 25.11\lambda^3 + 25.8\lambda^4) \qquad (1)$$

The resistance to the stable crack propagation may be characterized by the energy parameter RG, as well as the resistance to crack initiation by the critical energy release rate G_{1c} related to K_{1c}. The RG-value is equal to the energy requirement increment dW_R which is extended for crack surface increases bhd. The authors' defined average RG-values over all crack extension $(1-\lambda_o)h$ and doubled (analogous to Rice's formula for J-integral) area A_G and total descending branch.

$$RG = \frac{dW_R}{bhd\lambda} \cong \frac{2 A_G}{bh(1-\lambda_o)} \qquad (2)$$

On each λ_o-value 3 specimen-twins were tested for K_{1c} determination. The variation of the factor was equal to 2-17%. The tests did not reveal a definite dependence of K_{1c}-λ_o (Fig. 4). Thus, in further analysis, the K_{1c} values of each mix were averaged over all crack lengths. RG-values were obtained only for specimens without extenders (Fig. 5).

The two conclusions which follow from the obtained F-f and v-f curves are; a) F is inversely proportional to f^2, and, b) v increases with the growth of f linearly and does not depend on λ. Taking this into account, a further deformation relationship for beams and bending can be accepted using Fracture Mechanics (Fig. 2) :

$$tg(\theta/2) = 2f/L = \frac{c}{2r(1-\lambda)h} = \frac{v}{2(r-r\lambda+\lambda)h} \qquad (3)$$

Hence taking into account that the v-f plot has a slope approximately equal to 1 and 4h approximately equals L, the authors obtained limits for

factor r: r approximately equal to 1.   Alternatively, the turning center
of concrete beams in bending is coincident with the concentrated load
point.   From equation (2) a connection between the nett section size and
f is obtained.

$$1-\lambda = \frac{L\delta_c}{4h} \cdot \frac{1}{f} \tag{4}$$

$$\frac{d\lambda}{df} = \frac{L\delta_c}{4h} \cdot \frac{1}{f^2} \tag{5}$$

From Eq. (2) and (5):

$$RG = 8Ff^2/(bL\delta_c) \tag{6}$$

Hence, considering a critical opening displacement δ_c and stable crack
growth resistance RG as constant values of the material at the stage of
crack propagation, we have: F $1/f^2$, which is confirmed by experience.
Additionally, Eq. (6) shows interaction between δ_c, RG and the descending
part of the F-f diagram.   As the average RG-value may be determined from
the F-f curve, the δ_c value for stable crack growth may also be defined
from this curve :

$$\delta_c = (1-\lambda_0)Ff^2/A_G \tag{7}$$

Thus, knowing the crack kinetics from Eq. (4) it is possible to find the
K_{1c} value at stable crack propagation.   Thus, all characteristics of
crack growth resistance may be determined from the F-f diagram.

The resistances to crack initiation and propagation depend, by different
amounts, on the hcp content.   At increasing aggregate volume, K_{1c} values
achieve their maximum and thereafter decrease.   Analogously tensile
strength in bending similarly changes.   However, RG-values increase
monotonically and with increasing intensity for coarse aggregates.   It
is interesting that cracking resistance of the tested coarse concretes
did not practically depend on the kind of mortar matrix, but only by the
volume content of the coarse aggregate ϕ_{ca}.   This can be seen from the
practical coincidence of K_{1c} and the RG dependence on ϕ_{ca} (Fig. 6).

The structural influence on K_{1c} and RG may be explained by the following.
Previous to crack initiation, the formation of micro-cracks at grain-
matrix interface is noted, thus increasing the aggregate content makes
easier fracture process zone formations and K_{1c} decreases.   On the other
hand, a higher probability of matrix micro-cracking breaking in
aggregates yields an increased K_{1c}.   The application of these two
tendencies gives rise to a local maximum of K_{1c}.

Stable crack growth in heady concretes with aggregate strengths
significantly more than the matrix material allows the fracture surfaces
and RG values to grow monotonically as the aggregate content increases.

In compression tests, the fracture is caused by a certain concentration
of micro-cracks to be achieved which coalesce into macro-cracks.   Hence
the addition of aggregate reduces the material strength by increasing
the number of initial defects.

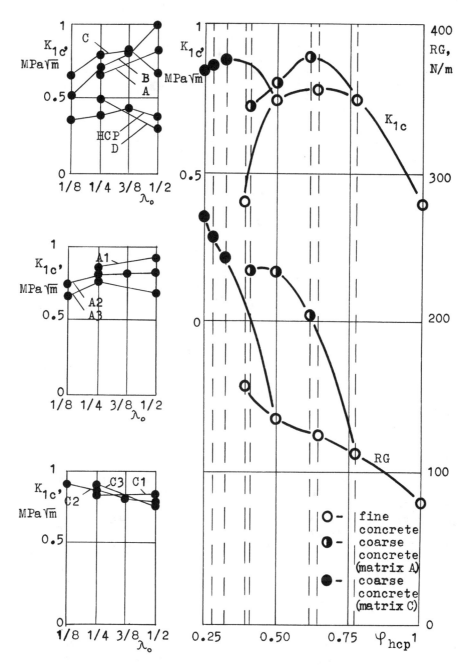

Figure 4. K_{1c} vs. λ_o.

Figure 5. K_{1c} and RG vs. hcp matrix content φ_{hcp}.

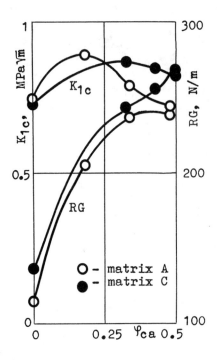

Figure 6. K_{1c} and RG vs. ca.

The fracture energy G_f is not a sufficiently accurate characteristic of concrete cracking resistance. For crack initiation and stable growth stages, it is important to use different characteristics of cracking resistance, e.g. K_{1c} and RG accordingly. The K_{1c} value has a maximum for certain aggregate contents. It allows the optimum design of concrete structures where unstable crack growth is possible. The descending part of the load-deflection curves allows the RG value to be defined, which increases with aggregate content and size owing to fractured surface developing around the aggregate particles.

REFERENCES

1. 50-FMC RILEM draft recommendation. Determination of the fracture energy of mortar and concrete by means of 3-point bend tests. Mater. et Const., 1985, 18, 285-90.

2. Kovler, K.L., Krasnovsky, R.O., Krol, I.S. and Tacher, M., Bend test method, A. s. 1397791 USSR, G 01 N 3/20, Bulletin of Discoveries and Inventions, 1988, No. 19, 170 (in Russian).

FRACTURE RESISTANCE EVALUATION OF STEEL FIBRE CONCRETE

Janusz KASPERKIEWICZ
Institute of Fundamental Technological Research
Polish Academy of Sciences
00-049 Warszawa, Swietokrzyska 21, POLAND
Åke SKARENDAHL
Swedish Cement and Concrete Research Institute,
100-44 Stockholm, SWEDEN

ABSTRACT

Methods of testing and evaluating SFRC (Steel Fibre Reinforced Concrete) made using different types of fibres (hooked-end, enlarged ends, indented and metallic glass), have been investigated in experiments on thin beams loaded in bending. The tests were aimed at evaluation of fracture behaviour of various kinds of thin layer SFRC, used e.g. in shotcrete.

The analysed magnitudes were among others the limit of proportionality, the modulus of rupture, and the toughness indices I(5) and I(10), the last two estimated following the ASTM Standard C 1018. Also the system of cracks has been recorded and analysed. The tested methodologies can be recommended for evaluation of SFRC materials. A strong reinforcing effect has been observed in case of metallic glass fibres which are much more efficient than the traditional metal fibres.

BACKGROUND, AIM AND SCOPE OF TESTS

Steel fibres are used in concrete mixes to improve various strength and deformation properties. In particular the behaviour of the material in the cracking stage can be improved. Crack control is of importance especially when moisture variations or temperature variations appear or when durability aspect is decisive. The application of the fibrous reinforcement is justified also when the production technology demands fast development of the cohesiveness.

Evaluation of the quality of steel fibre reinforced concrete (SFRC) is often connected with estimation of its flexural strength alone (f_f or MOR -

Modulus of Rupture, or maximum flexural stress). The relevant test gives however no information on the material behaviour in the later stage. More recent techniques of evaluating fibre concrete inform also on the stress at the first crack, and on the toughness, indicated by the shape of the area under the load-deflection curve. Such approach is used in ASTM C 1018 [1]. The stress at the limit of proportionality (σ_{LOP} – first crack stress), gives information about the quality of the matrix, while the toughness indices I(5) and I(10) describe the effect of the fibres.

The tests have been planned to check methods of investigating fracture resistance properties of thin SFRC layers fabricated e.g. by means of shotcreting. For a shotcrete layer as important as the carrying capacity of the layer is its toughness, i.e. the amount of energy absorbed in the fracture process. The test results should therefore include – among the others – also the whole force–deflection diagram. From the point of view of the durability of the material the structure of cracks in an element is even more important than its carrying capacity. The element with large number of tiny cracks is at the same carrying capacity 'better' than the element with one crack, single but large – a crack through which gases and liquids inducing corrosion can penetrate.

The aim of the experiments was to investigate a simple test which could supply adequate information about the behaviour of SFRC. The basic concept was to use the ASTM approach of describing the matrix properties through the first crack strength (at LOP) and the fibre reinforcing effect through a toughness index. In addition crack distribution should be evaluated.

The present tests were limited to rapid hardening cement mixes for practical reasons (time aspect). The strength development is not expected to influence the character of the fracture process.

MATERIALS AND PREPARATION OF SPECIMENS

For the tests designed have been 10 different SFRC mixes and three control mixes without reinforcement (plain matrix). The composite mixes were differing in the structure of fibrous reinforcement, the matrix being intentionally kept unchanged. For every batch the amounts of the components were chosen in such a way that the whole content of the mixing pan was used for producing a single testing plate – the problem of erroneous estimation of the fibre content in the mix was so avoided. From every plate 4 beams were sawn and the average dimensions of the tested beams were ca 27x117x350 mm.

TABLE 1
Nominal and actual average dimensions of the fibres [mm]

Type of fibres (denotation of the series)	Length nominal	actual	Cross-section ($d=d_{eq}$ or d x b) nominal	actual (CoV)	Equiv.diam. (d_{eq}) nominal	actual
Metallic (1,2)	30	30.9	0.025x1.5	0.025x1.59 (18.7%)	0.219	0.225
glass (3,4)	45	46.5	0.025x1.5	0.025x1.42 (5.5%)	0.219	0.2126
Dramix (5,6)	30	29.0	0.5	0.5 (0%)	0.5	0.5
Enl.E. (7,8,14)	18	18.2	0.6x0.6	0.627x0.627(14.7%)	0.677	0.7075
Hoerle (10)	28	27.5	0.5	0.57 (5.9%)	0.5	0.57

Equivalent fibre diameter is calculated here as: $d_{eq} = \sqrt{(4*A/\Pi)}$, the value of A being the actual fibre cross-section area.

All the mixes have been produced with the same type of cement (rapid hardening), at water-cement ratio of 0.51, using sand 0-2 mm. A liquid superplasticizer (P), melamine based, was added to the mixing water (W). To the mass of cement (C) added was 8% of silica (S). The water-cement ratio is calculated as: W/(C+S) .

After mixing the other components for about 3 min. in the vertical axis mixer the fibres were added to the mixing pan manually. This was possible due to small volumes of materials involved.

Four types of metal fibres were used in the tests. Their actual dimensions have been checked on some examples and are compared to the nominal dimensions in Table 1. Of these fibres the metallic glass fibres with amorphic microstructure have been produced by a special new technique described in [2], the other fibres were produced more conventionally (Dramix, Enlarged Ends and Hoerle fibres).

After the mechanical tests the broken half beams have been fully separated, and the visible parts of the fibres counted. It was then possible to estimate the actual value of α — the area corresponding to a single fibre in a cross-section. They have been compared with the expected (nominal) values calculated according to a formula from [3]:

$$\alpha = \Pi^2 . d_{eq}^2 / (8.V_f) \quad .$$

Here V_f is the volume content of the fibres. The results of this simple test prove that at the accepted method of the production of the specimens all the prepared fibres are placed in the SFRC mix with no rebound-type nor spilling losses. The coefficient of correlation obtained for collection of nominal and actual values of α was: $r = 0.904$. Also it seems permissible to use the nominal sizes of the fibrous reinforcement in the further analysis of the results, instead of the actual dimensions.

To characterize the system of reinforcement also other parameters have been calculated, such as p (perimeter of the cross-section of the fibre),

TABLE 2
Tested mixes and parameters of fibre systems

Series No.	Reinforcement vol.	l	d_{eq}	p	Fibre type	Bond coef. η	Parameters — α	$\eta.l.p$	$\eta.l/\alpha$
15	0.0%	0	0.000	0.000			100.000	0.000	0.000
0	0.0%	0	0.000	0.000			100.000	0.000	0.000
13	0.0%	0	0.000	0.000			100.000	0.000	0.000
1	0.3%	30	0.219	3.050	Met.gl.	0.80	19.723	73.200	1.217
2	0.6%	30	0.219	3.050	Met.gl.	0.80	9.862	73.200	2.434
3	0.3%	45	0.219	3.050	Met.gl.	0.80	19.723	109.800	1.825
4	0.6%	45	0.219	3.050	Met.gl.	0.80	9.862	109.800	3.651
5	1.0%	30	0.500	1.571	Dramix	1.30	30.843	61.269	1.264
6	2.0%	30	0.500	1.571	Dramix	1.30	15.421	61.269	2.529
14	1.0%	18	0.677	2.400	EE	1.30	56.544	56.160	0.414
7	2.0%	18	0.677	2.400	EE	1.30	28.272	56.160	0.828
8	4.0%	18	0.677	2.400	EE	1.30	14.136	56.160	1.655
10	2.3%	28	0.500	1.571	Hoerle	1.00	13.622	43.988	2.055

$\eta.1.p$, and $\eta.1/a$. Here and in the other places η is a bond efficiency coefficient, assumed: $\eta=1.0$ for plain fibres, $\eta=0.8$ for highly smooth metallic glass fibres, and $\eta=1.3$ for anchored fibres (Dramix or EE – Enlarged Ends fibres).

The symbols and compositions of all the mixes are shown in Table 2. During preparation of the specimens two tests of workability have been performed: the flow diameter test [4] and the slump test. The results are shown in the last columns of the Table 3 below. The coefficient of correlation between the two sets of workability data is r = 0.87.

After 24 hours the plates have been taken away from the fog room, marked, sawn into beams, and stored again in the fog-room until tested at the age of 7 days.

From accompanying tests on the plain matrix its 7-days bending strength was found to be 6.5 MPa, its compressive strength 36.4 MPa, and its density 2215 kg/m^3 (specimens 40x40x160 or 100x100x100 mm^3).

LOADING SYSTEM, PROGRAM OF OBSERVATIONS

All the specimens have been loaded in pure bending using loading supports with joints enabling self accommodation of the axial load in two

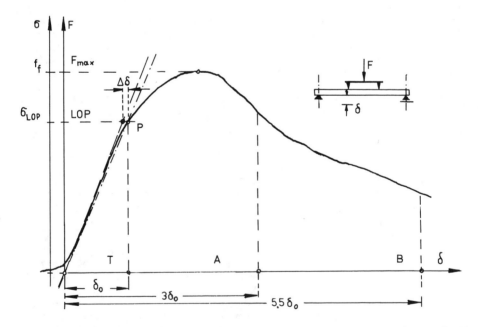

Figure 1. Definition of parameters used in the processing of the experimental data. Area A and B are measured by a planimeter. Area T is calculated as LOP.δ_o/2. Point P is determined by an offset from the tangent to the initial linear part of the curve: $\Delta\delta$ = 30 μm. The toughness indices are defined as: I(5)=(T+A)/T and I(10)=(T+A+B)/T.

perpendicular vertical planes. The beams were loaded immediately after removing from the fog room, in one monotonic cycle. The machine was controlled by a servo-system set to a constant rate of the deflection under the force. A typical measurement output is shown in Fig. 1, together with the explanation of the introduced parameters of the results.

Soon after the final unloading the specimens have been searched for the structure of cracking, The microcracks observed under a low power microscope were enhanced with a marker, and the specimens photographed.
When counting numbers of visible cracks they are counted only in the central region of each specimen, corresponding to the constant bending moment area. Cracks are calculated along four parallel lines drawn on the bottom surfaces of the specimens and the average of all the readings is taken as representative for the whole series (crack number).

TEST RESULTS

Main test results are presented in Table 3. Layout of the table corresponds to that of the Table 2. Each experimental value is an average of test results of 4 specimens. More detailed results can be found in [5]. In the Table 3 presented are for each series of specimens: observed maximum bending stress (f_f or MOR), limit of proportionality (LOP and σ_{LOP}), toughness indices I(5) and I(10), and crack number – mean value of the number of the cracks observed. Additionally presented are also values of the slump and flow diameter of the fresh mix.

In Fig. 2 shown are 'average' force-deflection diagrams. A diagram for one group of specimens is obtained by taking – for each deflection value – an arithmetic mean of the four corresponding readings of the bending force on different specimens. In the figure it is possible to observe different behaviour of the specimens due to differences in the structure of the fibrous reinforcement. It can be seen that at the right structure of the reinforcement it is possible to use one third only of the fibres content, getting at 0.3 % vol nearly the same result as normally at 1 % vol – compare series 1, 3 and 14. The 'strongest' material – the material of highest MOR value – has been obtained by using the largest amount of the reinforcement – series 8, which is however at quite unusual fibre content: 4%vol of EE fibres.

Comparing series 6, 7, 1, 3 and 2, it can be seen that approximately the same reinforcing effect on MOR can be obtained or by addition of 2% vol. of more common fibres (EE, Dramix), or by addition of 0.6% vol. (i.e. 3 times less !) of high bond strength metalic glass fibres. Yet from this point of view (of improved MOR) the series 1, 2, 3, 4 do not seem to be very much superior comparing to the plain matrix series (series: 0 and 13 – not shown in the figure).

Quite different picture is obtained in case of the fracture toughness properties. Here the small fibre content metalic glass fibres series – designated from 1 to 4 – are quite competitive compared to those with higher fibre contents of the other fibres. It is interesting that there must be a certain fibre length optimum in the metallic glass fibres reinforced concrete. The highest observed value of I(10) was obtained for series 2 – 0.6% vol of fibres of length 30 mm. Perhaps the similar content of longer fibres (of length 46 mm) generates too big matrix porosity.

TABLE 3
Average results in tested series

Series	MOR [MPa]	LOP [kN]	Tough.Indices I(5)	Tough.Indices I(10)	σ_{LOP} [MPa]	Avarage Crack Number	Slump [mm]	Flow [%]
15							160	107
0	5.20	1.43	1.29	1.29	5.19	1.38	150	107
13	4.87	1.28	1.47	1.53	4.88	1.13	165	112
1	5.43	1.49	4.18	7.13	5.25	2.50	120	102
2	5.90	1.52	5.27	10.42	5.22	5.13	75	82
3	4.55	1.34	4.13	7.39	4.52	2.50	90	97
4	5.00	1.37	3.99	8.63	4.85	3.00	25	62
5	5.17	1.49	3.75	6.78	5.05	2.06	100	102
6	5.68	1.44	4.53	9.39	5.07	4.69	90	
14	5.41	1.52	3.75	6.09	5.15	1.25	100	107
7	5.74	1.49	5.23	9.76	5.23	1.63	85	77
8	7.48	1.69	4.93	9.66	5.69	2.56	50	72
10	6.64	1.68	5.32	10.57	5.53	2.94	110	82
Average	5.589	1.478	3.987	7.387	5.136	2.564		
St.dev.	0.805	0.122	1.346	3.152	0.306	1.271		
CoV.	14.4%	8.3%	33.8%	42.7%	6.0%	49.6%		
Min.	4.55	1.28	1.29	1.29	4.52	1.13	25	62
Max	7.48	1.69	5.32	10.57	5.69	5.13	165	112

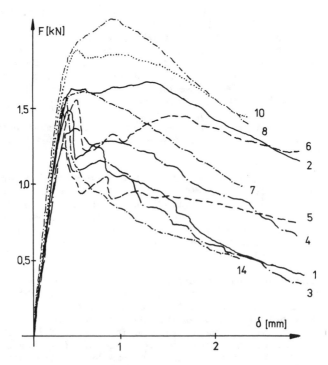

Fig.2. Comparison of averaged force-deflection diagrams for all the tested
series of beams.

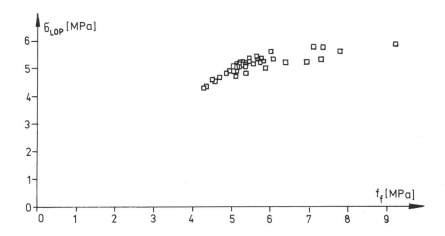

Figure 3. Nominal stresses in various specimens at the LOP loading level and at the maximum load (MOR).

Special investigation would be needed to clear this point out (perhaps some microscopic observations).

In Fig. 3 shown are observed for individual specimens values of σ_{LOP} and f_f. It can be seen that σ_{LOP} is influenced mostly by the quality of the matrix, while f_f (MOR) depends strongly on the effect of the fibrous reinforcement.

In Fig. 4 shown is – as an example – a picture of a crack system identified at the bottom surfaces of the beams from series 2. The beams are

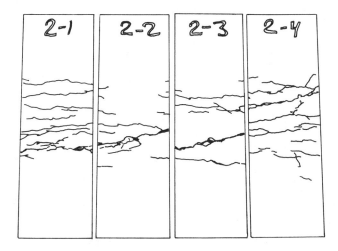

Figure 4. An example of the crack system observed in specimens of series 2.

placed in their original positions, i.e. as they were situated during casting. Observed can be the effect of continuation of the crack paths across several specimens. This seems to be an argument for the concept of the continuous inhomogeneity, proposed in [6]: cracked regions correspond to the regions of weaker matrix or lower fibrous reinforcement content. The observed lowest crack spacing was circa 15 mm (in series 2).

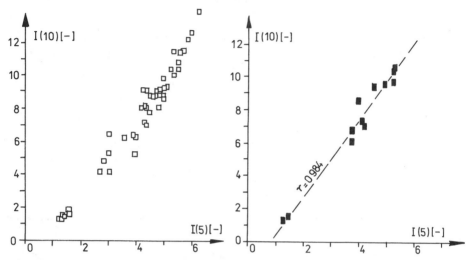

Figure 5. Values of toughness indices I(5) and I(10) as determined on individual specimens and as average values from the series.

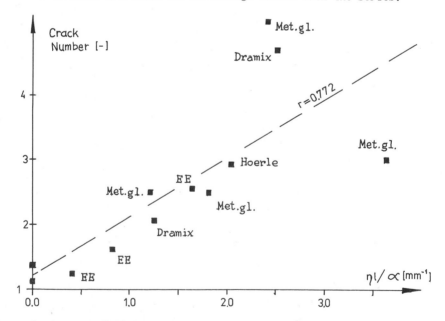

Figure 6. An example of the degree of cracking dependence on the nominal value of fibrous reinforcement structural parameter $\eta l/\alpha$.

The standard tested values of the both toughness indices I(5) and I(10) are compared in Fig. 5. There seems to be a certain nonlinear correlation, of still unknown meaning. The conclusion that there is some interdependence between I(5) and I(10) seems however to be limited to the narrow domain of the composite mixes tested in the present investigations, and will probably be untrue for other geometrical proportions of the specimens and for other mix compositions.

In Fig. 6 demonstrated is dependence of the average number of the observed cracks on parameter $\eta l/\alpha$. Similar dependence can be observed for toughness index I(10), where the coefficient of correlation for the same set of the results was found: $r = 0.727$.

CONCLUSIONS

The tested methodology seems adequate and can be recommended for the evaluation of fracture resistance of steel fibre reinforced concrete:

- the fracture process should be carried in pure bending, and recording the whole F–δ diagram, to a deflection of the beam at which the value of I(10) can be evaluated,

- after fracture the cracks should be identified on the tensile side of the beams, in the constant bending moment zone, and counted as an average value from at least 4 base lengths for each specimen; if possible the picture of the cracks should also be recorded.

The number of the cracks observed shows the material ability to multiple cracking. To be able to observe distribution of cracks in the tested specimens the third points loading system must be chosen, not the central point loading scheme, used in many tests of fibrous concrete composites.

The LOP level (Limit of Proportionality) is to be identified as a level at which the offset from the tangent to the linear part of the force-deflection diagram is circa 30 μm. The same approach can easily be adopted in analysing the test data with other shape of specimens, perhaps with another offset value. All the presented results should be checked from the point of view of the thickness of the test element. It is important for example to see whether the I(10) obtained in the tests described above is identical with I(10) obtained according to the ASTM test [1]. Due to more or less regular co-relation between I(10) and I(5) it is however of no importance which of the two indices is to be used. It is temporarily suggested to use I(10) which gives larger coverage to the specimen behaviour.

The conducted experiments demonstrated strong reinforcing effect of a particular fibre reinforcement – amorphous metallic glass fibres. Due to their special shape and high effectiveness the fibre content can be decreased 4–6 folds, compared to conventional SFRC mixes, still giving high fracture toughness. Mixing metallic glass fibres into a concrete (microconcrete) matrix is however more difficult than in case of the conventional fibres. The value of 0.6% vol. of 30 mm long fibres of such type seems to be maximum and – probably – also an optimum.

The general effect of fibrous reinforcement can be expressed by certain structural parameters. A promising example seems to be $\eta.l/\alpha$.

There is also an additional conclusion that at least sometimes the flow table test can be used instead of the slump test to estimate the workability. Such a conclusion needs however further experimental corroboration. The obtained results are at least not contradictory to the only found recently published results of the flow-slump test dependence [7].

ACKNOWLEDGEMENTS

The tests have been prepared and performed at the Cement and Concrete Institute in Stockholm (CBI), in the co-operation between CBI - Stockholm, the Åke Skarendahl AB - Djursholm, and the Institute of Fundamental Technological Research of the Polish Academy of Sciences in Warsaw. The help from all the co-operating institutions is kindly acknowledged.

REFERENCES

1. ASTM Standard C 1018-85, Standard test method for flexural toughness and first-crack strength of fiber reinforced concrete (using beam with third-point loading). Annual Book of ASTM Standards, vol. 04.02, 1985, pp.637-44.
2. de Guillebon, B. and Sohm J.M., Metallic glass ribbons - a new fibre for concrete reinforcement. In Developments in Fibre Reinforced Cement and Concrete, ed-s, R.N. Swamy, R.L. Wagstaffe, D.R. Oakley, RILEM TC 49-TFR, Sheffield 1986, vol. I, pp. 145-49.
3. Kasperkiewicz J., Analysis of idealized distributions of short fibres in composite materials. Bull.l'Acad.Pol.Sci., Ser.sci.techn., 1979, 27, No. 7, pp. 601-9.
4. BS 4551: 1980, Methods of testing: Mortars, screeds and plasters (Clause 12: Determination of flow), BSI 1980, Gr 8, 31 pp.
5. Kasperkiewicz J., Skarendahl Å., Influence of steel fibre data on toughness and cracking of fibre concrete. IFTR Reports (Prace IPPT), 1988, to be published, ca 40 pp.
6. Kasperkiewicz J., Fracture and crack propagation energy in plain concrete. Heron, 1986, 31, 2, 5-14.
7. Tanigawa Y., Mori H., Tsutsui K., Kurokawa Y., Estimation of rheological constants of fresh concrete by slump test and flow test. Trans. of the Japan Concrete Inst., vol.8, Tokyo 1986, pp.65-72.

TEST METHOD AND APPLICATION OF STEEL FIBRE REINFORCED CONCRETE

Zhao Guofan   Huang Chengkui
Dalian University of Technology,China

ABSTRACT

This paper is a summary of certain results from a series of tests on steel fibre reinforced concrete (SFRC) that had the following objectives: to recognize the specialities of test mothed for SFRC, to investigate the mechanical properties and the cost of SFRC with three types of fibres, the effect of steel fibres on the properties of lightweight concrete that is reffered to applications in high-rise buildings . The engineering characteristics, benefits, applications and potential of using steel fiber reinforced concrete are also presented in this paper.

INTRODUCTION

Considerable research efforts have been made on steel fibre reinforced concrete with the growth of its application in the last decade in China. In order to provide useful data for design and standardized testing to allow meaningful comparison of reported test results in the literature, the proposals for the trial edition of "Method of Test for SFRC" have been prepared. In this course a series of mechanical tests on SFRC has been carried out and it is found that most test methods for plain concrete are suitable for SFRC and some test procedures for SFRC are different from that for plain concrete.

Since SFRC is more and more widely used in engineering, the need of full understanding of the mechanism of this material with different types of steel fiber is becoming apparent . Many people have turned their attention from the pure and basic properties of SFRC to the engineering characteristics, benefits and applications of this material. This paper also gives a general introduction of SFRC with different types of steel fiber and to make a brief description of engineering applications and characteristics of SFRC in China.

In order to develop more applications of SFRC to high-rise building, the mechanical properties of SFRC with lightweight aggregate made from expanded shale has been studied experimentally.

TEST METHOD

Following suitable existing test methods, the authors have carried out many tests on mechanical properities of SFRC in recent years in the structural laboratory of Dalian University of Technology. It is recognized that most of the test methods for plain concrete [1,2] can be used in SFRC tests, such as compressive strength, splitting tensile strength, modulus of elasticity, shrinkage, creep and some other tests.

Some specialities of test methods for SFRC have been found as follows: In specimen preparation external vibration should be used, rodding is not acceptable, internal vibration may be used in some cases where the effect of fibre orientation and distribution caused by internal vibration on the test result is not significant. Slump is not a good indicator of workability and the Vebe procedure is a good way to get it. In compressive strength tests the standard specimens (150x150x150 mm cube or 150x150x300 mm prism) used conventionally in China are appropriate for SFRC ,sometimes smaller specimens (100x100x100mm, 100x100x300mm) may be used, but the effect of size of SFRC specimens on test results is more severe than that of plain concrete. According to the statistics of 120 specimens, the ratio of the compressive strength of 150x150x150 mm cube to that of 100x100x100 mm cube is 0.91 for SFRC [3] , but it is 0.95 for plain concrete [1]. For evaluating the engineering property of SFRC, the stress-strain curve in tension and compression and the load-deflection curve in flexure should be obtained. It is well known that testing equipment may be a problem for obtaning the above curves, so the testing machine with closed loop and high rigidity should be used in those tests. In China, conventional testing machines have been widely used in most laboratories, for avoiding abrupt failure and obtaining the descending portion of the curves, the testing machine should be stiffened by loading the specimens in parallel with two steel rods for tension and with four stiff springs for compression. In this manner, most tests might be successful. Potentionmeters or LVDT displacement transducers are prefered to dial gages or strain gages as transducers for measuring the deflection, elongation or strain and automatic dynamic data acquisition system with computer or X-Y recorder should be used to get data fast.

One of the properties of concrete which is improved by the addition of fibres is energy absorption, it can be represented by the toughness index. The toughness index is calculated as the area under the normalized load-deflection curve out to the reference deflecion Dr; $Ti=(L/Lp) \times (D/Dr) \times 100$, where Dr=span/150, Lp=the peak of load.

THE MECHANICAL PROPERTIES OF SFRC

The characteristics of SFRC with different types of fibre

Three types of fibres have been used in the tests: (1) Hooked
fibres made from sheared thin steel sheets in rectangular cross
section (H-fibre), (2) Melt extract fibres made from waste
steel and iron in kidney shaped cross section (M-fibre), (3)
Straight fibres produced by cutting the unraveled wire from
scrap or wornout steel cables and wire ropes (S-fibre),which
are much cheaper than the other two, the cost of S-fibres is
approximately 50% of that of M-fibres or 30% of that of H-
fibres. The aspect ratio (length to equivalent diameter) is
about 50 for H-fibre and M-fibre,75 for S-fibre. To allow
comparison of the properties of SFRC with different types and
contents of steel fibres, the matrix of all specimens is the
same for each group of tests.

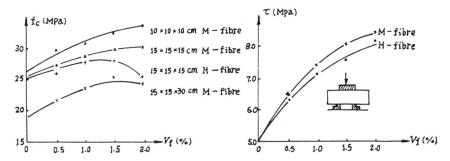

Fig.1 Relationship Between   Fig.2 Relationship Between
      Compressive Strength   Shear Strength and
      and Fibre Content   Fibre Content

Typical results of tests are shown in Fig. 1 to 7, which
indicate that all mechanical strengths except compressive
strength increase singnificantly with the increasing of steel
fibre content. Reffering to Fig.1, the compressive strength is
likely maximum at Vf=1.5% and the improvement ranges from 0 to
21 percent, if the content of steel fibre by volume is more
than 1.5%, the compressive strength decreases with the content
increasing because the workability and density of SFRC is
deteriorated.
The shear strength, splitting tensile strength and rupture
modulus (Fig.2,3 and 6) are improved by addition of 2% fibres
by 35 percent, 44 percent and 55 percent, respectively.

Referring to the property of beam-column joint, the bond
strength around reinforcing bar should be examined, test data
indicate that the improvements are more significant for
deformed bars and negligible for plain bars.

The improvement for the thoughness and post-elasticity property

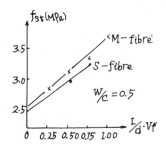

Fig.3 Typical Results of
Splitting Tensile
Test on SFRC

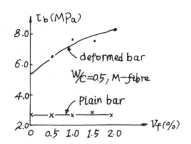

Fig.4 Ultimate Bond Strength
of Reinforcing Bars in
SFRC

of SFRC is more and more evident with the content of fibres
increasing (Fig.5 and 6), the thoughness index of SFRC with 2%
fibres is 41 times more than that of plain concrete.

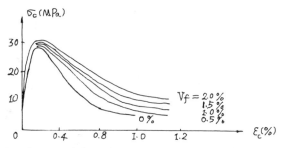

Fig.5 Stress-Strain Curves in Compresion for SFRC

On the basis of above mentioned properties, the melt extract
fibres are prefered because of their high surface area/volume
ratio, irregular contour and rough surface which improve the
adhesive and frictional bonds, and the optimal content is about
1.5 percent by volume (l/d Vf=.75). But considering the cost,
the cut wires made from waste steel wire rope are also
acceptable, and the economic content of fibres may be 1 to 1.5
percent by volume (l/d Vf=.75--1.00).

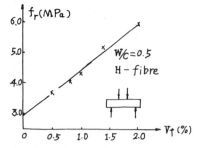

Fig.6 Modulus of rupture
as a Function of
Fibre Content

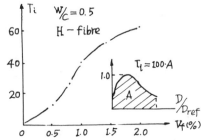

Fig.7 Tougthness Index of SFRC
as a Function of Fibre
Content

SFRC With Lightweight Aggregate

In the tests, lightweight concrete (density 1.95) contains river sand and aggregate made from expanded shale of ball shaped in diameter smaller than 10mm. The formula of matrix are the same for each series of steel fibre reinforced lightweight concrete specimens.

It has a strong appeal to the authors that the improvement in both compressive strength and other properties of lightweight concrete by addition of steel fibres is more significant than that of plain concrete. On the following data the emphasis should be put: By addition of steel fibres of 2.0% the improvements are 40-50 percent in compressive strength, 65 percent in splitting tensile strength, 95 percent in shear strength and 90 percent in rupture modulus, respectively. On the other hand, the failure behaviour of lightweight concrete is also improved by addition of steel fibres and appear in a more ductile manner.

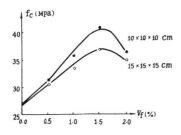
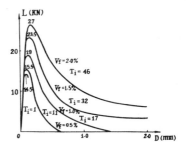

Fig.8 Compressive Strength of
       Lightweight   Concrete
       Improved by Addition of
       Steel fiber

Fig.9 Load-Deflection Curves
      in Flexure for SFRC
      with Lightweight Aggre-
      gate

The lightweight aggregate concrete reinforced by steel fibres will be beneficial to its uses in frames and shear walls in multistory buildings especially in beam-column joints and beams of coupled shear walls. In those cases, the shear strength, flexural strength and ductility of structures will be improved in a singnificant manner.

ENGINEERING APPLICATION OF SFRC

Because of the advantages of SFRC and the improvement of the manufacture technology of steel fibre, more and more people have focused or are now focusing their attention on the application of SFRC in China. In recently years SFRC has been widely used in some special engineering structures as high cracking resistance, impact resistance, anti-friction, abrasion resistance material in civil engineering structures in China, which include building, defensive work, highway, railway, hydraulic engineering, airport pavement and so on.

Use of SFRC as Excellent Impact Resistance Materials

Test results show that SFRC enhances not only the flexural and tensile strength, but also its toughness and impact strength to a great extent. Therefore it is much suitable for the construction of defensive works such as defensive gate which will subject to heavy impact load. Research on the SFRC defensive gate indicates that the performance of defensive gate is greatly improved. The resistance of SFRC defensive gate increases about 140% compared with the high strength concrete gate without reinforced with steel fibre if the gate weights are equal. The use of SFRC in defensive gate can greatly reduce the gate weight and improve the burst resistance, impact resistance,etc..

One of the most successful application of SFRC as impact resistance material is the use in the precast pile caps and pile shoes. In the foudation engineering,the piles are often adopted to deal with the weak foundation. It is needed to drive the pile into the designed depth in order to achieve the desired loading capacity. During the process of pile driving, the pile cap and pile shoe are often damaged by the repeated strong impact load used to drive the pile even before it reaches its designed depth. Therefore the SFRC is adopted to improve the performance of the pile. The experimental values show that the impact resistance of pile shoe and cap reinforced with steel fiber is 3-8 times higher than concrete pile and 1-4 times higher than reinforced concrete pile. Practical applications of SFRC shoes and caps in the constructions indicate that SFRC is very suitable. All the piles were driven to the designed depth smoothly without crack and damage. The detials of the piles reinforced with steel fibre are shown in Fig.10.

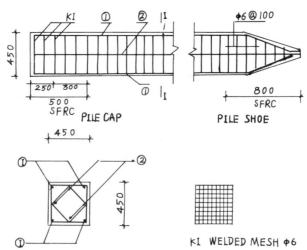

Fig.10 Details of Steel Fiber Reinforced
Pile Cap and Pile Head

On the joint of the track the sleepers in railway line are

often subjected to impact load, and horizontal thrust is also
loaded on the sleepers on the lines with small radius, which is
the main reason for the early failure of such sleepers. It is
not suitable to use the wooden sleepers because of the
shortage of the wood source . To settle this problem a new kind
of rail sleeper SF-J-2 was therefore developed by using the
steel fibre reinforced prestressed concrete as shown in Fig.11.

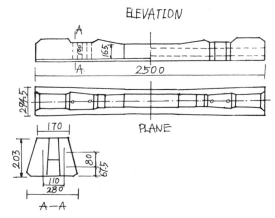

Fig.11 Detials of rail sleeper J-2

The practical application of this kind of rail sleepers in some
railway line in China since 1986 has shown the good performance
of steel fibre reinforced prestressed concrete.

Use of SFRC as Anti-Friction, Abrasion Resistance
Material in Hydraulic Structures

The concrete used in the hydraulic engineering is often
subjected to waterborne force of cavitation,erasion, impact and
friction which always cause the damage of such structures.
Therefore SFRC is much available for the construction of such
structural elements for its high impact and abrasion resistance
capacity.

In order to investigate the possibility of the application and
the performance of SFRC in hydraulic structures subjected to
friction and abrasion environment, a comparison test was
carried out on a two-level hydraulic power station on the Dadu
River installed and put into operation in 1965. This hydraulic
power station is Diversion type runoff hydroelectric power
plant, and the measured maximum grain diameter of the bed load
passing through the sluice gate is about 1.0-1.5 meter. After
one year operation, some of the scouring sluice board and apron
in the hydraulic power station were rushed into deep trough by
the passing of the bed load carried by the violent water
current with high speed . The maximum depth of the rubbed
trough is about 0.7 meter and the 28 mm diameter embedded steel
bars were completely broken. The comparison tests were carried
out by using many kinds of materials as anti-friction materials
such as epoxy resin concrete,furan concrete,steel fibre
reinforced high strength concrete and so on .The results

indicate that the use of SFRC as anti-friction materials in
hydraulic structures is much suitable with good performance and
low cost compared with other materials.

Recently in China the SFRC was adopted for the construction and
the repairment of many hydraulic structures for which the
impact resistance, anti-friction and abrasion resistance are
essential. The operation of such hydraulic structures shows
that the performance is significantly improved.

Use of the SFRC as High Cracking Resistance and High
Ductile Material

The greatest advantage of incorporating steel fibre into
concrete is the greatly increasing cracking resistance and
ductile capacity,so it can used in these structural objects
which requires high cracking resistance and ductile capacity
such as frame joint,hydraulic aqueduct, sluice gate and so on.

Beam-column joints for seismic resistance structures are
required to dissipate and absorb large amount of energy during
an earthquake.To meet such needs of seismic resistance, a mate
of intersecting bars and hoops in the joint are required which
makes the congestion and difficulty of casting concrete during
construction. To minimize the steel congestion, SFRC is used in
the beam-column joint region. In order to investigate the
performance of such joints, a series of tests have been carried
out. Details of the joint specimens are shown in Fig.12. Test
values show that it is by the addition of steel fiber that the
toughness, bond strength, crack strength,shear strengh,ultimate
strength of the joint is considerably improved. Therefore the
use of SFRC in joint region can solve the problem of steel
congestion and the difficulty of casting concrete.

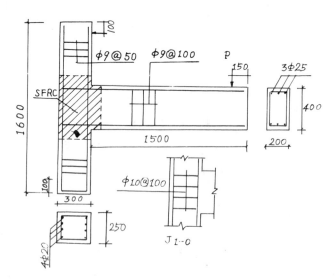

Fig. 12 Details of Beam-Column Joint Specimen

In civil engineering, using of concrete material in the construction of many structures is limited because of its low tensile strength. If the concrete structures are reinforced with steel fibre, the cracking strength and ultimate strength of concrete can be increased and the crack width and deformation of the structures can be reduced. Therefore SFRC is used instead of concrete in the constructions of many concrete structures limited by the low cracking strength of concrete or the excessive crack width or deformation.

During the construction of a concrete hydraulic aqueduct in Hunan Province, because of the low depth and long span of the aqueduct, the tensile stresses of the structure were great which may cause the damage because of crack of concrete. SFRC then was adopted to reinforce the aqueduct as shown in Fig.13. Test on such aqueduct showed that the crack and deformation were controlled.

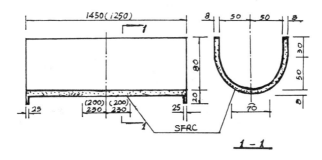

Fig.13   Cross Sections of Steel Fibre
Reinforced Hydraulic Aqueduct

SFRC was also used in the construction of lock gate and slot of sluice gate in hydraulic engineering to avoid the crack of concrete. For example, a lock gate in a dam in Zhejiang Province was previously a shell gate constructed with reinforced concrete. When the gate was put ioto operation, many cracks occured on the gate caused by excessive torsion deformation. In 1979 this gate was replaced by a new gate with the same gate size and reinforcement. During the casting of the gate, deformed steel fibers were added in concrete. After several years operation, it seems that SFRC shell gate works well with the torsion deformation being reduced from 3cm to 0.5cm.

In the design and construction of a catilever folded plate structure in Nanjing gym. SFRC was also used to reduce the crack width and the deformation. The steel fibre contents used were 100kg/m and 150kg/m , which improved the bending strength 1.5 to 2 times higher thah plain concrete, and the phenomenon of excessive crack width and deformation of the plate were controlled.

Other Application of SFRC

Except the applications of SFRC mentioned above, SFRC is also used to repair the deteriorated portion of concrete due to fire and long-time service, to stablize the rock slope, to construct and repair highway and airport pavement and so on. The interesting of application of SFRC for such purposes in China is very great , and SFRC is adopted for many construction since its appearance in China, for example, the use of steel fibre reinforced shotcrete to repair the deteriorated concrete caused by fire in a reinforced concrete basement in Wu Yang Steel Rolling Mill and a reinforced concrete bridge in Qin Huang Dao city . Many practical application of SFRC for the purpose of consolidating and stabilization have also been carried out in the building , mining, and tuning engineering all over the country. The practical use shows SFRC is the very suitable material for such purpose for its low cost time saving and good performance.

FUTURE DEVELOPMENT

It is certain that the application of SFRC in civil engineering will be greatly developed and SFRC will be adopted for many engineering structures. With more and more applications of SFRC ,the need of fully understanding the properties of SFRC is very urgent currently. Therefore many research works on the behaviour and applications of SFRC should be carried out in China to meet the requirment of the engineering application. Nowadays many research works such as the research on the standard test method of SFRC, Recommendation for the design and construction of SFRC joint and the Code for the design and construction of SFRC structures have been carried out in many universities and research institutes in China. But research works on the properties of SFRC,the manufacture technology and construction method should be more improved to reduce thr cost of steel fibre and ensure the good quality of construction of steel fibre reinforced concrete.

Although SFRC is adopted for many engineering structures the application of SFRC should be more widely developed, for example the use of steel fibre reinforced ferrocement structure,steel fiber reinforced high strength concrete, steel fibre reinforced light-weight concrete and the use of SFRC for offshore structure,floating dock,ship building and so on.

References

1. China Academy of Building Research, The standard of test method for plain concrete, Beijing, 1986

2. China Academy of Water Conservancy and hydraulic Engineering Research, The standard of test method for concrete in water conservancy and hydraulic Engineering, Beijing, 1982

3. Guan Liqiu and Zhao Guofan, A study on the mechanism of fibre reinforcement in short steel fibre concrete, RILEM Symposium FRC86

639

EFFECT OF PREPARATION CONDITIONS ON THE MECHANICAL PROPERTIES
OF LOW POROSITY HARDENED CEMENT PASTE

Lu Ping   Fu Jung   Shen Wei
Dept. of Materials Science and Engineering
Tongji University, Shanghai,200092, CHINA

ABSTRACT

Hardened cement paste (HCP) itself is a composite material at meso- and
microlevel. In this paper, the influence of compacting pressure, W/C ratio
and curing condition of a kind of very low porosity HCP --- Polymerized
Silicate Cement (PSC) during preparation on its mechanical behaviour has
been analysed in terms of GPC method in which peakwidening was revised with
the help of a computer. The non-evaporable water, the pore size distributions and the mechanical strength of all samples were measured.

Results show that for compacted cement paste bith porosity and degree
of hydration effect on its mechanical behaviour. The optimum W/C ratio during compaction is the maximum one under a compacting pressure. The optimum
compacting pressure is considered as that cooperation of the porosity of
HCP with its degree of hydration can achieve the optimum so that this material can come up to advanced mechanical behaviour. Furthermore, because of
improvement of its intrinsic propertirs the optimum compacting pressure of
PSC material may decrease in some degree in order to produce more hydrates.
Results also indicate that among several curing regulations, hydrothermal
curing condition in atmosphere is the best to the benefit of the improvement of its mechanical behaviour.

INTRODUCTION

Hardened cement paste (HCP) itself is recognized as a multiphase composite
material at meso- and microlevel. In 8th ICCC the authors published a paper
about study of intrinsic properties of HCP(1). As we know, the mechanical
behaviour of porous material firstly depends on its poro structure. It is
necessary to prepare HCP with lower porosity in order to investigate the
effect of intrinsic properties of HCP on its mechanical behaviour more directly. Since 1930s Freyssinet and LHermite et al had published several ar-
ticles of compacted cement paste(2,3), Lecznor and Barnoff measured the
compressive strength of compacted HCP(4). In recent 20 years Skalny(5) and
Bajza(6) reported the mechanical behaviour of compacted HCP, the used pressure was upto 50 --- 200 MPa, resulting in higher compressive strength

(about 200 MPa) of the HCP.

In general, the higher the degree of hydration of HCP is, the faster the development of its mechanical strength is, so the degree of hydration is an important parameter for HCP with low porosity, especially, for poly-merized silicate cement (PSC material), the purpose of which is to improve brittleness of cementitious materials. The higher degree of hydration can increase the degree of silicate polymerization at molecular level. But it is impossible to raise the degree of hydration of HCP with low porosity by a big margin, for its preparation conditions are limitted and very densed structure results in difficulty of curing in water.

This paper investigates the influence of the compacting pressure, W/C ratio and curing condition of PSC materials at room temperature on its me-chanical behaviour.

EXPERIMENTAL PROCEDURE

Starting materials

A works' ordinary Portland clinker mixed by gypsum of 3% by weight is used, ground to fineness of 6400 cm^2/g. According to Ref. 1 the cylindical spe-cimens of 12.7 12.7 mm are prepared. The adopted additives are 15% $FeSO_4$, $CuSO_4$ solutions and silane VTES ($CH_2=CH-Si(OC_2H_5)_3$). W/C ratio is varied in range 0.08 --- 0.17. The compacting pressure is 0.5, 1.0, 2.0 and 3.0 T/cm^2. The specimens are divided into 7 series, which are cured in water at 20°C, under hydrothermal (95°C) and autoclaved at 10 --- 30 atm, respec-tively. The age of all specimens is 28 days.

Testing methods

1). Physical properties: Compressive strength and splitting strength is measured, and non-evaporable water is detected. In addition, the pore size distributions of these samples are performed using mercury intrusion poro-simetry (MIP).

2). GPC method: By using GPC method the degree of silicate polymerization of TMS-derivatized samples is determined. The detail procedure was reported in Ref. 1. Because the resolution of GPC is not infinite, just as one of other chromatography, the peaks in graph are always widened. This phenome-non is usually shown by Tung eq. as follows(7):

$$f(v) = \int_{V_a}^{V_b} G(V,y)W(y)dy \qquad (1)$$

where $G(V,y)$ is peakwidening function, $W(y)$ is real molecular weight dis-tributions and $f(v)$ is apparent measured function.

In terms of least square method $f(v)$ function is simulated by follows

$$f(v) = (\frac{h}{\pi})^{\frac{1}{2}} \exp(-q^2(v-g_o)^2) \sum_{i=0} u_i(v-y_o)^i \qquad (2)$$

where h is resolution factor of GPC instrument, n is terms of polynomial, which can be arbitrarily choosen according to required precision. Solving

W(y) in eq. 1 by a computer, amending peakwidening phenomenon, we obtain
GPC graph as shown in Fig. 1. Each component in Fig. 1 can be transformed
to its percentage by area factors in Tab. 1.

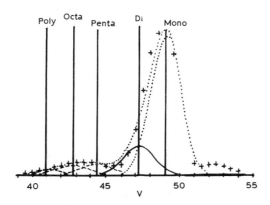

Fig. 1 GPC analysis of a typical TMS derivative

Tab. 1 Area factors of the peaks in Fig. 1

	mono–	di–	penta–	octa–	Q_{14}	Q_{23}	Q_{47}
area factor	0.073	0.093	0.109	0.116	0.120	0.122	0.124

RESULTS AND DISCUSSION

1. Optimum W/C ratio

The effect of W/C ratio on compacted HCP is different from that on ordinary
HCP. For the latter, the higher W/C ratio is not beneficial. Obviously,
this is because surplus water in HCP formed big capillaries which is not
beneficial to development of its microstructure. But for the former, the
lower W/C ratio is not enough to its completer hydration. There is no surplus water, the big capillary within paste is mainly formed from compacting
pressure.

During compaction of cement paste there is a maximum W/C ratio. Once
this value is exceeded, the water within paste is inevitably squeezed. The
test results of the maximum W/C ratio under the different compacting pressure are shown in Fig. 2, from which one can know the maximum W/C ratio
gradually reduced with the increase of compacting pressure, though the reduced rate is slower and slower. This is because within the region of high
compacting pressure, the cement particles have been close to denser heaped-
up state, the variation of its porosity vs. pressure is not so obvious.

As we know, W/C ratio is an important technological parameter of ce-

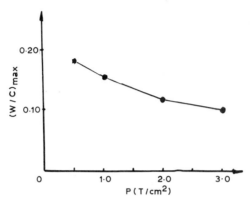

Fig. 2 Relationship of compacting pressure and the maximum W/C ratio

ment and concrete materials. For compacted HCP with low porosity, the optimum W/C ratio is obviously the maximum one under a compacting pressure.

2. The optimum compacting pressure

The test results of physical properties of the specimens are shown in Tab. 2, in which series A is ordinary low porosity HCP, series B is PSC material modified by Fe^{+2} ion, series C is that by Cu^{+2} ion and series D is that by VTES. From Tab. 2 one can know that under compacting pressure of 1 T/cm^2 the mechanical strength of series A is the highest. Under lower compacting pressure (such as specimen A7), higher initial porosity always results in higher ultimate porosity although its degree of hydration is also higher. The results of MIP indicate that the pore volume of A7 is 0.0397 cm^3/g, meanwhile, that of A4, A5 are 0.0296 cm^3/g and 0.0308 cm^3/g, respectively. On the other hand, under higher compacting pressure (such as specimen A4) its W/C ratio reduces and its initial porosity is so low that its further hydration is restrained to a certain degree. Comparing A4 with A5, one can know the strong dependence of mechanical strength of low porosity HCP on its degree of hydration. In fact, under a compacting pressure its porosity and degree of hydration are contradictory. Only under a proper compacting pressure the porosity of paste is in optimum cooperation with its degree of hydration, which can result in an optimum effect on its mechanical behaviour.

By using TMS technique the molecular weight distributions of the PSC samples of series B, C and D modified at molecular level are shown in Tab. 3. It can be seen from Tab. 3 that due to the addition of coordination complex ions and silane coupling agent, the degree of silicate polymerization in the samples increases, the intrinsic properties of the materials can be improved in some degree. Obviously, within region of low porosity the samples of series B, C and D have higher mechanical strength thanthat of series A, although the porosity of series B, C and D is the same as of series A. Meanwhile, as the mechanical strength of series A, B, C and D is equal to each other, the samples of series B, C and D may allow higher porosity than the samples of series A. Therfore, the optimum compacting pressure of PSC material may decrease in some degree, so that an new equilibrium between ultimate porosity of HCP and its degree of hydration can be reached.

Tab. 2   The physical properties of specimens

	compacting press.(T/cm²)	W/C	compress. str.(MPa)	tensile str.(MPa)	bonding water(%)
A1	3.0	0.10	175.8	9.8	10.58
A2	3.0	0.08	164.3	8.3	10.37
A3	2.0	0.12	182.1	11.0	12.19
A4	2.0	0.10	170.9	12.3	11.12
A5	1.0	0.15	200.6	11.3	12.20
A6	1.0	0.12	203.3	12.0	12.29
A7	0.5	0.17	141.3	7.5	14.01
B1	3.0	0.10	211.5	14.8	10.86
B2	3.0	0.08	201.7	11.7	11.32
B3	2.0	0.12	240.4	19.2	12.09
B4	2.0	0.10	289.7	20.6	13.55
B5	1.0	0.15	158.8	15.2	14.45
B6	1.0	0.12	143.3	14.7	14.19
C1	3.0	0.10	211.4	13.6	10.94
C2	3.0	0.08	220.0	15.4	12.43
C3	2.0	0.12	239.4	15.1	12.64
C4	2.0	0.10	274.3	15.7	12.73
C5	1.0	0.15	101.2	11.4	14.05
C6	1.0	0.12	192.2	14.1	14.89
D1	3.0	0.10	185.7	13.1	10.72
D2	3.0	0.08	189.9	14.4	10.78
D3	2.0	0.12	187.4	14.7	11.46
D4	2.0	0.10	212.3	13.6	11.91
D5	1.0	0.15	235.5	13.2	12.73
D6	1.0	0.12	227.0	11.5	11.83
D7	0.5	0.17	200.2	5.2	14.51

Tab. 3   Results of GPC measurement (%)

	Mn	Mw	mono-	di-	penta-	octa-	Q_{14}	Q_{23}	Q_{47}
A3	418	492	84.7	9.0	2.9	2.5	0.8	/	/
A5	421	502	82.7	10.6	3.1	2.7	0.9	/	/
B3	465	978	74.9	7.3	5.3	4.0	3.5	2.8	2.3
C3	422	565	86.0	5.8	0.6	5.2	2.0	0.5	/
D3	443	681	80.4	5.6	4.2	4.7	3.6	1.4	/
D5	450	702	77.5	8.0	4.0	5.0	3.9	1.5	/

It has to be pointed out that for both series B and C a lot of AFt phase formed in early stage hydration because sulphate solution were used, which reduces real W/C ratio during hydration of cement itself. In order to reach the same initial density as of series A, the compacting pressure of series B and C must increase.

Furthermore, the relationship of bonding water under different compacting pressure and ultimate porosity of the specimens is analysed as shown in Fig. 3. The line A1-A5, B1-B4, C1-C4 and D3-D5 from Fig. 3 indicates that owing to the decrease of compacting pressure, the increase of the maximum W/C ratio, ultimate degree of hydration of HCP increased and its ultimate porosity decreased. On the other hand, the line A5-A7, B4-B6, C4-C6 and D5-D7 indicates that as the compacting pressure is further decreased, it is impossible that higher initial porosity of HCP is counteracted by the increase of its degree of hydration. This ultimate porosity is still higher. It can be seen from Tab. 2 that the mechanical strength of HCP decreased. So the compacting pressure reflected by points A5, B4, C4 and D5 in Fig. 3 is the optimum.

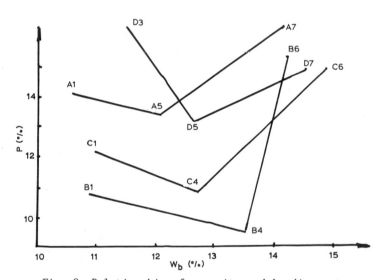

Fig. 3   Relationship of porosity and bonding water

It is further shown from Fig. 3 that the improvement of intrinsic properties of low porosity HCP can further increase its degree of hydration, decrease its porosity and thus develop its mechanical behaviour.

3.The optimum curing regulation

The physical properties of the specimens in various curing conditions are shown in Tab. 4. One can know from Tab. 4 that after changing curing condition, the degree of hydration of most specimens increases, their mechanical strengths also raise. Obviously, in used curing conditions the optimum curing regulation is the third type. Besides, the hydrothermal curing is better than ordinary curing, the autoclaved curing also is goodish. Comparing the second type with the third, one can find that the shorter precuring period is better to develop its microstructure.

Verbeck and Copeland(8) reported that steam can accelerate the hydra-

Tab. 4   Physical properties of the specimens in
various curing conditions

Curing condition	Symb.	Compress. str.(MPa)	Tensile str.(MPa)	Bonding water(%)
1. stand. curing to	A1	175.7	9.8	10.6
28 days	B1	211.5	14.8	10.9
	D1	185.7	13.1	10.7
2. stand. curing to	A1a	222.6	10.3	10.8
7d + hydrothermal	B1a	230.9	11.8	11.1
12h + stand. cu-	D1a	272.1	11.4	12.0
ring to 28 days				
3. stand. curing 24h	A1b	286.8	20.3	11.0
+ hydrothermal 12	B1b	290.7	20.8	13.2
h + stand. curing				
to 28 days				
4. stand. curing to	A1c	201.6	12.2	10.6
28d + autoclaved	B1c	223.3	15.2	11.0
10 atm 6h	D1c	231.0	17.2	11.3
5. stand. curing 7d	A1d	203.0	18.5	11.4
+hydrothermal 12h	B1d	263.1	19.2	11.7
+autoclaved 10atm	D1d	246.5	20.2	11.4
6h + stand.curing				
to 28 days				
6. stand. curing 24h	A1e	200.7	16.9	11.1
+autoclaved 10atm	B1e	213.9	17.7	11.5
6h + stand.curing	D1e	245.7	14.9	10.7
to 28 days				
7. stand. curing to	A1f	204.6	9.2	9.7
28d + autoclaved	B1f	251.1	15.3	10.3
30atm 6h	D1f	203.2	13.8	11.1

tion of cement particle, which formed a denser hydrate layer surrounding
unhydrated cement particle and preventing it from continuous hydration. It
results in development of heterogeneous microstructure. Especially, the
crystal hydrate of low porosity HCP formed in autoclaved condition is not
easy to transfer into near pores, thus more heterogenous microstructure is
developed. This result is indicated by SEM micrograph as shown in Fig. 4.
One can find from Fig. 4 that the hydrates of autoclaved specimen surround
the surface of unhydrated cement particle. In this microstructure pores
seem to be more. Meanwhile, the specimen in hydrothermal curing condition
has more hydrates which fill in the space beyond unhydrated cement particle
resulting in lower porosity.

Fig. 4 SEM micrograph
A). autoclaved 30atm B). hydrothermal curing

CONCLUSION

1. Both the porosity of compacted HCP with low porosity and its degree of hydration effect on its mechanical strength. If compacting pressure is too high, then its degree of hydration is too low. The optimum compacting pressure is considered as that the cooperation of the porosity of HCP with its degree of hydration can achieve the optimum so that this material can come to up advanced mechanical behaviour. The optimum W/C ratio thus is the maximum one under a compacting pressure.

2. Because the degree of silicate polymerization in CSH gel of PSC material increases, its intrinsic properties are improved. Comparing with ordinary low porosity HCP, the optimum compacting pressure of PSC material may be reduced that the more hydrates are produced.

3. Among several curing conditions the hydrothermal curing in atmosphere is the best to the benefit of the improvement of its mechanical behaviour.

ACKNOWLEDGEMENTS

The authors gretefully acknowledge the financial support of the National Natural Science Fund of China.

REFERENCES

1. Lu Ping and Huang Yiun-yuang, 8th Intern. Conf. on the Chem. Cem., Vol. 3, p. 343 (1986).

2. M. E. Freyssinet, Development in Concrete Making, Vol. 21, p. 207 (1936).

3. R. L'Hermite, Intern. Assoc. Congr. for Test. Mater., p. 332 (1937).

4. F. J. Lecznar and R. M. Barnoff, ACI Journal Proceedings, Vol. 57, p. 973 (1961).

5. J. Skalny and A. Bajza, ibid, Vol. 67, p. 221 (1970).

6. A. Bajza, V. Frano and J. Skalny, Zement-Kalk-Gips, Vol. 57, p. 400 (1968).

7. L. H. Tung, J. Appl. Polym. Sci., Vol. 10, p. 375 (1966).

8. G. Verbeck and L. E. Copeland, Paper SP32-I, PC4-2-2-5, PCA, Chicago, USA.

PROPERTIES OF POLYMER-MODIFIED MORTARS CONTAINING SILICA FUME

YOSHIHIKO OHAMA*, KATSUNORI DEMURA*, MASASHI MORIKAWA* and TAKAYUKI OGI**
* College of Engineering, Nihon University
Koriyama, Fukushima-ken, 963 Japan
** Hasegawa Komuten Co., Ltd., Tokyo, 115 Japan

ABSTRACT

This paper discusses the properties of polymer-modified mortars with silica fume. The polymer-modified mortars using a styrene-butadiene rubber latex and an ethylene-vinyl acetate emulsion are prepared with various polymer-cement ratios and silica fume contents, and tested for flexural and compressive strengths, adhesion, water absorption and permeation, water resistance, drying shrinkage, carbonation and chloride ion penetration. The effects of the polymer-cement ratio and silica fume content on the properties of the polymer-modified mortars are examined. In conclusion, the addition of the silica fume causes improvements in the compressive strength, resistance to water permeation and chloride ion penetration. However, the adhesion, resistance to water absorption and carbonation and drying shrinkage are not improved by silica fume addition.

INTRODUCTION

Polymer-modified mortars have widely been used as construction materials such as floorings, pavings, integral waterproofings, adhesives, decorative coatings and repairing materials because of their superior performance. In particular, they have recently received much attention as repairing materials for damaged reinforced concrete structures. Silica fume has recently been used for improving the mechanical properties of cement concrete [1]. The present paper deals with the properties of the polymer-modified mortars with the silica fume, which is added to the mortars for the purpose of improving their properties relating to the repairing materials. The polymer-modified mortars using a styrene-butadiene rubber latex, an ethylene-vinyl acetate emulsion and the silica fume were prepared with various polymer-cement ratios and silica fume contents, and tested for flexural and compressive strengths, adhesion, water absorption and permeation, water resistance, drying shrinkage, carbonation and chloride ion penetration. The effects of the polymer-cement ratio and silica fume content on the properties of the polymer-modified mortars are discussed.

MATERIALS

Cement, Aggregate and Admixture

Ordinary portland cement and Toyoura standard sand as specified in JIS (Japanese Industrial Standard) were used in all the mixes. Silica fume was used as an admixture. The properties of the silica fume are given in Table 1.

TABLE 1

Chemical compositions and physical properties of silica fume

Chemical Compositions (%)								
ig.loss	SiO_2	Al_2O_3	Fe_2O_3	CaO	MgO	Na_2O	K_2O	Total
3.0	90.6	1.2	1.1	0.3	0.8	0.45	1.97	99.4

Physical Properties		
Specific Gravity (20 °C)	Average Particle Size (μm)	Specific Surface Area (m^2/g)
2.2	0.1~0.3	28.32

Polymer Dispersions for Cement Modifiers

Commercial cement modifiers used were a styrene-butadiene rubber(SBR) latex and an ethylene-vinyl acetate (EVA)emulsion. Their basic properties are listed in Table 2. Before mixing, silicone-emulsion-type antifoamer was added to the polymer dispersions in a ratio of 0.7% of the silicone solids of the antifoamer to the total solids of the polymer dispersions.

TABLE 2

Properties of polymer dispersions

Type cf Polymer Dispersion	Total Solids (%)	Specific Gravity (20°C)	pH (20°C)	Viscosity (20°C,cP)
SBR	45.1	1.028	8.4	133
EVA	44.7	1.056	5.2	1440

TESTING PROCEDURES

Preparation of Specimens

According to JIS A 1171 (Method of Making Test Sample of Polymer-Modified Mortar in the Laboratory), polymer-modified mortars containing silica fume were mixed using the mix proportions given in Table 3. The polymer-cement ratio [P/(C+S)] was calculated on the basis of the total solids of the polymer dispersions (P), the content of the ordinary portland cement (C) and the content of silica fume (S). Mortar specimens having the desired shapes and sizes were molded, and then given a 2-day-20°C-80%R.H.-moist, 5-day-20°C-water and 21-day-20°C-50%R.H.-dry cure.

Strength Test

The cured mortar specimens 40x40x160mm were tested for flexural and compressive strengths according to JIS A 1172(Method of Test for Strength of Polymer-Modified Mortar).

TABLE 3
Mix proportions of mortars

Cement(C*+S**) : Sand (By Weight)	Silica Fume Content, S/(C+S) (%)	Type of Polymer Dispersion	Polymer-Cement Ratio, P/(C+S) (%)	Water-Cement Ratio, W/(C+S)(%)	Flow
1 : 3	0	—	0	77.5	160
		SBR	5	66.7	168
			10	68.6	170
			15	65.6	170
			20	62.9	172
		EVA	5	66.7	166
			10	63.7	172
			15	60.5	170
			20	57.2	168
	5	—	0	76.8	173
		SBR	5	67.2	168
			10	64.6	165
			15	59.8	168
			20	57.7	174
		EVA	5	73.3	170
			10	69.7	175
			15	65.0	166
			20	64.7	165
	10	—	0	79.0	172
		SBR	5	69.9	170
			10	67.1	166
			15	62.8	166
			20	57.9	167
		EVA	5	67.5	168
			10	63.2	165
			15	63.7	165
			20	62.4	167
	13	—	0	91.4	166
		SBR	5	78.9	166
			10	74.2	168
			15	70.4	168
			20	64.6	175
		EVA	5	77.8	165
			10	71.4	166
			15	70.5	167
			20	69.1	168

*;C:Ordinary portland cement.
**;S:Silica fume.

Adhesion Test

The cured mortar specimens adhered to the cement mortar plates with a cement:sand ratio of 1 : 2 (by weight) and a water-cement ratio of 65% were tested for adhesion in tension in accordance with JIS A 6915(Wall Coatings for Thick Textured Finishes). After adhesion test, the failed crosssections of the mortar specimens were observed for failure modes, which were classified into the following three types:

A:Adhesive failure(failure in the interface)
M:Cohesive failure in polymer-modified mortars
S:Cohesive failure in substrate (ordinary cement mortar)

The respective approximate rates of A, M and S areas in the total area of 10 on the failed crosssections are expressed as suffixes for A, M and S.

Water Absorption Test
The cured mortar specimens 40x40x160mm were immersed in water at 20°C for 48 hours, and their water absorption was determined by the method specified in JIS A 6203(Polymer Dispersions for Cement Modifiers). After water immersion, the mortar specimens were tested for flexural and compressive strengths to evaluate their water resistance.

Water Permeation Test
The cured specimens 150x40 mm were tested for water permeation according to JIS A 1404(Method of Test for Waterproof Agent of Cement for Concrete Construction).

Drying Shrinkage Test
The mortar specimens 40x40x160mm were used for measuring drying shrinkage. Immediately after they were given a 2-day-20°C-80%R.H.-moist and 5-day-20°C-water cure, their original length was measured. Then the mortar specimens were cured at 20°C and 50%R.H. for 28 days, and their drying shrinkage was measured by the comparator method specified in JIS A 1129(Methods of Test for Length Change of Mortar and Concrete).

Accelerated Carbonation Test
The cured mortar specimens 100x100x100mm were placed in a non-pressurizing carbonation test chamber for 28 days, in which temperature, humidity and CO_2 gas concentration were controlled to be 30°C, 60%R.H. and 5% respectively. After accelerated carbonation, the mortar specimens were split, and the split crosssections were sprayed with 1% phenolphthalein alcoholic solution. The depth of the rim of each crosssection changed to white color was measured with slide calipers as a carbonation depth as shown in Figure 1.

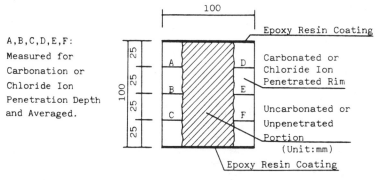

Figure 1. Crosssection of specimen after carbonation or chloride ion penetration test.

Chloride Ion Penetration Test

The cured mortar specimens 100x100x100mm were immersed in 2.5% NaCl
solution at 20°C for 28 days for chloride ion penetration. The test
solution was changed every 7 days up to an immersion period of 28 days.
After 28-day immersion, the mortar specimens were split, and the split
crosssections were sprayed with 0.1% sodium fluorescein and 0.1N silver
nitrate solutions. The depth of the rim of each crosssection changed to
white color was measured with slide calipers as a chloride ion (Cl^-)
penetration depth as shown in Figure 1.

TEST RESULTS AND DISCUSSION

Figure 2 represents the effects of polymer-cement ratio and silica fume
content on the water-cement ratio of polymer-modified mortars with a
constant flow of 170±5. The water-cement ratio of the polymer-modified
mortars having the constant flow is reduced with increasing polymer-

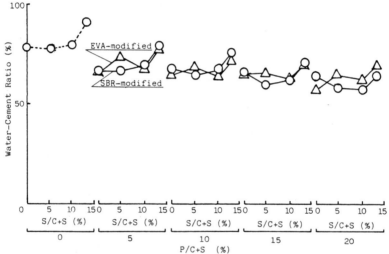

Figure 2. Effects of polymer-cement ratio and silica fume content on water-
cement ratio of polymer-modified mortars with constant flow of
170±5.

cement ratio regardless of the polymer type and silica fume content. At the
respective polymer-cement ratios, a difference in the water-cement ratio
between the polymer types or silica fume contents is hardly recognized up to
a silica fume content of 10%, and the water-cement ratio increases as the
silica fume content is raised to 13%.

Figure 3 shows the polymer-cement ratio and silica fume content vs.
flexural strength of polymer-modified mortars before and after water
immersion. In general, the flexural strength of the polymer-modified
mortars before and after water immersion increases with increasing
polymer-cement ratio. The flexural strength of unmodified mortars before
and after water immersion is slightly decreased with an increase in the
silica fume content. Irrespective of the polymer-cement ratio, the flexural

strength of SBR-modified mortars before water immersion increases with increasing silica fume content, and reaches the maximum at a silica fume content of 10 %. The flexural strength before water immersion of EVA-modified mortars with polymer-cement ratios of 5 and 10% decreases with increasing silica fume content. The flexural strength before water immersion of EVA-modified mortars with polymer-cement ratios of 15 and 20% increases with increasing silica fume content, and attains to the maximum at

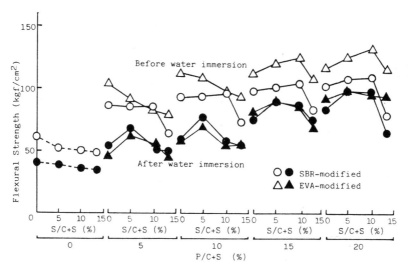

Figure 3. Polymer-cement ratio and silica fume content vs. flexural strength of polymer-modified mortars before and after water immersion.

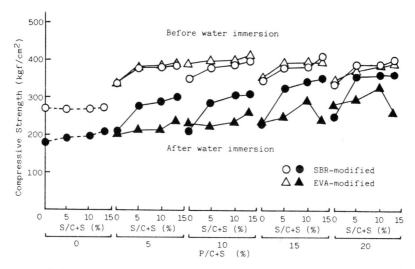

Figure 4. Polymer-cement ratio and silica fume content vs. compressive strength of polymer-modified mortars before and after water immersion.

a silica fume content of 10%. SBR- and EVA-modified mortars after water immersion give the maximum flexural strength at a silica fume content of 5% as the silica fume content increases at the respective polymer-cement ratios. A reduction in the flexural strength of the polymer-modified mortars after water immersion is improved with increasing polymer-cement ratio irrespective of the silica fume content.

Figure 4 exhibits the polymer-cement ratio and silica fume content vs. compressive strength of polymer-modified mortars before and after water immersion. The effect of the silica fume content on the compressive strength of unmodified mortars before water immersion is hardly recognized. The compressive strength of the unmodified mortars after water immersion is gradually increased with increasing silica fume content. The compressive strength of the polymer-modified mortars before water immersion is increased with raising polymer-cement ratio and silica fume content regardless of the polymer type. The compressive strength of SBR-modified mortars after water immersion is considerably increased with raising polymer-cement ratio and

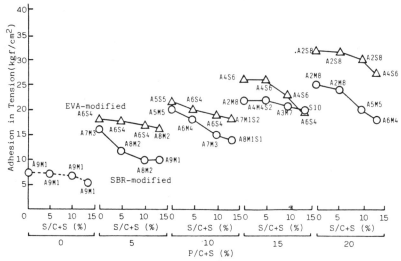

Figure 5. Adhesion in tension vs. polymer-modified mortars with variations of polymer-cement ratio and silica fume content.

silica fume content. The compressive strength after water immersion of EVA-modified mortars with polymer-cement ratios of 5 and 10% rises with an increase in the silica fume content. The compressive strength after water immersion of EVA-modified mortars with polymer-cement ratios of 15 and 20% is increased with an increase in the silica fume content, and reaches the maximum at a silica fume content of 10%. In general, a reduction in the compressive strength of the polymer-modified mortars after water immersion is also improved with raising polymer-cement ratio regardless of the silica fume content.

Figure 5 illustrates the adhesion in tension of polymer-modified mortars with variations of polymer-cement ratio and silica fume content. The adhesion in tension of the polymer-modified mortars is increased with an increase in the polymer-cement ratio, but decreased with raising silica fume content at the respective polymer-cement ratios.

The water absorption of polymer-modified mortars with variations of polymer-cement ratio and silica fume content is shown in Figure 6. In general, the water absorption of the polymer-modified mortars is markedly reduced with an increase in the polymer-cement ratio, but slightly increased with an increase in the silica fume content. The water absorption is affected to a lesser extent by the silica fume content at high polymer-cement ratio.

Figure 7 represents the water permeation of polymer-modified mortars with variations of polymer-cement ratio and silica fume content. Generally the water permeation of the polymer-modified mortars tends to decrease with increasing polymer-cement ratio and silica fume content. The effect of the polymer-cement ratio on the water permeation is more remarkable than that of

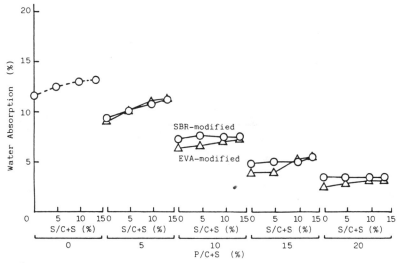

Figure 6. Water absorption vs. polymer-modified mortars with variations of polymer-cement ratio and silica fume content.

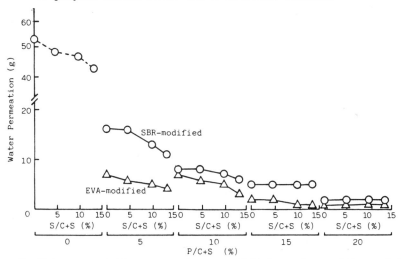

Figure 7. Water permeation vs. polymer-modified mortars with variations of polymer-cement ratio and silica fume content.

the silica fume content. The effect of the silica fume content is hardly recognized at the higher polymer-cement ratio.

Figure 8 shows the drying shrinkage of polymer-modified mortars with variations of polymer-cement ratio and silica fume content. Generally the

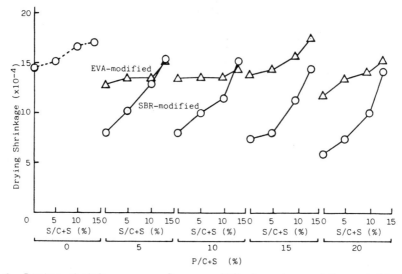

Figure 8. Drying shrinkage vs. polymer-modified mortars with variations of polymer-cement ratio and silica fume content.

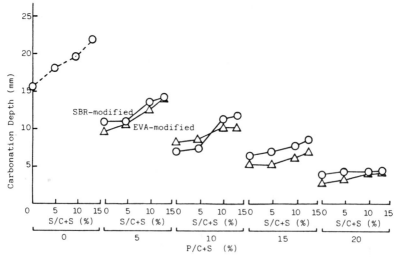

Figure 9. Carbonation depth of polymer-modified mortars with variations of polymer-cement ratio and silica fume content.

drying shrinkage of the polymer-modified mortars is inclined to decrease with increasing polymer-cement ratio. However, the drying shrinkage tends to considerably increase in a raise in the silica fume content, and this tendency is marked for SBR-modified mortars. The reasons for such large drying shrinkage may be explained to be due to the liberation of a large amount of water adsorbed by the silica fume with very large surface area.

Figure 9 represents the carbonation depth of polymer-modified mortars with variations of polymer-cement ratio and silica fume content. Regardless of polymer type, the carbonation depth of the polymer-modified mortars is markedly decreased with raising polymer-cement ratio, but increased with raising silica fume content.

Figure 10 illustrates the chloride ion penetration depth of polymer-modified mortars with variations of polymer-cement ratio and silica fume content. Irrespective of polymer type, the chloride ion penetration depth of the polymer-modified mortars is inclined to decrease with increasing polymer-cement ratio and silica fume content.

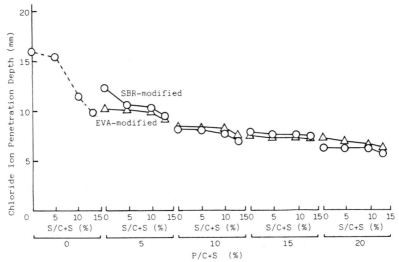

Figure 10. Chloride ion penetration depth of polymer-modified mortars with variations of polymer-cement ratio and silica fume content.

CONCLUSIONS

The properties of polymer-modified mortars are markedly improved with raising polymer-cement ratio regardless of polymer type and silica fume content. Generally the addition of silica fume causes improvements in the compressive strength, resistance to water permeation and chloride ion penetration of the polymer-modified mortars. However, the adhesion, resistance to water absorption and carbonation and drying shrinkage of the polymer-modified mortars are not improved by silica fume addition.

REFERENCE

1. Fly Ash, Silica Fume, Slag, and Natural Pozzolans in Concrete, Proceedings, Second International Conference, Madrid, Spain 1986, SP-91, Volume 1, ed.,V.M.Malhotra, American Concrete Institute, Detroit, 1986, 839p.

CALCULATION OF STRESSES IN LAMINATED MATERIALS UNDER TRANSVERSE BENDING

A.E. Bogdanovich, E.V. Yarve
Institute of Polymer Mechanics
Latvian SSR Academy of Sciences
Riga, USSR

ABSTRACT

The stress state analysis for multilayer beams manufactured
of anisotropic or orthotropic reinforced layers subjected to
transverse dynamic load is presented. The numerical realization of the proposed method is based on two-dimensional
spline approximation of the displacements across the beam
thickness and along the beam length. The numerical analysis
of transverse shear and normal stresses arising under short,
medium and long-time impulses is performed. The results illustrate high efficiency of the proposed method, obtained
even with a coarser division into sublayers of a highly non-
uniform through the thickness multilayer beams.

INTRODUCTION

The problem of calculation of the stress-strain state in multilayer and reinforced structural elements is becoming ever
more actual. The complexity of this problem is explicable by
the principal necessity of the structural specifics of such
structures to be taken into account. Namely, the mechanical
inhomogeneity of reinforced materials and the presence of
layers having different mechanical characteristics in laminates. These specific features require that special approaches to the numerical analysis should be elaborated. Thus, for
example, when solving the problem by the theory of elasticity in terms of displacements using the finite-element or fi-
nite-difference methods, certain additional procedures for

calculation of interfacial stresses are required. The use of a hybrid finite element technique with independent displacement and stress approximation leads to high-order systems of equations, thereby restricting the applicability of this method to dynamics. The problem of evolving efficient methods to be used in solving the dynamic problems in the theory of elasticity for structurally-inhomogeneous and, in particular, multilayer and reinforced media, remains open.

In the given paper a method for solving two-dimensional dynamic problem for a laminated rectangular packet consisting of orthotropic layers (Fig. I) by approximating the displacements by polynomial splines has been proposed. Independent approximations are used through the thickness of a packet (the z coordinate) and along the length (the x coordinate). In accordance with the approach of /I/, in the direction z each layer is divided (in a general case, nonuniformly) into sublayers, the boundaries between sublayers corresponding to coordinates z_i ($0 = z_0 < z_1 < \ldots < z_N = H$); $N = \sum\limits_{K=1}^{n} \ell_K$. On introducing the notation $I_S = \sum\limits_{i=1}^{s} \ell_i$ ($S = I, \ldots, n - I$) the horizontal line $z = z_{I_S}$ would represent the boundary surface between S- and $S + I$-th layers.

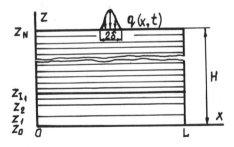

Figure I. The loading scheme and division into sublayers of a laminate.

The recurrent approach proposed for constructing the basic splines of an arbitrary power, being continually differentiated for the necessary number of times over a speci-

fied time interval, makes it possible to obtain continuous
displacements through the entire thickness of a laminate and
continuous (with respect to z) derivatives of displacements
through the thickness of each homogeneous layer. On the boundary surfaces between the layers having different elastic characteristics these derivatives can possess the first-kind disruptions. For a homogeneous layer, the approach outlined below will result in normalized B -splines.

SOLUTION PROCEDURE

Let us consider the procedure for constructing the basic
spline functions $\Phi_i (z)$, proposed in /2/, such that the displacements

$$u_x (x,z,t) = \sum_i U_i (x,t) \Phi_i (z) ;$$

$$u_z (x,z,t) = \sum_i W_i (x,t) \Phi_i (z)$$

(I)

for any selected functions $U_i (x ,t), W_i (x , t)$, being sufficiently smooth with respect to x , could be continually differentiated for all x and z from the region
$[z_0, z_N] \times [0, L]$, except the boundary surfaces between the
layers, at which $\frac{\partial u_x}{\partial z}$ and $\frac{\partial u_z}{\partial z}$ can possess the first-kind
disruptions. Let us introduce the following functions:

$$\mathcal{G}_i^- (z) = \begin{cases} 0, & z_0 \leq z < z_{i-1} ; \\ \frac{z - z_{i-1}}{z_i - z_{i-1}}, & z_{i-1} \leq z < z_i ; \\ 0, & z_i < z < z_N ; \end{cases}$$

(2)

$$\mathcal{G}_i^+ (z) = \begin{cases} 0, & z_0 \leq z < z_i ; \\ \frac{z_{i+1} - z}{z_{i+1} - z_i}, & z_i < z < z_{i+1} ; \\ 0, & z_{i+1} \leq z \leq z_N ; \quad i = 0, 1, \dots, N \end{cases}$$

and supplement (2) with the limiting relations $\mathcal{G}_i^-(z_i - 0) = 1$; $\mathcal{G}_i^-(z_i + 0) = 0$; $\mathcal{G}_i^+(z_i - 0) = 0$; $\mathcal{G}_i^+(z_i + 0) = 1$.

Let us proceed from the approximations of displacement derivatives $\dfrac{\partial u_x(x,z,t)}{\partial z}$, $\dfrac{\partial u_z(x,z,t)}{\partial z}$. For example, for $\dfrac{\partial u_x(x,z,t)}{\partial z}$ we have

$$\frac{\partial u_x}{\partial z} = \sum_{s=1}^{n-1} \left\{ U_{I_{s-1}}^+(x,t)\mathcal{G}_{I_{s-1}}^+(z) + \sum_{j=1}^{\ell_s-1} U_{j+I_{s-1}}(x,t) \times \right. \tag{3}$$

$$\left. \times \left[\mathcal{G}_{j+I_{s-1}}^+(z) + \mathcal{G}_{j+I_{s-1}}^-(z) \right] + U_{I_s}^-(x,t)\mathcal{G}_{I_s}^-(z) \right\}; \quad z_0 \le z \le z_N.$$

It is evident that for any selected $U_j^\pm$ and U_j the derivative $\dfrac{\partial u_x}{\partial z}$ is continuous at $z_{I_{s-1}} < z < z_{I_s}$ and

$$\frac{\partial u_x(x, z_{I_s} \pm 0, t)}{\partial z} = U_{I_s}^\pm(x,t).$$

In order that we could pass from $\dfrac{\partial u_x}{\partial z}$ to u_x, let us integrate (3) in the range from z_0 to z, and introduce designations $u_x(x, z_0, t) = U_{-1}(x, t)$. As a result we shall obtain

$$u_x(x, z, t) = U_{-1}(x, t)\Psi_{-1}(z) +$$

$$+ \sum_{s=1}^{n-1} \left\{ U_{I_{s-1}}^+(x,t)\Psi_{I_{s-1}}^+(z) + \sum_{j=1}^{\ell_s-1} U_{j+I_{s-1}}(x,t) \times \right. \tag{4}$$

$$\left. \times \left[\Psi_{j+I_{s-1}}^+(z) + \Psi_{j+I_{s-1}}^-(z) \right] + U_{I_s}^-(x,t)\Psi_{I_s}^-(z) \right\},$$

where

$$\Psi_{-1}(z) \equiv 1; \quad \Psi_j^\pm(z) = \int_{z_0}^{z} \mathcal{G}_j^\pm(z)\, dz; \quad j = 0, 1, \ldots, N. \tag{5}$$

The approximation (4) for any selected functions $U_{-1}(x,t)$, $U_{I_s}^\pm(x,t)$, $U_j(x,t)$; $j = 0, 1, \ldots, N$ $(j \ne I_s)$ will ensure the continuity of $u_x(x, z, t)$ at $z_0 \le z \le z_N$ and the continuity of $\dfrac{\partial u_x(x,z,t)}{\partial z}$ at $z \ne z_{I_s}$. On the contact surfaces between s- and $s+I$-th layers, the derivative $\dfrac{\partial u_x(x,z,t)}{\partial z}$ will possess the first-kind disruption. Renumbering the terms, let us express (4) in the standard form

$$u_x(x, z, t) = \sum_{j=-1}^{N+n-1} U_j(x, t) \Psi_j(z),$$ (6)

where

$$\Psi_j(z) = \begin{cases} \Psi_{I_{s-1}}^+, & j = I_{s-1} + S - 1; \\ \Psi_{j+1-s}^+ + \Psi_{j+1-s}^-, & I_{s-1} + S - 1 < j < I_s + S - 1; \\ \Psi_{I_s}^-, & j = I_s + S - 1; \end{cases}$$ (7)

$$U_j(x, t) = \begin{cases} U_{I_{s-1}}^+, & j = I_{s-1} + S - 1; \\ U_{j-s+1}, & I_{s-1} + S - 1 < j < I_s + S - 1; \\ U_{I_s}^-, & j = I_s + S - 1. \end{cases}$$

The basic functions (7) are inconvenient for practical purposes, because they do not comprise a local supporter. Their linear combinations

$$\Phi_j(z) = \frac{\Psi_{j-1}(z)}{\Psi_{j-1}(z_N)} - \frac{\Psi_j(z)}{\Psi_j(z_N)}, \quad j = 0, 1, \ldots, N+n-1;$$ (8)

$$\Phi_{N+n}(z) = \frac{\Psi_{N+n-1}(z)}{\Psi_{N+n-1}(z_N)},$$

are free of this imperfection. Having calculated $\Phi_j(z)$, we can easily make sure that $\Phi_j(z) \equiv 0$ at $z \leq z_{j-s-1}$ and $z \geq z_{j-s+2}$ for $I_{s-1} < j-s+1 < I_s$, i.e. the obtained piecewise-polynomial second-order basic functions have a local supporter of minimum length.

Substituting in (6) $\Psi_j(z)$ by $\Phi_j(z)$, we obtain the displacement approximation in the form

$$u_x(x, z, t) = \sum_{j=0}^{N+n} U_j(x, t) \Phi_j(z);$$ (9)

$$u_z(x, z, t) = \sum_{j=0}^{N+n} W_j(x, t) \Phi_j(z).$$

It is easy to check whether

$$u_x(x,z_0,t) = U_0(x,t) \; ; \; u_x(x,z_N,t) = U_{N+n}(x,t) \; ;$$

$$u_z(x,z_0,t) = W_0(x,t) \; ; \; u_z(x,z_N,t) = W_{N+n}(x,t), \tag{IO}$$

however, for the sake of calculating the displacement from the known U_j, W_j at the boundary surfaces between sublayers, the relations of the type (IO) are invalid.

If in the first stage of evolving the displacement approximation we had proceeded from the derivative $\frac{\partial^2 u_x(x,z,t)}{\partial z^2}$, then by double integrating over z , we might analogously have constructed the basic functions, provided the displacement approximation had been continually twice differentiated through the layer thickness.

Thus we have constructed m + I-st power basic splines which had been m times continually differentiated over a specified interval, except certain number of points, at which the m_I-th derivative ($m_I \le m$) possessed the first-kind disruption. The outlined procedure of evolving the approximation functions, defined by formulas (5), (7) and (8) can efficiently be employed in the division of the body along the remaining spatial coordinates.

The next stage involves the division of the two-dimensional region by straight lines $x = x_i$, $0 = x_0 < x_1 < ... < x_M$ = L into M rectangles inside each sublayer and approximation of the functions $U_i(x, t), W_i(x, t)$ by splines which are to be continually differentiated m times in all points x, $x_0 < x < x_M$, without any exception. Let us designate the basic splines by χ_j , $0 \le j \le M + m$. Each χ_j is a piecewise-polynomial function of m + I-st power. This function is other than zero only at $x_{j-m} < x < x_{j+1}$, i.e. it has a local supporter of minimum length. If the location of both ends of the supporter region satisfies the conditions $x_{j-m} > x_0$ and $x_{j+1} < x_M$, then χ_j coincides with the normalized B_{m+1} -spline /3/. If the above inequalities are disturbed, the usual technique involves introduction of fictitious nodes, lying outside the $[0,L]$ interval and constructing of normalized B -splines over the correspondingly widened inter-

val. Thereby the completeness of the system of basic functions has been confirmed /3/. To avoid the necessity of introducing fictitious nodes and treating B_{m+1} with the supporter being determined beyond the $[0, L]$ interval, we have
constructed special basic functions which differ from B -
splines at the edges of $[0, L]$ interval. It can be shown
that these functions will ensure the completeness of the system of basic functions, however, a comprehensive analysis of
this problem is beyond the scope of this paper.

In such a way, we shall express the full displacement
approximation in the rectangular region by two-dimensional
spline as follows

$$u_x(x, z, t) = \sum_{j=0}^{M+m} \sum_{i=0}^{N+n} U_{ij}(t) \chi_j(x) \phi_i(z) \; ; \tag{II}$$

$$u_z(x, z, t) = \sum_{j=0}^{M+m} \sum_{i=0}^{N+n} W_{ij}(t) \chi_j(x) \phi_i(z).$$

The representation (II) allows to solve a wide range of two-
dimensional static and dynamic plane stress and plane strain
problems for a laminated media. For example, it is possible
to consider loading with pressure pulses arbitrarily distributed over the lateral surfaces, including local impact loading, and study the various versions of boundary conditions
at the end faces. The calculation of the stress-strain states
in laminates, generated in the vicinity of rigid supports and
at the free edges becomes possible, too.

Let us consider as an example the problem of transverse
bending for a beam under normal load $q(x, t)$ which is
applied to the surface $z = H$. Having expressed the stresses
in the form

$$\sigma_{xx} = B_{11}(z)\frac{\partial u_x}{\partial x} + B_{13}(z)\frac{\partial u_z}{\partial z} \; ;$$

$$\sigma_{zz} = B_{31}(z)\frac{\partial u_x}{\partial x} + B_{33}(z)\frac{\partial u_z}{\partial z} \; ;$$

$$\sigma_{xz} = G_{13}(z)\left(\frac{\partial u_x}{\partial z} + \frac{\partial u_z}{\partial x}\right) \; ,$$

where $B_{ij}(z)$ are piecewise-constant over each layer stiffness characteristics, and employing the variational method
(taking (9) into account), we obtain the following system of
equations with respect to $U_i(x,t)$, $W_i(x,t)$:

$$\sum_{i=0}^{N+n} \left\{ -\frac{\partial^2 U_i}{\partial t^2} \int_0^H S(z)\, \phi_i \, \phi_m \, dz + \frac{\partial^2 U_i}{\partial x^2} \int_0^H B_{11}(z)\, \phi_i \, \phi_m \, dz + \right.$$

$$+ \frac{\partial W_i}{\partial x}\left[\int_0^H B_{13}(z)\frac{d\phi_i}{dz}\, \phi_m \, dz - \int_0^H G_{13}(z)\phi_i \frac{d\phi_m}{dz}\, dz \right] -$$

$$\left. - U_i \int_0^H G_{13}(z)\frac{d\phi_i}{dz} \frac{d\phi_m}{dz}\, dz \right\} = 0 ; \qquad (I2)$$

$$\sum_{i=0}^{N+n} \left\{ -\frac{\partial^2 W_i}{\partial t^2} \int_0^H S(z)\, \phi_i \, \phi_m \, dz + \frac{\partial^2 W_i}{\partial x^2} \int_0^H G_{13}(z)\, \phi_i \, \phi_m \, dz + \right.$$

$$+ \frac{\partial U_i}{\partial x}\left[\int_0^H G_{13}(z)\frac{d\phi_i}{dz}\, \phi_m \, dz - \int_0^H B_{13}(z)\phi_i \frac{d\phi_m}{dz}\, dz \right] -$$

$$\left. - W_i \int_0^H B_{33}(z)\frac{d\phi_i}{dz} \frac{d\phi_m}{dz}\, dz \right\} = \delta_{N+n}^m \, q(x,t) ;$$

$$m = 0, 1, \ldots, N+n.$$

In the case of boundary conditions

$$\frac{\partial U_i(x,t)}{\partial x}\bigg|_{x=0,L} = 0 ; \quad W_i(x,t)\bigg|_{x=0,L} = 0 ; \quad i=0,1,\ldots,N+n,$$

the solution of (I2) can be expressed as

$$U_i(x,t) = \sum_{K=1}^{\infty} U_{iK}(t) \cos\frac{\pi K x}{L} ;$$

$$\qquad (I3)$$

$$W_i(x,t) = \sum_{K=1}^{\infty} W_{iK}(t) \sin\frac{\pi K x}{L} .$$

After expressing the external load in the form of the Fourier
series

$$q(x,t) = \sum_{K} q_K Q(t) \sin\frac{\pi K x}{L} , \qquad (I4)$$

we shall obtain by substituting (I3) and (I4) into (I2) a separate system of ordinary differential equations with respect
to $U_{i\kappa}(t)$, $W_{i\kappa}(t)$ for each fixed number κ. The dimensionality of this system is 2 $(N + n + I)$. In the case of solving
this problem by two-dimensional spline approximation (II),
the dimensionality of the respective system of ordinary differential equations will be 2 $(N + n + I) \times (M + m - I)$.

ANALYSIS OF THE NUMERICAL RESULTS

Let us consider the results of calculating the transverse
stresses arising in three-ply and five-ply laminates under
normal surface load (I4). The time-dependent part of the load
$Q(t)$ is specified in the form of triangular impulse

$$Q(t) = \begin{cases} \dfrac{t}{t_o}, & t < t_o; \\ 2 - \dfrac{t}{t_o}, & t_o \le t \le 2t_o; \\ 0, & t > 2t_o. \end{cases} \tag{I5}$$

The load duration $2t_o$ in the calculations varied from tenths
to a few tens of T - value (T is the time of stress wave propagation through the thickness of a laminate). Of all the series (I4) only one harmonic corresponding to the number $\kappa = I$
or $\kappa = 5I$ was retained. In consecutive illustrations, the amplitude values of transverse stresses are presented. To find
the stresses in the particular $x = x_o$ section under the
load, representing one term of the series (I4), the given amplitude values should be multiplied by $\cos \dfrac{\pi\kappa x_o}{L}$ in the case
of σ_{xz} and by $\sin \dfrac{\pi\kappa x_o}{L}$ in the case of σ_{zz}.

The calculations were performed for laminates made up
of orthotropic and isotropic layers. In the first case, the
two-dimensional model imitates the $\{x, z\}$ section of the
cross-ply composite plate with unidirectionally reinforced
monolayers. The material of a monolayer is graphite-epoxy
composite having the following characteristics: $E_I = I.94.10^{II}$
N/m^2; $E_3 = 7.72.10^9$ N/m^2; $\nu_{I3} = 0.3$; $G_{I3} = 4.2I.10^9$ N/m^2;
$\S = I.63.10^3 kg/m^3$.

The isotropic laminates are made up of organic glass layers having the characteristics: $E = 6.10^{10} \text{N/m}^2$; $V = 0.3$; $\rho = 1.5.10^3 \text{kg/m}^3$ and adhesive layers with $E = 2.8.10^9 \text{N/m}^2$; $V = 0.33$; $\rho = 10^3 \text{kg/m}^3$. In all cases, $L = I$ m, $H = 10^{-2}$m were assumed.

Let us consider three-ply laminates under short-time load impulses. In Fig. 2 and 3, the values σ_{zz} / q_o in several consecutive time instants are shown. The direction of motion of the output impulse (I5) is indicated by arrows, as well as the direction of motion of the impulses passing through the boundary surfaces and being reflected from the boundaries. In the calculations, nonuniform division into sublayers, becoming more dense on approaching the boundary surfaces between layers, was made. The total number of sublayers was 30, each layer being divided into I0 sublayers. The location of boundaries between sublayers are indicated on the z axis. It was shown that for an isotropic three-ply laminate (Fig. 2) the relative magnitude of σ_{zz} / q_o disruptions on the boundary surfaces ($z = z_{I0}$ and $z = z_{20}$) in all time instants did not exceed 0.002.

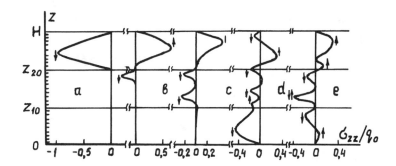

Figure 2. The σ_{zz} / q_o variance with z for a three-ply glass-adhesive-glass packet at the time instants $t = 2t_o$ (a), $4t_o$(b), $8t_o$(c), IIt_o(d), $I2t_o$(e) at $t_o = T/II$, $K = I$.

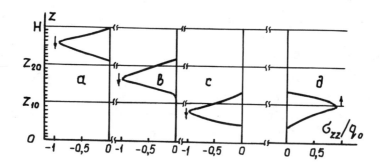

Figure 3. The σ_{zz}/q_o variance with z for a three-ply graph-ite-epoxy packet $[0^\circ/90^\circ/0^\circ]$ at the time instants $t = 2t_o(a)$, $4t_o(b)$, $6t_o(c)$, $10t_o(d)$ at $t_o = T/7$, $K = I$.

As is clear from Fig. 3, in the case of an orthotropic three-ply laminate with a $[0^\circ/90^\circ/0^\circ]$ layup (reinforcement in the direction of the x axis) the distortion of the input impulse (I3) was not observed upon passing through the interfaces; the reflected impulses were not produced. This can be attributed to the fact that for all the layers in this laminate the transversal moduli E_3 and G_{13} were equal.

The results obtained for the load harmonically distributed along the x coordinate allows us to determine the stress σ_{zz} under short-time local pressure impulses with high accuracy. Let us illustrate this fact on the example of the local load

$$q(x,t) = q_o\, Q(t)\, \beta(x),$$

where

$$\beta(x) = \begin{cases} [(x - L/2)^2 - \varepsilon^2]^2/\varepsilon^4, & |x - L/2| \le \varepsilon; \\ 0, & |x - L/2| > \varepsilon. \end{cases} \tag{I6}$$

The respective problem was solved by using the two-dimensional spline-approximation (II). The value $\varepsilon = 0.02L$ was assumed. Over a half-length of the package 27 internal nodes have been selected, I9 of which are inside $[\frac{L}{2}-\varepsilon, \frac{L}{2}+\varepsilon]$ interval, the remaining nodes are disposed nonuniformly — the distance

between the adjacent nodes decreases according to geometric
progression law on approaching the middle of the package. The
loading time t_o is the same as for Figs. 2 and 3.

The obtained results showed that in $x = L/2$ section the
distribution of 6_{zz} over the z coordinate at the time instants, indicated in the captions to Figs. 2 and 3, coincides
with the distributions presented in the mentioned figures.
Besides, it turned out that in order to find 6_{zz} in any sec-
tion x_o, $|x_o - L/2| < \varepsilon$, it is enough to multiply the 6_{zz} distri-
butions for $x = L/2$ section by $\beta(x_o)$. The stress 6_{zz} out of
the loading zone is negligible.

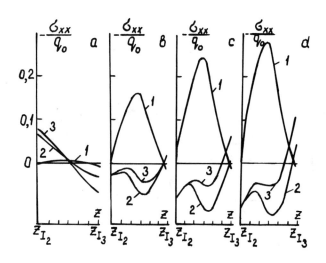

Figure 4. The dependencies of $6_{xx}/q_o$ on z through the upper
layer of glass-adhesive-glass package at the time
instants $t = 2t_o$ (curve I), $8t_o$ (2), $12t_o$ (3) at
$x = 0.5L$ (a); $0.495L$ (b); $0.49L$ (c); $0.48L$(d).

In Fig. 4, the dependencies $6_{xx}(z)$ in the upper layer
of the glass-adhesive-glass package for several x sections
are shown. At $t = 2t_o$, these dependencies are similar to the
$6_{zz}(z)$ function, while the magnitudes of 6_{xx} are close to
the magnitudes of 6_{zz} multiplied by the Poisson's ratio $V = 0.3$.
In $x_o = L/2$ section 6_{xx} reaches its maximum, whereas in other x_o,
$|x_o - L/2| < \varepsilon$, sections the magnitudes of this stress can be obtained by multiplying its value by $\beta(x_o)$ at $x_o = L/2$. Out of
the loading zone 6_{xx} at $t = 2t_o$ is fairly small. At $t = 8t_o$

and $t = 12 t_o$ it becomes more notable, thus substantiating the growing role of the process of flexural wave propagation along the x coordinate.

Further let us treat the stresses arising in three-ply and five-ply packages subjected to comparatively long-time impulses. The σ_{xz}/q_o and σ_{zz}/q_o dependencies on z coordinate for the case of three-ply isotropic package are presented in Fig. 5. It is assumed that $N = 18$, each layer comprising 6 sublayers; the boundaries between sublayers are marked on the z axis.

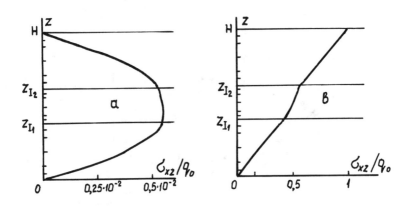

Figure 5. The dependencies of σ_{xz}/q_o (a) and σ_{zz}/q_o (b) on z for a three-ply glass-adhesive-glass packet at the time instant $t = t_o$ at $t_o = 100$ T, $K = 1$.

The diagrams for transverse shear stresses across the thickness of two five-ply packages, loaded with more prolonged impulses, are presented in Fig. 6. The division was as follows: $N = 30$; 6 sublayers making up a layer.

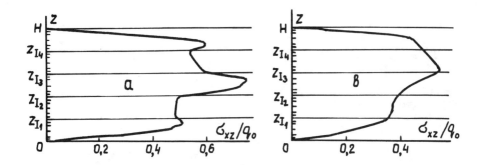

Figure 6. The dependencies of σ_{xz}/q_0 on z for a five-ply glass-adhesive-glass-adhesive-glass (a) and $[0^\circ/90^\circ/0^\circ/90^\circ/0^\circ]$ (b) packets at the time instant $t = t_0$ at $t_0 = 2$ T, $K = 5I$.

CONCLUSIONS

The obtained results confirm that the proposed solution procedure based on approximating the displacements in respect to z coordinate according to (9) and in respect to x coordinate according to (II) or (I3), allows high accuracy of results to be attained, even with comparatively coarser division into sublayers of the medium having highly nonuniform mechanical characteristics through the thickness. It should be emphasized that in the case of fixed division this accuracy was maintained both for very short and long-time impulses, as well as over a wide range of K -values.

The conclusion can be drawn from the results that in the packages built up of one-type elastic layers, which differ only by in-plane fiber layup, the short impulses propagate through the thickness many times, without undergoing practically any changes (Fig. 3). At the same time, the introduction of the layers, causing strong nonuniformity of transverse elastic characteristics, opens up vast possibilities for governing the nonstationary processes in laminated materials.

REFERENCES

I. Bogdanovich, A.E. and Yarve, E.V., Stress analysis in multilayered beams under transverse dynamic bending. Mechanics of Compos. Mater., 1983, 19, 604-616 (Transl. from Russian).

2. Bogdanovich, A.E. and Yarve, E.V., Numerical solution for the problem of two-dimensional transient deformation of layered media. Mekh. Kompozitn. Mater., 1988, N I, 36-44 (in Russian).

3. Korneichuk, N.P. Splines in the Theory of Approximation, Nauka Publ., Moscow, 1984, 352 p.

INDEX OF CONTRIBUTORS